AF767267

CHEMICAL REACTOR DESIGN AND CONTROL

The Wiley Bicentennial–Knowledge for Generations

Each generation has its unique needs and aspirations. When Charles Wiley first opened his small printing shop in lower Manhattan in 1807, it was a generation of boundless potential searching for an identity. And we were there, helping to define a new American literary tradition. Over half a century later, in the midst of the Second Industrial Revolution, it was a generation focused on building the future. Once again, we were there, supplying the critical scientific, technical, and engineering knowledge that helped frame the world. Throughout the 20th Century, and into the new millennium, nations began to reach out beyond their own borders and a new international community was born. Wiley was there, expanding its operations around the world to enable a global exchange of ideas, opinions, and know-how.

For 200 years, Wiley has been an integral part of each generation's journey, enabling the flow of information and understanding necessary to meet their needs and fulfill their aspirations. Today, bold new technologies are changing the way we live and learn. Wiley will be there, providing you the must-have knowledge you need to imagine new worlds, new possibilities, and new opportunities.

Generations come and go, but you can always count on Wiley to provide you the knowledge you need, when and where you need it!

William J. Pesce
President and Chief Executive Officer

Peter Booth Wiley
Chairman of the Board

CHEMICAL REACTOR DESIGN AND CONTROL

WILLIAM L. LUYBEN
Lehigh University

AIChE®

BICENTENNIAL 1807 WILEY 2007 BICENTENNIAL

WILEY-INTERSCIENCE
A JOHN WILEY & SONS, INC., PUBLICATION

Published by John Wiley & Sons, Inc., Hoboken, New Jersey
Published simultaneously in Canada.

Library of Congress Cataloging-in-Publication Data:

Luyben, William L.
Chemical reactor design and control/William L. Luyben.
p. cm.
Includes index.
ISBN 978-0-470-09770-0 (cloth)
1. Chemical reactors—Design and construction. I. Title.
TP157.L89 2007
600'.2832--dc22 2006036208

MIX
Paper from responsible sources
FSC
www.fsc.org
FSC® C013604

Dedicated to 40 classes of
Lehigh Chemical Engineers

CONTENTS

PREFACE

Chemical reactors are unquestionably the most vital parts of many chemical, biochemical, polymer, and petroleum processes because they transform raw materials into valuable chemicals. A vast variety of useful and essential products are generated via reactions that convert reactants into products. Much of modern society is based on the safe, economic, and consistent operation of chemical reactors.

In the petroleum industry, for example, a significant fraction of our transportation fuel (gasoline, diesel, and jet fuel) is produced within process units of a petroleum refinery that involve reactions. Reforming reactions are used to convert cyclical saturated naphthenes into aromatics, which have higher octane numbers. Light C4 hydrocarbons are alkylated to form high-octane C8 material for blending into gasoline. Heavy (longer-chain) hydrocarbons are converted by catalytic or thermal cracking into lighter (shorter-chain) components that can be used to produce all kinds of products. The unsaturated olefins that are used in many polymerization processes (ethylene and propylene) are generated in these reactors. The polluting sulfur components in many petroleum products are removed by reacting them with hydrogen.

The chemical and materials industries use reactors in almost all plants to convert basic raw materials into products. Many of the materials that are used for clothing, housing, automobiles, appliances, construction, electronics, and healthcare come from processes that utilize reactors. Reactors are important even in the food and beverage industries, where farm products are processed. The production of ammonia fertilizer to grow our food uses chemical reactors that consume hydrogen and nitrogen. The pesticides and herbicides we use on crop fields and orchards aid in the advances of modern agriculture. Some of the drugs that form the basis of modern medicine are produced by fermentation reactors. It should be clear in any reasonable analysis that our modern society, for better or worse, makes extensive use of chemical reactors.

Many types of reactions exist. This results in chemical reactors with a wide variety of configurations, operating conditions, and sizes. We encounter reactions that occur in solely the liquid or the vapor phase. Many reactions require catalysts (homogeneous if the catalyst is the same phase as the reactants or heterogeneous if the catalyst has a different phase). Catalysts and the thermodynamic properties of reactants and products can lead to multiphase reactors (some of which can involve vapor, multiple liquids, and solid phases). Reactions can be exothermic (producing heat) or endothermic (absorbing heat). An example of the first is the nitration of toluene to form TNT. A very important example of the second is steam–methane reforming to produce synthesis gas.

Reactors can operate at low temperature (e.g., C4 sulfuric acid alkylation reactors run at 10°C) and at high temperatures (hydrodealkylation of toluene reactors run at 600°C). Some reactors operate in a batch or fed-batch mode, others in a continuous mode, and still others in a periodic mode. Beer fermentation is conducted in batch reactors. Ammonia is produced in a continuous vapor-phase reactor with a solid "promoted" iron catalyst.

The three classical generic chemical reactors are the batch reactor, the continuous stirred-tank reactor (CSTR), and the plug flow tubular reactor (PFR). Each of these reactor types has its own unique characteristics, advantages, and disadvantages. As the name implies, the *batch reactor* is a vessel in which the reactants are initially charged and the reactions proceed with time. During parts of the batch cycle, the reactor contents can be heated or cooled to achieve some desired temperature–time trajectory. If some of the reactant is fed into the vessel during the batch cycle, it is called a "fed-batch reactor." Emulsion polymerization is an important example. The reactions conducted in batch reactors are almost always liquid-phase and typically involve slow reactions that would require large residence times (large vessels) if operated continuously. Batch reactors are also used for small-volume products in which there is little economic incentive to go to continuous operation. In some systems batch reactors can provide final product properties that cannot be achieved in continuous reactors, such as molecular weight distribution or viscosity. Higher conversion can be achieved by increasing batch time. Perfect mixing of the liquid in the reactor is usually assumed, so the modeling of a batch reactor involves ordinary differential equations. The control of a batch reactor is a "servo" problem, in which the temperature and/or concentration profiles follow some desired trajectory with time.

The CSTR reactor is usually used for liquid-phase or multiphase reactions that have fairly high reaction rates. Reactant streams are continuously fed into the vessel, and product streams are withdrawn. Cooling or heating is achieved by a number of different mechanisms. The two most common involve the use of a jacket surrounding the vessel or an internal coil. If high conversion is required, a single CSTR must be quite large unless reaction rates are very fast. Therefore, several CSTRs in series are sometimes used to reduce total reactor volume for a given conversion. Perfect mixing of the liquid in the reactor is usually assumed, so the modeling of a CSTR involves ordinary differential equations. The control of a CSTR or a series of CSTRs is often a "regulator" problem, in which the temperature(s) and/or concentration(s) are held at the desired values in the face of disturbances. Of course, some continuous processes produce different grades of products at different times, so the transition from one mode of operation to another is a servo problem.

The PFR tubular reactor is used for both liquid and gas phases. The reactor is a long vessel with feed entering at one end and product leaving at the other end. In some applications the vessel is packed with a solid catalyst. Some tubular reactors run adiabatically (i.e., with no heat transferred externally down the length of the vessel). The heat generated or consumed by the reaction increases or decreases the temperature of the process material as it flows down the reactor. If the reaction is exothermic, the adiabatic temperature rise may produce an exit temperature that exceeds some safety limitation. It may also yield a low reaction equilibrium constant that limits conversion. If the reaction is endothermic, the adiabatic temperature change may produce reactor temperatures so low that the resulting small chemical reaction rate limits conversion.

In these cases, some type of heat transfer to or from the reactor vessel may be required. The reactor vessel can be constructed like a tube-in-shell heat exchanger. The process fluid flows inside the tubes, which may contain catalyst, and the heating/cooling medium is on the shell side. Variables in a PFR change with both axial position and time, so the modeling of a tubular reactor involves partial differential equations. The control of a PFR can be quite challenging because of the distributed nature of the process (i.e., changes in temperature and composition variables with length and sometime radial position). Tubular reactor control is usually a regulator problem, but grade transitions can lead to servo problems in some processes.

The area of reactor design has been widely studied, and there are many excellent textbooks that cover this subject. Most of the emphasis in these books is on steady-state operation. Dynamics are also considered, but mostly from the mathematical standpoint (openloop instability, multiple steady states, and bifurcation analysis). The subject of developing effective stable closedloop control systems for chemical reactors is treated only very lightly in these textbooks. The important practical issues involved in providing reactor control systems that achieve safe, economic, and consistent operation of these complex units are seldom understood by both students and practicing chemical engineers.

The safety issue is an overriding concern in reactor design and control. The US Chemical Safety Board (CSB) published a report in 2002 in which they listed 167 serious incidents involving uncontrolled chemical reactivity between 1980 and 2001. There were 108 fatalities as a result of 48 of these incidents. The CSB has a number of reports on these and more recent incidents that should be required reading for anyone involved in reactor design and control. In 2003 the American Institute of Chemical Engineers published *Essential Practices for Managing Chemical Reactivity Hazard*, which is well worth reading.

There are hundreds of papers dealing with the control of a wide variety of chemical reactors. However, there is no textbook that pulls the scattered material together in a cohesive way. One major reason for this is the very wide variety in types of chemistry and products, which results in a vast number of different chemical reactor configurations. It would be impossible to discuss the control of the myriad of reactor types found in the entire spectrum of industry. This book attempts to discuss the design and control of some of the more important generic chemical reactors.

The development of stable and practical reactors and effective control systems for the three types of classical reactors are covered. Notice that "reactors" are included, not just control schemes. Underlying the material and approaches in this book is my basic philosophy (theology) that the design of the process and the process equipment has a much greater effect on the successful control of a reactor than do the controllers that are hung on the process or the algorithms that are used in these controllers. This does not imply that the use of models is unimportant in reactor control, since in a number of important cases they are essential for achieving the desired product properties.

The basic message is that the essential problem in reactor control is temperature control. Temperature is a dominant variable and must be effectively controlled to achieve the desired compositions, conversions, and yields in the safe, economic, and consistent operation of chemical reactors. In many types of reactors, this is achieved by providing plenty of heat transfer area and cooling or heating medium so that dynamic disturbances can be handled. Once temperature control has been achieved, providing base-level stable operation, additional objectives for the control system can be specified. These can be physical property specifications (density, viscosity, molecular weight distribution, etc.) or economic objectives (conversion, yield, selectivity, etc.).

The scope of this book, like that of all books, is limited by the experience of the author. It would be impossible to discuss all possible types of chemical reactors and presumptuous to include material on reactors with which I have little or no familiarity. Despite its limitations, I hope the readers find this book interesting and useful in providing some guidance for handling the challenging and very vital problems of chemical reactor control.

The many helpful comments and suggestions of Michael L. Luyben are gratefully acknowledged.

William L. Luyben

CHAPTER 1

REACTOR BASICS

In this chapter we first review some of the basics of chemical equilibrium and reaction kinetics. We need to understand clearly the fundamentals about chemical reaction rates and chemical equilibrium, particularly the effects of temperature on rate and equilibrium for different types of reactions. Reactions are generally catagorized as exothermic (releasing energy) or endothermic (requiring energy), as reversible (balance of reactants and products) or irreversible (proceeding completely to products), and as homogeneous (single-phase) or heterogeneous (multiphase).

One major emphasis in this book is the focus of reactor design on the control of temperature, simply because temperature plays such a dominant role in reactor operation. However, in many reactors the control of other variables is the ultimate objective or determines the economic viability of the process. Some examples of these other properties include reactant or product compositions, particle size, viscosity, and molecular weight distribution. These issues are discussed and studied in subsequent chapters.

Many polymer reactions, for example, are highly exothermic, so the temperature control concepts outlined in this book must be applied. At the same time, controlling just the temperature in a polymer reactor may not adequately satisfy the economic objectives of the plant, since many of the desired polymer product properties (molecular weight, composition, etc.) are created within the polymerization reactor. These key properties must be controlled using other process parameters (i.e. vessel pressure in a polycondensation reactor or chain transfer agent composition in a free-radical polymerization reactor).

Many agricultural chemicals (pesticides, fungicides, etc.), for another example, are generated in a series of often complex batch or semibatch reaction and separation steps. The efficacy of the chemical often depends on its ultimate purity. Operation and control of the reactor to minimize the formation of undesirable and hard-to-separate byproducts

Chemical Reactor Design and Control. By William L. Luyben

then become of urgent priority. Trajectories of reactor and feed process conditions must be developed and followed to ensure the economic success of the enterprise.

Returning now to the issue of reactor temperature control, we can generally state that reactors with either substantially reversible or endothermic reactions seldom present temperature control problems. Endothermic reactions require that heat be supplied to generate products. Hence, they do not undergo the dangerous phenomenon of "runaway" because they are self-regulating, that is, an increase in temperature increases the reaction rate, which removes more heat and tends to decrease the temperature.

Reversible reactions, even if they are exothermic, are also self-regulating because an increase in temperature decreases the chemical equilibrium constant. This reduces the net reaction rate between the forward and reverse reactions and limits how much product can ultimately be generated.

We also can generally state that major temperature control problems can and often do occur when the reactions are both exothermic and irreversible. These systems are not inherently self-regulatory because an increase in temperature increases the reaction rate, which increases temperature even further. The potential for reactor runaways is particularly high if the reactor is operating at a low level of conversion. The large inventory of reactant provides plenty of "fuel" for reaction runaway. These concepts will be quantitatively studied in later chapters.

Probably the most important aspect of reactor design and control for a substantial number of industrial processes involves heat transfer, that is, maintaining stable and safe temperature control. Temperature is the "dominant variable" in many chemical reactors. By *dominant variable*, we mean it plays a significant role in determining the economics, quality, safety, and operability of the reactor. The various heat transfer methods for chemical reactors are discussed in a qualitative way in this chapter, while subsequent chapters deal with these issues in detail with several illustrative quantitative examples.

> The key element in temperature control of chemical reactors is to provide sufficient heat transfer surface area or some other heat removal mechanism so that dynamic disturbances can be safely handled without reactor runaways.

In this chapter the design and operation of the three types of classical reactors are discussed. Their advantages and disadvantages, limitations, and typical application areas are also enumerated.

The final subject discussed in this chapter is the issue of reactor scaleup. Moving from a laboratory test tube in a constant temperature bath to a 20-L pilot plant reactor to a 200,000-L commercial plant reactor involves critical design and control decisions. One major problem is the reduction of the heat transfer area relative to the reactor volume (and heat transfer duty) as we move to larger reactors. This has an important effect on temperature control and reactor stability.

Another major problem with scaleup involves mixing within the reactor. The larger the reactor, the more difficult it potentially becomes to ensure that the entire contents are well mixed and at uniform conditions (if that is the reactor type) or that the contents remain distributed and not mixed (if that is the reactor type). Mixing is typically achieved using internal agitators. Gas sparging is also used to achieve mixing in systems that involve a gaseous feedstream. Mixing also affects the heat transfer film coefficient between the vessel wall and the process liquid. Therefore it impacts the ability to measure and control temperature effectively. For a given total reactor volume, the physical dimensions of the reactor vessel (the ratio of diameter to height) affect both the heat transfer area and the level of mixing. All these issues are discussed in several examples in subsequent chapters.

1.1 FUNDAMENTALS OF KINETICS AND REACTION EQUILIBRIUM

The rate at which a chemical reaction occurs in homogeneous systems (single-phase) depends primarily on temperature and the concentrations of reactants and products. Other variables, such as catalyst concentration, initiator concentration, inhibitor concentration, or pH, also can affect reaction rates. In heterogeneous systems (multiple phases), chemical reaction rates can become more complex because they may not be governed solely by chemical kinetics but also by the rate of mass and/or heat transfer, which often play significant roles.

1.1.1 Power-Law Kinetics

If we consider the *irreversible* reaction with two reactants forming a product

$$\mathrm{A} + \mathrm{B} \longrightarrow \mathrm{C} \tag{1.1}$$

the overall rate of reaction $\mathcal{R}$ can be viewed as the moles of component A being consumed per unit time per unit volume. Sometimes reaction rates are based per mass of catalyst present. Of course, by stoichiometry in this system, the moles of component B consumed have to equal the moles of A, along with the moles of component C produced. If component B had a stoichiometric coefficient of 2, then the rate of consumption of B would be twice that for A.

The overall reaction rate has a temperature dependence governed by the specific reaction rate $k_{(T)}$ and a concentration dependence that is expressed in terms of several concentration-based properties depending on the suitability for the particular reaction type: mole or mass concentration, component vapor partial pressure, component activity, and mole or mass fraction. For example, if the dependence is expressed in terms of molar concentrations for components A(C_A) and B(C_B), the overall reaction rate can be written as

$$\mathcal{R} = k_{(T)} C_\mathrm{A}^{\alpha} C_\mathrm{B}^{\beta} \tag{1.2}$$

where the exponents α and β are the "order" of the reaction for the respective two reactants. The actual reaction mechanism determines the form of the kinetic expression. More than one mechanism can give the same rate expression. Only in elementary reaction steps is the reaction order equal to the stoichiometry. The concept of a single rate-controlling step is often used in the development of kinetic expressions.

The temperature-dependent specific reaction rate $k_{(T)}$ is represented by the Arrhenius equation

$$k_{(T)} = k_0 e^{-E/RT} \tag{1.3}$$

where k_0 is a constant called the *preexponential factor*, E is the activation energy (typical units are kcal/mol, kJ/kmol, or Btu/lb · mol), R is the ideal-gas constant (in suitable units that depend on the units of E and T), and T is the absolute temperature [in K (degrees Kelvin) or °R (degrees Rankine)].

The k_0 preexponential factor is a large positive number (much greater than one) and has units that depend on the concentration units and the order of the reaction with respect to each component. The exponential term in Eq. (1.3) is a small positive number. Its minimum value is zero (when E/RT is infinite at very low absolute temperatures because of the negative sign in the exponential). Its maximum value is unity (when E/RT is zero at very high temperatures). Therefore at low temperature the E/RT term becomes large, which makes the exponential small and produces a low specific reaction rate. Conversely, at high temperature the E/RT term becomes small, which makes the exponential approach unity (in the limit as temperature goes to infinity, the exponential term goes to one). Thus the specific reaction rate *increases* with increasing temperature.

Clearly the rate of change of $k_{(T)}$ with temperature depends on the value of the activation energy. Figure 1.1 compares the relative rates of reaction as a function of activation

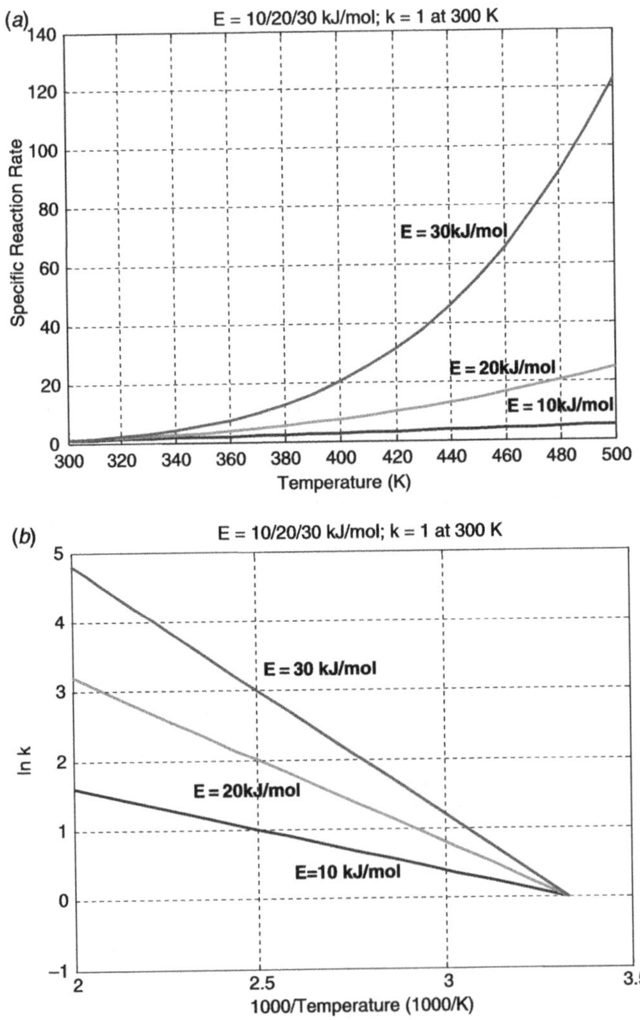

Figure 1.1 Effect of activation energy on temperature dependence of reaction rate.

energy and temperature. The activation energies are 10, 20, and 30 kJ/mol, and the reaction rates are calculated relative to a rate of unity at 300 K. Reactions with low activation energies are relatively insensitive to temperature, whereas reactions with high activation energies are quite sensitive to temperature. This can be seen by comparing the slopes of the lines for the relative reaction rates versus $1/T$. With an activation energy of 10 kJ/mol, the change in reaction rate from 300 to 500 K is much less than the change at an activation energy of 30 kJ/mol. Also, we see that the sensitivity of reaction rate to temperature is relatively greater at lower than at higher temperatures. Both of these observations play a role in the control of temperature in a chemical reactor.

The main point of the discussion above is

> Specific reaction rates always increase as temperature increases and the higher the activation energy, the more sensitive the reaction rate is to temperature.

Now we consider the reversible reaction where we do not achieve complete conversion of the reactants:

$$\mathrm{A} + \mathrm{B} \Leftrightarrow \mathrm{C} \tag{1.4}$$

We can express the forward reaction rate in terms of molar concentrations of reactants C_A and C_B that are dependent on the reaction orders α and β

$$\Re_F = k_{F(T)} C_\mathrm{A}^\alpha C_\mathrm{B}^\beta \tag{1.5}$$

with the specific rate

$$k_{F(T)} = k_{OF} e^{-E_F/RT} \tag{1.6}$$

The reverse reaction rate can also be written in terms of the molar concentration of product C_C dependent on the reaction order γ

$$\Re_R = k_{R(T)} C_\mathrm{C}^\gamma \tag{1.7}$$

with the specific rate

$$k_{R(T)} = k_{OR} e^{-E_R/RT} \tag{1.8}$$

The net overall reaction rate is the difference between the forward and the reverse

$$\Re = \Re_F - \Re_R = k_{F(T)} C_\mathrm{A}^\alpha C_\mathrm{B}^\beta - k_{R(T)} C_\mathrm{C}^\gamma \tag{1.9}$$

Under conditions of chemical equilibrium, the net overall reaction rate is zero, which leads to the relationship between the forward and reverse specific reaction rates and the chemical equilibrium constant (K_EQ) for the reaction:

$$K_\mathrm{EQ} = \frac{C_\mathrm{A}^\alpha C_\mathrm{B}^\beta}{C_\mathrm{C}^\gamma} = \frac{k_{F(T)}}{k_{R(T)}} \tag{1.10}$$

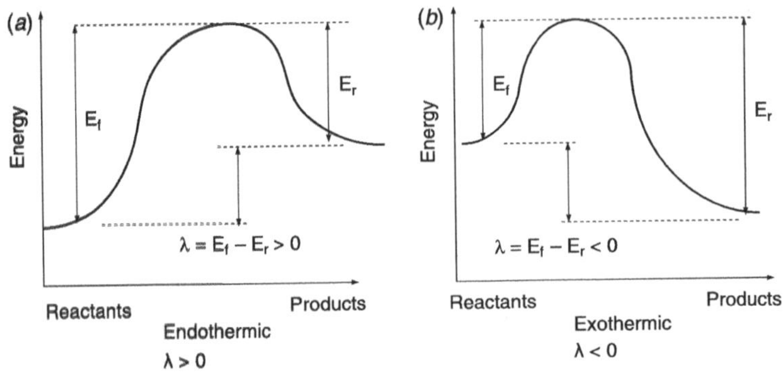

Figure 1.2 Energy change from reactants to products.

$$K_{EQ} = \frac{k_{F(T)}}{k_{R(T)}} = \frac{k_{0F}e^{-E_F/RT}}{k_{0R}e^{-E_R/RT}} = \frac{k_{0F}}{k_{0R}}e^{(E_R-E_F)/RT} \tag{1.11}$$

Just as the specific reaction rates k_F and k_R depend only on temperature, the same is true for the chemical equilibrium constant K_{EQ}. This temperature dependence is governed by the difference between the activation energies of the reverse and forward reactions. We can visualize the relative change in energy from reactants to products as shown in Figure 1.2. If the activation energies of forward and reverse reactions are equal, the equilibrium constant is independent of temperature. If the activation energy of the reverse reaction E_R is greater than the activation energy of the forward reaction E_F, then we release energy going from reactants to products. For this case, the numerator in the exponential term in Eq. (1.11) is positive; therefore as temperature *increases* the exponential term becomes smaller, and the equilibrium constant *decreases*. If the difference between the activation energies is the opposite (with E_F larger than E_R), then we require energy going from reactants to products. For this case, the numerator is negative, which means that the exponential term becomes larger as temperature increases, and the equilibrium constant *increases*.

The van't Hoff equation in thermodynamics gives the temperature dependence of the chemical equilibrium constant

$$\frac{d(\ln K_{EQ})}{dT} = \left(\frac{\lambda}{RT^2}\right) \tag{1.12}$$

where λ is the heat of reaction. This equation shows that the sign of the heat of reaction determines whether the equilibrium constant increases or decreases with increasing temperature. Exothermic reactions have negative heats of reaction, so the equilibrium constant *decreases* with increasing temperature:

The chemical equilibrium constant of a reversible exothermic reaction decreases as temperature increases.

Endothermic reactions have positive heats of reaction, so the equilibrium constant of a reversible endothermic reaction increases with increasing temperature.

Differentiating Eq. (1.11) with respect to temperature and combining with Eq. (1.12) give the relationship between the activation energies and the heat of reaction λ:

$$E_F - E_R = \lambda \tag{1.13}$$

From the previous discussion about the temperature sensitivity of reaction rate as a function of activation energy, we can understand why the chemical equilibrium constant of an exothermic reversible reaction decreases with increasing temperature. An exothermic reaction has a negative heat of reaction, since the activation energy of the reverse reaction exceeds that of the forward reaction. As temperature increases, the reverse reaction increases relatively more rapidly than the forward reaction, which means that at chemical equilibrium we have relatively more reactants than products and a lower equilibrium constant.

We note that particular catalysts or initiators used in chemical reactors change only the effective specific reaction rate and *do not* change the value of the chemical equilibrium constant.

1.1.2 Heterogeneous Reaction Kinetics

Power-law kinetic rate expressions can frequently be used to quantify homogeneous reactions. However, many reactions occur among species in different phases (gas, liquid, and solid). Reaction rate equations in such heterogeneous systems often become more complicated to account for the movement of material from one phase to another. An additional complication arises from the different ways in which the phases can be contacted with each other. Many important industrial reactors involve heterogeneous systems. One of the more common heterogeneous systems involves gas-phase reactions promoted with porous solid catalyst particles.

One approach to describe the kinetics of such systems involves the use of various resistances to reaction. If we consider an irreversible gas-phase reaction A $\rightarrow$ B that occurs in the presence of a solid catalyst pellet, we can postulate seven different steps required to accomplish the chemical transformation. First, we have to move the reactant A from the bulk gas to the surface of the catalyst particle. Solid catalyst particles are often manufactured out of aluminas or other similar materials that have large internal surface areas where the active metal sites (gold, platinum, palladium, etc.) are located. The porosity of the catalyst typically means that the interior of a pellet contains much more surface area for reaction than what is found only on the exterior of the pellet itself. Hence, the gaseous reactant A must diffuse from the surface through the pores of the catalyst pellet. At some point, the gaseous reactant reaches an active site, where it must be adsorbed onto the surface. The chemical transformation of reactant into product occurs on this active site. The product B must desorb from the active site back to the gas phase. The product B must diffuse from inside the catalyst pore back to the surface. Finally, the product molecule must be moved from the surface to the bulk gas fluid.

To look at the kinetics in heterogeneous systems, we consider the step of adsorbing a gaseous molecule A onto an active site s to form an adsorbed species As. The adsorption rate constant is k_a. The process is reversible, with a desorption rate constant k_d:

$$\mathrm{A} + \mathrm{s} \underset{k_d}{\overset{k_a}{\longleftrightarrow}} \mathrm{As} \tag{1.14}$$

Since we are dealing with gaseous molecules, we usually write the rate of adsorption in terms of the partial pressure of A (P_A) rather than molar concentration. The net rate of adsorption and desorption is

$$r = k_a' P_A C_S - k_d' C_{AS} \tag{1.15}$$

where C_S is the concentration of open active sites and C_{AS} is the concentration of sites occupied by an adsorbed molecule of A. The total number of sites (C_T) is fixed and is the sum of the open and occupied sites:

$$C_T = C_S + C_{AS} \tag{1.16}$$

If we define θ as the fraction of total sites covered by the adsorbed molecules, then

$$\theta = \frac{C_{AS}}{C_T} \tag{1.17}$$

We can rewrite these equations and combine constant parameters into the following rate expression:

$$r = k_a P_A (1 - \theta) - k_d \theta \tag{1.18}$$

At equilibrium the net rate is zero, and we can define an adsorption equilibrium constant (K_A) to produce the following expressions that define what is typically called *Langmuir isotherm behavior*:

$$\begin{aligned} K_A &= \frac{k_a}{k_d} \\ \theta &= \frac{K_A P_A}{1 + K_A P_A} \end{aligned} \tag{1.19}$$

Figure 1.3 shows a plot of θ versus partial pressure for various values of the adsorption equilibrium constant. These show that as the equilibrium constant increases for a given pressure, we increase the surface fraction covered, up to a value of 1. As the pressure increases, we increase the fraction of the surface covered with A. But we have only a finite amount of catalyst surface area, which means that we will eventually reach a point where increasing the partial pressure of A will have little effect on the amount that can be adsorbed and hence on the rate of any reaction taking place. This is a kind of behavior fundamentally different from that of simple power-law kinetics, where increasing the reactant concentration always leads to an increase in reaction rate proportional to the order in the kinetic expression.

We now consider the irreversible reaction A → B, where both components are gaseous and the reaction occurs on a solid catalyst. We can consider three steps to the mechanism: the adsorption of reactant A onto the surface (assumed to be reversible), the transformation of A into B on the catalyst surface (assumed to be irreversible), and finally the desorption

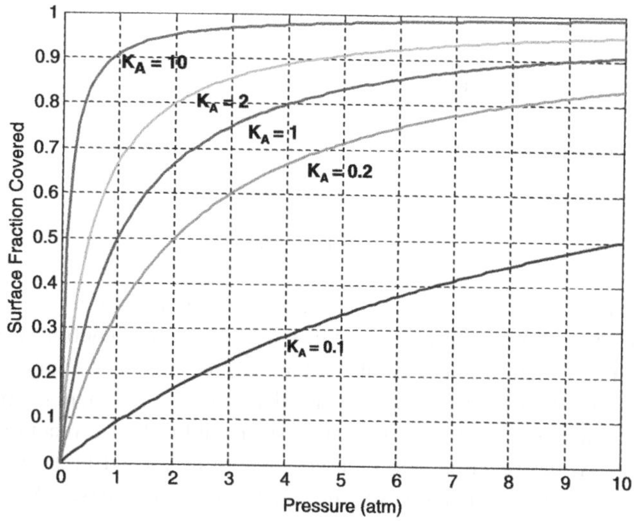

Figure 1.3 Langmuir isotherms for heterogeneous systems.

of product B from the surface (assumed to be reversible):

$$\begin{aligned} &\mathrm{A} + \mathrm{s} \underset{k_d^{\mathrm{A}}}{\overset{k_a^{\mathrm{A}}}{\longleftrightarrow}} \mathrm{As} \\ &\mathrm{As} \xrightarrow{k_{sr}} \mathrm{Bs} \\ &\mathrm{Bs} \underset{k_a^{\mathrm{B}}}{\overset{k_d^{\mathrm{B}}}{\longleftrightarrow}} \mathrm{B} + \mathrm{s} \end{aligned} \tag{1.20}$$

The assumption of which step is slowest governs the form of the final kinetic expression. For the purposes of this simple example, we assume that the second step is the slowest and is first-order with respect to the adsorbed A species. Therefore the rate r is determined by a rate constant and the concentration of A absorbed on the surface (C_{AS}) according to standard power-law kinetics:

$$r = k_{sr} C_{\mathrm{AS}} \tag{1.21}$$

We can write the absorption equilibrium coefficients for A and B in terms of their partial pressures (P_{A} and P_{B}) and the concentration of open sites (C_{S}):

$$\begin{aligned} K_{\mathrm{A}} &= \frac{C_{\mathrm{AS}}}{P_{\mathrm{A}} C_{\mathrm{s}}} = \frac{k_a^{\mathrm{A}}}{k_d^{\mathrm{A}}} \\ K_{\mathrm{B}} &= \frac{C_{\mathrm{BS}}}{P_{\mathrm{B}} C_{\mathrm{s}}} = \frac{k_a^{\mathrm{B}}}{k_d^{\mathrm{B}}} \end{aligned} \tag{1.22}$$

The total concentration of sites is a constant (C_T) and is the sum of open and occupied sites. We can express this in terms of the equilibrium constants under the assumption that the transformation step is the slowest:

$$C_T = C_s + C_{AS} + C_{BS} = C_s(1 + K_A P_A + K_B P_B) \tag{1.23}$$

We can write the overall reaction rate as

$$r = \frac{k_{(T)} P_A}{1 + K_A P_A + K_B P_B} \tag{1.24}$$

where $k_{(T)}$ is a kinetic rate constant that is a function of temperature.

For this assumed mechanism of what is an irreversible overall reaction, we observe that the reaction rate is a function not only of the partial pressure of reactant A but also the partial pressure of product B. The reaction rate decreases as we increase the amount of B because it occupies active sites on the catalyst and inhibits the reaction. At a given partial pressure of A, the reaction rate is largest when the partial pressure of B goes to zero. As the concentration of B increases, the reaction rate decreases. When the partial pressure of A is small and the term $K_A P_A + K_B P_B$ is much less than one, the reaction rate turns into first-order power-law kinetics that depends on P_A. In the limit of large partial pressures of A, the rate no longer depends on the concentration of A and becomes only a constant value equal to k/K_A. Figure 1.4 shows the reaction rate normalized by (k/K_A) for various values of P_B as a function of P_A. When the value of K_A is large compared with K_B (as shown in Fig. 1.4*a*), the reaction rates are relatively large and do not depend as much on P_B. This is because more of reactant A is adsorbed onto active sites of the catalyst. Since the transformation of adsorbed A to adsorbed B is the slowest step, the higher concentration of adsorbed A increases the reaction rate. On the other hand, when the value of K_A is small compared with K_B (Fig. 1.4*b*), the reaction rates are much slower and depend more on P_B. This is caused by the large concentration of adsorbed B on the active catalyst sites inhibiting the reaction.

The general forms of rate expressions in heterogeneous systems can have concentration or partial pressure dependences in both numerator and denominator along with various exponents. In heterogeneous reactors, it is not unusual to derive kinetic expressions that are more complicated than just a power-law expression. This, of course, has implications on how the reactor is controlled and the potential for runaway in exothermic systems. In some cases, where kinetics are very fast relative to mass transfer rates, the reactor behavior is governed by mass transfer and the variables that affect it.

1.1.3 Biochemical Reaction Kinetics

One special type of heterogeneous reactor involves biological systems with enzymes or microorganisms that convert some organic starting material into chemicals, pharmaceuticals, foodstuffs, and other substances. The conversion of sugar into alcohol via fermentation represents historically one of the oldest types of chemical reactors for the production of beer and wine. In fermentation, a reactant such as glucose (typically called the *substrate* S) is converted into a product P by the action of a microorganism or by the catalytic effect of an enzyme produced by a microorganism.

We can view an enzyme as a biological catalyst, and as such it leads to kinetic rate expressions that are of similar form to those derived in heterogeneous reaction

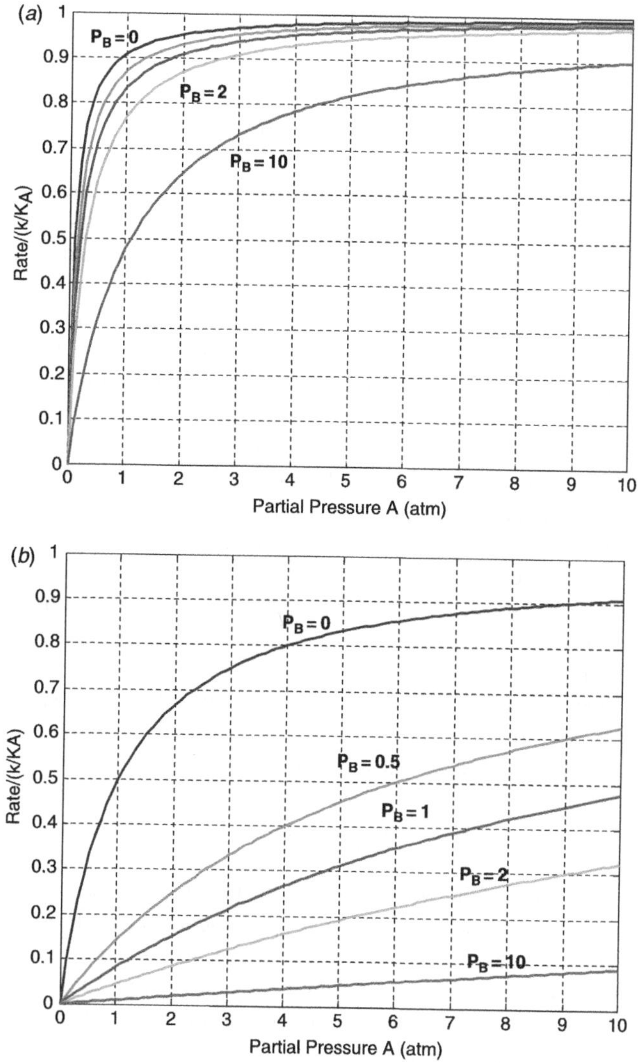

Figure 1.4 Normalized reaction rate as a function of P_A and P_B, $k = 1$; (*a*) $K_A = 10$, $K_B = 1$; (*b*) $K_A = 1$, $K_B = 2$.

systems. The Michaelis–Menton kinetic expression is one standard formulation used in enzyme-catalyzed fermentation. It assumes that the substrate and enzyme (E) form a complex (ES) via a reversible reaction. The enzyme–substrate complex is assumed to be very reactive and goes on to form the product in an irreversible reaction:

$$\begin{aligned} \mathrm{S} + \mathrm{E} &\leftrightarrow \mathrm{ES} \\ \mathrm{ES} &\longrightarrow \mathrm{P} \end{aligned} \tag{1.25}$$

The reaction rate for the formation of product (r_P) can then be derived from certain assumptions to take the following form that represents observed experimental behavior reasonably well:

$$r_P = \frac{kC_{E0}C_S}{C_S + K_M} \tag{1.26}$$

Here k is the rate constant for the irreversible reaction, C_{E0} is the total enzyme concentration, C_S is the substrate concentration, and K_M is the Michaelis–Menton constant. Both k and K_M may be functions of pH, temperature, and other properties of the fermentation medium. From this kinetic expression, we see that at high substrate concentrations the rate of product formation is independent of C_S and is approximately equal to kC_{E0}. This is due to the presence of a limited amount of enzyme, which is required for the reaction to proceed, and adding more substrate under these conditions will not cause the reaction rate to increase further. At low substrate concentrations, the rate of product formation becomes first-order with respect to C_S. Under these conditions the substrate concentration becomes the determinant for product formation, and increasing C_S produces a proportional increase in rate. The rate is also proportional to the total enzyme concentration under all conditions of substrate concentration.

Substrate can also be converted into product in fermenters by cells or microbes or "bugs," which not only act as the reaction catalyst but also reproduce themselves to promote further reaction. The substrate fed to the cell biomass supplies carbon, hydrogen, and oxygen to the organisms. The substrate is also the energy source for the cells and goes into maintaining their existence and into growing new cells. Sometimes, such as in wastewater treatment, we want the cells to break down the substrate and generate carbon dioxide and water. In other cases, such as yeast production, we are after the cells themselves. Further, such as in chemical or pharmaceutical production, we often want the cells to take the substrate and produce a desired "product" that is one part of the organism's biochemical pathway.

Fermentations may be aerobic when the cells must be in the presence of an O_2 environment or anaerobic when they cannot. Water is the standard fermentation medium and is also one of the products, as is carbon dioxide, which is removed from the liquid and leaves in a vapor product stream since it may have a negative effect on the cells. Other nutrients or media (sources of nitrogen, phosphorus, minerals, vitamins, etc.) typically must be supplied to keep the organisms happy and healthy.

Since many biochemical reactions and their stoichiometry are not well understood, we often find a more empirical approach to the quantitative assessment of the kinetics. Mass concentration units (e.g., g/L) are often used along with yield coefficients to calculate the distribution of products formed and the amount of substrate consumed. In the absence of any inhibition effects and in the presence of an infinite supply of substrate, the rate of cell growth r_X is autocatalytic, that is, it depends only on the concentration of cells (C_X), and the more cells we have, the higher the growth rate. The cell biomass is typically represented by X:

$$r_X = \mu_{\max} C_X \tag{1.27}$$

Here $\mu_{\max}$ is the nomenclature for the maximum cell growth rate [typically in h^{-1} (reciprocal hours)] and C_X is the mass concentration of cells (g/L). Hence the cell growth rate initially is exponential with time (called the *exponential* growth phase).

The value of μ_{max} depends on temperature and pH. Different organisms operate in different optimal temperature and pH ranges. Once we go beyond the boundaries of these ranges (either too low or too high), the organism behavior changes significantly and cannot be represented by the same kinetics.

Unfortunately, the cell growth rate is limited or inhibited by a number of factors. First is the limitation created by the substrate S or some other nutrient. The Monad kinetic model is typically used to represent the behavior of such biochemical systems according to the following equation:

$$\mu = \mu_{max} \frac{C_S}{K_S + C_S} \tag{1.28}$$

Here μ is the specific cell growth rate (h^{-1}), C_S is the substrate concentration, and K_S is a constant that is equal to the value of substrate concentration at half the maximum cell growth rate. This model produces the relationship shown in Figure 1.5. At low substrate concentrations, the growth rate is first-order with respect to C_S. At high substrate concentrations, the specific growth rate is independent of concentration. The value of K_S determines how quickly we reach the maximum specific growth rate.

The reaction rate for cell growth then becomes

$$r_X = \mu C_X = \mu_{max} \frac{C_S\, C_X}{K_S + C_S} \tag{1.29}$$

In addition to the depletion of substrate (or a lack of oxygen in aerobic systems), the cell growth rate can also be slowed by inhibition caused by products generated by the cells themselves (the desired product or byproducts such as carbon dioxide). This is incorporated into the kinetic rate expression by the effect of an inhibition term:

$$r_X = \mu_{max} \frac{C_S\, C_X}{K_S + C_S} \left[1 - \frac{C_P}{C_{P,\,max}}\right]^{n_P} \tag{1.30}$$

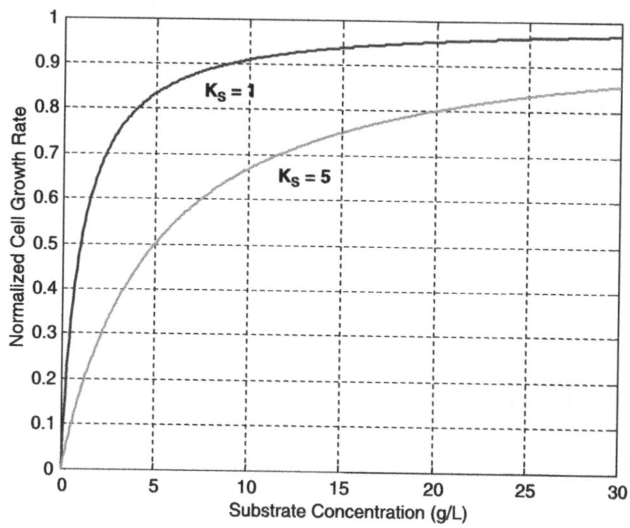

Figure 1.5 Normalized cell growth rate as a function of substrate concentration.

Here C_P is the product concentration, $C_{P,max}$ is the maximum product concentration when cell growth stops, and n_P is the order of inhibition. At low values of C_P, the inhibition term plays no significant role. However, as the product concentration increases, the cell growth rate begins to decrease until the biomass concentration eventually reaches a plateau (what is called the "stationary" phase). From there, the fermentation broth is typically harvested before the cells start to die and the biomass concentration starts to decrease.

Empirically determined yield factors are typically used to relate the mass of cells produced per unit mass of substrate consumed (Y_{XS}) and the mass of product generated per unit mass of biomass produced (Y_{PX}).

Fermentation reactors generally produce heat, so temperature control is an important issue for such reactors. This becomes more aggravated at larger scales when the surface area of a cooling jacket may not be large enough in relation to the volume. Because sterility is a key requirement for successful fermentations, there is a strong reluctance to insert anything, such as cooling coils, into the fermenter itself. The biological nature of the cells, however, has direct consequences on their sensitivity to a temperature runaway. The temperature in the fermenter can increase only so much before it begins to place physiological stress on the cells, which then slows their growth rate and the heat generation rate. If the temperature rises too much, it may be fatal for the organisms. So, temperature control is one key part of effective fermenter control, but the control of other variables, such as pH, vessel backpressure, agitation rate, substrate concentration, and dissolved oxygen concentration (for aerobic systems), is also essential.

1.1.4 Literature

This section has presented a brief review of some of the important kinetic concepts encountered in reactor analysis, modeling, and control. These concepts must be understood within the context of how they affect reactor temperature control and other aspects of reactor control. We recognize that many excellent reference books on chemical reaction engineering are available. These books cover the topic of kinetics and a host of other reactor design concepts in extensive depth. Our intention is not to attempt to provide anything like the scope of that material, so we assume some familiarity with it. A short list of excellent reference books includes

K. G. Denbigh, *Chemical Reactor Theory*, Cambridge University Press, 1965.

H. S. Fogler, *Elements of Chemical Reaction Engineering*, 2nd edition, Prentice-Hall, 1992.

H. F. Rase, *Chemical Reactor Design for Process Plants*, Wiley, 1977.

O. Levenspiel, *Chemical Reaction Engineering*, 3rd edition, Wiley, 1999.

S. M. Walas, *Reaction Kinetics for Chemical Engineers*, McGraw-Hill, 1959.

K. R. Westerterp, W. P. M. van Swaaij, and A. A. C. M. Beenackers, *Chemical Reactor Design and Operation*, Wiley, 1984.

1.2 MULTIPLE REACTIONS

It is most unusual for only a single desired reaction to occur in a chemical reactor. Nature is typically not that generous and exacts a penalty that takes the form of side reactions generating undesired impurity components. The side reactions can involve other transformations of the reactant species (in *parallel* with the desired reaction) or further transformations of the desired product species (in *series* with the desired reaction). Typically we encounter some combination of both types. We discuss each of these schemes in some detail here because they often play a critical role in understanding the behavior of the reactor, how it has to be operated, and also how it can be controlled.

They also have a major impact on the design of the entire process. To suppress undesirable side reactions, it is often necessary to operate the reactor with a low concentration of one of the reactants and an excess of other reactants. These must be recovered in a separation section and recycled back to the reaction section.

1.2.1 Parallel Reactions

The first reaction type is when the reactants form, not just the desired products, but also other undesired products in parallel with the main reaction. We want to show here the implications of parallel reactions, so we consider a simple *batch isothermal* reactor at constant volume:

$$\begin{aligned} \mathrm{A} &\xrightarrow{k_\mathrm{B}} \mathrm{B} \\ \mathrm{A} &\xrightarrow{k_\mathrm{C}} \mathrm{C} \end{aligned} \tag{1.31}$$

Assuming first-order kinetics, we can express the change with time in the concentrations of reactant A (C_A) and products B (C_B) and C (C_C):

$$\frac{dC_\mathrm{A}}{dt} = -(k_\mathrm{B} + k_\mathrm{C})\, C_\mathrm{A} \tag{1.32}$$

$$\frac{dC_\mathrm{B}}{dt} = k_\mathrm{B}\, C_\mathrm{A} \tag{1.33}$$

$$\frac{dC_\mathrm{C}}{dt} = k_\mathrm{C}\, C_\mathrm{A} \tag{1.34}$$

The kinetic rate constants are k_B and k_C. We can analytically solve these differential equations, assuming that we start at time zero with only reactant A ($C_{\mathrm{A}0}$):

$$\frac{C_{\mathrm{A}(t)}}{C_{\mathrm{A}0}} = e^{-(k_\mathrm{B}+k_\mathrm{C})t} \tag{1.35}$$

$$\frac{C_{\mathrm{B}(t)}}{C_{\mathrm{A}0}} = \frac{k_\mathrm{B}}{k_\mathrm{B}+k_\mathrm{C}}\left[1 - e^{-(k_\mathrm{B}+k_\mathrm{C})t}\right] \tag{1.36}$$

$$\frac{C_{\mathrm{C}(t)}}{C_{\mathrm{A}0}} = \frac{k_\mathrm{C}}{k_\mathrm{B}+k_\mathrm{C}}\left[1 - e^{-(k_\mathrm{B}+k_\mathrm{C})t}\right] \tag{1.37}$$

Figure 1.6 shows the normalized concentrations as functions of time during the batch for different values of the two rate constants k_B and k_C. The higher the value of k_B compared with k_C, the more of product B is generated. This is expressed in terms of the selectivity (S) to the desired product, which we preferentially want to be as large as possible:

$$S = \frac{C_\mathrm{B}}{C_\mathrm{C}} \tag{1.38}$$

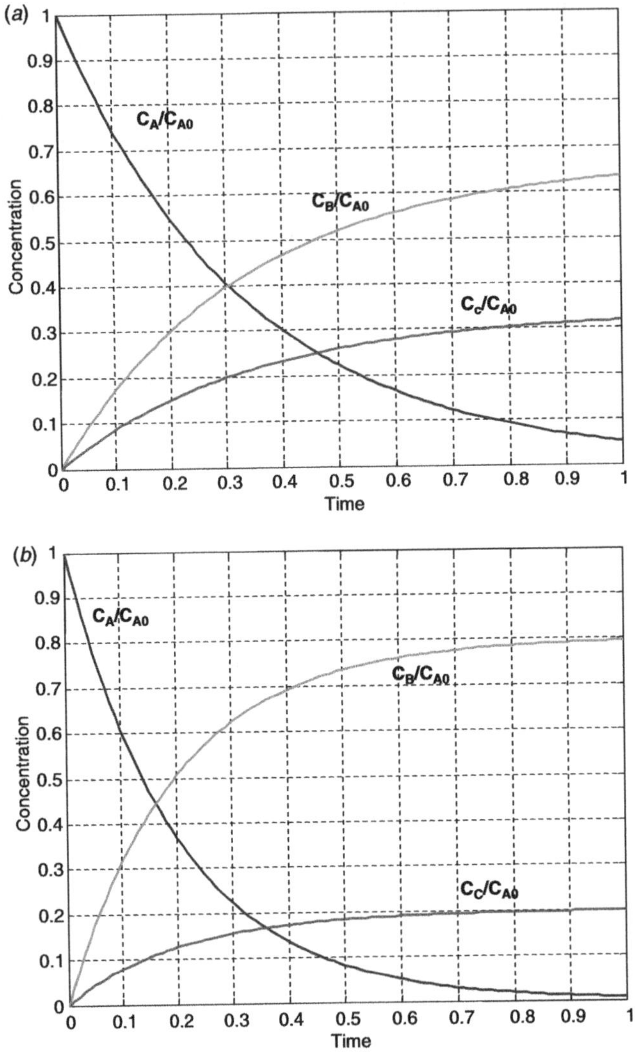

Figure 1.6 Concentrations for parallel reactions: (*a*) $k_B = 2$, $k_C = 1$; (*b*) $k_B = 4$, $k_C = 1$.

One aspect of optimizing the operation of a batch reactor is establishing the temperature such that the selectivity is as high as possible. If the activation energies of the two reactions are different, changing temperature shifts the ratio of the rates.

If the chemistry involves two reactants, selectivity is affected by the concentrations of the reactants. For example, supposed that there are two parallel reactions in which C is the desired product and D is the undesired product:

$$\begin{aligned} \mathrm{A} + \mathrm{B} &\xrightarrow{k_C} \mathrm{C} \\ \mathrm{A} &\xrightarrow{k_D} \mathrm{D} \end{aligned} \tag{1.39}$$

Keeping the concentration of A low in the reactor and the concentration of B high in the reactor will help improve the yield of the desired product. An important industrial example of this type of system is the production of isooctane from the reaction of isobutene and isobutane. The isobutene can react with itself to form polymer, so a large excess if isobutane is used and the concentration of isobutene is kept small by distributing the fresh feed among a number of reactors.

1.2.2 Series Reactions

The second reaction type involves reactants forming products, but then the products undergo further reaction in series with the main reaction. We want to show here the implications of series reactions, so we consider a simple batch isothermal reactor at constant volume:

$$\mathrm{A} \xrightarrow{k_\mathrm{B}} \mathrm{B} \xrightarrow{k_\mathrm{C}} \mathrm{C} \tag{1.40}$$

Assuming first-order kinetics, we can write the change with time in the concentrations of reactant A (C_A) and products B (C_B) and C (C_C):

$$\frac{dC_\mathrm{A}}{dt} = -k_\mathrm{B}\, C_\mathrm{A} \tag{1.41}$$

$$\frac{dC_\mathrm{B}}{dt} = k_\mathrm{B}\, C_\mathrm{A} - k_\mathrm{C}\, C_\mathrm{B} \tag{1.42}$$

$$\frac{dC_\mathrm{C}}{dt} = k_\mathrm{C}\, C_\mathrm{B} \tag{1.43}$$

The kinetic rate constants are k_B and k_C. We can solve these differential equations analytically, assuming that we start at time zero with only reactant A ($C_{\mathrm{A}0}$).

$$\frac{C_{\mathrm{A}(t)}}{C_{\mathrm{A}0}} = e^{-k_\mathrm{B} t} \tag{1.44}$$

$$\frac{C_{\mathrm{B}(t)}}{C_{\mathrm{A}0}} = \frac{k_\mathrm{B}}{k_\mathrm{C} - k_\mathrm{B}}\left[e^{-k_\mathrm{B} t} - e^{-k_\mathrm{C} t}\right] \tag{1.45}$$

$$\frac{C_{\mathrm{C}(t)}}{C_{\mathrm{A}0}} = 1 - \frac{C_{\mathrm{A}(t)}}{C_{\mathrm{A}0}} - \frac{C_{\mathrm{B}(t)}}{C_{\mathrm{A}0}} \tag{1.46}$$

Figure 1.7 shows typical composition profiles. Notice that there is a peak in the C_B at some point in time. The higher the value of k_B relative to k_C, the higher the peak in C_B and the earlier in the batch the peak occurs. The batch should be stopped when the peak occurs if we wish to maximize selectivity. Thus batch time is an important operating parameter for series reactions. This is not the case for parallel reactions. Reactor temperature should be adjusted to favor k_B.

If series reactions are conducted in a CSTR, the concentrations in the reactor can be adjusted to influence selectivity and conversion. Because the production of the undesirable product C depends on the concentration of the desired product B, this concentration should be kept small. The reactor can be operated with low conversion (small concentration of B).

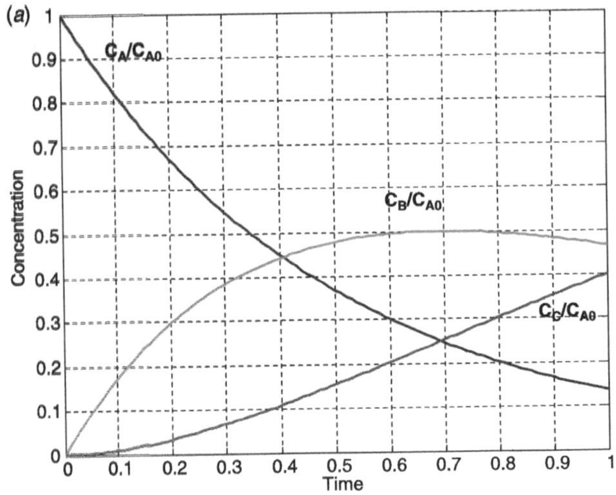

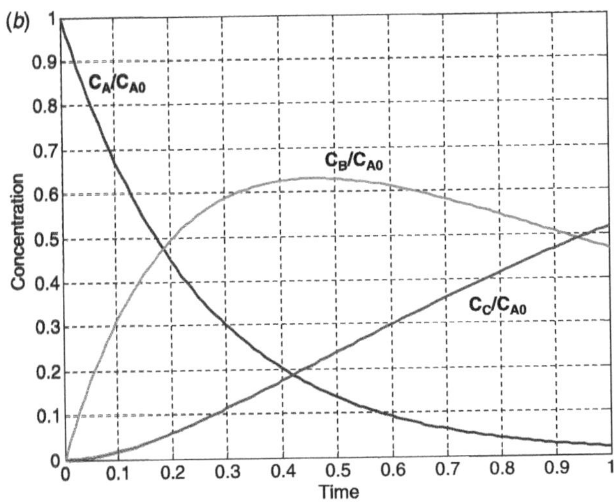

Figure 1.7 Concentrations for series reactions: (*a*) $k_B = 2$, $k_C = 1$; (*b*) $k_B = 4$, $k_C = 1$.

Of course, this means that the concentration of A is large, so recovery and recycle of unreacted A is required to make the process economical.

In the case where there are two reactants, one of which is involved in an undesirable series reaction ($A + B \rightarrow C$ and $C + B \rightarrow D$), the concentration of B in the reactor can be kept small to improve selectivity. An important industrial example of this type of series reactions is in the production of ethylbenzene. The desired reaction is the formation of ethylbenzene from ethylene and benzene. The undesirable reaction is the formation of diethylbenzene from ethylene and ethylbenzene. To suppress this second series reaction, the concentration of ethylene is kept low and an excess of benzene is employed, which must be recovered and recycled.

This classical tradeoff between selectivity and recycle is considered in several examples in subsequent chapters.

1.3 DETERMINING KINETIC PARAMETERS

The many preexponential factors, activation energies and reaction order parameters required to describe the kinetics of chemical reactors must be determined, usually from laboratory, pilot plant, or plant experimental data. Ideally, the chemist or biologist has made extensive experiments in the laboratory at different temperatures, residence times and reactant concentrations. From these data, parameters can be estimated using a variety of mathematical methods. Some of these methods are quite simple. Others involve elegant statistical methods to attack this nonlinear optimization problem. A discussion of these methods is beyond the scope of this book. The reader is referred to the textbooks previously mentioned.

In many practical applications, the engineer often has only plant performance data to use to backcalculate kinetic parameters. Data of this type are seldom extensive enough to permit precise calculation of all parameters since the plant normally operates in a fairly narrow window of operating conditions. However, useful simplified kinetics and parameters can often be determined that describe the major kinetics inside this region. Extrapolation outside the region from which the data has been obtained is very risky.

1.4 TYPES AND FUNDAMENTAL PROPERTIES OF REACTORS

In this section we discuss in a qualitative way the *classical* types of reactors: batch, continuous stirred-tank reactor (CSTR), and plug flow reactor (PFR). Our purpose is to point out the features of each that impact the ease or difficulty of their temperature control.

These classical reactors are idealizations of real industrial reactors. Perfect mixing is assumed in classical batch and CSTR reactors, but mixing is never perfect in an agitated vessel, no matter how intense the mixing. No axial mixing and no radial gradients (plug flow) are assumed in the classical PFR tubular reactor, but the flow patterns in a real tubular reactor are never without some axial mixing and differences in flow velocities and properties at different radial positions. However, the classical idealizations are usually close enough to reality so that they can be used for studying both steady-state design and the dynamic control of chemical reactors.

1.4.1 Continuous Stirred-Tank Reactor

Figure 1.8 shows a vessel with an agitator for mixing, a jacket that surrounds the vessel for heating or cooling, feedlines entering the vessel and a liquid product stream exiting from the bottom. The liquid in the reactor is assumed to be perfectly mixed, that is, with no radial, axial, or angular gradients in properties (temperature and composition). The product stream has a composition and a temperature that are exactly the same as the contents of the liquid throughout the vessel. This is always true, both under steady-state conditions and dynamically at any point in time.

This characteristic of a CSTR immediately generates an inherent weakness of the CSTR type of reactor, that is, the concentration of reactant in the vessel is the same as the concentration of reactant in the product. The concentration of reactant is inversely related to conversion.

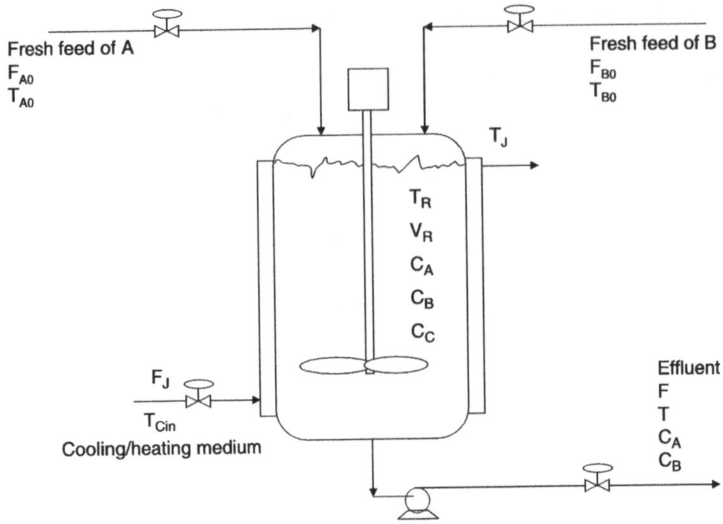

Figure 1.8 CSTR with jacket.

Fractional conversion χ defined as

$$\chi = \frac{C_{A0} - C_A}{C_{A0}} \tag{1.47}$$

If a high conversion is desired, the reactant concentration must be small. But the reaction rate depends directly on the reactant concentration. It also depends on the reactor volume. So, if a high conversion desired, the reactor must be large to compensate for the small reactant concentration. Thus a single CSTR is seldom used if high conversion is desired. Of course, using several CSTRs in series is one way to reduce the total reactor volume because only the last vessel will have the small reactant concentration.

We will develop detailed steady-state and dynamic mathematical models of CSTRs in Chapters 2 and 3 with several types of reactions and quantitatively explore the effect of kinetic and design parameters on controllability. For the moment, let us just make some qualitative observations. There are several features of a CSTR that impact controllability:

1. A variety of methods and configurations can be used for heat transfer. These are described in Section 1.5. Since heat transfer is one of the key issues in reactor control, the CSTR is usually more easily controlled than a tubular reactor. It is physically difficult to adjust the heat removal down the length of a tubular reactor.
2. The temperature of the feed has some effect on controllability, but it is much less important in a CSTR than in a tubular reactor, as discussed in Section 1.4.3. If heat is being removed from the reactor, a feed that is at a lower temperature than the temperature in the reactor will reduce the heat transfer requirements.
3. Conversion is the fraction of a reactant that is fed to the reactor that reacts in the reactor. The level of conversion in a CSTR has a very significant impact on its stability and controllability. This is discussed in detail in Chapter 2. A high conversion means a small reactant concentration in the reactor vessel, so there is little "fuel"

available to permit a reactor runaway. On the other hand, a low conversion means that there is plenty of reactant available to react. If the reaction is exothermic and irreversible, a reactor temperature runaway can more easily occur in a CSTR operating with low reactant conversion than in one operating with high reactant conversion. In addition to affecting reactant concentration, the design conversion affects reactor size. Low conversion means a smaller reactor. This small reactor has less heat transfer area if an external jacket or an internal coil is used, which has a negative impact on controllability.

1.4.2 Batch Reactor

The classical batch reactor is a perfectly mixed vessel in which reactants are converted to products during the course of a batch cycle. All variables change dynamically with time. The reactants are charged into the vessel. Heat/or catalyst is added to initiate reaction. Reactant concentrations decrease and product concentrations increase with time. Temperature or pressure is controlled according to some desired time trajectory. Batch time is also a design and operating variable, which has a strong impact on productivity.

Temperature profiles are established so that conversion and yield objectives are achieved while not exceeding heat transfer capacity limitations. These optimum temperature profiles depend on the chemistry. For example, if the reaction is reversible and exothermic, the temperature profile may ramp up to a high temperature to get the reactions going and then drop off with time to avoid the decrease in the chemical equilibrium constant at high temperature. If the reaction is reversible and endothermic, the temperature profile would rise to the highest possible temperature as quickly as possible because the chemical equilibrium constant increases with temperature.

If all the reactants are charged to the reactor, the reactant concentrations are initially large, which means that the reaction rate is high and the heat transfer load is high at the beginning of the batch cycle unless the temperature is kept low. The initial high reactant concentration problem can be avoided by using a "fed-batch reactor." Some material is initially charged to the reactor, but most of the reactant is fed during the course of the batch cycle. This causes the volume of the liquid in the reactor to increase with time, so volume as well as compositions and temperatures are all time-varying.

Several special features of a batch reactor impact control:

1. The process is inherently time-varying. There is no steady state. This causes process parameters to change with time, which means that controller parameters may have to change with time. Control strategies such as "gain scheduling" (changing controller gain and integral time) are frequently required in batch reactor control.
2. Rigorous nonlinear models must be used in analyzing batch reactors because of the changing process parameters. Continuous reactors operate around some steady-state level, so linear models are sometime adequate for establishing controller tuning constants.
3. Selecting the best time–temperature trajectory is a challenging dynamic optimization problem with constraints. There are rigorous nonlinear programming approaches to this problem, but there are also some more simple and practical methods that can be employed, as discussed in Chapter 4.
4. All the heat transfer configurations used on CSTRs can be applied to batch reactors.

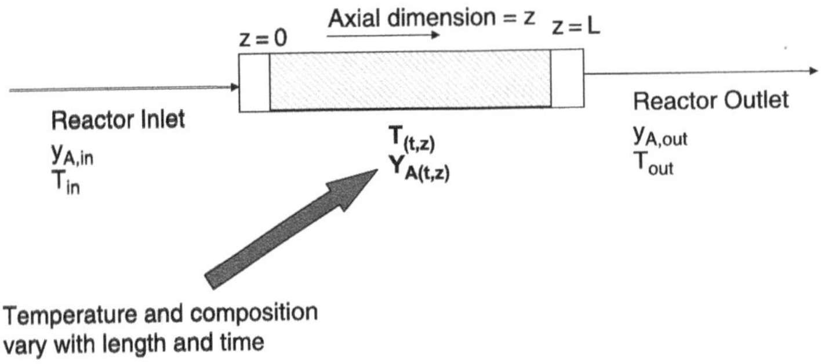

Figure 1.9 Adiabatic tubular reactor.

Mathematical models of batch reactors and control strategies are developed in Chapter 4. Both classical batch and fed-batch reactors are discussed using numerical examples.

1.4.3 Tubular Plug Flow Reactor

Figure 1.9 gives a sketch of a typical adiabatic tubular reactor. The major distinguishing characteristic of tubular reactors is their distributed-parameter nature, that is, variables change with physical dimensions as well as with time. The classical plug flow reactor assumes that the reactor vessel is cylindrical, that fluid flows down the length of the reactor with a flat velocity profile, that no axial mixing occurs, and that no radial gradients exist in temperature or compositions.

The tubular reactor can be an empty vessel if no catalyst is used. If a solid catalyst is required, the vessel is packed with catalyst, either in a bed or inside tubes. The dynamic behavior of the reactor is significantly affected by the presence of catalyst in the reactor because the thermal capacitance of the catalyst is usually greater than that of the process fluid, particularly if the system is gas-phase. The temperatures of both the process fluid and the catalyst change with time. Of course, under steady-state conditions, the two temperatures are equal at any axial position.

There are several modes of operation of tubular reactors:

1. *Adiabatic.* There is no heat transfer to or from the reactor. Temperature and compositions change with length. Since there is no heat transfer, there are no radial gradients in temperature. The adiabatic temperature change depends on the per-pass conversion in the reactor and the amount of material fed to the reactor and its heat capacity. The adiabatic temperature change is small if conversion is low. If the feed contains materials (inerts or product components) that do not reactant, this material can serve as a "heat sink" to reduce the adiabatic temperature change. The sensible heat of this material can soak up some of the heat of reaction. Of course, this nonreacting material usually has to be recovered and recycled, so this mode of operation increases the capital and energy costs of the separation section of the plant.

2. *With Heat Transfer.* The tubular reactor is constructed in a similar way as a tube-in-shell heat exchanger or a fired furnace. Process fluid flows inside the tubes and is cooled or heated by the heat transfer medium within the shell. Radial temperature gradients are inherent in tubular reactors with heat transfer, so the plug flow assumption

is less accurate. These radial gradients depend strongly on tube diameter and fluid properties and fluid velocities. The larger the tubes, the larger the radial temperature gradients. One standard reactor design and development procedure is to study the system and/or catalyst in a single tube in the laboratory or pilot plant and then use multiple tubes of the same diameter in parallel in the plant reactor. The furnace or reactor used in steam–methane reforming to produce synthesis gas (a mixture of hydrogen, carbon monoxide, and carbon dioxide) is an important example. The furnace has multiple parallel tubes that are heated by burning fuel to provide the required heat to drive the endothermic reactions at a high temperature level. The problem of flow maldistribution among a large number of parallel tubes presents further potential complications.

3. *Adiabatic with Intermediate Heat Transfer.* Many tubular reactor systems use a series of adiabatic reactors with heating or cooling between the reactor vessels. For example, naphtha reforming has endothermic reactions of removing hydrogen from saturated cyclical naphthene hydrocarbons to form aromatics. The process has multiple adiabatic reactors with fired furnaces between the reactors to heat the material back up to the required reactor inlet temperature.

4. *Adiabatic with "Cold-Shot Cooling."* Some exothermic reactions are conducted in vessels with multiple beds of catalyst, which operate adiabatically (temperature increases through the bed). At the exit of each bed, a cold stream is mixed with the hot stream leaving the bed to bring the temperature back down to the desire inlet temperature for the downstream bed. This cold stream is typically some of the feedstream that has been bypassed around the reactor feed preheating system.

All of these alternatives are discussed in Chapter 5.

The control of tubular reactors is probably the most difficult of all reactor systems. The reasons for this difficulty and the special features of tubular reactors are summarized below:

1. The distributed nature of the process leads to complex dynamic responses in which axial changes in variables can sometimes result in counter-intuitive dynamic behavior. One example of this is the "wrongway" response that occurs in some adiabatic packed-bed tubular reactors. A decrease in the reactor inlet temperature will eventually result in a lower reactor exit temperature. But there may be a transient *increase* in the exit temperature. This is caused by the colder feed decreasing the temperature in the front end of the reactor. The lower reaction rate consumes less reactant, so the reactant concentration increases at locations further down the reactor. The solid catalyst packing is still hotter at this location because of its thermal capacitance. So the combination of higher reactant concentration and higher temperature causes a rapid reaction rate at locations further down the length of the reactor, which raises the temperature above the normal steady-state value that will eventually be established.

2. Temperature and composition transients move in waves down the length of the reactor, and this can lead to limit cycles when the reactor is part of a complete plant with feed preheating and recycle streams.

3. Tubular reactors often have high-temperature limitations because of the occurrence of undesirable reactions, catalyst degradation, or materials of construction. This means that the maximum temperature anywhere in the reactor cannot exceed this limit. An exothermic reaction in an adiabatic reactor produces a maximum temperature at the exit under steady-state conditions. An exothermic reaction in a cooled reactor can

have the maximum temperature at some intermediate axial position or at the end. Both the magnitude and the location of the "peak" temperature vary with the design of the system and the age of the catalyst. They also vary with the operation of the system as disturbances occur. Controlling this peak temperature requires that multiple temperature measurements must be used down the length of the reactor for its detection.

4. Feed temperature is a very important design parameter in tubular reactors. A low feed temperature results in low reaction rates. So a long reactor is required to achieve the desired level of conversion. A high feed temperature results in a high exit temperature. If the reaction is exothermic and if there is a maximum temperature limitation, the per-pass conversion may have to be reduced or more "heat sink" material may have to be fed to lower the temperature rise. This usually means higher recycle flowrates with the associated higher capital and energy costs of the downstream separation system.

5. In cooled or heated tubular reactors the heat transfer options are limited. It is mechanically very difficult to change the temperature of the cooling or heating medium with axial position. The usual configuration is an essentially constant temperature of the heat transfer medium down the length of the reactor. In systems requiring cooling at fairly high temperatures, steam is generated on the shell side of the reactor to remove heat. The steam temperature is the same at any axial position. In systems requiring heating, burning fuel or condensing high-pressure steam is used at an essentially constant temperature at any axial position. If a heat transfer medium is used that does change in temperature down the length of the reactor, the available design parameters are the direction of flow (co-current or countercurrent flow of the cooling medium with respect to the direction of the process flow), the inlet temperature of the heat transfer medium and its flowrate. All of these must be balanced so that the desired temperatures and conversions are achieved. In systems requiring cooling at very high temperatures, molten salt is sometimes used as the heat removal medium.

The need to reduce energy consumption and reuse the exothermic heat of reaction so that we achieve a certain inlet temperature in a tubular reactor often leads to the use of a feed-effluent heat exchanger (FEHE). This can create some challenging control problems. Consider an exothermic reaction occurring in an adiabatic tubular reactor that has an inlet temperature of 450 K and an exit temperature of 500 K. One inefficient way to achieve the inlet temperature is to use a hot utility (steam or combustion of fuel) to heat up the cold feed. Energy can be saved by using the hotter reactor effluent stream in a FEHE. We study the dynamic problems that occur in reactor–FEHE systems like this in Chapter 4 and show that the positive feedback of energy can produce an openloop unstable process. The system can be made closedloop-stable by the use of an inlet temperature controller that bypasses cold material around the heat exchanger and mixes it with the heated stream to achieve the desired inlet reactor temperature. Chapter 7 contains a quantitative discussion of the interesting steady-state and dynamic tradeoff between energy and controllability in this type of system.

1.5 HEAT TRANSFER IN REACTORS

Batch and CSTR reactors can be cooled or heated in a variety of ways, which accounts in part for their superior controllability compared to tubular reactors. Figure 1.10*a*–1.10*f* show several of these alternatives.

The use of a jacket surrounding the reactor vessel is probably the most common method for providing heat transfer because it is relatively inexpensive in terms of equipment capital cost (see Fig. 1.10*a*). If heating is required, steam is condensed in the jacket or a hot heat transfer fluid stream is fed to the jacket. If cooling is required, a cooling medium is fed to the jacket. For moderate reactor temperatures (between 50 and 80°C), cooling water at 30°C is typically used. For lower temperature reactors, a cold refrigeration stream (brine) is used.

For reactor temperatures between 80 and 130°C, a tempered water or oil cooling medium is used. Plain cooling water should not be used because the large temperature difference between the reactor and the cooling medium leads to dynamic control problems. This is illustrated quantitatively in Chapter 2. It occurs because the temperature difference can be changed by only a small amount, which means that the heat removal rate cannot be changed much. Therefore the magnitude of the dynamic upsets that can be handled is quite limited.

For reactor temperatures above 130°C, steam can be generated in the jacket at a suitable pressure (to provide a 30–50°C temperature differential between the steam and the reactor; see Fig. 1.10*b*). Reactors operating at very high temperatures usually employ a molten salt for heat removal.

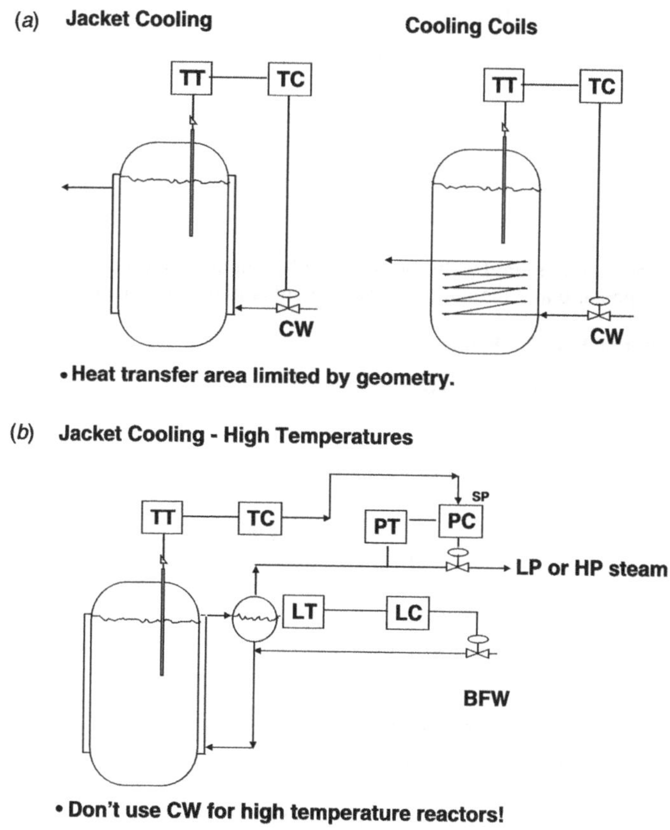

Figure 1.10 Reactor heat transfer methods.

(*c*) **Jacket Cooling; Circulating Water System; Direct Addition of CW**

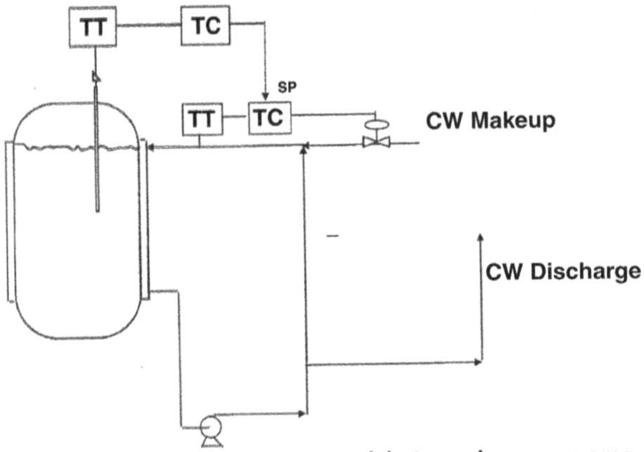

- **Gives better temperature control, but requires more area.**

Jacket Cooling; Circulating Water System; Heat Exchanger and Bypass

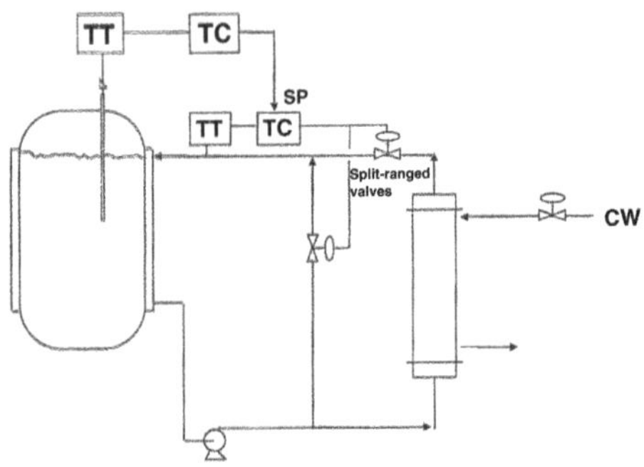

- **Can use tempered water or oil for high temperature reactor**
- **Bypassing and blending avoids dynamics of heat exchanger**

(*d*) **External Heat Exchanger**

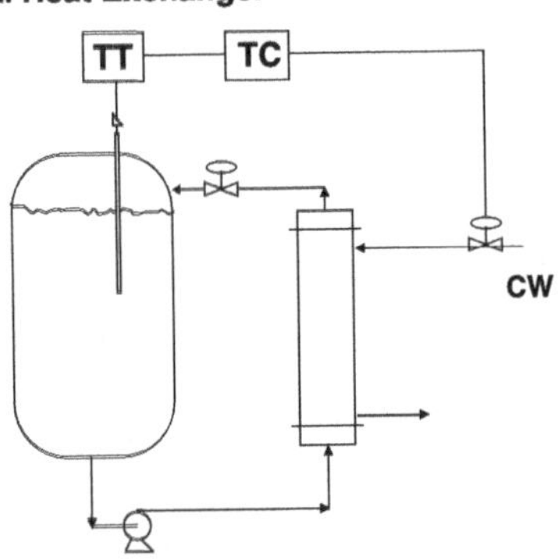

- **Area independent of reactor size.**
- **Must be able to pump reaction mass.**

Figure 1.10 *Continued.*

(*e*) **External Heat Exchanger with Blending and Cascade Control**

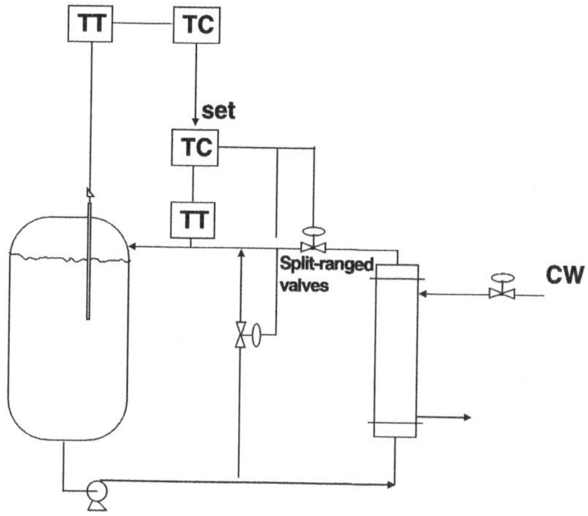

(*f*) **Autorefrigeration**

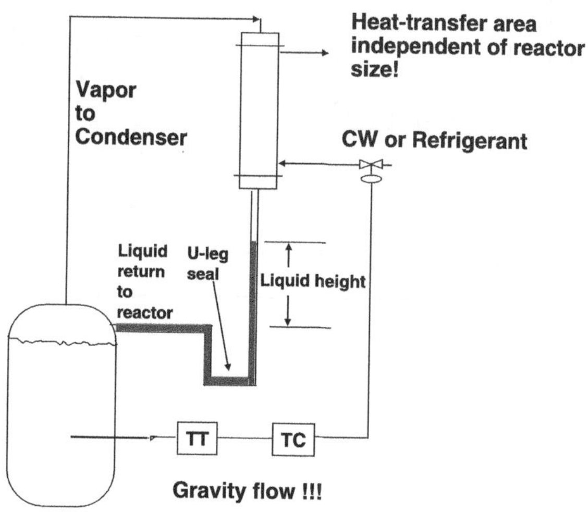

Figure 1.10 *Continued.*

Instead of using a once-through system of coolant through the jacket or coil, circulating cooling water systems are frequently used, as shown in Figure 1.10*c*. The circulation rate is large, so all the coolant in the jacket is essentially at the same temperature. The flowrate is fixed, so the heat-transfer coefficient does not change with the flowrate of make-up water, as is the case with a once-through system. This results in tighter reactor temperature control. Circulating water systems will be used in many of the examples in this book.

These circulating coolant systems come in two flavors. As shown in the top flowsheet in Figure 1.10*c*, the cold make-up water can be added directly into the circulating loop as needed to control the temperature of the circulating water. The hot water is removed to keep the inventory in the circulating loop constant. As shown in the bottom flowsheet in Figure 1.10*c*, an external heat exchanger that is cooled by some cooling medium can be used in the circulating loop to provide cooling. The material in the circulating loop can be tempered water or oil if reactor temperature is too high for cooling water. The bypassing and blending of hot and cold streams give very tight temperature control of the circulating fluid.

Another effective alternative is to circulate the reaction liquid through an external heat exchanger as shown in Figure 1.10*e* and 1.10*d*. This system has the advantage of being able to provide whatever heat transfer area is desired. With jacket and coil systems, the heat transfer area is limited by the physical dimension of the reactor vessel. With the external heat exchanger, the heat transfer area can be easily set at any desired level. As discussed in Section 1.6, this feature avoids some of the problems of reactor scaleup. Of course, the scheme requires pumping the liquid in the reactor, which can be undesirable in some processes with toxic materials (that can leak in pump seals), or with materials that are sensitive to shearing (some biological systems) or can foul the small tubes in the external heat exchanger.

The final heat removal scheme discussed here is called *autorefrigeration* or *evaporative cooling* (Fig. 1.10*f*). The pressure in the reactor is adjusted so that the liquid can boil if

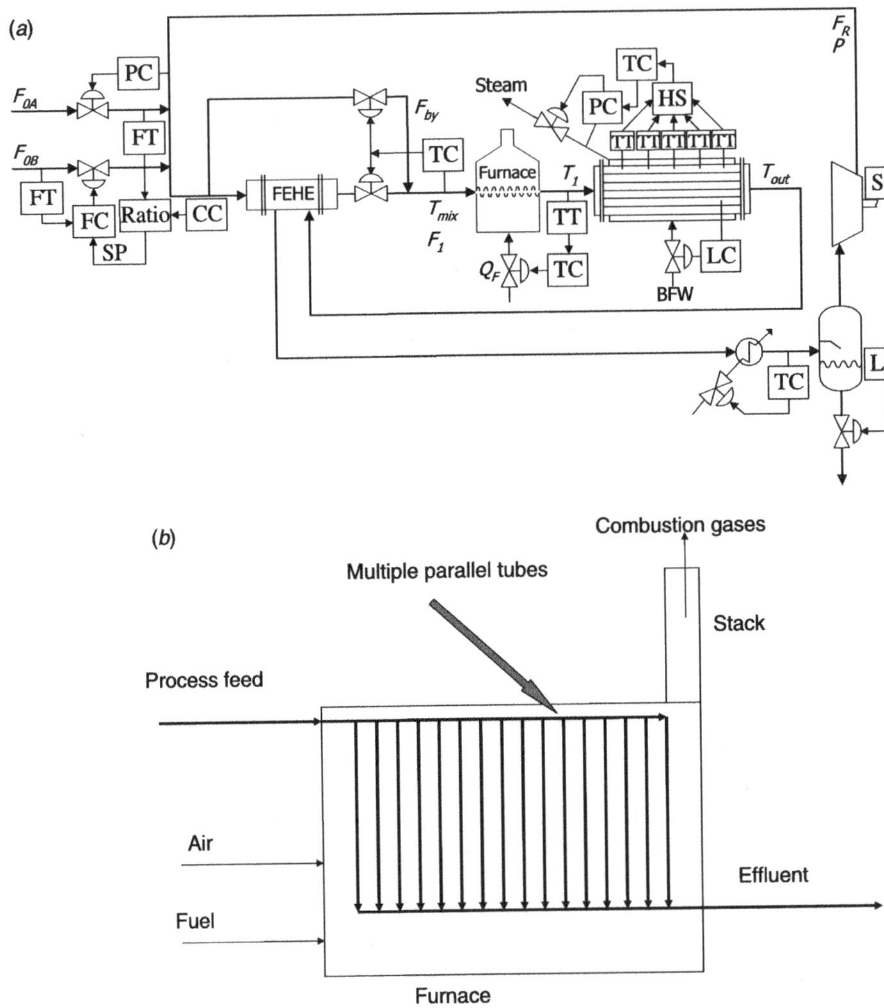

Figure 1.11 (*a*) Cooled tubular reactor; (*b*) heated tubular reactor.

temperature increases. The latent heat of vaporization of this phase change removes heat from the reactor. The vapor flows up to a condenser, and the condensed liquid is returned back to the reactor. The system has several desirable features. The boiling facilitates mixing of the liquid and prevents hotspots. The heat transfer area can be as large as desired, so scaleup problems are reduced.

However, this method also has several undesirable features. In the usual configuration, gravity with a U-leg seal is used to get the liquid to flow from the condenser back into the reactor, which is at a higher pressure (provides the driving force for the vapor to flow from the reactor to the condenser). The hydraulics of these gravity flow systems can lead to severe control problems. For example, if the vapor rate increases by 50%, the pressure drop in the vapor line increases by 225% (1.5 squared is 2.25). This means that the height of liquid in the U-leg must more than double. A large upset can back liquid up into the condenser and cover heat transfer area, which limits heat removal and can lead to a temperature runaway. Autorefrigerated systems must be designed for very small vapor pressure drop and enough elevation difference between the reactor and the condenser to provide a height of liquid that will handle the worst-case situation.

The other disadvantage of this type of system is that the temperature in the condenser is lower than in the reactor because of both the lower pressure and the more volatile components that are in the vapor phase. This means that a lower-temperature coolant must be used in the condenser compared to what could be used in a jacket- or coil-cooled system.

Many tubular reactors are operated adiabatically because of the problems in providing heat transfer. Figure 1.11*a* shows a complete gas-phase reaction process with a high-temperature tubular reactor that is cooled by generating steam. Figure 1.11*b* shows a fuel-fired furnace being used as a tubular reactor.

1.6 REACTOR SCALEUP

One of the most challenging aspects of chemical engineering is the problem of scaling up a process unit from a small laboratory or pilot plant to a large commercial size. Reactors are perhaps one of the more difficult to deal with. In this section we show quantitatively what the heat transfer scaleup problem is for a CSTR.

Suppose that a pilot plant reactor has a volume of 0.019 m^3 [5 gal (gallons)] and is fed 3.506 g/s of feed with a density of 801 kg/m^3 and a temperature of 294 K. The reaction is $A \rightarrow B$, which takes place in the liquid phase at a reactor temperature of 333 K. The concentration of reactant in the feed is 8.01 $kmol/m^3$ and the specific reaction at 333 K is $2.409 \times 10^{-4}\ s^{-1}$. With the given feedflow, reactor volume, and temperature, the reactant concentration of the product stream is 3.926 $kmol/m^3$ (conversion is 51%).

The reactor is cooled by a circulating cooling water system. The heat of reaction is 69.71×10^6 J/kmol. The heat that must be transferred to the jacket is 817 W. If an aspect ratio (height to diameter) of 2 is assumed, the diameter and jacket area can be calculated:

$$\text{Volume} = 0.019\,\text{m}^3 = \left(\frac{\pi D^2}{4}\right)(L) = \left(\frac{\pi D^2}{4}\right)(2D) = \frac{\pi D^3}{2} \Rightarrow D = 0.229\,\text{m}$$

$$\text{Jacket area} = \pi DL = \pi D(2D) = 2\pi D^2 = 0.3295\,\text{m}^2$$

Assuming an overall heat transfer coefficient of 851 W K^{-1} m^{-2}, the required temperature differential between the reactor and the jacket is only 2.9 K, giving a jacket temperature of 330 K. If the supply cooling water temperature is 294 K, the cooling water makeup flowrate is 5.43 g/s.

If the flowrate of the makeup cooling water were made very large, the jacket temperature could be reduced to almost 294 K, which would give a differential temperature of 333 − 294 = 39 K. But we are using only 2.9 K of the potential driving force. Thus there is plenty of cooling "muscle" available to achieve very tight temperature control.

Now suppose that we design a plant-scale reactor that is 1000 times larger in volume (19 m^3). The feed flowrate is 1000 times larger (3.506 kg/s), and the required heat transfer is 1000 times larger (817 kW).

Assuming the same aspect ratio ($L/D = 2$), the diameter is 10 times larger (2.29 m), which gives a heat transfer area that is only 100 times larger (32.95 m^2). If the overall heat transfer coefficient is the same as in the pilot plant (we will come back to this issue in Chapter 2), the required temperature differential between the reactor and jacket increases by a factor of 10 (jacket temperature is 304 K instead of 330 K). The flowrate of makeup cooling water (19.54 kg/s) increases by a factor of 4000.

In this large reactor the temperature differential driving force under design conditions is 333 − 304 = 29 K. The largest it can ever be is 333 − 294 = 39 K. Since we are using a large fraction of this maximum differential temperature, there is less "muscle" available. We demonstrate quantitatively in Chapter 3 the deterioration in temperature control as a larger fraction of the maximum differential is used.

One issue that should be mentioned here is the selection of the aspect ratio. We have used $L/D = 2$ in the example above, and this is commonly used for reactors. But why not use a larger aspect ratio? The larger the aspect ratio, the larger the heat transfer area for a given volume, which would improve temperature control. For example, the 19 m^3 reactor would have a diameter of 1.822 m and a heat transfer area of 41.71 m^2 if an aspect ratio of 4 were used. Using an aspect ratio of 10 gives a diameter of 1.446 m and a heat transfer area of 52.55 m^2. These areas should be compared with the smaller 32.95 m^2 in the $L/D = 2$ design.

There are two factors that explain why aspect ratios between 1 and 2 are frequently used for reactors. The first is the capital cost. The weight of metal required to build a reactor of a fixed volume is minimized using an aspect ratio of ~ 1. The second consideration is mixing. It becomes more difficult to achieve good mixing as the aspect ratio increases. More details are provided in Chapter 2.

1.7 CONCLUSION

Several fundamental concepts have been reviewed in the chapter, and some of the important characteristics of the different types of reactors have been discussed. Most of the treatment in this chapter has been qualitative in nature so as to convey basic ideas.

The next two chapters delve into the dynamics and control of CSTR systems in a much more detailed and quantitative way. The effects on controllability of a variety of parameter values (specific reaction rates, throughput, heats of reaction, and heat transfer coefficients), heat removal schemes, and design conversion levels are studied.

CHAPTER 2

STEADY-STATE DESIGN OF CSTR SYSTEMS

In this chapter we study the steady-state design of perfectly mixed, continuously operating, liquid-phase reactors. The effects of a wide variety of reaction types, kinetics, design parameters, and heat removal schemes are explored. The important effects of design conversion and design temperature on heat transfer area and other process parameters are quantitatively studied.

The important message we are trying to convey in this chapter is that steady-state design results can be used to yield valuable information that gives reasonably reliable indications of how effectively the reactor can be dynamically controlled. A parameter for assessing controllability at the steady-state design stage will be presented. This parameter indicates the fraction of the maximum available temperature driving force that is actually needed under steady-state design conditions. For example, when a circulating jacket cooling water system is used, the design value of the temperature difference is the reactor temperature minus the *jacket* temperature. The maximum temperature differential is the reactor temperature minus the coolant *supply* temperature. A large fraction indicates a design in which only limited increases in the rate of heat removal are possible, and thus reactor temperature control may be poor.

Several important types of reactions are considered in the following sections. The equations describing each of these systems are developed. The steady-state design of CSTRs with these reactions are discussed, using Matlab programs for hypothetical chemical examples and the commercial software Aspen Plus for a real chemical example.

2.1 IRREVERSIBLE, SINGLE REACTANT

Let us begin with the simplest possible example. Although this example may seem unrealistic and trivial, we will show that it provides a lot of insight into the thermal

Chemical Reactor Design and Control. By William L. Luyben

effects of important parameters. Moreover, these effects carry over into more complex chemical reactions and systems. The first-order, exothermic irreversible reaction takes place in a CSTR:

$$A \rightarrow B \tag{2.1}$$

The reaction rate is

$$\mathfrak{R} = kC_A = C_A k_0 \exp\left(\frac{-E}{RT_R}\right) \tag{2.2}$$

where $\mathfrak{R}$ = rate of consumption of reactant A (kmol $s^{-1}m^{-3}$)
k = specific reaction rate (s^{-1})
C_A = concentration of reactant A in reactor (kmol/m^3)
k_0 = preexponential factor (s^{-1})
E = activation energy (J/kmol)
R = 8314 J $kmol^{-1}$ K^{-1}
T_R = reactor temperature (K)

The reactor steady state is described by three algebraic balances: a total mass balance, a component balance (since there are only two components), and an energy balance:

Total mass balance (kg/s):

$$\rho_0 F_0 = \rho F \tag{2.3}$$

Component A balance (kmol A/s):

$$\begin{aligned} F_0 C_{A0} &= FC_A + V_R \mathfrak{R} \\ &= FC_A + V_R k C_A \end{aligned} \tag{2.4}$$

Reactor energy balance (J/s):

$$\rho_0 c_{p0} F_0 T_0 = \rho c_p F T_R + \lambda V_R \mathfrak{R} + Q \tag{2.5}$$

where ρ_0 = density of the feedstream (kg/m^3)
F_0 = flowrate of feed (m^3/s)
ρ = density of product stream (kg/m^3)
F = flowrate of product (m^3/s)
C_{A0} = concentration of reactant A in the feed (kmol/m^3)
V_R = volumetric holdup of liquid in reactor (m^3)
c_{p0} = heat capacity of feed (J kg^{-1} K^{-1})
T_0 = temperature of feed (K)
c_p = heat capacity of product (J kg^{-1} K^{-1})
λ = heat of reaction (J/kmol)
Q = rate of heat removal from liquid in reactor (J/s)

2.1.1 Jacket-Cooled

The Q term in Eq. (2.5) depends on the heat removal scheme used. If a circulating jacket water system is used, the jacket is essentially at one temperature T_J, and the heat transfer rate depends on the jacket area, the overall heat transfer coefficient and the differential temperature driving force

$$Q = UA_J(T_R - T_J) \tag{2.6}$$

where U = overall heat transfer coefficient (W K^{-1} m^{-2})
A_J = jacket heat transfer area (m^2) = πDL
D = reactor diameter (m)
L = reactor length (m)

The consumption of the cooling medium is calculated from an energy balance around the perfectly mixed jacket at temperature T_J. Constant physical properties of the cooling medium are assumed

$$F_J \rho_J c_J T_{C,\text{in}} = F_J \rho_J c_J T_J - Q \tag{2.7}$$

where F_J = flowrate of coolant (m^3/s)
ρ_J = density of coolant (kg/m^3)
c_J = heat capacity of coolant (J kg^{-1} K^{-1})
$T_{C,\text{in}}$ = supply temperature of cooling medium (K)

The conversion χ of reactant A is given by

$$\chi = \frac{C_{A0} - C_A}{C_{A0}} \tag{2.8}$$

The typical reactor design situation is to be given the feed conditions (flowrate, temperature, and composition), the kinetic information, and the desired conversion. The problem is to determine the temperature and the size of the reactor.

In general, the temperature of the reactor is established in a number of different ways that depend very strongly on the chemistry and the kinetics. For the simple irreversible reaction studied in this section, in which the only issue is to achieve the desired conversion, it would appear that the reactor temperature should be made as high as possible. This would give the largest specific reaction rate and therefore the smallest reactor size, thus minimizing capital investment. However, as we show below, there are dynamic controllability considerations that must be factored in when selecting reactor temperature. For the more complex reactions considered in later sections (such as reversible, consecutive, or simultaneous reactions), in which issues of both conversion and yield are important, the selection of reactor temperature must consider the production of undesirable products as well as reactant conversion.

A high-temperature limit may be established by several factors such as catalyst degradation and materials of construction (excessive corrosion); undesirable side reactions that occur are high-temperature or excessive cracking or polymerization.

TABLE 2.1 Base Case Parameters for Irreversible Exothermic Reaction

Preexponential factor k_0	s^{-1}	20.75×10^6
Activation energy E	J/kmol	69.71×10^6
Process molecular weight	kg/kmol	100
Process densities ρ_0 and ρ	kg/m^3	801
Coolant density ρ_J	kg/m^3	1000
Process heat capacities c_{p0} and c_p	$J\ kg^{-1}\ K^{-1}$	3137
Coolant heat capacity c_J	$J\ kg^{-1} K^{-1}$	4183
Heat of reaction λ	J/kmol	-69.71×10^6
Feed temperature T_0	K	294
Feed flowrate F_0	kmol/s	4.377×10^{-3}
Feed composition C_{A0}	$kmol/m^3$	8.01
Inlet coolant temperature $T_{C,in}$	K	294
Overall heat transfer coefficient U	$W\ m^{-2}\ K^{-1}$	851
Aspect ratio (L/D)	—	2

For a numerical base case, the kinetic and process parameters given in Table 2.1 are selected. Reactors with several design values of conversion and over a range of temperatures are sized. The purpose is to see the effect of these parameters on the size of the reactor and its heat transfer area. The effects of changes in the base case parameters, such as feed flowrate, heat of reaction, and overall heat transfer coefficient, will also be explored. Densities and heat capacities are assumed to be constant.

Results for Base Case Parameters Results are shown in Figure 2.1 for the base case feed flowrate of 4.377×10^{-3} m^3/s, which corresponds to an annual production rate of 110,000 metric tons. The effects of the design reactor temperature T_R on various equipment sizes and operating conditions are given for three values of the design conversion: 50, 80, and 90%. The corresponding values of the reactant concentration in the CSTR (and in the product stream) are 4.00, 1.60, and 0.801 $kmol/m^3$, respectively.

As temperature increases for a given conversion, reactor size decreases, as expected. The heat transfer rate also decreases because there is a greater contribution of the sensible heat of the cooler feed as reactor temperature increases.

The behaviors of the jacket temperature and the coolant flowrate are quite interesting for the 50 and 80% conversion cases. At low reactor temperatures, the jacket temperature is low. As reactor temperature increases, the jacket temperature increases for a while but then begins to decrease. This occurs because the smaller reactors have less heat-transfer area, so a larger differential temperature driving force is needed. Despite the fact that the reactor temperature is higher, the much larger required ΔT forces the jacket temperature lower. The flowrate of the cooling medium becomes very large as the jacket temperature approaches the supply coolant temperature of 294 K.

As the design conversion χ increases, reactor size increases, which provides more heat transfer area. Although the heat transfer rate increases, the jacket temperature is higher because the larger area requires a smaller ΔT driving force.

These results indicate that there may be an operability limit to raising reactor temperature for the purpose of reducing capital investment. This limit can be expressed as the point where the differential temperature driving force between the reactor and the jacket

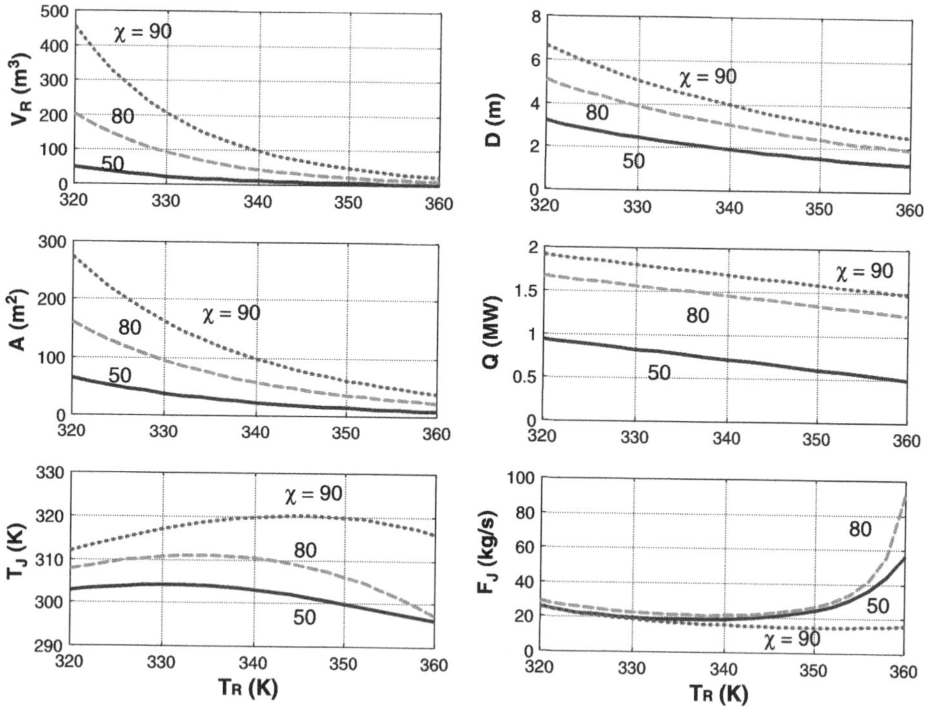

Figure 2.1 Effect of temperature and conversion.

becomes too large a fraction of the total available ΔT (the difference between the reactor temperature and the temperature of the *supply* coolant). We define this ratio as the reactor stability index (RSI):

$$\text{RSI} = \frac{\text{design } \Delta T}{\text{maximum } \Delta T} = \frac{T_R - T_J}{T_R - T_{C,\text{in}}} \tag{2.9}$$

For example, with a 50% design conversion, a reactor temperature of 353 K gives a reactor volume of 4.36 m^3 with a heat transfer area of 12.4 m^2 that requires a jacket temperature of 298.8 K. This is only 4.8 K above the supply coolant temperature of 294 K. The required coolant flowrate is very large (28.8 kg/s). The fraction of the total available ΔT is

$$(\text{RSI})_{353} = \frac{T_R - T_J}{T_R - T_{C,\text{in}}} = \frac{353 - 298.8}{353 - 294} = 0.92 \tag{2.10}$$

This very large RSI means that there is little additional ΔT driving force to be able to increase heat removal rates in order to handle upsets. Therefore temperature control is expected to be poor. This prediction will be quantitatively confirmed by a dynamic analysis in Chapter 3.

If a reactor temperature of 320 K is used, the reactor volume is over 10 times greater (50.5 m^3) with a heat transfer area of 63.5 m^2 that requires a jacket temperature of

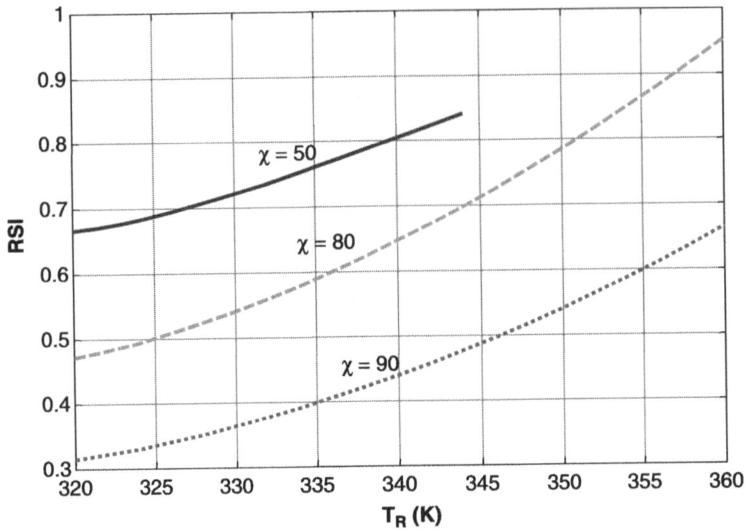

Figure 2.2 Effect of temperature and conversion on reactor stability index.

302.7 K. The fraction of the total available ΔT is

$$(\mathrm{RSI})_{320} = \frac{T_R - T_J}{T_R - T_{C,\mathrm{in}}} = \frac{320 - 302.7}{320 - 294} = 0.66 \tag{2.11}$$

This RSI indicates that the reactor should be able to handle upsets better than the reactor designed for 353 K operation. However, depending on the magnitude of the disturbances and other parameters, particularly the activation energy, even an RSI of 0.66 may not be adequate. These issues are studied in Chapter 3.

Figure 2.2 gives the values of RSI for three design conversions as functions of reactor design temperature. It is clear that higher temperature and lower conversion increase the value of RSI and therefore imply worse control.

Effect of Throughput Figures 2.3 and 2.4 show what happens when the feed flowrate is increased by 50% from the base case production rate. The 50% conversion design requires very large coolant flowrates for reactor temperatures above 340 K. At this temperature the reactor volume of 16.2 m^3 with a heat transfer area of 29.8 m^2 requires a jacket temperature of 297.6 K. The fraction of the total available ΔT is

$$(\mathrm{RSI})_{340} = \frac{T_R - T_J}{T_R - T_{C,\mathrm{in}}} = \frac{340 - 297.6}{340 - 294} = 0.92 \tag{2.12}$$

There is very little additional ΔT driving force to handle upsets, so we anticipate that temperature control will be poor. For a design reactor temperature of 320 K, the reactor is much larger (75.8 m^3) and the heat transfer area is also much larger (83.3 m^2). However, the jacket temperature is still quite low (300.2 K) because of the low

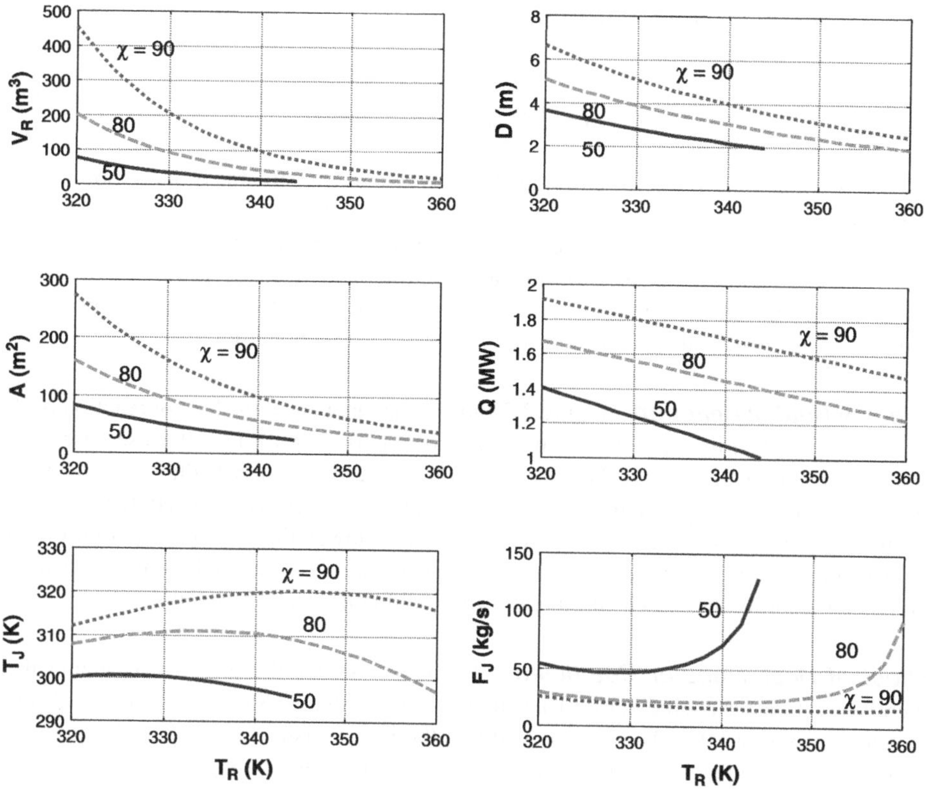

Figure 2.3 Effect of temperature and conversion; 50% more feed.

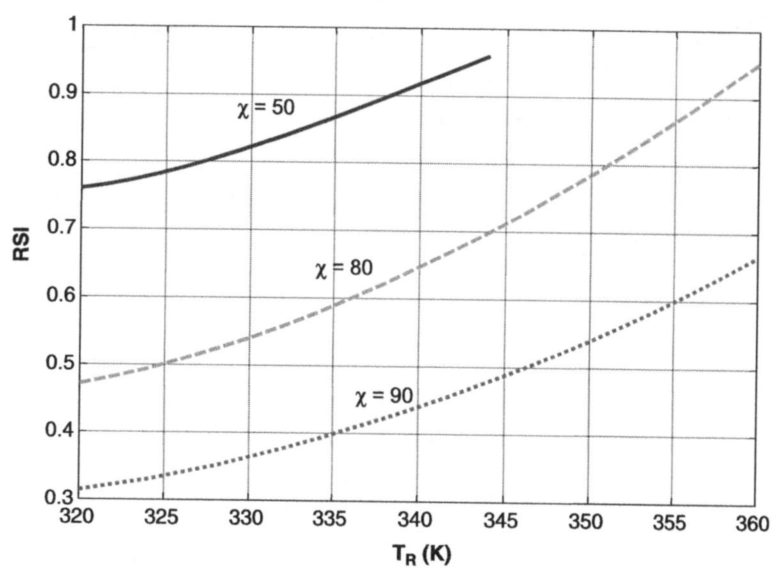

Figure 2.4 Reactor stability index with 50% higher feed flowrate.

reactor temperature. The resulting RSI is

$$(\text{RSI})_{320} = \frac{T_R - T_J}{T_R - T_{C,\text{in}}} = \frac{320 - 300.2}{320 - 294} = 0.76 \tag{2.13}$$

This is better than the 340 K design, but the fraction is still fairly large so control performance may not be acceptable. Notice that the RSI for the higher feed flowrate is larger than that for the base case feed flowrate at the same temperature and conversion (0.76 vs. 0.66). This indicates that increasing throughput makes reactor control more difficult. The fundamental reason for this degradation of control with production rate is the decrease in the area-to-volume ratio as vessel size increases.

Effect of Heat of Reaction Figure 2.5 shows what happens if the heat of reaction is 10 or 20% higher than the base case value. The conversion is 80% for all these cases. As expected, the heat of reaction has no effect on the reactor volume, diameter, or area. These parameters are set by throughput, temperature, and conversion. Higher heats of reaction require higher heat transfer rates Q, which lower the jacket temperature and increase the cooling water flowrate. As shown in Figure 2.6, the result is an increase in the reactor stability index, which indicates more difficult control problems.

Effect of Heat Transfer Coefficient Figure 2.7 shows the effect of reducing the overall heat transfer coefficient U from the base case value of 851 W m^{-2} K^{-1}with the conversion fixed at 80%. The reactor size and the heat transfer rate do not change with U, but jacket temperature and cooling water flowrate are strongly affected. As U

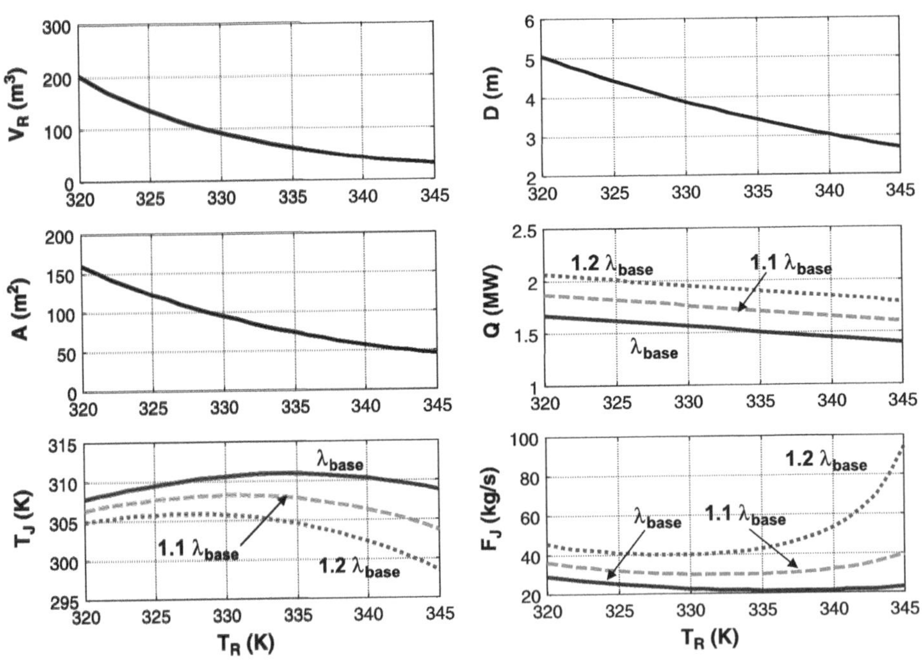

Figure 2.5 Effect of heat of reaction; $\chi = 80\%$.

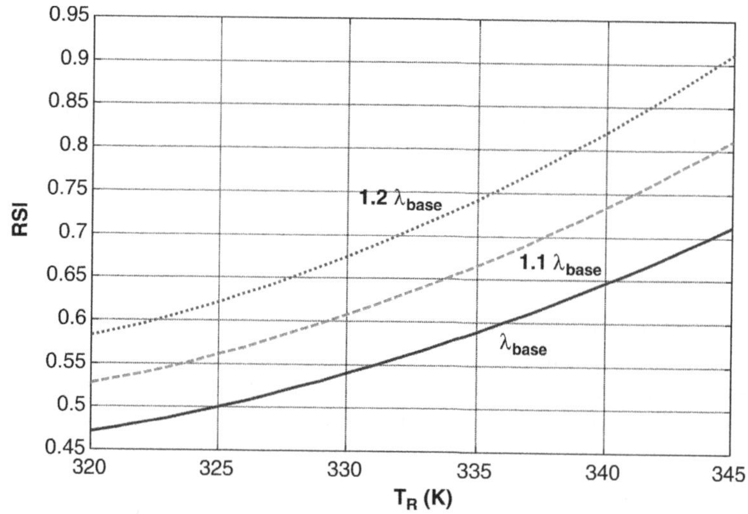

Figure 2.6 Effect heat of reaction on RSI.

becomes small, lower jacket temperatures are required. This adversely affects the reactor stability index (increasing the values of RSI).

Since the heat transfer coefficient is such an important parameter, it might be appropriate to discuss how to estimate its value in a CSTR with different types of cooling equipment. Detailed coverage of this issue is presented in most heat transfer books. For

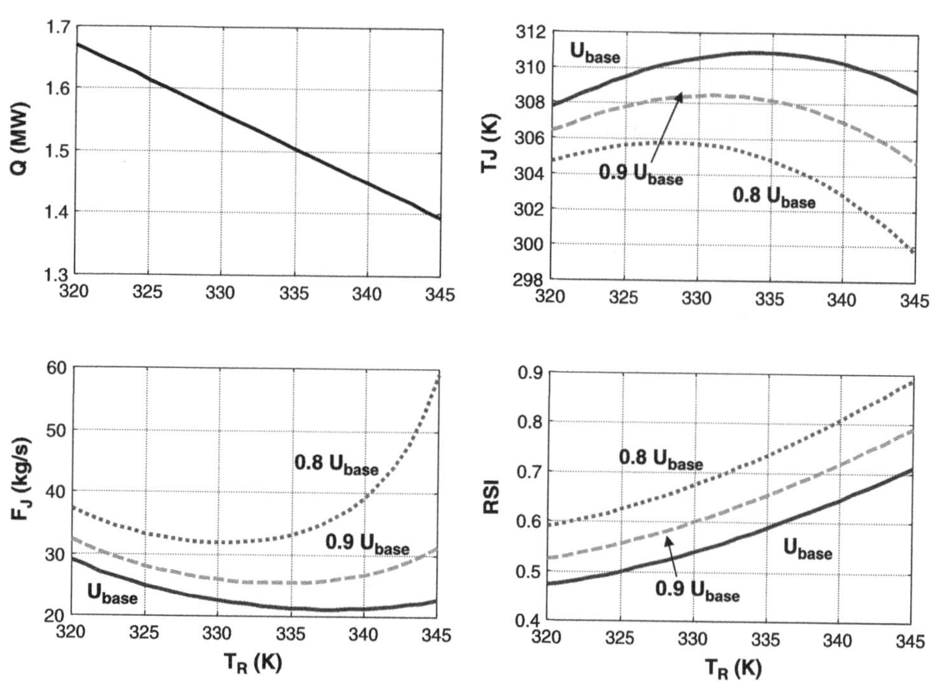

Figure 2.7 Effect of heat transfer coefficient.

example, Geankoplis reviews methods for estimating the film coefficient at the inside wall of an agitated vessel, on the outside of a cooling coil and on the inside a pipe or jacket.[1]

There are three resistances/coefficients that must be considered in a jacket-cooled CSTR. There is a film coefficient h_{in} at the inside wall of the vessel, a thermal conductivity k_m of the metal walls and a film coefficient h_{out} at the outside surface of the wall:

$$\frac{1}{U} = \frac{1}{h_i} + \frac{A_i}{A_o h_o} + \frac{L_m}{A_i k_m} \tag{2.14}$$

Unless the wall is very thick, the resistance of conduction through the wall is usually small. The film coefficient in the jacket depends on the velocity of coolant through the jacket to the 0.8 power. This is one of the reasons for using a circulating water system in which the velocity is not only high but also constant. It does not vary with the flowrate of the makeup cold water. If the flow of the cooling water through the jacket is once-through, the velocity is small and the outside film coefficient is small. Significant changes in the outside film coefficient can occur. In addition, the dynamics of the jacket vary with the coolant flowrate. Both of these potential problems make the circulating water system superior to the once-through system, as the numerical example given below illustrates.

The inside film coefficient h_i of an agitated vessel depends on the type and degree of agitation. Geankoplis[1] gives parameter values, which have been drawn from many sources, for a jacketed CSTR. The basic equation is

$$\frac{h_i D_T}{k} = a\left(\frac{(D_A)^2 N\rho}{\mu}\right)^b \left(\frac{c_p\mu}{k}\right)^{1/3} \left(\frac{\mu}{\mu_W}\right)^m \tag{2.15}$$

where D_T = diameter of tank
k = thermal conductivity of the process liquid
D_A = diameter of agitator
N = revolutions per second of agitator
ρ = density of process liquid
μ = viscosity of process liquid at temperature in vessel
μ_W = viscosity of process liquid at wall temperature
c_p = heat capacity of process liquid

The parameters a, b, and m depend on the type of agitator and the presence or absence of baffles. For example, the values of these parameters for a flat-blade agitator with baffles are $a = 0.74$, $b = \frac{2}{3}$, and $m = 0.14$. A similar equation can be used for a CSTR with an internal cooling coil (see footnote 1).

The outside film coefficient h_o in the jacket can be estimated from the Dittus–Boelter equation for flow through a pipe if a suitable equivalent diameter is used:

$$\frac{h_o D_{eq}}{k} = 0.027\left(\frac{V D_{eq}\rho}{\mu}\right)^{0.8} \left(\frac{c_p\mu}{k}\right)^{1/3} \left(\frac{\mu}{\mu_W}\right)^{0.14} \tag{2.16}$$

[1] C. J. Geankoplis, *Transport Processes and Unit Operations*, 3rd edition, Prentice-Hall, 1993, pp. 238, 300, 302.

Here, the equivalent diameter for an annular space is the distance between the vessel wall and the jacket wall. This same equation can be used to estimate the inside film coefficient in a cooling coil.

A numerical example might be useful to illustrate some of the issues discussed above. The base case reactor is used with a design conversion of 80% and reactor temperature of 320 K. The diameter of the reactor is 5 m, and the jacket area is 160 m^2. With a perfectly mixed circulating cooling water system, the jacket temperature is 304 K and the flowrate of makeup cooling water is 50.8 kg/s. The physical properties of the reaction liquid are

$$
\begin{aligned}
\rho &= \text{density} = 801\ \text{kg/m}^3 \\
c_p &= \text{heat capacity} = 3137\ \text{J kg}^{-1}\text{K}^{-1} \\
k &= \text{thermal conductivity} = 0.15\ \text{J s}^{-1}\ \text{K}^{-1}\ \text{m}^{-1} \\
\mu_{320} &= \text{viscosity at 320 K} = 2 \times 10^{-3}\ \text{kg m}^{-1}\text{s}^{-1} \\
\mu_{305} &= \text{viscosity at 305 K} = 3 \times 10^{-3}\ \text{kg m}^{-1}\ \text{s}^{-1}
\end{aligned}
$$

The diameter of the impeller is 2.5 m in this 5-m vessel and rotates at 0.5 revolution per second (rps). Substituting these values into Eq. (2.15) gives a value of $h_i = 1082\ \text{J s}^{-1}\ \text{K}^{-1}\ \text{m}^{-2}$:

$$\frac{h_i(5.50)}{0.15} = 0.74\left(\frac{(2.5)^2 0.5(801)}{2 \times 10^{-3}}\right)^{2/3}\left(\frac{3137(2 \times 10^{-3})}{0.15}\right)^{1/3}\left(\frac{2 \times 10^{-3}}{3 \times 10^{-3}}\right)^{0.14}$$

Assuming a jacket clearance of 0.1 m and using a circulation rate 50 times the makeup cooling water flow give a jacket water velocity of 1.6 m/s and a film coefficient of $h_o = 2045\ \text{J s}^{-1}\ \text{K}^{-1}\ \text{m}^{-2}$.

$$\frac{h_o(0.1)}{0.1} = 0.027\left(\frac{1.6(0.1)(1000)}{0.8 \times 10^{-3}}\right)^{0.8}\left(\frac{4183(0.8 \times 10^{-3})}{0.1}\right)^{1/3}\left(\frac{0.8 \times 10^{-3}}{0.65 \times 10^{-3}}\right)^{0.14}$$

The equivalent diameter of the annular jacket geometry is the jacket clearance. If the wall thickness is 0.003 m and the thermal conductivity of the metal is 45 J s^{-1} K^{-1} m^{-2}, the overall heat transfer coefficient for the circulating water system is $U = 700\ \text{J s}^{-1}\ \text{K}^{-1}\ \text{m}^{-2}$.

If a once-through cooling water system is used, the velocity through the jacket is very small (0.032 m/s), which gives such a small $h_o = 68\ \text{J s}^{-1}\ \text{K}^{-1}\ \text{m}^{-2}$ that the design is unfeasible. Reducing the jacket clearance from 0.1 to 0.01 m, which is probably too small for a 5-m-diameter vessel, gives a $h_o = 682\ \text{J s}^{-1}\ \text{K}^{-1}\ \text{m}^{-2}$.

If an internal cooling coil is used (as discussed in the next section) with a 0.0763 m diameter, the velocity in the tube is 3.7 m/s and the film coefficient is $h_o = 3200\ \text{J s}^{-1}\ \text{K}^{-1}\ \text{m}^{-2}$, giving a $U = 800\ \text{J s}^{-1}\ \text{K}^{-1}\ \text{m}^{-2}$. It is clear that the once-through cooling water system is by far the most inferior of the three systems.

Effect of Aspect Ratio The ratio of the length to the diameter of the reactor is a design parameter. We use an aspect ratio of 2 in the examples given in this book. Values from 1 to 4 can be found in literature examples.

There are several issues involved in selecting an aspect ratio. The cost of the vessel depends on the amount of metal required. The weight of metal in a cylindrical vessel is minimized by using an aspect ratio is about 1. Achieving good mixing becomes more difficult as the aspect ratio is increased above 2.

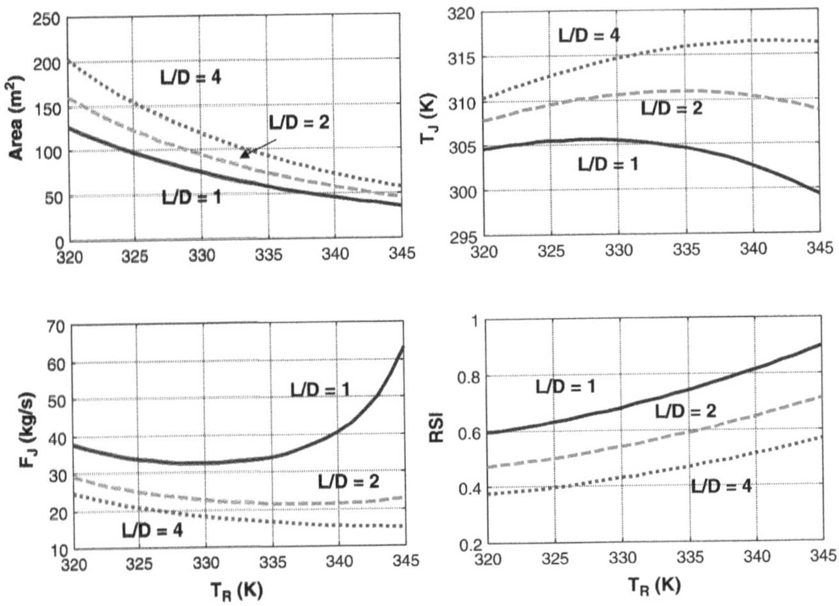

Figure 2.8 Effect of aspect ratio.

However, the circumferential wall heat transfer area increases as the aspect ratio increases. So, from a heat transfer, dynamic stability perspective, a large aspect ratio is desirable. Figure 2.8 shows the effect of aspect ratio on the various design and operating parameters for a design conversion of 80%. Increasing L/D increases heat transfer area, which decreases the required ΔT driving force and raises jacket temperature. The reactor stability index improves substantially.

A reasonable engineering compromise of these competing effects is to use an aspect ratio of ~2.

Another important consideration regarding the aspect ratio is its effect on mixing. It becomes more difficult to achieve good mixing as the aspect ratio increases. An interesting paper by Himmelsbach et al.[2] discusses some of these problems. They point out that the trend is to build reactors with larger volumes. However, wall thickness is strongly dependent on vessel diameter, and overland transportation of vessels over ~ 4.6 m in diameter is difficult because of road bridge dimensions and overhead electric line heights. This means that vessels with larger aspect ratios are used as vessel volume increases. New techniques for achieving good mixing in vessels with larger L/D ratios are discussed in their paper.

Effect of Reaction Order All the designs up to this point have assumed that the reaction is first-order in the concentration of the reactant C_A. Many reactions have reaction rates that depend on the reactant concentration to a power other than 1:

$$\mathcal{R} = k(C_A)^n \tag{2.17}$$

[2]W. Himmelsbach, D. Houlton, D. Ortlieb, and M. Lovallo. New advances in agitation technology for exothermic reactions in very large reactors, *Chem. Eng. Sci.* **61**, 3044–3052 (2006).

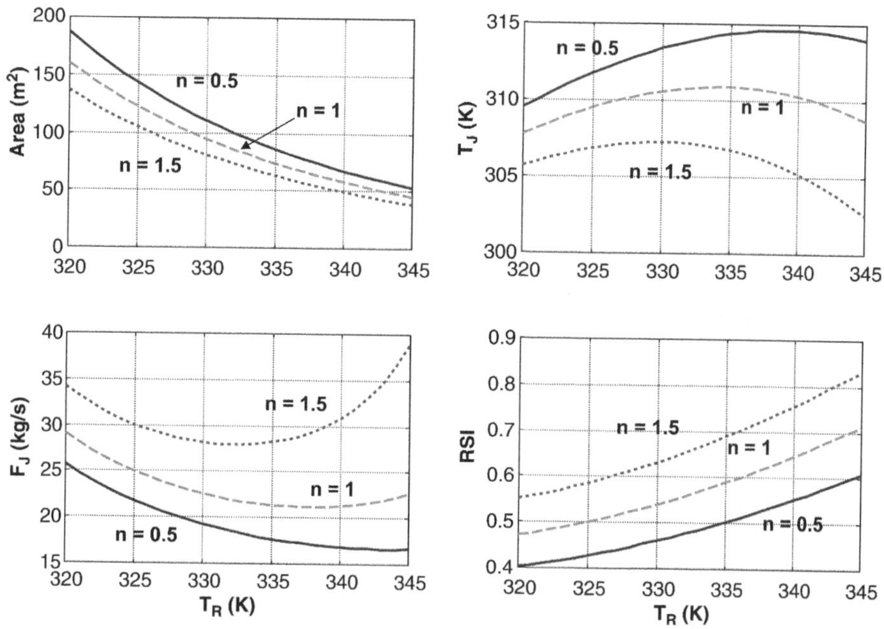

Figure 2.9 Effect of reaction order.

Since the concentration (kmol/m^3) is usually greater than 1, the reaction rate is larger if $n > 1$ and smaller if $n < 1$. The resulting reactor volumes are smaller and larger, respectively, in the two cases. The heat transfer area increases as the reactor volume increases, so the required ΔT driving force decreases. This translates into a more controllable reactor. Figure 2.9 illustrates these effects. Results for reaction orders of 0.5, 1, and 1.5 are shown. The $n = 0.5$ case has more area, higher jacket temperatures and lower RSI.

In later sections, reactions with two reactants are discussed in which the reaction rate depends on the concentrations of both reactants:

$$\mathcal{R} = k7(C_A)^{\alpha}(C_B)^{\beta} \tag{2.18}$$

The relative concentrations of the reactants, as well as the reaction orders, affect the reaction rate and the reactor volume.

Overview The message in these examples is that reactors in this exothermic irreversible reaction system should not necessarily be designed for the maximum temperature. Operability issues should be considered. Designing for a lower temperature gives a larger reactor with more heat transfer area that is more controllable.

Of course, the reactor that is designed at a lower temperature can be operated at a higher temperature. The large area will give a small ΔT, and controllability will be even better than when operated at the design temperature:

> The key to reactor temperature control is to provide an excess of heat transfer area so that disturbances can be handled.

```
% cstrjacket.m
% Irreversible reaction A = B
%
% Given conversion and parameter values, program calculates
% Steady-state design (volume,jacket area, coolant flow and temp)
%
clear
ca0=8.01;k0=20.75e6;e=69.71e6;t0=294;tcin=294;u=851;lambda=-69.71e6;
roe=801;m=100;cp=3137;cj=4183;roej=1000;

% Set conversion = 50% and feed rate and temperature
% ********************
conversion =0.5;f=4.377e-3;
ca=ca0*(1-conversion)
tr=320;
% ***********************
 k=k0*exp(-e/tr/8314);
vr=f*(ca0-ca)/k/ca; d=(2*vr/pi)^0.3333;
areaj=2*pi*d^2;q=(ca0-ca)*f*(-lambda)-cp*roe*f*(tr-t0);
tj=tr-q/u/areaj;vj=0.3333*areaj;
% fj is in kg/sec
fj=q/(cj*(tj-tcin));
deltatemp=tr-tj;
% to print results for one case
 tr,vr,d,areaj,q,tj,fj,deltatemp/(tr-294)
```

Figure 2.10 Matlab program for CSTR with jacket.

This conclusion has been inferred by looking at steady-state design results. In Chapter 3 dynamic simulation will quantitatively demonstrate the improved controllability of reactors with plenty of heat transfer area. Figure 2.10 gives a simple Matlab program that calculates the required reactor size, heat transfer area, and cooling water flowrate for a CSTR with jacket cooling, given the feed flowrate, feed composition, feed temperature, reactor temperature, desired conversion, cooling water supply temperature, physical properties, and kinetic parameters.

2.1.2 Internal Coil

If an internal cooling coil is used, the cooling medium flows in plug flow through the coil. The temperature differential driving force for heat transfer is a log mean average of the differential temperatures at the two ends of the coil

$$\begin{aligned}\Delta T_{\text{in}} &= T_R - T_{C,\text{in}} \\ \Delta T_{\text{out}} &= T_R - T_{C,\text{out}} \\ (\Delta T)_{LM} &= \frac{\Delta T_{\text{in}} - \Delta T_{\text{out}}}{\ln\left(\dfrac{\Delta T_{\text{in}}}{\Delta T_{\text{out}}}\right)}\end{aligned} \tag{2.19}$$

where $T_{C,\text{in}}$ is the supply temperature of the cooling medium (K) and $T_{C,\text{out}}$ is the temperature of the cooling medium leaving the coil (K).

The energy balance on the coil is given in Eq. (2.20):

$$F_J \rho_J c_J T_{C,\text{in}} = F_J \rho_J c_J T_{C,\text{out}} - Q \tag{2.20}$$

There are several issues to be resolved in the design of a cooling coil regarding the physical layout of the coil in the vessel. The diameter of the pipe used to make the coil must be specified. In addition, the diameter of the spiral loop and the spacing between the loops must be specified. Should one coil be used, or should several coils be installed in parallel?

The coil takes up volume in the vessel, so the size of the reactor will have to be larger to provide the same holdup for reaction. If the coil volume is too high a fraction of the total volume, the mixing in the reactor may be adversely affected.

If a small-diameter pipe is used, the heat transfer area per unit of coil volume will be large. However, the pressure drop through the coil will be large unless multiple parallel passes are used. Parallel passes can present flow distribution problems. Small-diameter tubes also present mechanical strength problems.

Obviously the diameter of the loop must smaller than the diameter of the vessel, and if the diameter of the loop is too small, it will interfere with the agitator.

The spacing between the flights of the loop cannot be too small, or mixing and heat transfer will be degraded.

The following coil specifications are used in this study:

1. The pipe diameter is 0.0763 m (3 in.).
2. The loop diameter constitutes 80% of the diameter of the vessel.
3. The spacing between the loop flights is the diameter of the pipe.
4. Only one coil is used.
5. The reactor vessel has an aspect ratio (L/D) of 2.

With these specifications, the reactor–coil system can be designed for a given conversion. The variable V_R is the required volume of liquid in the vessel, excluding the volume occupied by the internal coil. It is calculated from Eq. (2.4) for specified conversion, feed conditions, and reactor temperature.

Step 1. Make an initial guess of the diameter of the vessel.

$$D_{\text{guess}} = \left(\frac{2V_R}{\pi}\right)^{1/3} \tag{2.21}$$

Step 2. Calculate the length of the vessel:

$$L_{\text{vessel}} = 2D_{\text{guess}} \tag{2.22}$$

Step 3. Calculate the number of coil loops that will fit into a vessel with this length:

$$N_{\text{loops}} = \frac{L_{\text{vessel}}}{2D_{\text{coil}}} \tag{2.23}$$

Step 4. Calculate the length of the coil:

$$L_{\text{coil}} = 0.8D_{\text{guess}}(\pi)N_{\text{loops}} \tag{2.24}$$

Step 5. Calculate the volume of the coil:

$$V_{\text{coil}} = \frac{\pi(D_{\text{coil}})^2}{4} L_{\text{coil}} \tag{2.25}$$

Step 6. Calculate the total volume of the vessel, which is the required volume of the reaction mass plus the volume of the coil:

$$V_{\text{total}} = V_{\text{coil}} + V_R \tag{2.26}$$

Step 7. Calculate the diameter of a vessel with this volume:

$$D_{\text{calc}} = \left(\frac{2V_{\text{total}}}{\pi}\right)^{1/3} \tag{2.27}$$

Step 8. If D_{calc} is sufficiently close to D_{guess}, continue. Otherwise go back to step 2. Direct substitution works quite effectively in the iteration.

Step 9. Calculate the heat transfer area of the coil:

$$A_{\text{coil}} = \pi D_{\text{coil}} L_{\text{coil}} \tag{2.28}$$

Once the physical size of the vessel and coil have been determined, a second iterative calculation is required to determine how much coolant is required to remove the known amount of heat.

The temperature driving force for heat transfer between the reaction liquid at temperature T_R and the coil is the log-mean temperature difference given in Eq. (2.19). The heat transfer rate is given by

$$Q_{\text{coil}} = UA_{\text{coil}}\Delta T_{LM} \tag{2.29}$$

We know Q_{coil}, U, A_{coil}, T_R and $T_{C,\text{in}}$. Combining the two equations above gives one equation in one unknown, the temperature of the coolant leaving the coil $T_{C,\text{out}}$. However, the log term precludes an analytic solution, so an iterative "interval halving" solution method is used.

1. A value of $T_{C,\text{out}}$ is guessed. It must be less than T_R.
2. The value of ΔT_{LM} is calculated.
3. Then a value of the heat transfer rate is calculated: $Q_{\text{calc}} = UA_{\text{coil}}\Delta T_{LM}$.
4. If this calculated Q_{calc} is sufficiently close to the known value of Q_{coil}, the iteration has converged. If not, continue.
5. If Q_{calc} is greater than Q_{coil}, increase the guessed value of $T_{C,\text{out}}$.
6. If Q_{calc} is less than Q_{coil}, decrease the guessed value of $T_{C,\text{out}}$.

Note that the temperature approach on the "cold end" is known ($T_R - T_{C,\text{in}}$).

Figure 2.11 gives a Matlab program that performs these sizing calculations. Results for the base case feedrate, a 50% conversion, and a 320 K reactor temperature yield a coolant exit temperature of 314 K and a log-mean temperature difference of 13.5 K. The heat transfer area of the coil is 81.44 m^2, compared to a jacket area of 63.4 m^2. The vessel

```
% Program "cstrcoil.m"
% Irreversible reaction A = B
% Cooling coil only
% Coil diameter is 0.0763 m (3 inches)
% Coil loop diameter is 0.8 of vessel diameter
% Loop spacing is one coil diameter
% Given conversion and parameter values, program calculates
% (1) Steady-state design (volume,jacket area, coolant flow and temp)
% (2) Responses to step increases of various sizes in feed rate
clear
ca0=8.01;k0=20.75e6;e=69.71e6;t0=294;tcin=294;u=851;lambda=-69.71e6;
roe=801;m=50;cp=3137;cj=4183;roej=1000;dcoil=0.0763;
% Set conversion = 50% and feed rate and temperature
% *********************
conversion =0.5;f=4.377e-3;
ca=ca0*(1-conversion);
tr=320;
k=k0*exp(-e/tr/8314);
vr=f*(ca0-ca)/k/ca;
% ****************************
% Iterative loop to find diameter using direct substitution
% Guess diameter from vessl with no coil
 dguess=(2*vr/pi)^0.3333;
 error=10;
while error>0.011;
lvessel=2*dguess;
nloops=lvessel/2/dcoil;lcoil=0.8*dguess*pi*nloops;
vcoil=lcoil*pi*(dcoil^2)/4;
vtot=vr+vcoil;dvessel=(2*vtot/pi)^0.3333;
error=abs(dvessel-dguess);dguess=dvessel;
end
% Print results
tr,ca,f,vr,nloops,lcoil,vcoil,dvessel,areacoil=lcoil*dcoil*pi
q=(ca0-ca)*f*(-lambda)-cp*roe*f*(tr-t0);
% *********************************
% Iterative calculation for tcout using log mean temperature difference
%    using interval halving
% Make initial guess of tcout
tcout=tr-1;
flagm=-1;flagp=-1;dtcout=2;loop=0;errorq=1000;
while errorq>10;
  dt1=tr-tcin;dr2=tr-tcout;dtlm=(dt1-dt2)/log(dt1/dt2);qcalc=u*areacoil*dtlm;
  errorq=abs(q-qcale); loop=loop+1;
    if loop>50;q,qcalc,tcout
    error('LMTD loop'); end;
  if qcalc>q;if flagp>0;dtcout=dtcout/2;end
    tcout=tcout+dtcout;flagm=1;if tcout>tr;tcout=tr-0.01;end;en
    if qcalc<q;if flagm>0;dtcout=dtcout/2;end
    tcout=tcout-dtcout;flagp=1;end
end
fcoil=q/(cj*(tcout-tcin)),tcout,q,dtlm
% Iterative calculation for tcout using log mean temperature difference
%    using interval halving
% Make initial guess of tcout
tcout=tr-1;
flagm=-1;flagp=-1;dtcout=2;loop=0;errorq=1000;
while errorq>10;
  dt1=tr-tcin;dr2=tr-tcout;dtlm=(dt1-dt2)/log(dt1/dt2);qcalc=u*areacoil*dtlm;
  errorq=abs(q-qcale); loop=loop+1;
    if loop>50;q,qcalc,tcout
    error('LMTD loop'); end;
  if qcalc>q;if flagp>0;dtcout=dtcout/2;end
    tcout=tcout+dtcout;flagm=1;if tcout>tr;tcout=tr-0.01;end;en
    if qcalc<q;if flagm>0;dtcout=dtcout/2;end
    tcout=tcout-dtcout;flagp=1;end
end
fcoil=q/(cj*(tcout-tcin)),tcout,q,dtlm
```

Figure 2.11 Matlab program for CSTR with coil.

with a coil is slightly larger in diameter (3.212 m vs. 3.18 m) since the coil volume is 1.55 m^3 and the reactive liquid volume is 50.53 m^3. The 0.0763-m-diameter coil pipe has 42 loops with a total length of 340 m.

Now the coolant flowrate can be calculated:

$$F_{\text{coil}} = \frac{Q_{\text{coil}}}{c_C(T_{C,\text{out}} - T_{C,\text{in}})} \tag{2.30}$$

The coolant flowrate for this case is 11.1 kg/s. This should be compared with the coolant flowrate required in the jacket-cooled configuration of 25.8 kg/s, which is over twice as much.

Clearly using a coil provides some advantages over a jacket; the principal one is more heat transfer area. In Chapter 3 we demonstrate this dynamic advantage.

2.1.3 Other Issues

There are several other aspects about CSTRs with exothermic reactions that should be mentioned at this point. The first involves the temperature of the feed. The colder the feed, the less heat must be transferred from the reactor. So control would be expected to be improved. However, as we will see in Chapter 3, a cold feed can produce some interesting dynamics; for instance, an increase in feed flowrate initially *decreases* reactor temperature because of the sensible-heat effect. But as the reactant concentration in the reactor increases, the temperature eventually increases. A reactor temperature runaway can result if the cold feed "quenches" the reaction and reactant concentration builds to a very high level before the reaction "lights off."

The second item is the concentration of the reactant in the feed. We considered the case in which the feed is pure reactant A and found a heat removal rate for a given conversion and reactor temperature. However, suppose that the feed were a mixture of reactant A and product B. Now for the same feedrate and conversion, there is less of A to react so the heat transfer requirements are lower. This indicates one method of improving reactor controllability, which is to reduce reactant feed composition by diluting the feed with some nonreactive component. Of course, the downside of this approach is that there must be more material to recycle, which increases capital and energy costs.

The third item is other ways to increase heat transfer area. They include circulating reactor liquid through an external heat exchanger or autorefrigeration. The dynamics of both of these alternatives will be studied in Chapter 3. As earlier mentioned, for a jacketed reactor, the heat transfer area can be increased by using larger aspect ratios.

2.2 IRREVERSIBLE, TWO REACTANTS

A lot of insight into the CSTR control problem was gained in the previous section by considering only a very simple reaction. The importance of heat transfer area was clearly demonstrated. Thermal effects are dominant in reactor control, and these depend on net reaction rate, which is essentially set by conversion. Issues of yield and selectivity arise when more complex chemistry is involved, but these are not as important in terms of reactor controllability.

Now let us consider a irreversible reaction involving *two* reactants. This will allow us to see some of the problems of balancing the stoichiometry of the reaction. For every molecule of reactant A fed to the overall process (reactor plus separation section), exactly one (and only one) molecule of reactant B must be fed. Having two reactants also gives an additional degree of freedom in both the steady-state design and the dynamic control because the concentrations of the two reactants in the reactor can be adjusted to give the same reaction rate over a wide range of the individual reactant compositions.

The reaction considered is an exothermic, irreversible reaction, which is first-order in both reactants. There are two reactants and only one product, so there is a change in the number of moles entering and leaving the reactor. Mass is conserved, but moles are not:

$$A + B \rightarrow C \tag{2.31}$$

The reaction rate is first-order in the two reactants:

$$\Re = kC_{\mathrm{A}}C_{\mathrm{B}} = C_{\mathrm{A}}C_{\mathrm{B}}k_0 \exp\left(\frac{-E}{RT_R}\right) \tag{2.32}$$

There are two fresh feedstreams of pure reactants entering the reactor with flowrates F_{A0} and F_{B0} and compositions C_{A0} and C_{B0}, respectively.

It is important to note that the units of the specific reaction rate k are no longer reciprocal second (s^{-1}) but have changed to $\mathrm{m}^3\ \mathrm{s}^{-1}\ \mathrm{kmol}^{-1}$ because of the product of the two compositions.

2.2.1 Equations

There are now three components, so a mass balance, two component balances, and an energy balance are required to describe the system:

Total mass balance (kg/s):

$$\rho_{\mathrm{A0}}F_{\mathrm{A0}} + \rho_{\mathrm{B0}}F_{\mathrm{B0}} = \rho F \tag{2.33}$$

Component A balance (kmol A/s):

$$\begin{aligned} F_0C_{\mathrm{A0}} &= FC_{\mathrm{A}} + V_R\Re \\ &= FC_{\mathrm{A}} + V_RkC_{\mathrm{A}}C_{\mathrm{B}} \end{aligned} \tag{2.34}$$

Component B balance (kmol B/s):

$$\begin{aligned} F_0C_{\mathrm{B0}} &= FC_{\mathrm{B}} + V_R\Re \\ &= FC_{\mathrm{B}} + V_RkC_{\mathrm{A}}C_{\mathrm{B}} \end{aligned} \tag{2.35}$$

Reactor energy balance (J/s):

$$\rho_{\mathrm{A0}}c_{p\mathrm{A0}}F_{\mathrm{A0}}T_{\mathrm{A0}} + \rho_{\mathrm{B0}}c_{p\mathrm{B0}}F_{\mathrm{B0}}T_{\mathrm{B0}} = \rho c_pFT_R + \lambda V_R\Re + Q \tag{2.36}$$

2.2.2 Design

This system has two reactants, so the rate of reaction depends on the product of the two concentrations $C_A C_B$ in the reactor. An infinite number of combinations of the C_A and C_B can give the same reaction rate. Therefore one of the important design issues is the selection of the "best" pair of concentrations. The upper graph in Figure 2.12 gives a plot of the possible combinations of the two concentrations that give the same rate of reaction of reactants A and B. The parameters of the numerical case are given in Table 2.2.

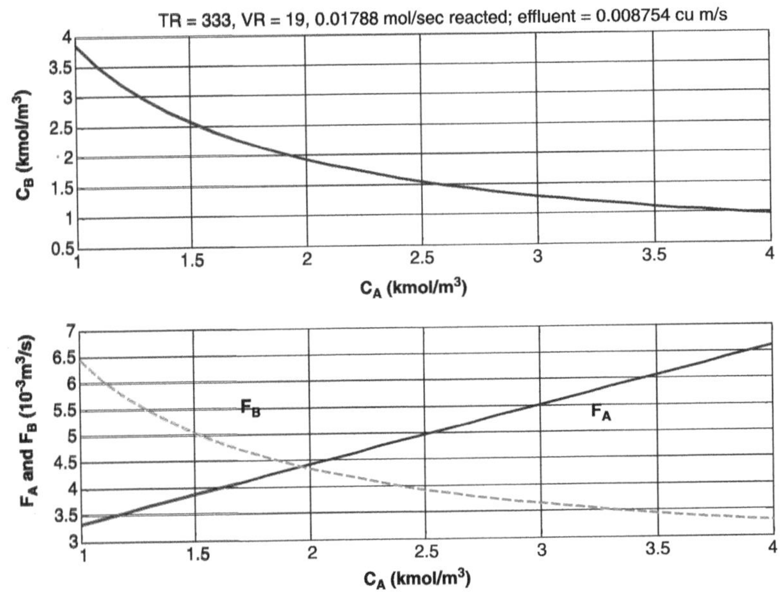

Figure 2.12 Two reactant concentrations.

TABLE 2.2 Irreversible Exothermic Reaction Parameters with Two Reactants

Reactor effluent F	m^3/s	8.754×10^{-3}
Feed composition C_{A0} and C_{B0}	$kmol/m^3$	8.01
Reactor volume	m^3	19
Reactor temperature	K	333
Specific reaction rate k at 333 K	$m^3\ s^{-1}\ kmol^{-1}$	2.4434×10^{-4}
Preexponential factor k_0	$m^3\ s^{-1}\ kmol^{-1}$	2.1045×10^{7}
Activation energy E	J/kmol	69.71×10^{6}
Molecular weights of A and B	kg/kmol	100
Molecular weight of C	kg/kmol	200
Densities	kg/m^3	801
Coolant density ρ_J	kg/m^3	1000
Heat capacities c_{pA0}, c_{pB0}, and c_p	$J\ kg^{-1}\ K^{-1}$	3137
Coolant heat capacity c_J	$J\ kg^{-1}\ K^{-1}$	4183
Heat of reaction λ	J/kmol	-69.71×10^{6}
Rate of production of C	kmol/s	0.01788
Feed temperature T_0	K	294
Inlet coolant temperature $T_{C,in}$	K	294

The reactor temperature (333 K) and volume (19 m^3) are fixed. The other parameter that is held constant is the volumetric flowrate of the reactor effluent (0.008754 m^3/s). The rate of reaction is 0.01788 kmol/s for all reactor compositions. The product of the two reactor compositions $C_A C_B$ is constant at 3.8514 $kmol^2/m^6$.

The flowrates of the two feedstreams to the reactor change with each set of feed compositions. These are shown in the lower graph in Figure 2.12. As the concentration of C_A increases (and C_B decreases), the flowrate of the component A feed to the reactor increases, while the flowrate of component B decreases. The importance of this is its effect on the overall flowsheet of the plant. Suppose that the unreacted A and B in the reactor effluent must by recovered in a separation system and recycled back to the reactor. Figure 2.13 gives the resulting plantwide flowsheets for two different designs with two different pairs of C_A and C_B.

The relative volatilities of the three components are assumed to be

$$\alpha_A > \alpha_C > \alpha_B \tag{2.37}$$

so two distillation columns are required. In the traditional "direct" separation sequence, the unreacted A is recycled from the top of the first distillation column, and the unreacted B is recycled from the bottom of the second column. Since the production rate of C is the same in both designs, the distillate product from the second column is 0.01788 kmol/s in both flowsheets. In addition, the flowrates of the fresh feedstreams into the plant (not the feedstreams to the reactor itself) are each 0.01788 kmol/s in both flowsheets.

The two processes differ in the flowrates of the two recycles. A high-C_A reactor composition design has a large recycle of A and a small recycle of B. The reverse is true for a low-C_A reactor composition design. Looking at the distillation system, we might expect that less energy is involved in recycling B because it goes out the bottom of the distillation

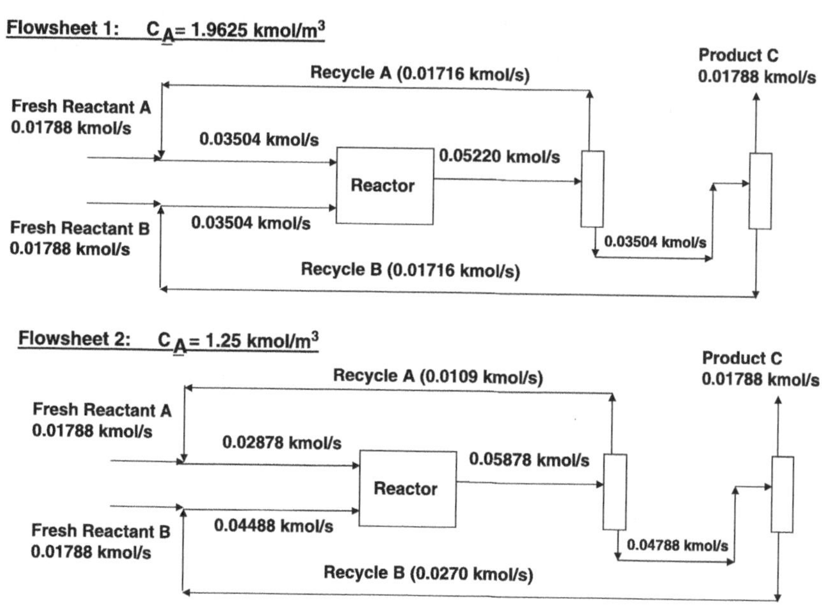

Figure 2.13 Two flowsheets.

columns, while recycling A requires that it be vaporized and taken overhead in the first column. Therefore, the best economic design would favor a smaller C_A design. Note that if the cost of recycling is about the same for both reactants, the value of C_A that minimizes the sum of the two reactor feedstreams, as shown in Figure 2.12, is about 1.9625 kmol/m^3, which is the point where the two concentrations are equal.

The discussion above illustrates that a reaction involving two reactants provides an additional design degree of freedom, which can be used to affect steady-state economics. As we will illustrate in Chapter 3, this feature also provides a parameter that impacts dynamic controllability. To reduce the problem of reactor runaway in a highly exothermic reaction with a strong temperature dependence, the designer can adjust the design to have a very small amount of one of the reactants present in the reactor. This "limiting reactant" gets chewed up quickly if the temperature starts to take off, which limits the reaction rate. This composition effect provides an inherent degree of self-regulation that compensates for the strong temperature dependence of the specific reaction rate k.

Of course, there is no guarantee that the steady-state economic optimum set of reactor compositions is the best set in terms of reactor stability. This is one of the many classical examples of the inherent engineering tradeoff between steady-state economics and dynamic controllability that occurs in many processes.

2.3 REVERSIBLE EXOTHERMIC REACTION

The next type of reaction considered is an exothermic reversible reaction with two reactants and two products:

$$A + B \Longleftrightarrow C + D \tag{2.38}$$

The reaction takes place in a CSTR that has two feedstreams, F_{A0} and F_{B0} (m^3/s), with compositions C_{A0} and C_{B0} (kmol/m^3). The net reaction rate is the difference between the forward and reverse reactions. We assume first-order dependence on compositions:

$$\begin{aligned} \mathfrak{R} &= k_F C_A C_B - k_R C_C C_D \\ &= C_A C_B k_{0F} \exp\left(\frac{-E_F}{RT_R}\right) - C_C C_D k_{0R} \exp\left(\frac{-E_R}{RT_R}\right) \end{aligned} \tag{2.39}$$

There are four components in this system, so four component balances can be written:

Component A balance (kmol A/s):

$$\begin{aligned} F_{A0} C_{A0} &= F C_A + V_R \mathfrak{R} \\ &= F C_A + V_R (k_F C_A C_B - k_R C_C C_D) \end{aligned} \tag{2.40}$$

Component B balance (kmol B/s):

$$F_{B0} C_{B0} = F C_B + V_R (k_F C_A C_B - k_R C_C C_D) \tag{2.41}$$

Component C balance (kmol C/s):

$$0 = F C_C + V_R (-k_F C_A C_B + k_R C_C C_D) \tag{2.42}$$

Component D balance (kmol D/s):

$$0 = FC_{\mathrm{D}} + V_R(-k_F C_{\mathrm{A}} C_{\mathrm{B}} + k_R C_{\mathrm{C}} C_{\mathrm{D}}) \tag{2.43}$$

Assuming that we know the feed conditions, the reactor volume, and the temperature (and therefore the forward and reverse specific reactions rates k_{F} and k_R), there are five unknowns in the four equations above: F, C_{A}, C_{B}, C_{C}, and C_{D}. Therefore an additional equation is needed. If the densities of all the components are the same, the volumetric flowrate of the effluent will be equal to the volumetric flowrates of the two feeds:

$$F_{\mathrm{A0}} + F_{\mathrm{B0}} = F \tag{2.44}$$

There are now four equations and four unknowns. But the solution of these simultaneous nonlinear algebraic equations is difficult. It would be even more difficult if the reactions were not first-order.

It is quite easy to use the *fsolve* function in Matlab to solve for the four unknowns. Figure 2.14 gives a program that solves for the four reactor compositions given reactor temperature, reactor volume, feed conditions and kinetics.

The design of a CSTR with a reversible reaction involves finding the temperature that maximizes conversion. Since the reaction is exothermic, the activation energy of the

```
% Program "cstrreverse.m"
% Uses fsolve to find steady-state composition of reversible reaction in a
%        CSTR of volume vr at temperature tr
% Set Keq at 333 K
keq=10,kor=((49.3e-6)/keq)*exp(139.4e6/8314/333)
% Set VR and TR
    vr=100;tr=350;
kf=4.239e6*exp(-69.7e6/tr/8314);
kr=kor*exp(-139.4e6/8314/tr);
fa0=0.004377;ca0=8.01;fb0=0.004377;cb0=8.01;
flow=fa0+fb0;
parameters=[vr,kf,kr,fa0,ca0,fb0,cb0];
% Initial guesses of ca, cb, cc and cd
xo=[2 2 2 2]';
options=optimset('MaxFunEvals',1000,'TolFun',1e-8);
% Use fsolve to solve four nonlinear algebraic equations
[x,fval]=fsolve(@reverse,xo,options,parameters);
ca=x(1,1);cb=x(2,1);cc=x(3,1);cd=x(4,1);
conversion=(fa0*ca0-flow*ca)/(fa0*ca0)
ca,cb,cc,cd
```

```
% function "reverse.m"
% Use fsolve to find solution of 4 nonlinear algebraic equations
%
function f=reverse(x,parameters)
ca=x(1,1);cb=x(2,1);cc=x(3,1);cd=x(4,1);
vr=parameters(1);kf=parameters(2);kr=parameters(3);
fa0=parameters(4);ca0=parameters(5);fb0=parameters(6);cb0=parameters(7);
flow=fa0+fb0;
ratef=kf*ca*cb;
rater=kr*cc*cd;
% Four equations
f(1,1)=fa0*ca0-flow*ca-vr*(ratef-rater);
f(2,1)=fb0*cb0-flow*cb-vr*(ratef-rater);
f(3,1)=-flow*cc-vr*(-ratef+rater);
f(4,1)=-flow*cd-vr*(-ratef+rater);
```

Figure 2.14 Matlab program for reversible reactions.

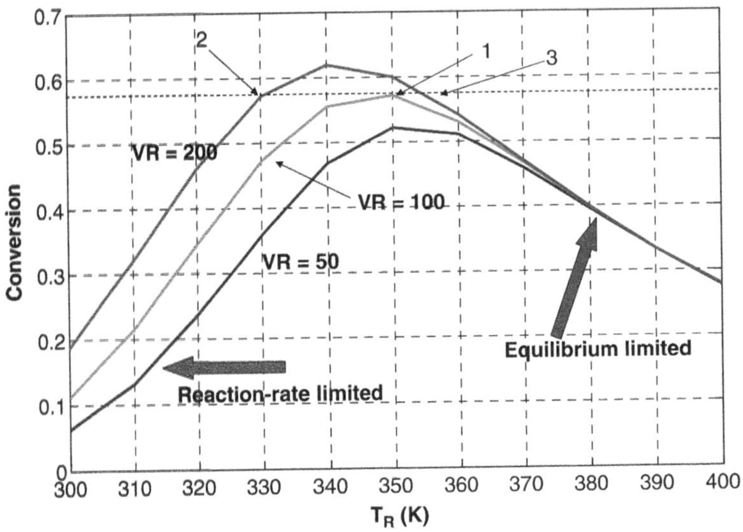

Figure 2.15 Conversion versus temperature for reversible reactions.

reverse reaction is greater than the activation energy of the forward reaction. The higher the temperature, the faster the reaction, but the lower the equilibrium constant. At low temperatures, the conversion is low because of small specific reaction rates. At high temperatures, the conversion is also low, this time because of a low equilibrium constant. These effects are illustrated in Figure 2.15, in which the conversions achieved in CSTR reactors of different sizes are plotted versus temperature. The kinetic parameters used in the specific numerical case are given in Table 2.3.

The smallest reactor that can achieve a 57% conversion in a single CSTR has a volume of 100 m^3 and operates at 350 K (point 1 in Figure 2.15). This conversion could be achieved in a larger (200 m^3) reactor, which could operate at either of two temperatures (points 2 and 3 in Figure 2.15). Obviously capital investment is minimized if the smallest reactor is selected.

The heat transfer area in this large reactor should provide good dynamic controllability because of the small ΔT driving force required. Even with less area, the system is inherently self-regulating because of the reversible reactions.

TABLE 2.3 Reversible Exothermic Reaction Kinetic Parameters

Forward-specific reaction rate k at 333 K	$m^3\ s^{-1}\ kmol^{-1}$	49.3×10^{-6}
Reverse-specific reaction rate k at 333 K	$m^3\ s^{-1}\ kmol^{-1}$	4.93×10^{-7}
Forward preexponential factor k_{0F}	$m^3\ s^{-1}\ kmol^{-1}$	4.239×10^{6}
Reverse preexponential factor k_{0R}	$m^3\ s^{-1}\ kmol^{-1}$	3.631×10^{16}
Forward activation energy E_F	J/kmol	69.71×10^{6}
Reverse activation energy E_R	J/kmol	139.4×10^{6}
Heat of reaction λ	J/kmol	-69.71×10^{6}
Feed flowrate F_{A0}	m^3/s	4.377×10^{-3}
Feed flowrate F_{B0}	m^3/s	4.377×10^{-3}
Feed composition C_{A0}	$kmol/m^3$	8.01
Feed composition C_{B0}	kmol m^3	8.01

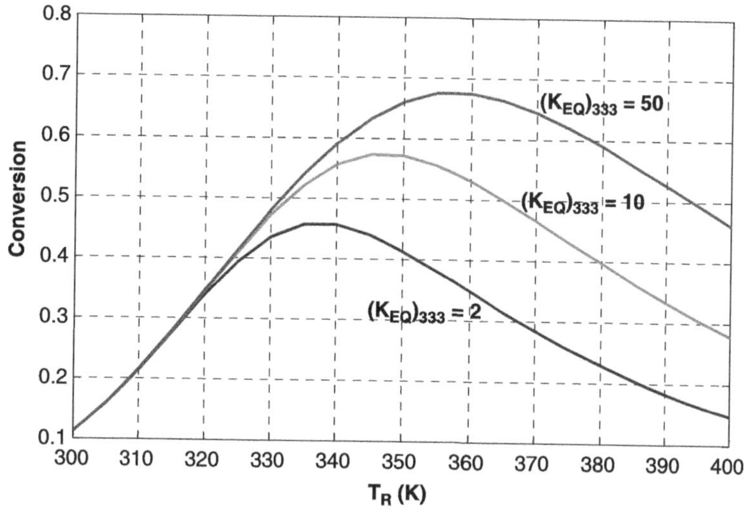

Figure 2.16 Conversion versus temperature for different K_{eq}; $V_R = 100$.

The equilibrium constant at 333 K in the numerical example considered above is

$$(K_{EQ})_{333} = \frac{(k_F)_{333}}{(k_R)_{333}} = \frac{49.3 \times 10^{-6}}{4.93 \times 10^{-6}} = 10 \tag{2.45}$$

Figure 2.16 shows the effect of temperature on conversion for different values of $(K_{EQ})_{333}$ with reactor volume fixed at 100 m^3.

2.4 CONSECUTIVE REACTIONS

Reactor design becomes more challenging when yield as well as conversion must be considered. One common situation in which this arises is when there are consecutive irreversible reactions such as the following:

$$\begin{aligned} A + B &\longrightarrow C \\ C + B &\longrightarrow D \end{aligned} \tag{2.46}$$

The desired product is C. The undesired product is D. There are many important industrial examples of this type: chlorination, oxidation, and nitration of a variety of hydrocarbons. The specific reaction rates for the first and second reactions are k_1 and k_2, respectively.

The "conversion" of reactant A is defined as

$$\text{Conversion of A} = \frac{\text{moles A fed} - \text{moles A leaving reactor}}{\text{moles A fed}} \tag{2.47}$$

There are several ways to define "yield" (or "selectivity") of the desired product C, as shown in Eq. (2.48). One is on the basis of the amount of A fed. The other is on the basis of the amount of A that has reacted.

$$
\begin{aligned}
\text{Yield 1} &= \frac{\text{moles C produced}}{\text{moles A fed}} \\
\text{Yield 2} &= \frac{\text{moles C produced}}{\text{moles A reacted}}
\end{aligned}
\tag{2.48}
$$

The desirable product C is produced by the first reaction whose rate depends on the concentrations of A and B in the reactor. But C is consumed by the second reaction whose rate depends on the concentrations of C and B in the reactor.

$$
\begin{aligned}
\mathfrak{R}_C &= -k_1 C_A C_B + k_2 C_C C_B \\
&= -C_A C_B k_{01} e^{-E_1/RT_R} + C_C C_B k_{02} e^{-E_2/RT_R} \\
\mathfrak{R}_D &= -k_2 C_C C_B \\
&= -C_C C_B k_{02} e^{-E_2/RT_R}
\end{aligned}
\tag{2.49}
$$

where the reaction rates $\mathfrak{R}_C$ and $\mathfrak{R}_D$ are rates of consumption.

It is obvious that the way to design this reactor is to keep the concentrations of C and B small so as to keep the second reaction rate small. This implies that a large excess of reactant A should be used in the reactor, which will dilute the C and B concentrations. It will also help to drive the first reaction because of the large value of C_A despite the small value of C_B. With this design, the per-pass conversion of A will be small, but the yield of C per mole of A reacted will be large.

The downside of this design is that the excess A must be recovered and recycled, which means high capital and energy costs. However, the resulting improvement in the yield of C is typically well worth the added cost. This is particularly true when the undesirable product D is toxic, corrosive, explosive or an environmental pollutant that is difficult to dispose of. Environmental and safety concerns have pushed the designs of many chemical processes to include several large recycle streams so that the yields of desirable products are increased and the yields of undesirable products are decreased. These recycle streams increase the difficulty of the plantwide control problem.

There are four components, so four component balances can be written.

Component A balance (kmol A/s):

$$F_{A0} C_{A0} = F C_A + V_R k_1 C_A C_B \tag{2.50}$$

Component B balance (kmol B/s):

$$F_{B0} C_{B0} = F C_B + V_R (k_1 C_A C_B + k_2 C_C C_B) \tag{2.51}$$

Component C balance (kmol C/s):

$$0 = F C_C + V_R (- k_1 C_A C_B + k_2 C_C C_B) \tag{2.52}$$

Component D balance (kmol D/s):

$$0 = FC_{\mathrm{D}} - V_R k_2 C_{\mathrm{C}} C_{\mathrm{B}} \tag{2.53}$$

The assumption of constant liquid densities means the volumetric flowrate of the reactor effluent F is the sum of the two volumetric feed flowrates.

The solution of these simultaneous nonlinear algebraic equations is difficult, but the Matlab program given in Figure 2.17 uses *fsolve* to find the concentrations of the components in the reactor effluent. The parameters given are reactor volume, reactor temperature, and feed conditions. Figure 2.18 gives a plot of conversion of A and yield of C as functions of reactor temperature. The numerical example uses the parameters given in Table 2.4. Reactor volume is 200 m^3.

The results shown in Figure 2.18 reveal some interesting effects of reactor temperature. Because the second reaction has a higher activation energy, its specific reaction rate increases more rapidly with temperature than the first reaction. Therefore the production of the undesirable product D is small at temperatures below 320 K, but increases at higher temperatures. The concentration of the desired component C reaches a maximum at about 332 K. This may or may not be the best temperature at which to operate the reactor. It depends on the economic value or the difficulty of disposing of

```
% Program "cstrconsecutivebasic.m"
% Uses fsolve to find steady-state compositions of consecutive reactions
kratio=10,ko2=((49.3e-6)/kratio)*exp(139.4e6/8314/333)
vr=200;tr=340;
k1=4.239e6*exp(-69.7e6/tr/8314)
k2=ko2*exp(-139.4e6/8314/tr)
fa0=0.004377;ca0=8.01;fb0=0.004377;cb0=8.01;cc=2;flow=fa0+fb0;
parameters=[vr,k1,k2,fa0,ca0,fb0,cb0];
% Initial guesses of ca, cb, cc, cd
xo=[2 2 2 .1]';
options=optimset('MaxFunEvals',1000);
% Use fsolve to solve four nonlinear algebraic equations
[x,fval]=fsolve(@consecutive,xo,options,parameters);
ca=x(1,1);cb=x(2,1);cc=x(3,1);cd=x(4,1);
ca,cb,cc,cd
conversion=(fa0*ca0-flow*ca)/(fa0*ca0)
yield1=flow*cc/(fa0*ca0)
yield2=flow*cc/(fa0*ca0-flow*ca)
```

```
% function "consecutive.m"
% Use fsolve to find solution of 4 nonlinear algebraic equations
function f=consecutive(x,parameters)
ca=x(1,1);cb=x(2,1);cc=x(3,1);cd=x(4,1);
vr=parameters(1);k1=parameters(2);k2=parameters(3);
fa0=parameters(4);ca0=parameters(5);fb0=parameters(6);cb0=parameters(7);
rate1=k1*ca*cb;rate2=k2*cc*cb;
flow=fa0+fb0;
% Four equations
f(1,1)=fa0*ca0-flow*ca-vr*rate1;
f(2,1)=fb0*cb0-flow*cb-vr*(rate1+rate2);
f(3,1)=-flow*cc+vr*(rate1-rate2);
f(4,1)=-flow*cd+vr*(rate2);
```

Figure 2.17 Matlab program for consecutive reactions.

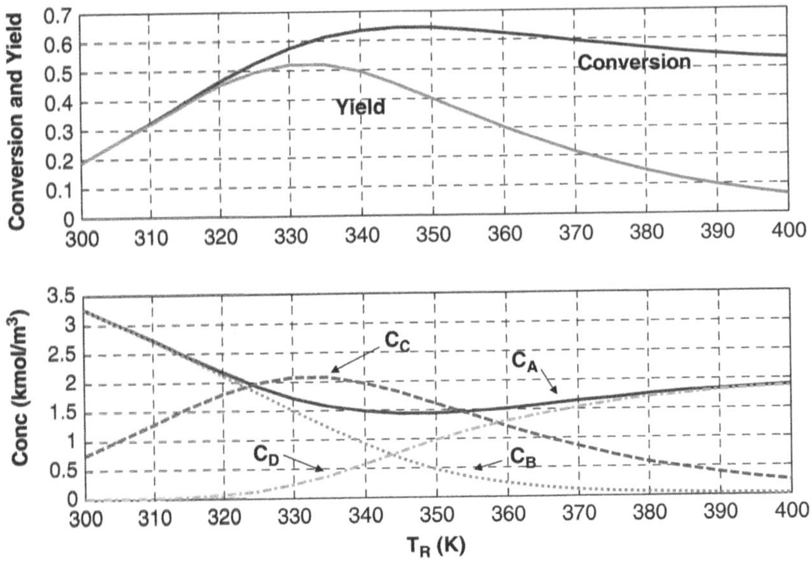

Figure 2.18 Effect of temperature; consecutive reactions.

the undesirable component D because at 332 K the concentration of D is far from negligible (~0.3 kmol/m^3).

Note that conversion of A increases as temperature increases up to about 350 K. Above this reactor temperature, the conversion of A begins to decrease slightly. This occurs because the second reaction consumes so much B in reacting with C that not enough is left to react with all the A. The concentration C_A increases slightly at higher temperatures.

The steady-state economic design of a process with this type of reaction requires consideration of the effects of reactor size and temperature on the entire plant. High recycle flowrates of A and a large reactor operating at a low temperature will suppress the production of D. But this will require a large capital investment in the reactor and separation sections of the plant and consume significant energy.

TABLE 2.4 Kinetic Parameters for Consecutive and Simultaneous Reactions

First specific reaction rate k_1 at 333 K/m^3	s^{-1} kmol^{-1}	49.3×10^{-6}
Second specific reaction rate k_2 at 333 K	m^3 s^{-1} kmol^{-1}	4.93×10^{-7}
First preexponential factor k_{01}	m^3 s^{-1} kmol^{-1}	4.239×10^6
Second preexponential factor k_{02}	m^3 s^{-1} kmol^{-1}	3.631×10^{16}
First activation energy E_1	J/kmol	69.71×10^6
Second activation energy E_2	J/kmol	139.4×10^6
Heat of reaction λ	J/kmol	-69.71×10^6
Feed flowrate F_{A0}	m^3/s	4.377×10^{-3}
Feed flowrate F_{B0}	m^3/s	4.377×10^3
Feed composition C_{A0}	kmol/m^3	8.01
Feed composition C_{B0}	kmol m^3	8.01
Reactor volume V_R	m^3	200

2.5 SIMULTANEOUS REACTIONS

Reactors in which simultaneous reactions occur are quite similar to the consecutive reactors case in that yield of the desired product is important. But since the same reactants are involved in both reactions, reactant concentrations cannot be as readily adjusted to affect yield. Yield of the desired product depends primarily on reactor temperature and on the kinetics of the two reactions.

$$\begin{aligned} A + B &\longrightarrow C \\ A + B &\longrightarrow D \end{aligned} \tag{2.54}$$

If the desired product is C, reactor temperature should be selected so that the specific reaction rate k_1 of the desired reaction is large compared to the specific reaction rate k_2 of the undesired reaction. If the orders of reactions are different in terms of their dependence on the two reactants, the concentrations can also be adjusted to favor the desired reaction:

$$\begin{aligned} \mathcal{R}_A &= \mathcal{R}_B = (k_1 + k_2) C_A C_B \\ &= C_A C_B \left[k_{01} e^{-E_1/RT_R} + k_{02} e^{-E_2/RT_R} \right] \\ \mathcal{R}_C &= -k_1 C_A C_B \\ \mathcal{R}_D &= -k_2 C_A C_B \end{aligned} \tag{2.55}$$

To illustrate some of these issues, we take a numerical example with the parameters given in Table 2.4. There are four components, so four component balances can be written as follows:

Component A balance (kmol A/s):

$$F_{A0} C_{A0} = F C_A + V_R (k_1 + k_2) C_A C_B \tag{2.56}$$

Component B balance (kmol B/s):

$$F_{B0} C_{B0} = F C_B + V_R (k_1 + k_2) C_A C_B \tag{2.57}$$

Component C balance (kmol C/s):

$$0 = F C_C - V_R k_1 C_A C_B \tag{2.58}$$

Component D balance (kmol D/s):

$$0 = F C_D - V_R k_2 C_A C_B \tag{2.59}$$

The Matlab program given in Figure 2.19 solves these nonlinear simultaneous equations for given reactor volume, temperature, and feeds. Figure 2.20 gives results

```
% Program "cstrsimultaneousbasic.m"
% Uses fsolve to find steady-state compositions of simultaneous reactions
kratio=10,ko2=((49.3e-6)/kratio)*exp(139.4e6/8314/333)
vr=200;tr=340;
k1=4.239e6*exp(-69.7e6/tr/8314)
k2=ko2*exp(-139.4e6/8314/tr)
fa0=0.004377;ca0=8.01;fb0=0.004377;cb0=8.01;cc=2;flow=fa0+fb0;
parameters=[vr,k1,k2,fa0,ca0,fb0,cb0];
% Initial guesses of ca, cb, cc, cd
xo=[2 2 2 .1]';
options=optimset('MaxFunEvals',1000);
% Use fsolve to solve four nonlinear algebraic equations
[x,fval]=fsolve(@simultaneous,xo,options,parameters);
ca=x(1,1);cb=x(2,1);cc=x(3,1);cd=x(4,1);
ca,cb,cc,cd
conversion=(fa0*ca0-flow*ca)/(fa0*ca0)
yield1=flow*cc/(fa0*ca0)
yield2=flow*cc/(fa0*ca0-flow*ca)
```

```
% function "simultaneous.m"
% Use fsolve to find solution of 4 nonlinear algebraic equations
function f=simultaneous(x,parameters)
ca=x(1,1);cb=x(2,1);cc=x(3,1);cd=x(4,1);
vr=parameters(1);k1=parameters(2);k2=parameters(3);
fa0=parameters(4);ca0=parameters(5);fb0=parameters(6);cb0=parameters(7);
rate1=k1*ca*cb;rate2=k2*ca*cb;
flow=fa0+fb0;
% Four equations
f(1,1)=fa0*ca0-flow*ca-vr*(rate1+rate2);
f(2,1)=fb0*cb0-flow*cb-vr*(rate1+rate2);
f(3,1)=-flow*cc+vr*rate1;
f(4,1)=-flow*cd+vr*rate2;
```

Figure 2.19 Matlab program for simultaneous reactions.

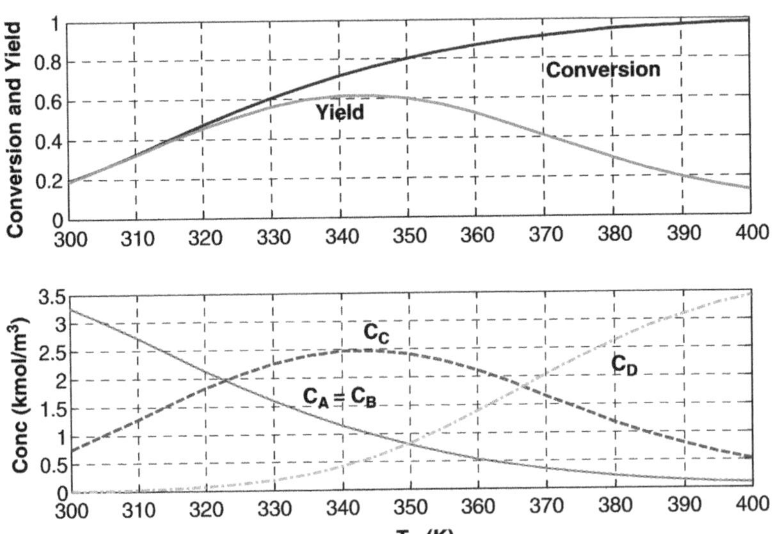

Figure 2.20 Effect of temperature; simultaneous reactions.

for a 200-m^3 reactor with volumetric flowrates of 4.377×10^{-3} m^3/s of each fresh feed. The maximum concentration of C occurs at 343 K, but there is a significant amount of D produced at this temperature. The optimum design may be a larger reactor operating at a lower temperature so that the production of D is kept very small.

2.6 MULTIPLE CSTRs

Up to this point we have looked at a single CSTR and have pointed out some of this reactor's problems, particularly if high conversions are desired. Alternative flowsheets with multiple reactors are frequently used to reduce capital investment, improve yields and to reduce inventories of dangerous chemicals.

One of the recent trends in chemical plant design is the push for "intensification." This is defined as making inventories of material in a plant small so that safety and pollution risks are reduced in the event of equipment failure. As we illustrate in this section, the use of multiple CSTRs can significantly reduce the total reactor volume required, which means a more "intensified" process. However, a word of caution is appropriate here. As we indicated earlier in this chapter, large reactors have large heat transfer areas and may give more stable operation that small reactors. Dynamic upsets and poor control can lead to unsafe situations and atmospheric pollution (venting of safety valves and rupture disks). Therefore you should keep in mind that intensification is only one aspect of plant design and operation. As with most issues in engineering (and in life), there are trade-offs that must be made among the various important considerations.

2.6.1 Multiple Isothermal CSTRs in Series with Reaction A → B

The first multiple-reactor system we study is the reaction considered in Section 2.1, which is the irreversible reaction A → B. The design equation for the volume of a single CSTR is given in Eq. (2.60) in terms of conversion χ, specific reaction rate k, and feed flowrate F:

$$V_R = \frac{F\chi}{k(1-\chi)} \tag{2.60}$$

We specify the desired conversion to be quite high (98 %), so a single CSTR is large. Using the numerical parameters given in Table 2.1, a reactor temperature of 350 K gives a specific reaction rate of 8.185×10^{-4} s^{-1}. A feedrate of 4.377×10^{-3} m^3/s requires a reactor with volume 262 m^3 and diameter 5.5 m, using an aspect ratio of 2. The process shown at the top of Figure 2.21 gives the design conditions. The heat transfer area is large (190 m^2), so the ΔT driving force between the reactor at 350 K and the jacket at 339 K is only 11 K. This reactor should be very controllable since only 20% of the total available ΔT is utilized at design conditions (inlet cooling water temperature is 294 K):

$$\text{RSI} = \frac{350-339}{350-294} = 0.20 \tag{2.61}$$

Now suppose that two tanks in series are used. We assume that they are of equal size and operate at the same 350 K temperature. The design equation for the volume of two

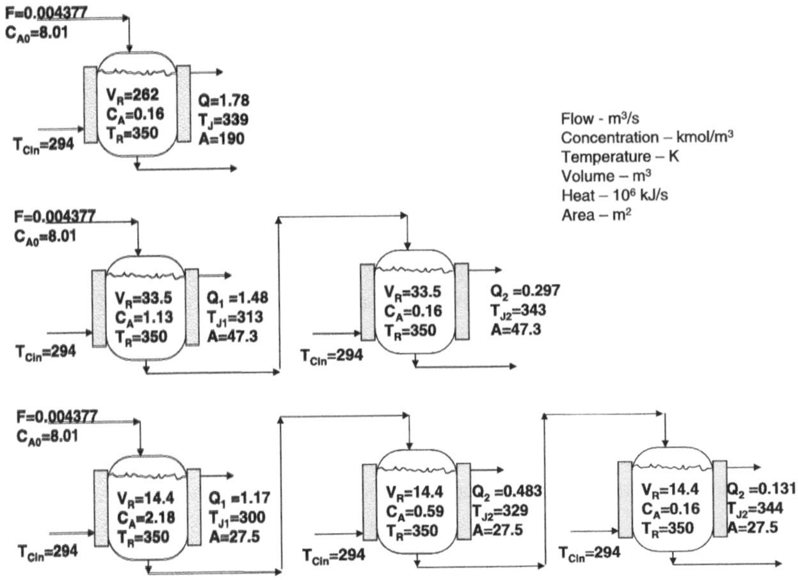

Figure 2.21 Flowsheet for multiple CSTRs.

CSTRs in series of equal volume is given as

$$V_{R1} = V_{R2} = \frac{F\lfloor 1 - \sqrt{1-\chi}\rfloor}{k\sqrt{1-\chi}} \tag{2.62}$$

For the same 98% conversion, the volume of each of the two tanks is only 32.5 m^3 with a 2.74 m diameter. Thus the total volume has been reduced by a factor of 4 from 262 m^3 in the case of one CSTR to only 65 m^3 when two CSTRs are run in series. This is a tremendous decrease in capital investment and represents a great intensification.

However, let us look at some of the details about this two-reactor design (shown as the middle process in Fig. 2.21). The concentration of reactant in the first reactor is 1.13 kmol/m^3. This gives a large reaction rate, so the heat transfer rate (1.48×10^6 J/s) is almost as large as that in the single CSTR design (1.78×10^6 J/s). But the two small reactors have much less heat transfer area (47.3 m^2) than does the single large reactor (190 m^2). The ΔT driving force between the reactor at 350 K and the jacket must increase to compensate for this relative reduction in area. The resulting jacket temperature in the first reactor is 313 K, giving a ΔT of 37 K, which is 66% if the total available (250 − 294 K). The reactor stability index is 3 times that of the 1-CSTR process:

$$\text{RSI}_2 = \frac{350 - 313}{350 - 294} = 0.66 \tag{2.63}$$

So the first reactor will not be able to handle disturbances as well as the single large reactor design. This is quantitatively demonstrated in Chapter 3.

The second of the two reactors in the 2-CSTR design has a low reactant concentration (0.16 kmol/m^3), which gives a small reaction rate and a small heat transfer rate (0.297×10^6 J/s). Therefore the ΔT is much smaller, and the jacket temperature

(343 K) is much higher. The second reactor in the series should be very controllable. It is the first reactor that could produce temperature runaways. It has all the ingredients: irreversible exothermic reaction with sufficient reactant present to fuel the runaway because of the conversion is only 85% in the first reactor.

If the 2-CSTR design has more potential control problems than the 1-CSTR design, we would expect the 3-CSTR design to be even worse. This is precisely what occurs. The volume of each reactor is further reduced (14.4 m^3 with diameter 2.09 m), so the total reactor volume of the three vessels is 43 m^3, which is only 16% of the volume of the reactor in the 1-CSTR design. Clearly this is a very significant reduction in the holdup of reactive material. Capital investment would be lower even though using three vessels is more expensive than using one vessel if their total volume were the same.

The design equation for three equal-volume CSTRs in series is

$$V_{R1} = V_{R2} = V_{R3} = \frac{F\left[1 - (1-\chi)^{1/3}\right]}{k(1-\chi)^{1/3}} \tag{2.64}$$

The process shown at the bottom of Figure 2.21 gives the conditions for this design. The reactant concentration in the first reactor is even higher (2.18 kmol/m^3), so the reaction rate is large despite the smaller volume. This produces a required heat transfer rate that is still quite large (1.16×10^6 J/s) relative to the smaller heat transfer area (27.5 m^2). The result is a very large ΔT in the first reactor (jacket temperature is 300 K), which is 89% of the total available:

$$\mathrm{RSI}_3 = \frac{350 - 300}{350 - 294} = 0.89 \tag{2.65}$$

This reactor is shown to be uncontrollable in Chapter 3.

These steady-state design results indicate that dynamic controllability of a multiple-CSTR process could be poor if the reactions are irreversible and exothermic.

If the reactor temperatures are reduced, the vessels become larger and more heat transfer area is available. So the multiple CSTR system controllability could be improved by designing them for lower reactor temperatures. Of course, this would reduce the savings in capital investment compared to the single CSTR and negatively impact the desire for intensification. As an example, suppose the temperatures in the 2-CSTR process are set at 340 K instead of 350 K. The vessels increase from 32.5 to 65.7 m^3 with heat transfer areas that increase from 47.3 to 75.7 m^2. The percentage of the total ΔT driving force in the critical first reactor decreases from 66 to 54%, indicating a more controllable system.

2.6.2 Multiple CSTRs in Series with Different Temperatures

Another alternative is to operate the reactors at different temperatures. The sizing equations have to be modified to reflect that the specific reactions rates are different in the different reactors. For a 2-CSTR process, the quadratic equation given in Eq. (2.66) must be solved for reactor volume V_R. The parameters k_1 and k_2 are the specific reaction rates at the two reactor temperatures T_{R1} and T_{R2}, respectively.

$$\left(\frac{V_R}{F}\right)^2 (k_1 k_2) + \left(\frac{V_R}{F}\right)(k_1 + k_2) + \left(1 - \frac{1}{1-\chi}\right) = 0 \tag{2.66}$$

As a numerical example, suppose that we run the first reactor at 325 K and the second at 350 K. The resulting design has vessel volumes of 64.5 m^3 with heat transfer areas of 74.8 m^2. However, the percentage of the total ΔT driving force in the critical first reactor decreases only to 61%. This approach appears to be less effective than lowering both reactor temperatures.

2.6.3 Multiple CSTRs in Parallel

Instead of running the reactors in series, they could be arranged to operate in parallel. The advantage of this flowsheet is somewhat subtle because the concentration of the reactant in the reactors would be the same low concentration required in a single reactor for the same conversion. The total volume of the two vessels is exactly the same as the volume in one big vessel. Building two vessels is more expensive than building one vessel, so it is not obvious what the advantage of the parallel operation is.

The improvement comes from the lower feed flowrate to each reactor. This gives smaller reactors that have a higher ratio of heat transfer area to volume. To illustrate this, we use the same numerical case studied above. Now the 4.377×10^{-3} m^3/s of feed is split between two parallel reactors, each operating at 350 K with 98% conversion. The required reactor volume in each vessel is 131 m^3, giving a total of 262 m^3, which is the same volume as one reactor. However, the heat transfer area in each is 120 m^2, giving a total of 240 m^2, which is larger than the 190 m^2 of the single large vessel. The heat transfer rate in each reactor is half that of the single reactor. The result is a jacket temperature of 341 K in the two parallel reactors, compared to 339 K in the single reactor. Thus less of the total available ΔT is consumed, and operability should be improved.

2.6.4 Multiple CSTRs with Reversible Exothermic Reactions

In a system with reversible exothermic reactions, intuition tells us that operating a series of CSTRs operating at different temperatures should improve performance. The first reactor could have a high temperature to get the reactions going, but the latter reactors could have lower temperatures to avoid the chemical equilibrium constant limitation associated with high temperatures. In Section 2.3 we showed that there is an optimum temperature for a single CSTR that maximized conversion. In this section we first see whether using a series of CSTRs operating at the same temperature improves performance. Then we see if operating at different temperatures improves conversion.

Reactors with Same Volume Figure 2.22 gives a Matlab program for solving the equations describing two CSTRs operating in series with reversible reactions. We assume that the two reactors have the same volume. The temperatures in the two reactors are T_{R1} and T_{R2}. The two fresh feedstreams enter the first vessel and produce an effluent containing all four components. This is fed to the second reactor. Two Matlab functions ("reverse1" and "reverse2") are used.

Figure 2.23*a* shows how conversion is affected by reactor size and the temperature of the second reactor. The temperature in the first reactor is set at 350 K (the temperature that maximizes conversion in a single 100-m^3 CSTR). There is a second-reactor temperature that maximizes conversion for a given reactor size. The maximum conversion increases as reactor size increases, and the temperature at which it occurs shifts to lower values. For example, two 200-m^3 reactors can achieve a 72.4% conversion if the second reactor is operated at 328 K. Two 100-m^3 reactors can achieve a 68.7% conversion if the

```
% Program "multiplecstrreversebasic.m"
% Two CSTR's in series with reversible reaction at different temperatures
%  Each reactor has same volume
clear
keq=10,kor=((49.3e-6)/keq)*exp(139.4e6/8314/333)
% Set VR, TR1 and TR2
    vr=100;tr1=350;tr2=320;
%*********************************************
%  First reactor
kf1=4.239e6*exp(-69.7e6/tr1/8314);kr1=kor*exp(-139.4e6/8314/tr1);
fa0=0.004377;ca0=8.01;fb0=0.004377;cb0=8.01;flow1=fa0+fb0;
parameters=[vr,kf1,kr1,fa0,ca0,fb0,cb0];
% Initial guesses of ca1, cb1, cc1 and cd1
xo=[2 2 2 2]';
options=optimset('MaxFunEvals',1000);
[x,fval]=fsolve(@reverse1,xo,options,parameters);
ca1=x(1,1);cb1=x(2,1);cc1=x(3,1);cd1=x(4,1);
conversion1=(fa0*ca0-flow1*ca1)/(fa0*ca0);
ca1,cb1,cc1,cd1,conversion1,pause
% ****************************************************
%  Second reactor
kf2=4.239e6*exp(-69.7e6/tr2/8314);kr2=kor*exp(-139.4e6/8314/tr2);
flow2=flow1;
parameters=[vr,kf2,kr2,flow2,ca1,cb1,cc1,cd1];
% Initial guesses of ca2, cb2, cc2 and cd2
xo=[2 2 2 2]';
[x,fval]=fsolve(@reverse2,xo,options,parameters);
ca2=x(1,1);cb2=x(2,1);cc2=x(3,1);cd2=x(4,1);
conversion2=(fa0*ca0-flow2*ca2)/(fa0*ca0);
ca2,cb2,cc2,cd2,conversion2
```

```
% function "reverse1.m"
% Use fsolve to find solution for first reactor in series
function f=reverse1(x,parameters)
ca1=x(1,1);cb1=x(2,1);cc1=x(3,1);cd1=x(4,1);
vr=parameters(1);kf1=parameters(2);kr1=parameters(3);
fa0=parameters(4);ca0=parameters(5);fb0=parameters(6);cb0=parameters(7);
flow=fa0+fb0;
ratef=kf1*ca1*cb1;rater=kr1*cc1*cd1;
% Four equations
f(1,1)=fa0*ca0-flow*ca1-vr*(ratef-rater);
f(2,1)=fb0*cb0-flow*cb1-vr*(ratef-rater);
f(3,1)=-flow*cc1-vr*(-ratef+rater);
f(4,1)=-flow*cd1-vr*(-ratef+rater);
```

```
% function "reverse2.m"
% Use fsolve to find solution for second reactor in series
function f=reverse2(x,parameters)
ca2=x(1,1);cb2=x(2,1);cc2=x(3,1);cd2=x(4,1);
vr=parameters(1);kf2=parameters(2);kr2=parameters(3);
flow2=parameters(4);ca1=parameters(5);cb1=parameters(6);cc1=parameters(7);cd1=parameters(8);
ratef=kf2*ca2*cb2;rater=kr2*cc2*cd2;
% Four equations
f(1,1)=flow2*ca1-flow2*ca2-vr*(ratef-rater);
f(2,1)=flow2*cb1-flow2*cb2-vr*(ratef-rater);
f(3,1)=flow2*cc1-flow2*cc2-vr*(-ratef+rater);
f(4,1)=flow2*cd1-flow2*cd2-vr*(-ratef+rater);
```

Figure 2.22 Reversible reactions in two CSTRs.

second reactor is operated at 333 K. These should be compared with the 62% conversion attainable is a single 200-m^3 reactor operating at 340 K.

Obviously three or more stages could be used to achieve additional improvements in conversion. At some point the incremental increase in conversion cannot justify the increase in capital investment and operating complexity of using multiple vessels.

Reactors with Different Volumes There may be some advantage in having reactors of different sizes, in addition to operating them at different temperatures. This question is explored by examining various combinations. Figure 2.23*b* shows how conversion depends on the temperature and volume in the second reactor. The first reactor temperature is constant at 350 K and its volume is 100 m^3. Assuming that we want to design at the peak in the conversion curve and the design conversion is 70%, the smallest V_{R2} that can achieve this conversion is about 150 m^3 at a temperature in the second reactor of 332 K.

Figure 2.23*c* gives results for a range of temperatures T_{R1} and various volumes V_{R1} in the first reactor. The temperature in the second reactor is fixed at 332 K and the volume of the second reactor $(V_{R2})^{opt}$ that gives the desired 70% conversion is found by an iterative procedure. The graph on the left gives the total reactor volume $(V_{R1} + V_{R2})$ for three

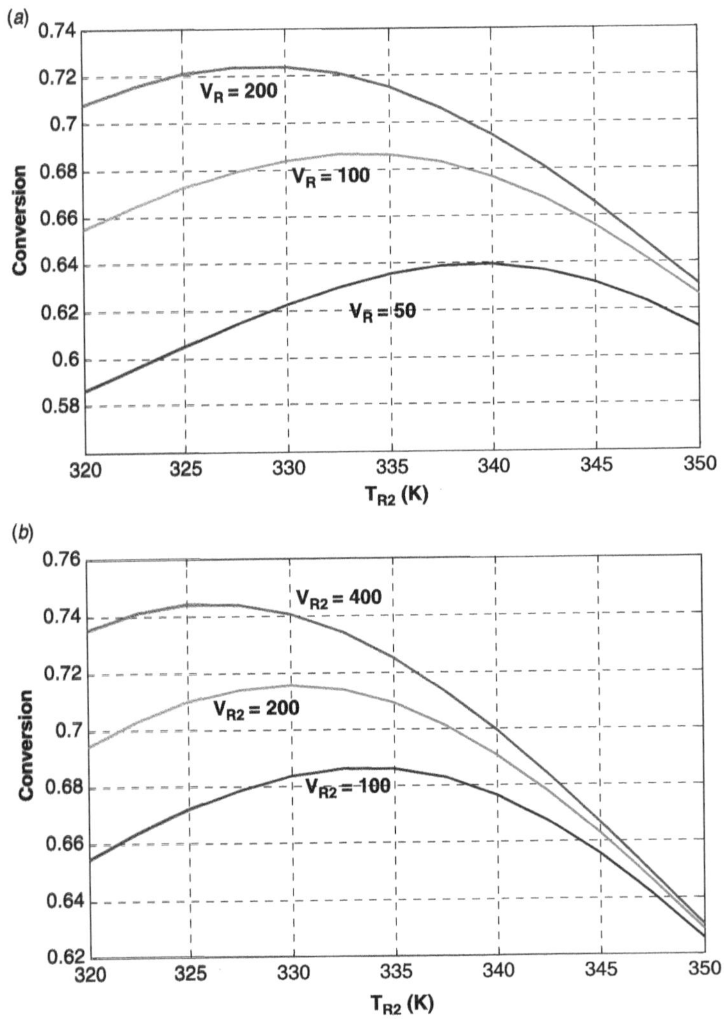

Figure 2.23 (*a*) Effect of second reactor temperature—reversible reactions, $T_{R1} = 350$ K; (*b*) effects of second reactor temperature and size—reversible reactions, $T_{R1} = 350$ K, $V_{R1} = 100$; (*c*) total reactor volume—reversible reactions, $T_{R2} = 332$ K, conversion = 70%.

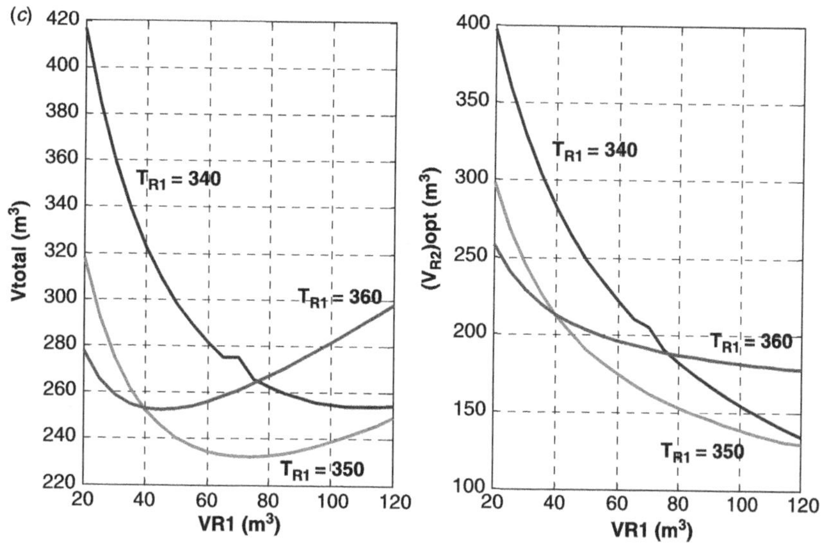

Figure 2.23 *Continued.*

values of T_{R1}. The temperature in the first reactor that minimizes the total volume is 350 K. The corresponding total volume is 230 m^3, 70 m^3 in the first reactor, and 160 m^3 in the second reactor (see the graph on the right).

2.7 AUTOREFRIGERATED REACTOR

All the reactors considered in previous sections had only liquid phases with jacket cooling or coil cooling. There is no consideration of the vapor phase in the reactor, although the reactor liquid composition and temperature certainly set the reactor pressure unless there is an inert component in the system.

There are many reactors in industry that use evaporative cooling. The liquid in the reactor boils to remove reaction heat. The vapor leaving the reactor is condensed in an external heat exchanger, and the liquid is returned to the reactor. Clearly the vapor phase is important in these autorefrigerated reactors. In addition to chemical kinetics, the vapor–liquid equilibrium properties influence the design of the reactor–condenser system.

To illustrate the steady-state design of an autorefrigerated system, a numerical example is developed in this section. The reaction is exothermic, irreversible and occurs in the liquid phase:

$$A + B \longrightarrow C \tag{2.67}$$

The process, kinetic, and phase-equilibrium parameters are given in Table 2.5. There is a single feedstream F_0 (m^3/s) with concentrations of the reactants C_{A0} and C_{B0} (kmol/m^3). A slight excess of reactant B is fed to the reactor, so the conversion is specified in terms of this reactant:

$$\chi = \frac{F_0 C_{B0} - F C_B}{F_0 C_{B0}} \tag{2.68}$$

TABLE 2.5 Autorefrigerated Reactor Parameters

Reactor temperature	K	353
Feed temperature	K	294
Cooling water inlet temperature	K	294
Specific reaction rate k at 353 K	$m^3\ s^{-1}\ kmol^{-1}$	0.001806
Preexponential factor k_0	$m^3\ s^{-1}\ kmol^{-1}$	37.35×10^6
Activation energy E	J/kmol	69.71×10^6
Heat of reaction λ	J/kmol	-69.71×10^6
Process heat of vaporization	J/kmol	23.24×10^6
Feed flowrate F_0	m^3/s	7.812×10^{-4}
Feed composition C_{A0}	$kmol/m^3$	8.804
Feed composition C_{B0}	$kmol\ m^3$	7.203
Molecular weights M_A and M_B	kg/kmol	50
Molecular weight M_C	kg/kmol	100
Process liquid density	kg/m^3	801
Cooling water density	kg/m^3	1000
Process heat capacity	$J\ kg^{-1}\ K^{-1}$	3137
Cooling water heat capacity	$J\ kg^{-1}K^{-1}$	4183
Overall heat transfer coefficient	$W\ K^{-1}\ m^{-2}$	851
Pressure drop in vapor line	kPa	10
Vapor pressure constants[a]	A_{VP}	B_{VP}
A	14.9287	−2825.7
B	14.2347	−2825.7
C	13.3187	−2825.7

[a]Pressure in kPa and temperature in K.

If the conversion is specified, the concentration of B in the reactor liquid phase can be calculated:

$$C_B = C_{B0}(1 - \chi) \tag{2.69}$$

Since the consumption of A is the same as the consumption of B, the concentration of A in the reactor liquid phase is given as

$$C_A = C_{A0} - \chi C_{B0} \tag{2.70}$$

Assuming that the feed contains no product component C, the concentration of C can be calculated:

$$C_C = \chi C_{B0} \tag{2.71}$$

If the temperature in the reactor T_R is specified, the specific reaction rate k can be calculated; now all the information is available to determine the reactor size:

$$V_R = \frac{FC_C}{kC_A C_B} \tag{2.72}$$

Figures 2.24*a* and 2.24*b* give the process conditions and equipment sizes for 90 and 80 % conversion cases. The reactor volume in the 90% conversion case is 1.68 m^3, giving a diameter of 1.02 m if the aspect ratio is 2.

The pressure in the reactor is determined by a bubblepoint calculation at the known temperature and liquid composition. Equation (2.73) is used to convert from

(a)

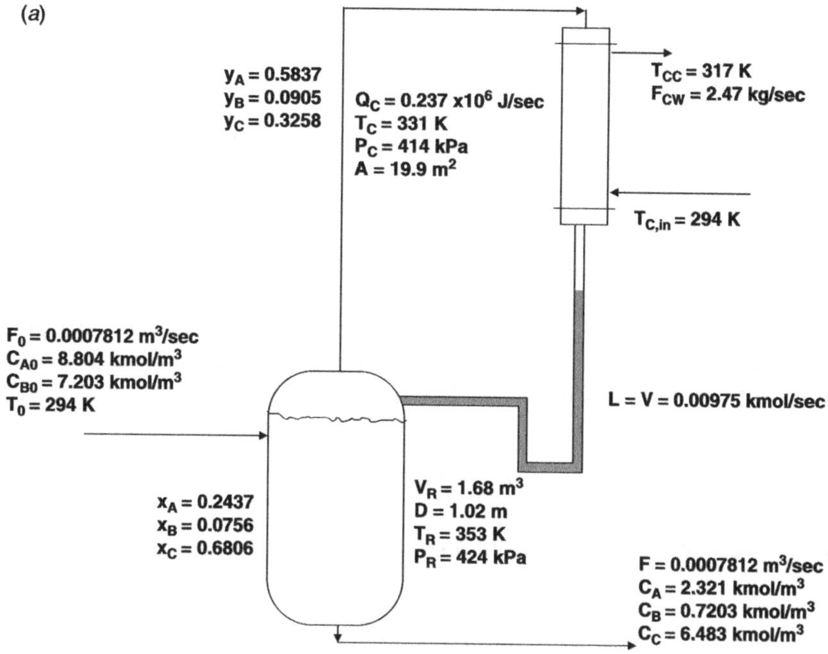

(b)

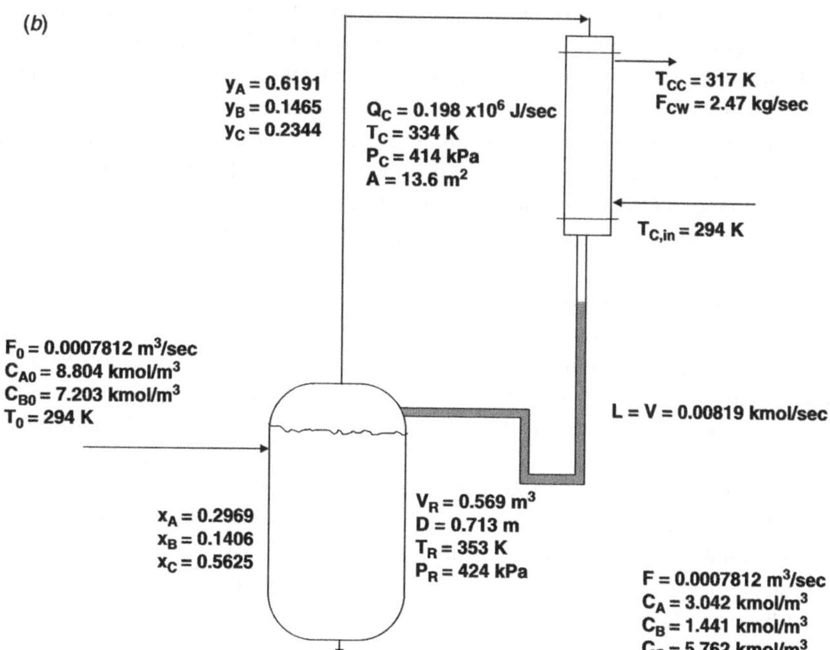

Figure 2.24 Autorefrigerated reactors with (*a*) 90% conversion and (*b*) 80% conversion.

concentrations to mole fractions:

$$x_j = \frac{C_j}{C_A + C_B + C_C} \tag{2.73}$$

The pressure in the reactor P_R is calculated explicitly from Eq. (2.74) assuming ideal vapor–liquid behavior

$$P_R = \sum_{j=A}^{C} x_j P_j^S \tag{2.74}$$

where P_j^S is the vapor pressure of component j (kPa). A two-parameter vapor pressure equation is used. The values of the constants A_{VP} and B_{VP} are given in Table 2.5. Pressures are in kPa and temperature is in degrees Kelvin.

$$\ln P_j^S = A_{VP,j} + \frac{B_{VP,j}}{T} \tag{2.75}$$

The composition of the vapor in equilibrium with the liquid in the reactor is

$$y_j = \frac{x_j P_j^S}{P_R} \tag{2.76}$$

The conditions of the liquid phase in the reactor are now known. Next, the rate of vaporization and the conditions in the condenser must be calculated. The pressure in the condenser must be lower than the pressure in the reactor to make the vapor flow to the condenser. A 10 kPa pressure drop is assumed. This sets the condenser pressure P_C. Assuming a total condenser, the composition of the liquid leaving the condenser $x_{C,j}$ is the same as the vapor entering y_j. So the pressure and the liquid composition are known in the condenser. An iterative bubblepoint calculation is required to determine the condenser temperature T_C. The Newton–Raphson method works well because an analytical derivative of the temperature-dependent vapor pressures can be used. The Matlab program given in Figure 2.25 performs the design sizing calculations given the feed conditions, reactor temperature, and desired conversion.

Notice in Figure 2.24*a* that the temperature in the condenser (331 K) is lower than the temperature in the reactor (353 K). This occurs for two reasons: (1) the pressure in the condenser is lower and (2) the vapor leaving the reactor is richer in the lighter components. These results illustrate one of the inherent disadvantages of autorefrigeration. The temperature of the energy sink is lower than when direct cooling of the reactor liquid is employed. In some cases this means that refrigeration must be used in the condenser, which makes cooling costs much higher than when cooling water can be used.

The flowrate of the vapor to the condenser V (and the liquid returned to the reactor) is calculated by first using an overall energy balance around the entire system to find the condenser heat removal Q_C

$$Q_C = kV_R C_A C_B(-\lambda) + F\rho c_p(T_0 - T_R) \tag{2.77}$$

$$V = \frac{Q_C}{Mc_p(T_R - T_C) + \Delta H_V} \tag{2.78}$$

```
% Program "autodesign.m"
% vapor pressure coefficients for components A, B and C
% In SI units
clear
a=[14.9287 14.2347 13.3187];b=[-2825.7 -2825.7 -2825.7];
% Sizes reactor given tr and conversion and feed
f=.0007812;ca0=8.804;cb0=7.203;tr=353;conv=0.9;
k=37.35e6*exp(-69.71e6/8314/tr);
cb=cb0*(1-conv),ca=ca0-conv*cb0,cc=conv*cb0,vr=f*cc/(k*ca*cb)
dr=(2*vr/pi)^0.3333,ajacket=2*pi*dr^2
% calculate mole fraction of liquid
x(1)=ca/(ca+cb+cc);x(2)=cb/(ca+cb+cc);x(3)=cc/(ca+cb+cc);
% Calculate reactor pressure and vapor composition
 for j=1:3;ps(j)=exp(a(j)+b(j)/tr);end
 pr=0;for j=1:3;pr=pr+ps(j)*x(j);end
 for j=1:3;y(j)=ps(j)*x(j)/pr;xc(j)=y(j);end
 pr,x,y,pause
 pc=pr-10;
 % Bubblepoint calculation in condenser using Newton-Raphson
 tc=tr-15;error=10;
 while error>.0001
    for j=1:3;ps(j)=exp(a(j)+b(j)/tc);end
    pcalc=0;for j=1:3;pcalc=pcalc+ps(j)*xc(j);end
    error=abs(pcalc-pc);
    func=pcalc-pc;
    % Analytical derivative
    dfunc=0;for j=1:3;dfunc=dfunc-x(j)*ps(j)*b(j)/(tc^2);end
    tc=tc-func/dfunc;
 end
 tc
 % Overall energy balance
 roe=800;cp=3137;lambda=69.71e6;dhvap=23.24e6;t0=294;u=851;tcc=317;
 qc=k*vr*ca*cb*(lambda)+f*roe*cp*(t0-tr)
 v=qc/((cp*roe*(tr-tc)/50)+dhvap)
 area=qc/(tc-tcc)/u
 fcw=qc/(tcc-294)/4183
```

Figure 2.25 Matlab program for autorefrigerated reactor.

where M is the average molecular weight of the vapor (assumed to be 50 kg/kmol) and ΔH_V is the heat of vaporization (23.24×10^6 J/kmol).

The design is completed by finding the heat transfer area required in the condenser

$$A = \frac{Q_C}{U(T_C - T_{CC})} \tag{2.79}$$

where U is the overall heat transfer coefficient (851 W m^{-2} K^{-1}) and T_{CC} is the temperature of the cooling water in the condenser (317 K), which is assumed to be perfectly mixed using a circulating system. This temperature is an important design parameter and has a profound effect on reactor controllability because it establishes the area of the condenser. Designing for a high T_{CC} gives more heat transfer area because it corresponds to a smaller ΔT driving force. We will demonstrate in Chapter 3 that dynamic stability requires large areas.

The cooling water flowrate F_{cw} (kg/s) is determined from Eq. (2.80). The temperature of the makeup cooling water is $T_{C,\text{in}}$ (294 K) and its heat capacity is c_{cw} (4183 J kg^{-1} K^{-1}):

$$F_{\text{cw}} = \frac{Q_C}{c_{\text{cw}}(T_{CC} - T_{C,\text{in}})} \tag{2.80}$$

Figure 2.24 shows the effect of different design levels of conversion on the sizes of the reactor and the condenser. A 90% conversion design produces a 1.68-m^3 reactor and 19.9 m^2 of area in the condenser. An 80% conversion design yields a much smaller (0.569 m^3) reactor but only a slightly smaller condenser (13.6 m^2). It is interesting to compare these heat transfer areas with the area that is available if jacket cooling were used. In the 90% conversion design, the vessel diameter is 1.02 m and the jacket area is 6.56 m^2. In the 80% conversion design, the vessel diameter is 0.713 m and the jacket area is 3.19 m^2. The condenser areas are a factor of 3 or 4 larger than the jacket area.

These two designs use a condenser cooling water temperature T_{CC} of 317 K. If this is changed to 322 K, the condenser areas increase to 31.0 and 19.3 m^2, respectively, in the two cases. The dynamic controllability of these alternative designs will be studied in Chapter 3.

2.8 ASPEN PLUS SIMULATION OF CSTRs

The reactions considered in previous sections have involved hypothetical components A, B, C, and D for which arbitrary physical properties and kinetics could be selected to illustrate various phenomena. Simple Matlab programs can be easily generated for these systems.

To study a real chemical system, engineers typically use one of several commercial process simulators. Both steady-state and dynamic simulation software is available for rigorous simulation of several types of reactors. In this section we go through the details of using Aspen Plus to simulate a CSTR in which real chemicals are reacted. The step-by-step procedure is given for defining components, specifying reaction type, providing kinetic parameters, and selecting alternative types of heat removal systems.

The specific chemistry used to illustrate the use of Aspen Plus is the reaction of ethylene (E) with benzene (B) to form the desired product ethylbenzene (EB). There is a consecutive reaction that produces an undesirable product diethylbenzene (DEB). A third reaction combines benzene and diethyl benzene to form ethylbenzene:

$$\begin{aligned} \mathrm{E} + \mathrm{B} &\longrightarrow \mathrm{EB} \\ \mathrm{E} + \mathrm{EB} &\longrightarrow \mathrm{DEB} \\ \mathrm{DEB} + \mathrm{B} &\longrightarrow \mathrm{EB} \end{aligned} \tag{2.81}$$

The reactions occur in the liquid phase and are assumed to be irreversible. The reaction rates of the three reactions are assumed to be those given here:

$$\begin{aligned} \mathcal{R}_1 &= (C_\mathrm{E})(C_\mathrm{B})(1.528 \times 10^6)e^{(-71.13 \times 10^6)/RT} \\ \mathcal{R}_2 &= (C_\mathrm{E})(C_\mathrm{EB})(2.778 \times 10^4)e^{(-83.68 \times 10^6)/RT} \\ \mathcal{R}_3 &= (C_\mathrm{DEB})(C_\mathrm{B})(0.4167)e^{(-62.76 \times 10^6)/RT} \end{aligned} \tag{2.82}$$

The units of $\mathcal{R}$ are kmol s^{-1} m^{-3}. Concentrations have units of kmol/m^3. Activation energies have units of J/kmol. Temperature is in degrees Kelvin.

There are two feedstreams to the reactor: 0.2 kmol/s of pure ethylene and 0.4 kmol/s of benzene. The excess of benzene is used to keep the ethylene concentration low so that the formation of DEB is suppressed.

The steps in setting up the Aspen Plus simulation are outlined below. The rigorous "RCSTR" model is used, which requires specifying reactions and kinetic parameters. An alternative, which is useful in some systems with reversible reactions, is the "RGIBBS" reactor module. Kinetic parameters are not required. Chemical equilibrium compositions are calculated for given feed and reactor temperature and pressure. If the forward and reverse reactions are known to be fast, so that the reactor effluent is at equilibrium conditions, the RGIBBS reactor provides a simple way to model a reactor. In Chapter 3 we will illustrate how this type of reactor can incorporate some approximate dynamics for developing control systems.

2.8.1 Simulation Setup

We click on *Start* and select *Programs, Aspen Tech, Aspen Engineering Suite, Aspen Plus 2004*, and *Aspen Plus User Interface*. The window shown in Figure 2.26 opens. A *Blank Simulation* is selected, and clicking *OK* opens a blank flowsheet shown in Figure 2.27.

At the bottom of the window are page tabs for various units and operations. Clicking the *Reactor* page tab produces a display of alternative types of reactors (see Fig. 2.28). The arrow just to the right of the *RCSTR* type is clicked. Select one of the icons and move the cursor to the blank flowsheet. Clicking on the flowsheet inserts a CSTR (see Fig. 2.29). Multiple CSTRs can be installed by continuing to click on the flowsheet. To stop adding units, click on the arrow at the far left near the bottom of the window. This cancels the input mode. The reactor block can be renamed by clicking the icon on the flowsheet, right-clicking, and selecting *Rename*.

Now we need to install feed and product lines with valves and a pump. Clicking the *Pressure Changers* page tab produces a display of compressors, pumps, valves, and other components (see Fig. 2.30). Clicking the appropriate icon and inserting the device on the flowsheet produces the flowsheet shown in Figure 2.31 with three valves and one pump. The valves and pump have been renamed.

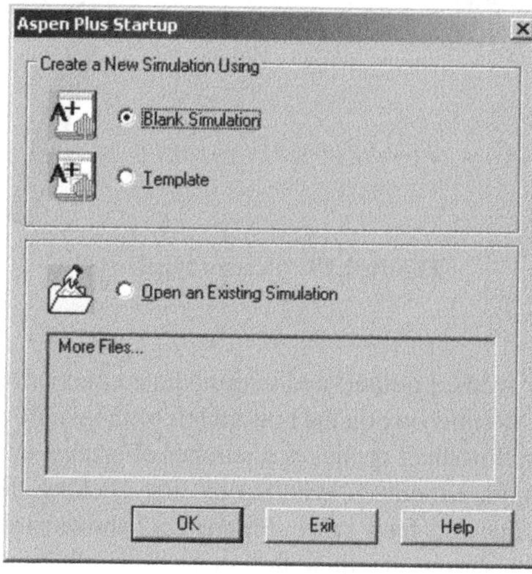

Figure 2.26 Open Aspen Plus simulation.

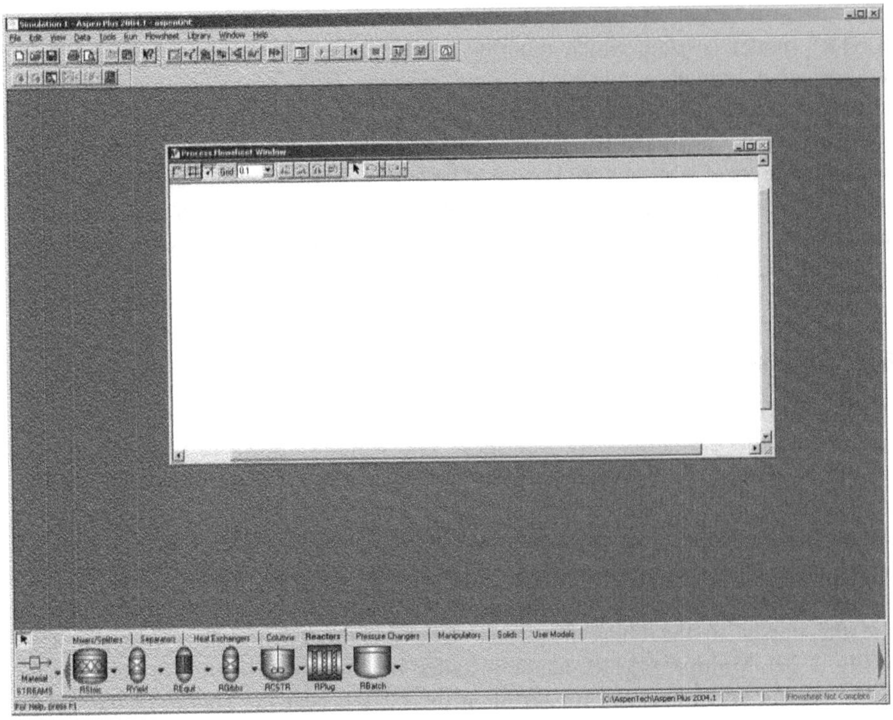

Figure 2.27 Aspen Plus window with flowsheet.

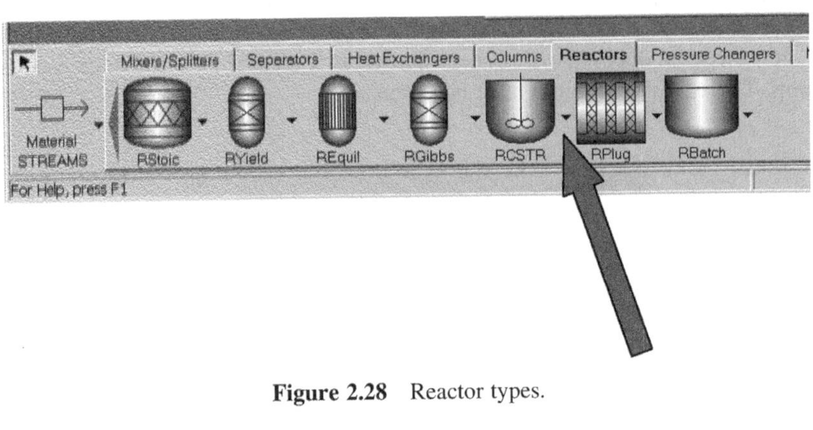

Figure 2.28 Reactor types.

To install streams for feeds, product, and intermediate connections, click the arrow to the right of the *Material Streams* box on the bottom left of the window and select *Material*. Moving the cursor to the flowsheet produces a number of arrows on the inlets and outlets of the various blocks. A feedstream is installed by first clicking the flowsheet and then clicking the arrow pointing to a feed valve. Figure 2.32 shows stream "1" connected to valve "V1." Figure 2.33 shows the final flowsheet with all lines installed and some streams renamed for clarity. Save the file in an appropriate directory.

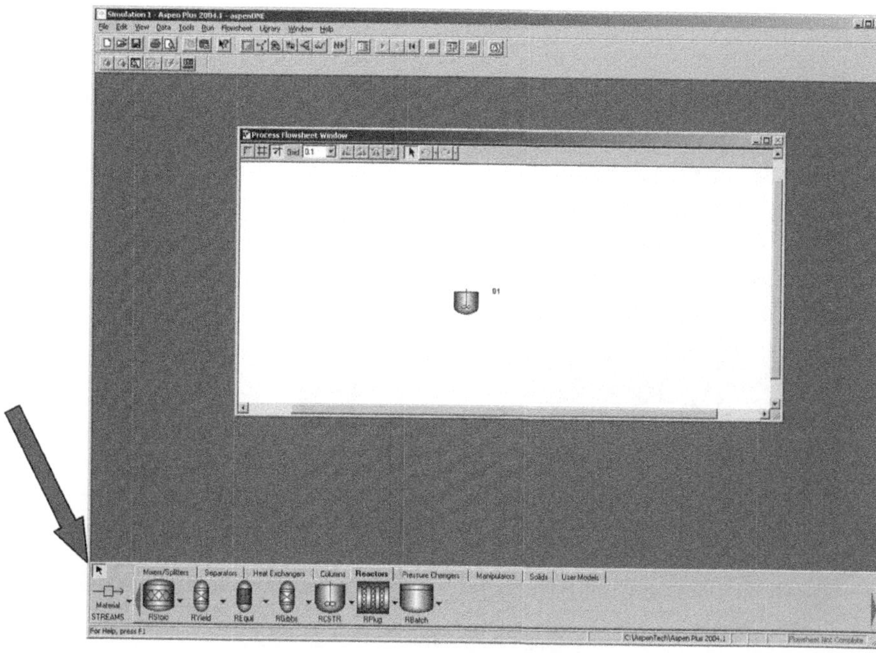

Figure 2.29 Placing RCSTR on flowsheet.

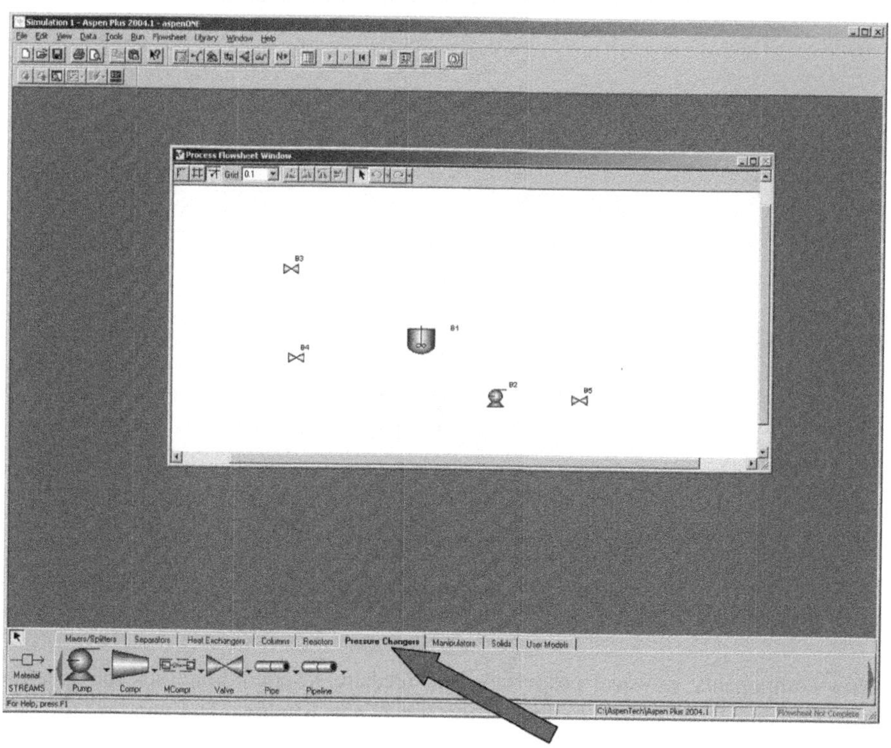

Figure 2.30 Installing pump and valves.

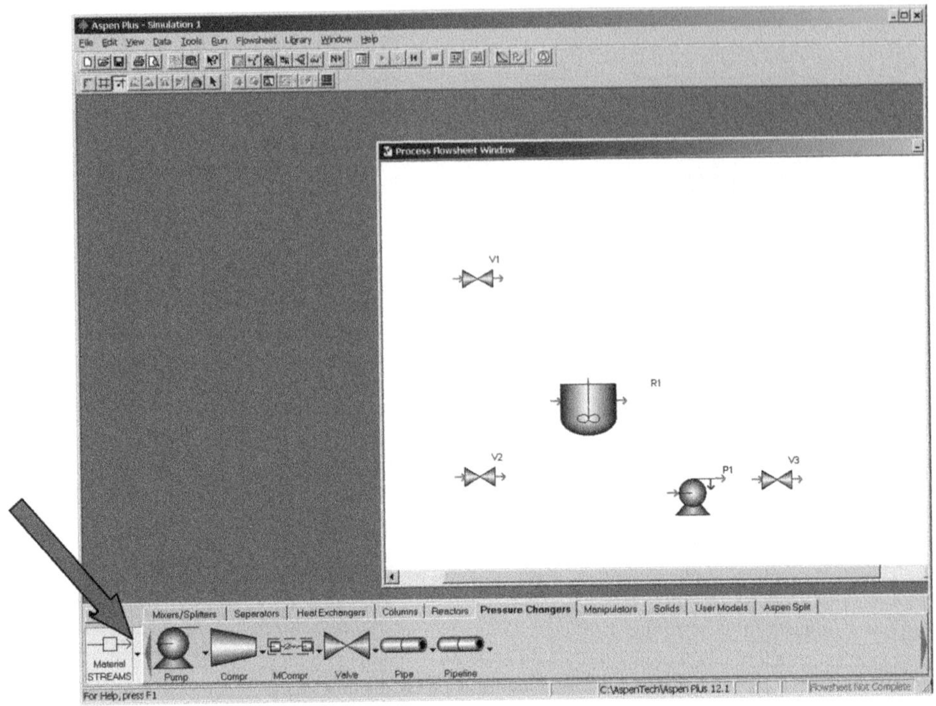

Figure 2.31 Flowsheet with equipment installed.

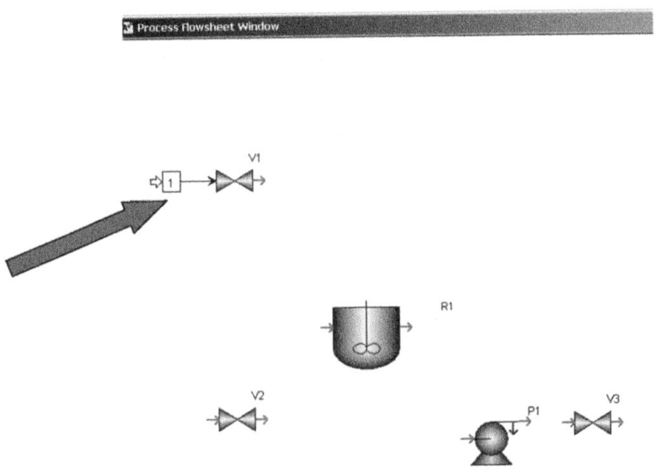

Figure 2.32 Installing material stream.

Now components, physical properties, and feedstream conditions must be specified. Select *Data* and *Setup* on the toolbar at the top of the window. The window shown in Figure 2.34 opens. Almost all aspects of the simulation can be accessed from this *Data Browser* window. The units to use in the simulation are specified in the *Units of*

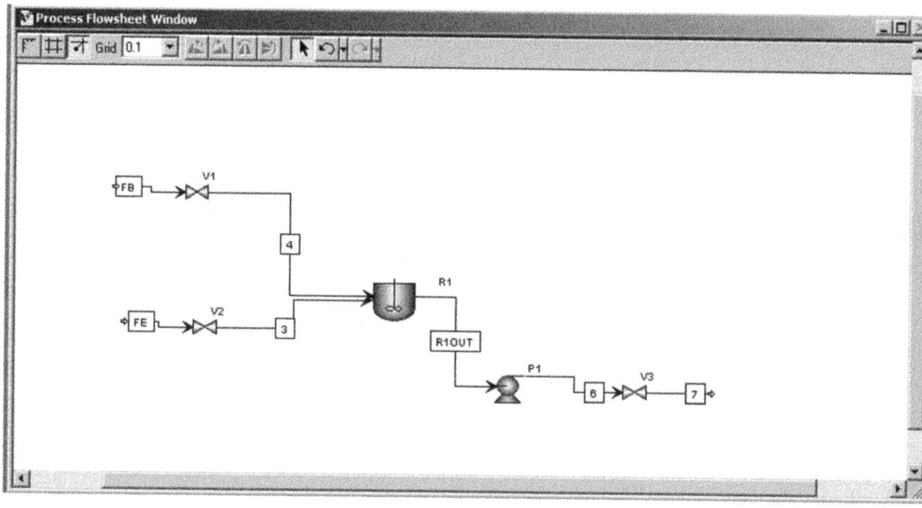

Figure 2.33 Final flowsheet.

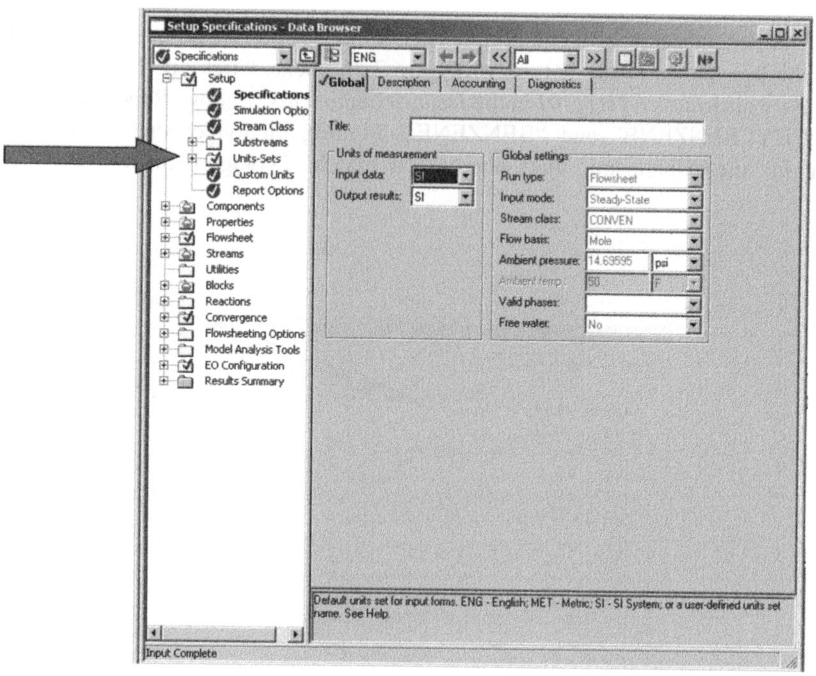

Figure 2.34 Specifying units.

measurement section in the middle of the window. The SI units have been selected. Clicking *Report options* and selecting the *Stream* page tab opens the window shown in Figure 2.35 on which reporting of mole fractions can be selected (it is not a default).

The chemical components to be used in the simulation are specified by clicking *Components, Specifications*, and *Find*. The window shown in Figure 2.36 opens. Type

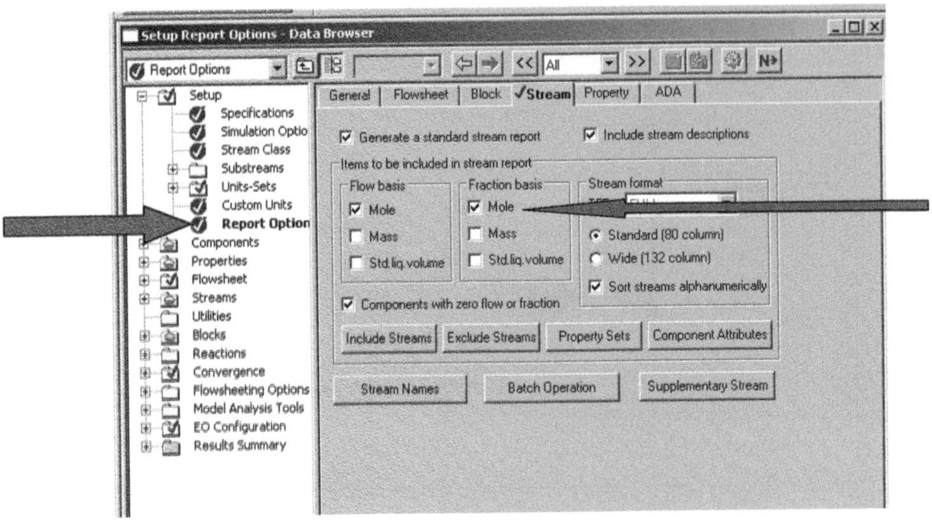

Figure 2.35 Defining report items.

in the component name or formula and click *Find now*. For example, if "ethylene" is typed, the box shown at the bottom of Figure 2.36 opens. Click on "ETHYLENE" and click *Add*. This inserts *ETHYL-01* in the *Define components* window. Repeat the procedure for "ETHYLBENZENE" and "BENZENE." Finding diethylbenzene is not so easy. Typing in "diethylbenzene" does not give a match. Typing in its formula ($C_{10}H_{14}$)

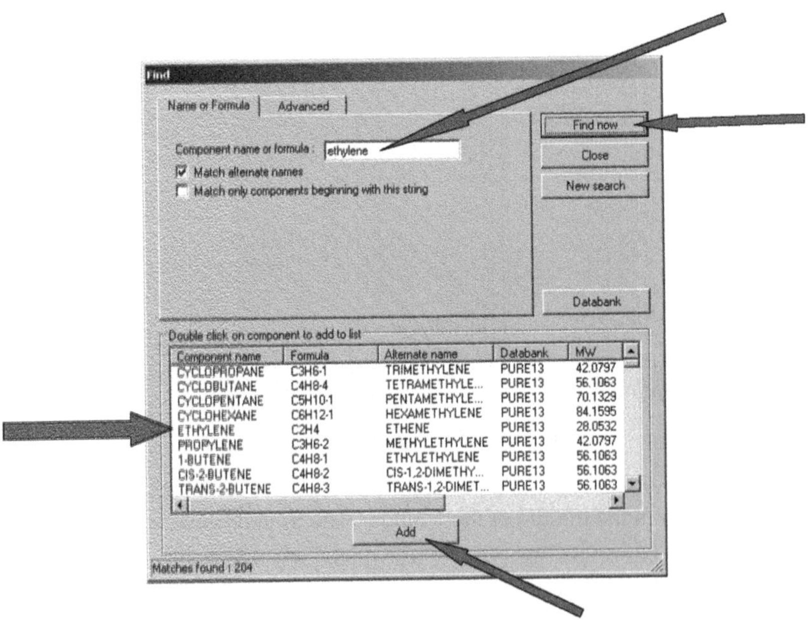

Figure 2.36 Selecting component ethylene.

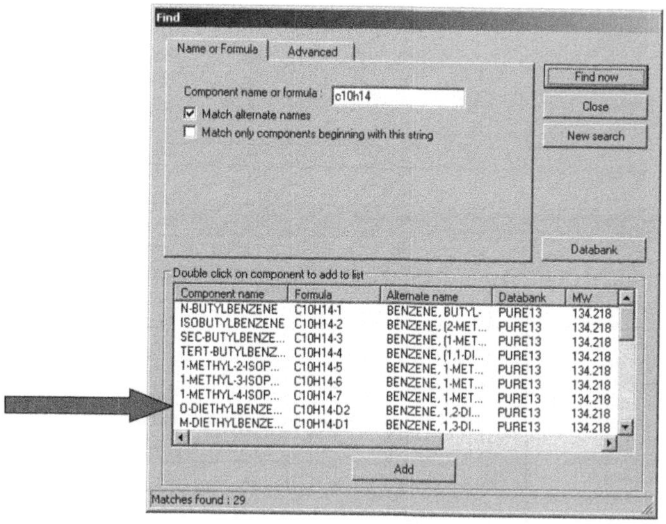

Figure 2.37 Selecting diethylbenzene.

opens the window given in Figure 2.37 on which one of the isomers can be selected ("O-DIETHYLBENZENE") and added to the list.

Figure 2.38 shows the list of four components. The names of the components in the *Component ID* list can be changed to something more convenient by highlighting, typing in

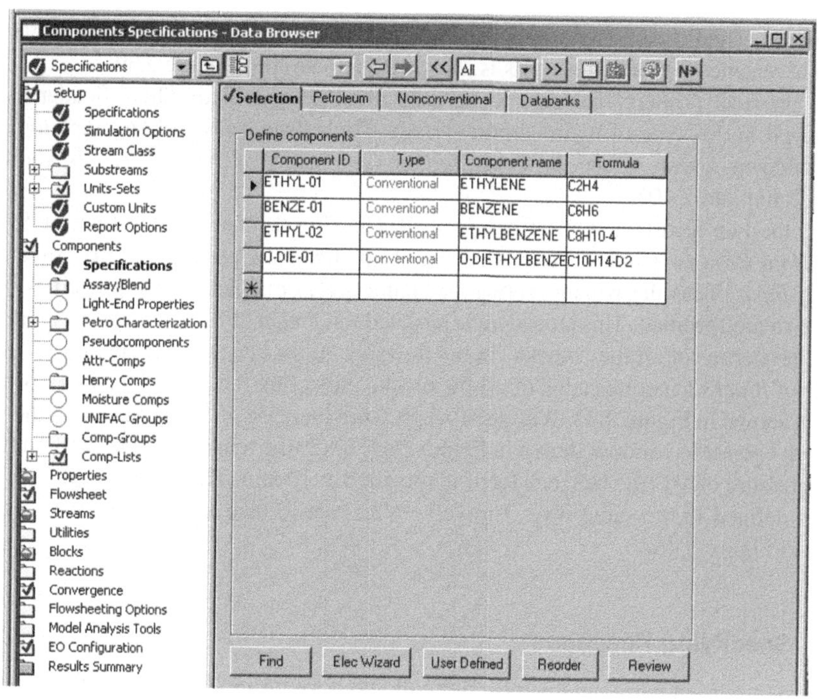

Figure 2.38 All components selected.

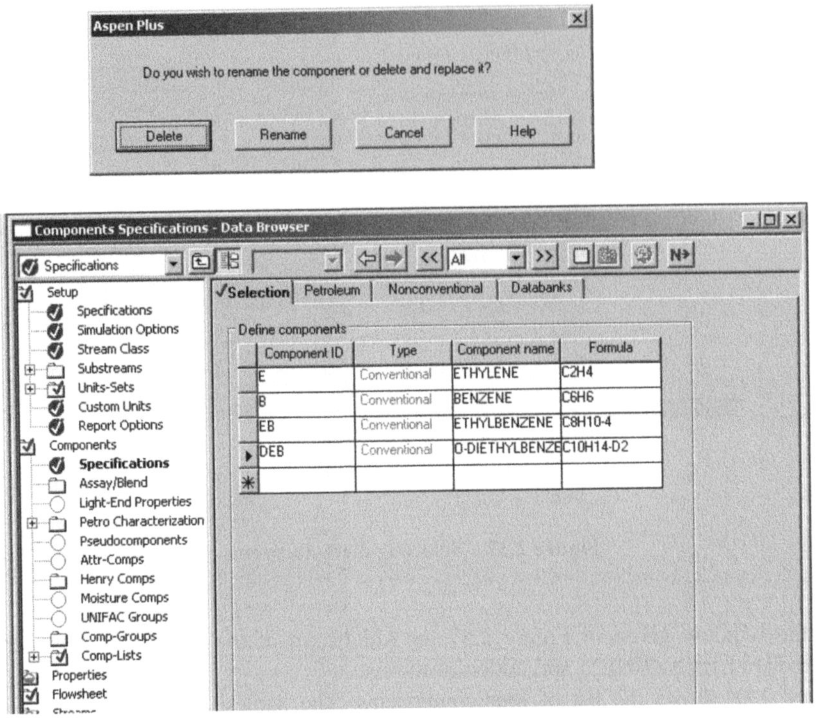

Figure 2.39 Renaming components.

the new name, and clicking *Rename* when the window shown at the top of Figure 2.39 opens. The final renamed list of components is shown at the bottom of Figure 2.39.

The physical property package to be used must be specified. Chao–Seader usually works well in this type of hydrocarbon system. Click *Properties* and *Specifications*. Use the dropdown arrow in the *Base method* box to select "CHAO-SEA" as shown in Figure 2.40*a* and 2.40*b*.

Next the two feedstreams must be specified. Clicking the "+" box in front of *Streams* in the *Data Browser* window produces a list of all streams. Opening the "FB" stream and clicking *Input* opens the window shown in Figure 2.41, on which all the properties of this feedstream are specified. This procedure is repeated for stream "FE" as shown in Figure 2.42.

The properties of all the "blocks" in the flowsheet must be set. Clicking on the " + " to the left of *Blocks* generates a list of all the blocks. Selecting the pump block P1 opens the window shown in Figure 2.43. We specify a pressure increase of 5 atm. Selecting the valve V1 block opens the window shown in Figure 2.44. Since the reactor pressure is 10 atm, the outlet pressure of V1 (the benzene feed) is specified at 10 atm. The valve V2 (the ethylene feed) is defined in the same way. The valve V3 is set to take a 3 atm pressure drop, as shown in Figure 2.45.

2.8.2 Specifying Reactions

The most important part (and the most difficult and error prone) is setting up the reactions. Click *Reactions* on the left column of the *Data Browser* window and click *Reactions*

(a)

(b)

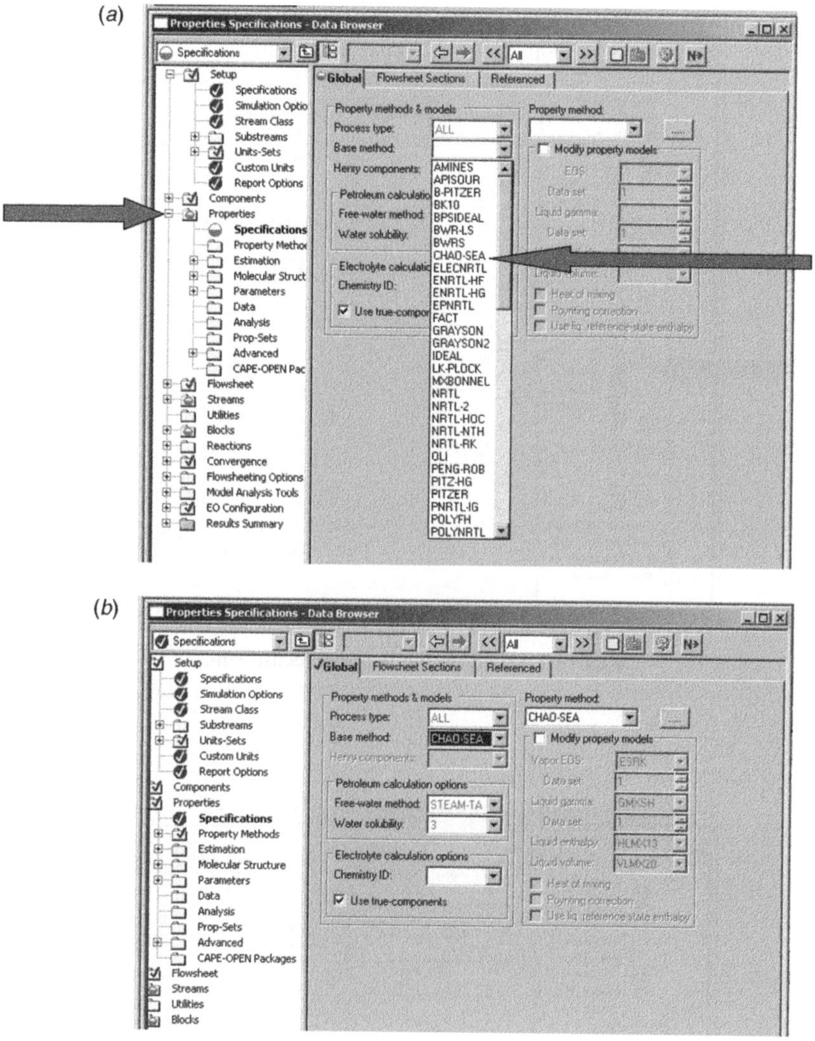

Figure 2.40 (*a*) Selection physical property method; (*b*) Chao–Seader method chosen.

again, as shown in Figure 2.46. To set up a new reaction, click the *New* button. The little window shown in Figure 2.47 opens on which a reaction name (R-1) is specified and the type of reaction is selected. The three reaction rates are given in power-law form [Eq. (2.82)], so *POWERLAW* is selected. Then select *Input* and click *New*, as shown in Figure 2.48. The window shown in Figure 2.49 opens. In the upper right corner, select *Kinetic*. Note that this is *Reaction No.* 1, as shown in the upper left corner. The reaction set R-1 will contain all three reactions (numbers 1, 2, and 3).

Click the box under *Reactant Components* and use the dropdown menu to select component "E" (ethylene). Set the stoichiometric *Coefficient* at "-1" (reactant is consumed) and the *Exponent* at "1" (first-order reaction) (see Fig. 2.50). Repeat for the other reactant "B" (benzene). In a similar way, define the product as "EB" with a coefficient of 1

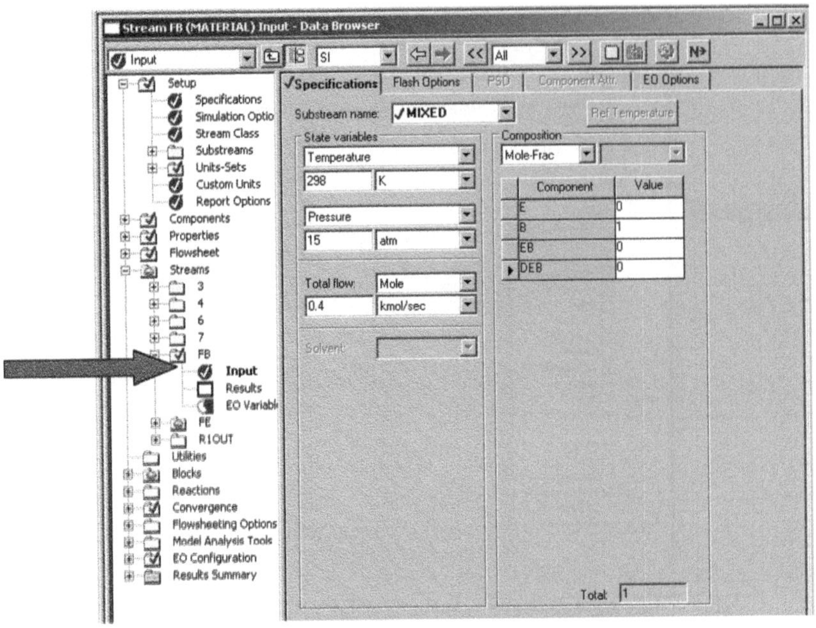

Figure 2.41 Defining benzene feedstream.

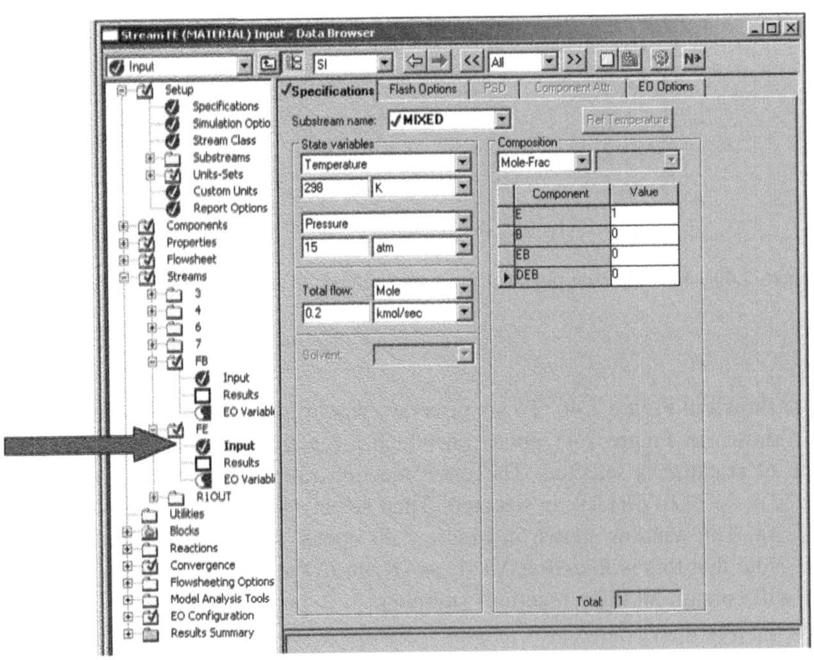

Figure 2.42 Defining ethylene feedstream.

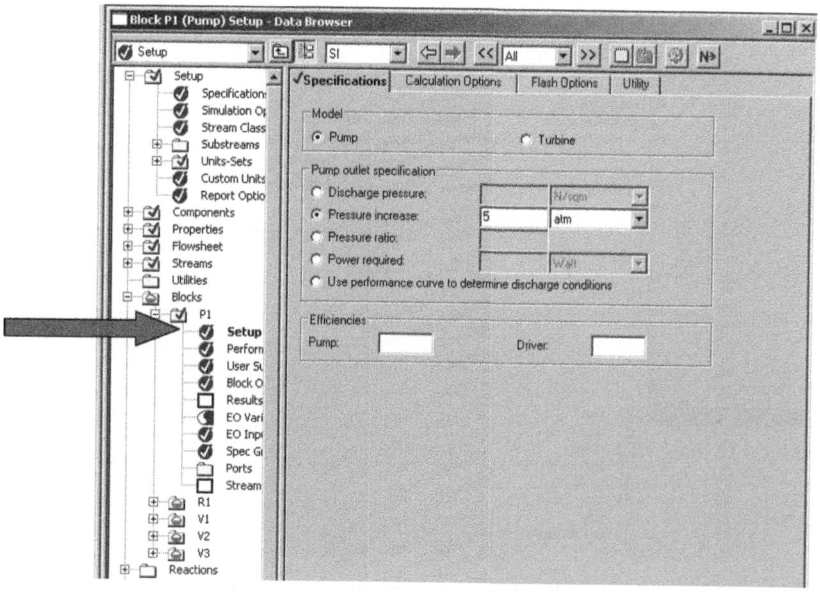

Figure 2.43 Defining pump head.

(produced in the reaction). The *Exponent* is left blank. Then click *Close*. The other two reactions are set up in the same way (see Fig. 2.51).

Clicking the *Kinetics* page tab opens the window shown in Figure 2.52. The dropdown arrow on the first line lets you select one of the three reactions. The kinetic parameters for reaction "1" are shown in Figure 2.52. The *Reacting Phase* is selected as *Liquid*. The *Rate basis* is selected as *React (vol)*. The concentration *[Ci] Basis* is selected as

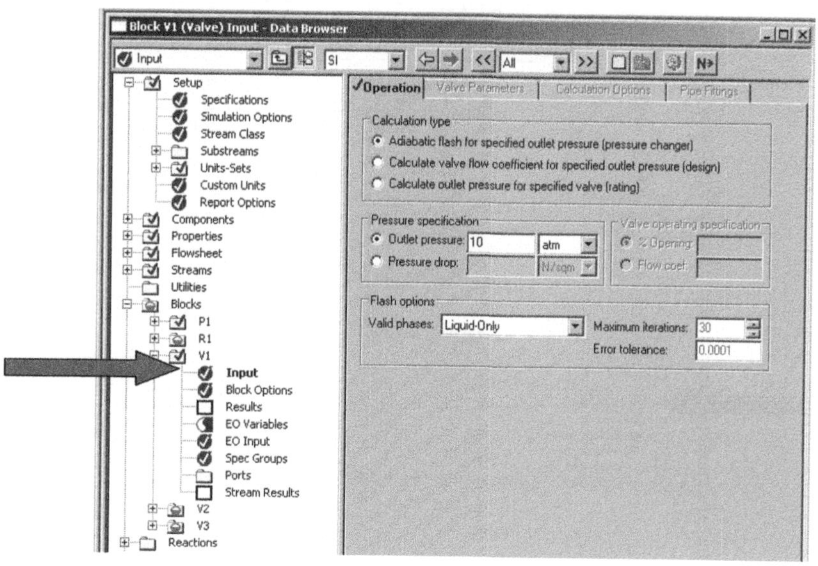

Figure 2.44 Setting feed valve properties.

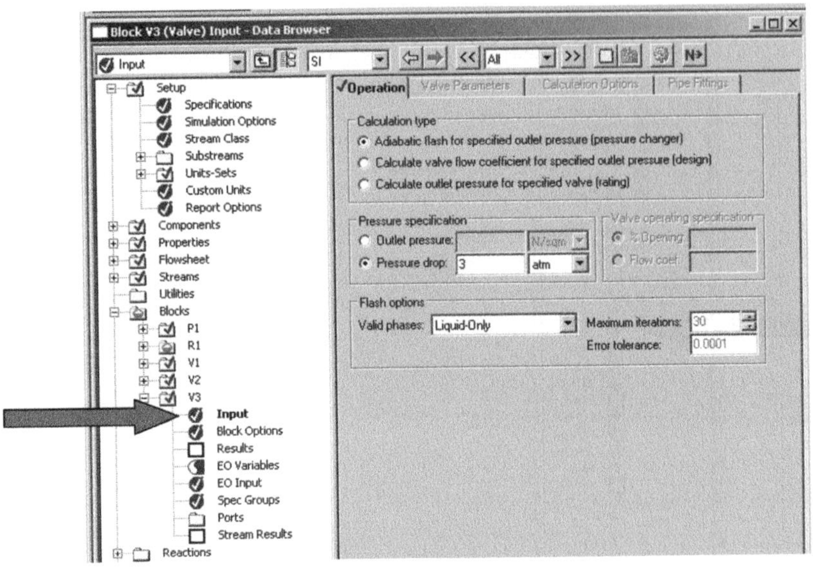

Figure 2.45 Specifying product valve parameters.

Molarity. The values for k and E are entered. The activation energy can specified in several different types of units, but the k must have units of kmol s^{-1} m^{-3}. The kinetic parameters of the two other reactions are set up in the same way (see Figs. 2.53 and 2.54). The reactions are now all specified and can be used in the reactor block (R1).

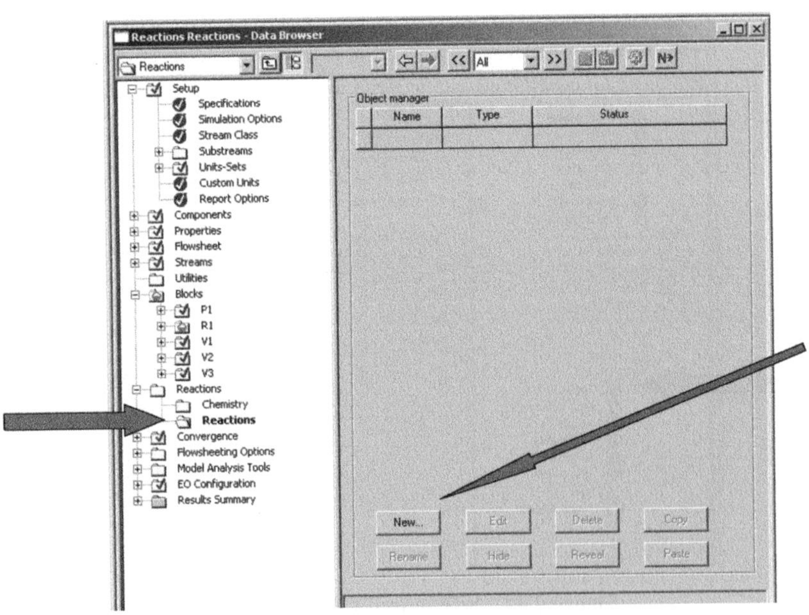

Figure 2.46 Setting up reactions.

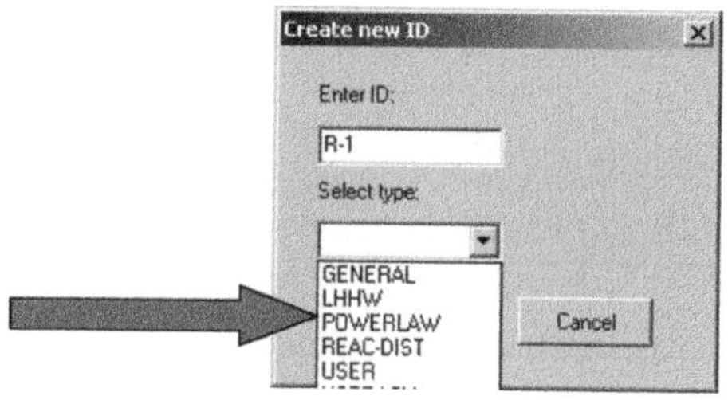

Figure 2.47 Power-law reaction rates.

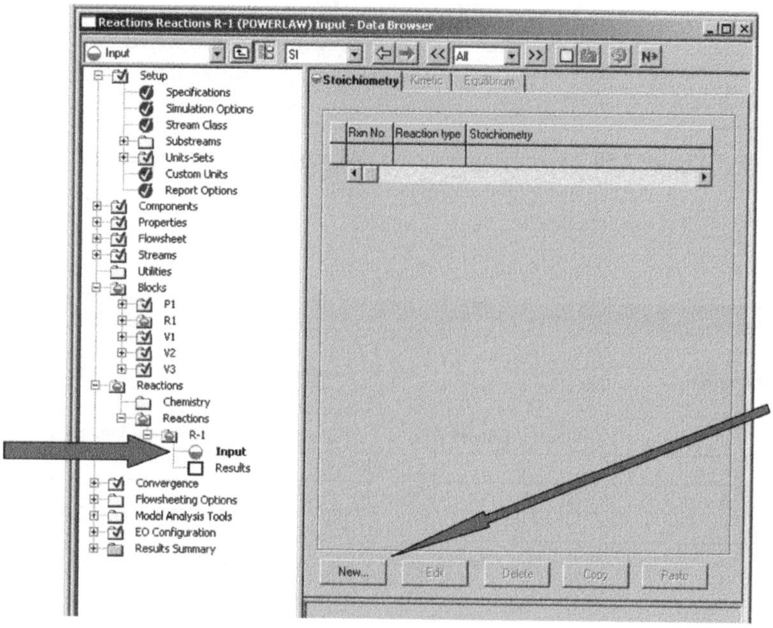

Figure 2.48 Input reactions.

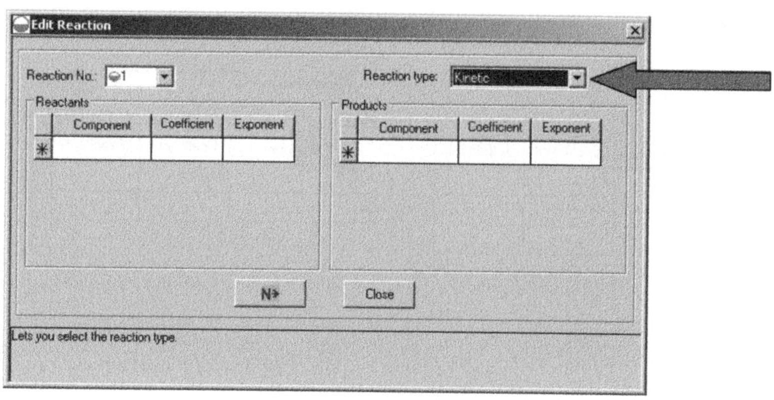

Figure 2.49 Specifying reaction type.

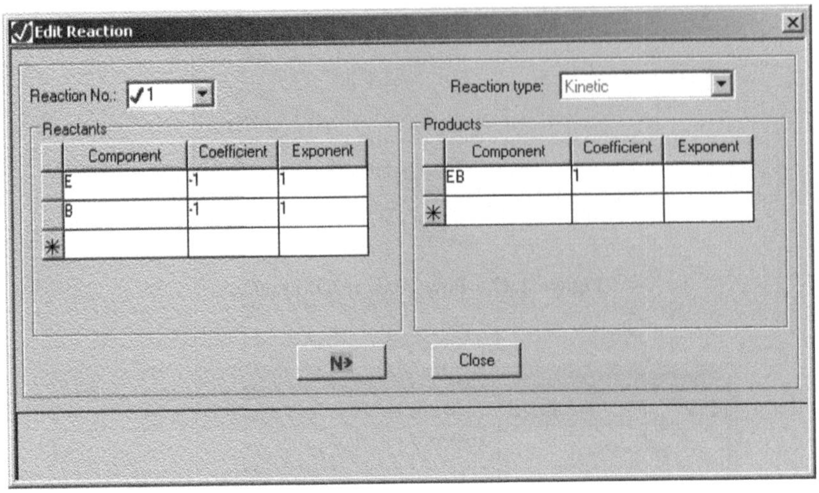

Figure 2.50 Specifying reactants and products for first reaction.

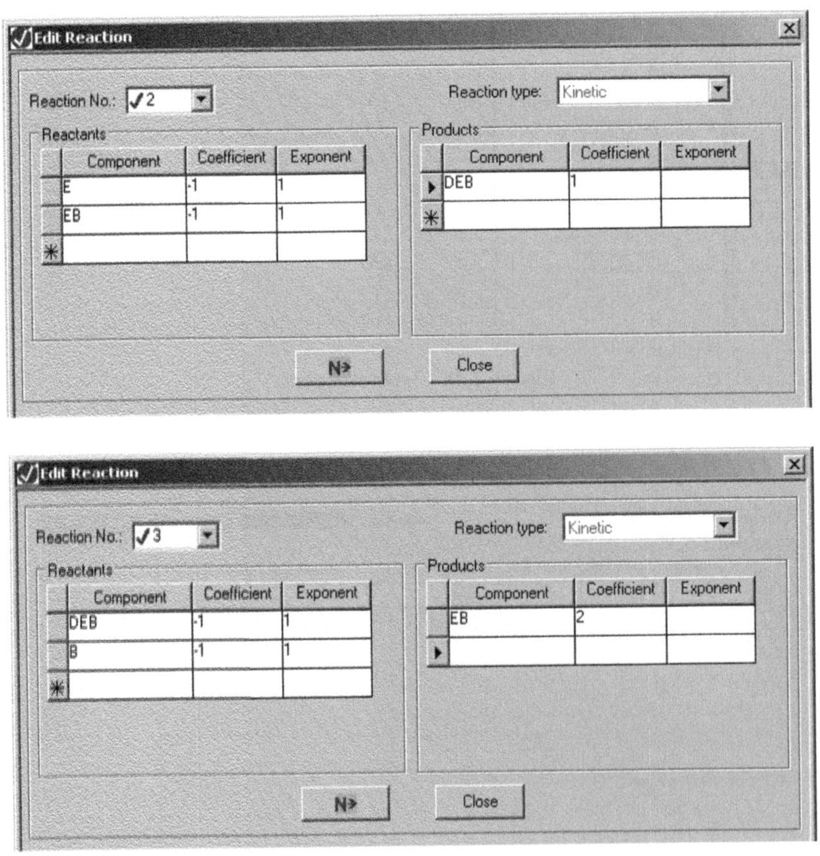

Figure 2.51 Specifying reactants and products for second and third reactions.

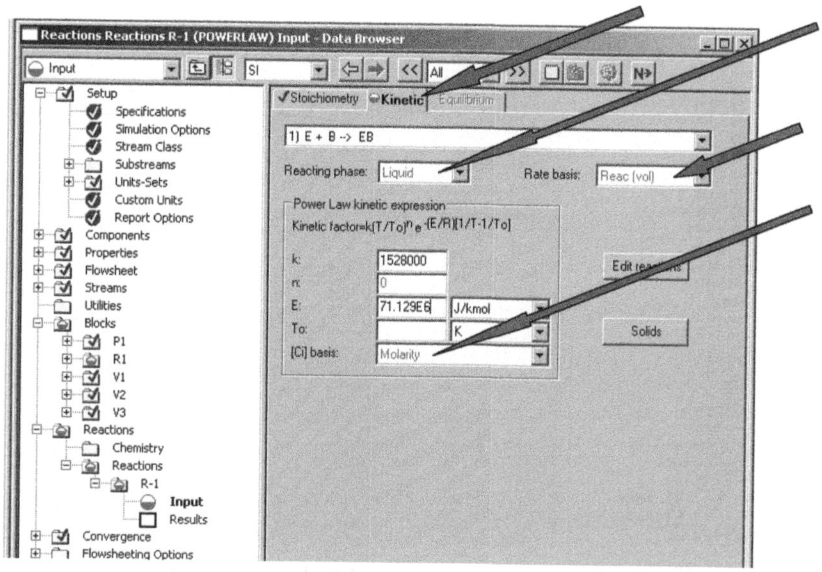

Figure 2.52 Specifying kinetic parameters for reaction 1.

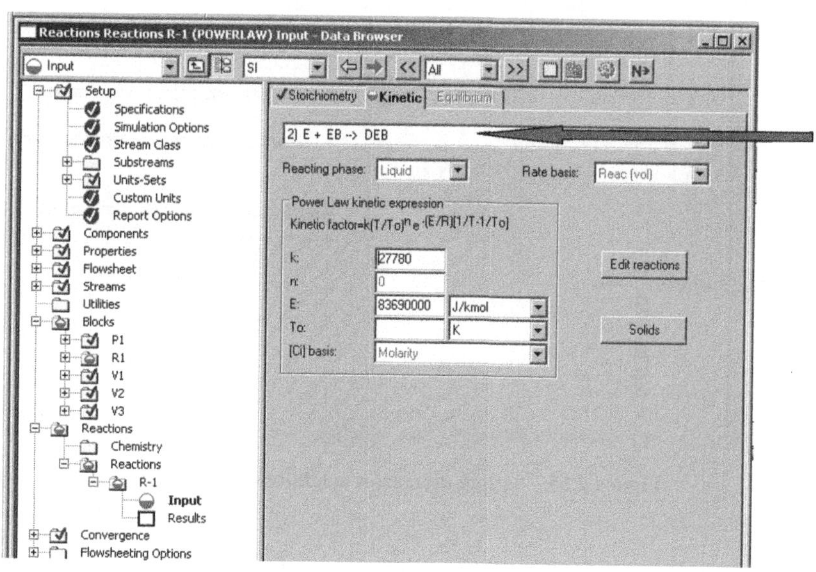

Figure 2.53 Specifying kinetic parameters for reaction 2.

2.8.3 Reactor Setup

Click on the reactor block and then on *Setup*. The window shown in Figure 2.55 opens on which the operating temperature and pressure are set, as well as the reactor volume and the valid phases (*Liquid-Only*). Instead of setting temperature, other options could be selected such as reactor heat transfer rate. Selecting the *Reactions* page tab under *Setup* opens the

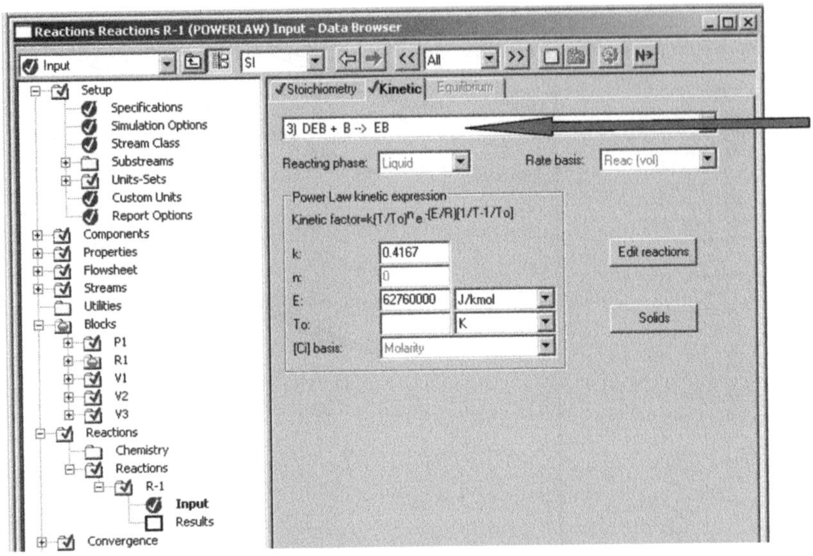

Figure 2.54 Specifying kinetic parameters for reaction 3.

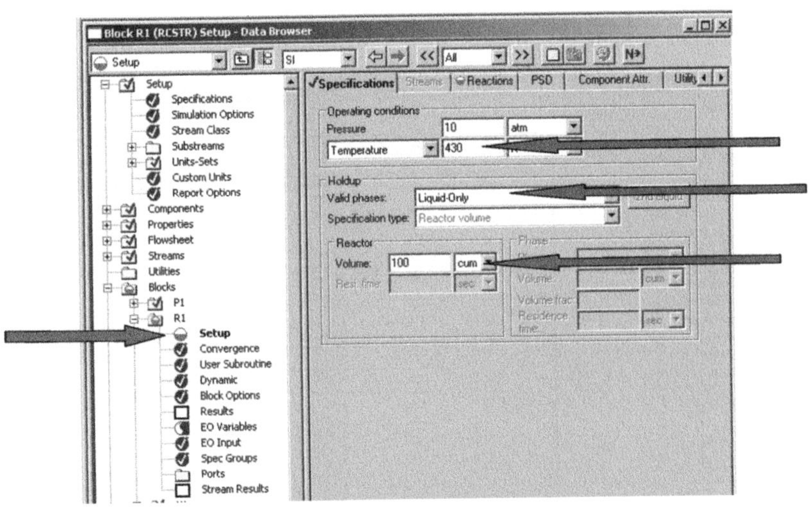

Figure 2.55 Setting up reactor conditions and size.

window shown in Figure 2.56. The R-1 reaction is moved from the left column (*Available reaction sets*) to the right column (*Selected reaction sets*) using the arrows between the two columns.

The simulation is now completely specified and is ready to run. Clicking the blue *N* button opens the window shown in Figure 2.57. Click the *OK* button and the simulation runs. The inlet and outlet streams of the reactor can be checked by clicking *Stream Results*, as shown in Figure 2.58. At the 430 K temperature and with a 100 m^3 reactor volume, the conversion of ethylene is $(0.2 - 0.006912)/0.2 = 0.965$. The production of DEB is quite small (0.5 % of ethylene fed). Since the activation energy of the DEB

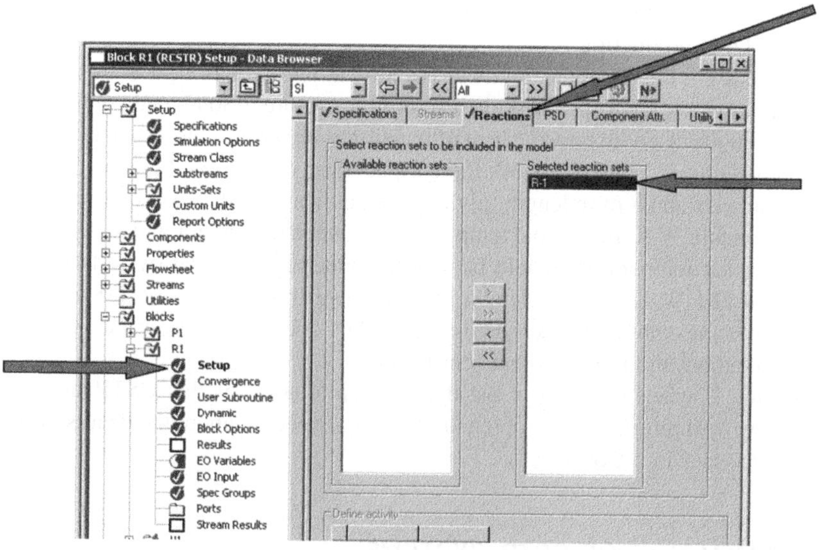

Figure 2.56 Selecting reaction R-1.

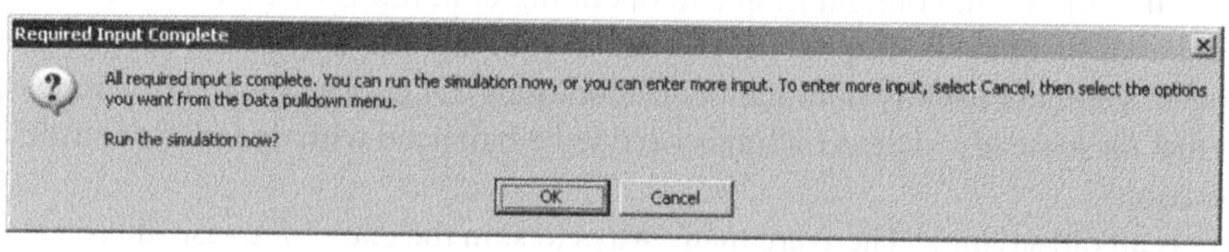

Figure 2.57 Running the simulation.

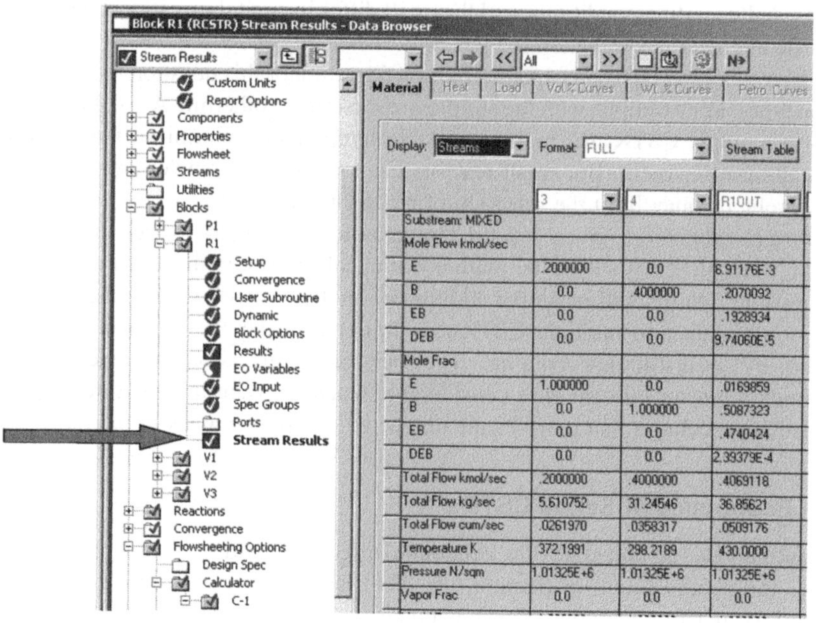

Figure 2.58 Stream results.

reaction is larger than the EB reaction, operating the reactor at higher temperatures would increase the DEB production.

Before we leave this example, let us take a look at the issue of heat transfer. In setting up the simulation, we have specified the reactor temperature (430 K) and volume (100 m^3) but have said nothing about how the heat of reaction is removed. The simulation calculates a heat removal rate of 12.46×10^6 W. If the aspect ratio of the vessel is 2, a 100-m^3 vessel is 4 m in diameter and 8 m in length, giving a jacket heat transfer area of 100.5 m^2. If we select a reasonable 30 K differential temperature between the reactor and the coolant in the jacket, the jacket temperature would be 400 K. Selecting a typical overall heat transfer coefficient of 851 W K^{-1} m^{-2} gives a required heat transfer area of 488 m^2, which is almost 5 times the available jacket area. Aspen Plus does not consider the issue of area. It simply calculates the required heat transfer rate.

In Chapter 3, when we discuss the dynamic simulation of this CSTR using Aspen Dynamics, we will return to the problem and be more specific about the details of realistic heat transfer issues in a CSTR.

2.9 OPTIMIZATION OF CSTR SYSTEMS

The economic optimum configuration and operating conditions of CSTR systems depend on a number of parameters. Capital investment in equipment is important, but energy costs and product yield are usually dominant considerations. Of course, the major theme of this book is that these steady-state economics need to be balanced with dynamic controllability considerations.

There is an old saying: "There are many ways to skin the cat." Likewise, there are many different configurations of equipment and operating conditions that can take the same feedstreams and produce exactly the same products. These flowsheets will differ in topology and operating conditions, and they will differ in capital investment and operating costs. The engineer has to determine which is "best."

We have already given some qualitative consideration of this question for several of the reaction types. For example, with the simple irreversible, exothermic A $\rightarrow$ B reaction conducted in a single CSTR, the highest possible temperature-minimized reactor size and hence capital investment. But small reactors have small heat transfer areas, so dynamic control problems may limit the selection of reactor temperature.

However, even for a given reactor temperature, we would expect that the use of several CSTRs in series should be more economical up to a point. Eventually the point of diminishing returns will make the addition of another vessel unattractive.

As an alternative to multiple CSTRs we might consider the use of a single CSTR followed by a distillation column. The per-pass conversion of reactant can be low, giving a reactor effluent with considerable reactant. Then this mixture is separated in a distillation column that recycles the unreacted component back to the reactor.

2.9.1 Economics of Series CSTRs

To quantitatively demonstrate the economic effects of adding more reactors, let us consider the results found in Section 2.6. The three flowsheets considered have one, two, or three reactors in series. Reactor temperatures are 350 K, and the conversion is 98%. The specific reaction rate k_{350} is 8.185×10^{-4} s^{-1}.

The respective volumes are 262 m^3 for the 1-CSTR process, 32.5 m^3 in each of the two vessels in the 2-CSTR process, and 14.4 m^3 in each of the three vessels in the 3-CSTR process. The vessel diameters are 5.5, 2.74, and 2.09 m, respectively.

What is the capital investment for each of these flowsheets? The cost ($) of a plain stainless-steel pressure vessel of length L (m) and diameter D (m) is

$$\text{Cost} = 17{,}640(L)^{1.066}(D)^{0.802} \tag{2.83}$$

We assume that the reactor is constructed from fairly exotic material and has heat transfer equipment (jacket or coil) and an agitator. So the capital cost is estimated at 10 times the basic vessel cost. The total capital costs of the reactors in the three different flowsheets are $7,429,000, $4,043,000, and $3,657,000 for the 1-CSTR, 2-CSTR, and 3-CSTR processes, respectively. You can see that the reduction in cost between the 2-CSTR and the 3-CSTR processes is quite small. The cost of a 4-CSTR process could be somewhat higher because of having more vessels, even if each is somewhat smaller.

2.9.2 Economics of a Reactor–Column Process

An alternative to the series of CSTRs is to have one reactor followed by a distillation column that recycles reactant back to the reactor. The flowsheet of this process is shown in Figure 2.59. The reactor can be small because the concentration of the reactant in the reactor can be large. The per-pass conversion of reactant is not equal to the overall conversion.

Of course, there is additional capital investment in the distillation column and its associated heat exchangers (condenser and reboiler) and most importantly, there will be an energy cost to provide heat to the reboiler. We neglect any energy cost in the

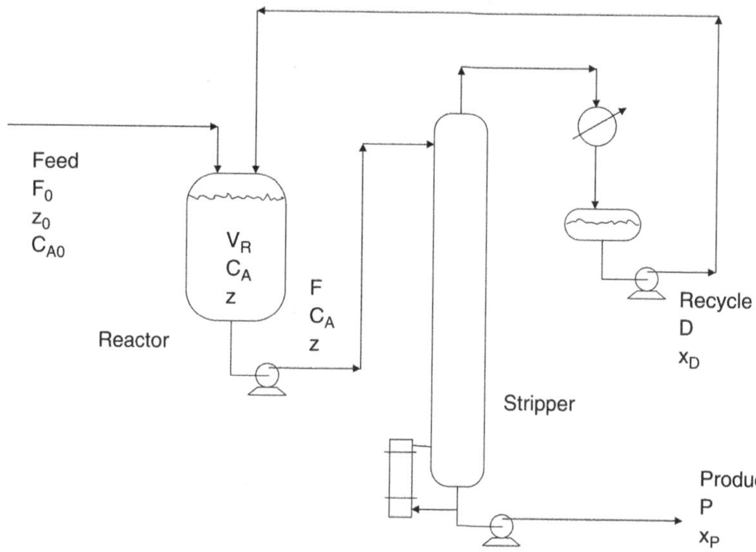

Figure 2.59 Reactor–stripper flowsheet.

multiple-reactor system because we assume inexpensive cooling water is used in all the reactors.

This reactor–column process has several degrees of freedom that can be adjusted to satisfy some economic objective. The most important is the size of the reactor. The smaller the reactor, the larger the column and the recycle flow.

The same kinetic case considered in the previous section is used in the reactor. The reactor effluent F (with composition z) is fed to a stripping column that sends the lighter component A out the top as a recycle stream and the heavier component B out the bottom as the final product stream. A stripping column is used, as opposed to a full column with reflux, because previous studies have shown that a stripper is the optimum economic design. Since the overhead is not a product stream, it does not have to have a high purity. The vapor going overhead is condensed and pumped back to the reactor at a molar flowrate D and composition x_D. The liquid rate in the stripper is F and the vapor rate is D.

The vapor–liquid equilibrium (VLE) is assume to be described by a constant relative volatility between components A and B of $\alpha = 1.5$. On each tray in the column, the compositions (mole fraction A) of the liquid x and vapor y are related as follows:

$$y = \frac{\alpha x}{1 + (\alpha - 1)x} \tag{2.84}$$

The design of the stripper is based on finding the pinch point between the VLE line and the operating line. If the feed is saturated liquid, this occurs at $y^{\max}$:

$$y^{\max} = \frac{\alpha z}{1 + (\alpha - 1)z} \tag{2.85}$$

Then the overhead composition is set to be greater than z by 80% of the difference between $y^{\max}$ and z, as shown in Figure 2.60:

$$x_D = y_{NT} = z + 0.8(y^{\max} - z) \tag{2.86}$$

This criterion is similar to using a reflux ratio of 1.2 times the minimum reflux ratio in a full distillation column. It provides a reasonable compromise between the number of trays and the vapor boilup required. The slope of the resulting operating line is the liquid-to-vapor ratio F/D:

$$\frac{F}{D} = \frac{x_D - x_P}{z - x_P} \tag{2.87}$$

Molar concentrations are converted into mole fractions, and volumetric flowrates are converted to molar flowrates for the tray-to-tray calculations in the column. The fresh feedstream F_0 is 0.03506 kmol/s with a composition $z_0 = 1$ mole fraction A. Since the reaction is equimolar (one mole of A produces one mole of B), the molar flowrate of the bottoms from the column P is equal to the fresh feed flowrate F_0. The overall conversion is set at 98%, so the concentration of reactant in the column bottoms (the product stream P) is $x_P = 0.02$ mole fraction A.

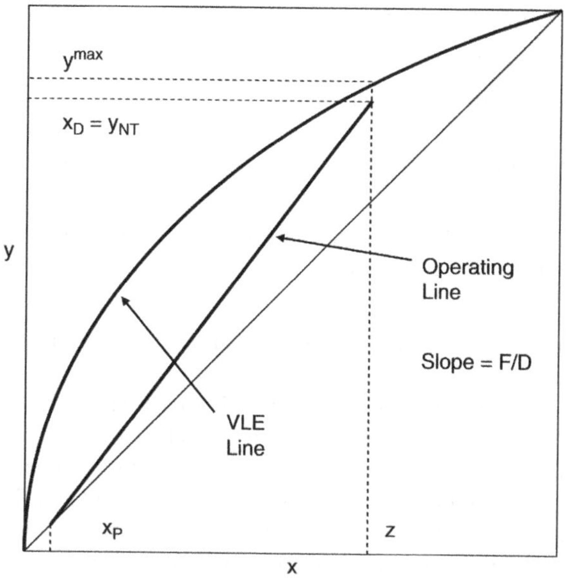

Figure 2.60 McCabe–Thiele diagram of stripper.

The component balance around the column can be used to solve for the overhead flowrate D once the F/D ratio is known from the following equation:

$$zF = Dx_D + Px_P$$
$$z\left[\frac{F}{D}\right]D = Dx_D + Px_P \tag{2.88}$$
$$D = \frac{Px_P}{\left[\frac{F}{D}\right]z - x_D}$$

The composition (kmol/m^3) of reactant in the reactor C_A, which is used in the kinetic evaluation, is converted into mole fraction A using

$$z = \frac{C_A}{8.01} \tag{2.89}$$

The rate of consumption of reactant in the reactor is equal to the conversion χ times the molar fresh feed of pure A:

$$V_R k C_A = \chi F_0 \tag{2.90}$$

The design parameters that are known are χ, k, F_0, z_0, and x_P. The stepwise procedure for finding the "best" reactor size is

1. Pick a reactor size V_R (m^3).
2. Calculate C_A from Eq. (2.90) and z from Eq. (2.89).

3. Calculate y^{max} from Eq. (2.85).
4. Set the overhead composition ($x_D = y_{NT}$) using Eq. (2.86).
5. Calculate the F/D ratio from Eq. (2.87).
6. Calculate D from Eq. (2.88).
7. Perform tray-to-tray calculations starting at x_P in the bottom and using the VLE relationship and a component balance on each successive tray.
8. The required number of trays is reached when the composition of the vapor leaving that tray is equal to (or greater than) x_D.

The Matlab program given in Figure 2.61 performs these calculations for a given value of reactor volume V_R.

A range of values of reactor sizes is explored. For each value the capital costs of the reactor, the column and the heat exchangers are calculated using the cost and sizing relationships given in Table 2.6. The recycle stream is assumed to be 322 K.

The energy cost and the total annual cost (TAC) are also calculated. The reactor volume that minimizes TAC is the optimum economic steady-state design.

Figure 2.62 gives results over a range of reactor volumes. The reactor volume that minimizes TAC ($1,117,000 per year) is 20 m^3, giving a recycle flowrate D of 0.1241 kmol/s. Figure 2.63 gives the values of variables and parameters for the 10 m^3 reactor process. The reactor concentration is $C_A = 2.099$ kmol/m^3 ($z = 0.262$ mole fraction A). The column has 16 trays. The distillate composition is 0.3304 mole fraction A. Energy cost is $427,400 per year. The capital cost of the column is $123,000. The capital cost of the

```
% Program "cstrstripper.m"
%  Set reactor volume; alpha=1.5, 98% conversion, 350 K
% fresh feed = 0.03506 kmol/sec (0.004377 cu m/sec)
clear
tr=350,conv=.98,factor=10;ca0=8.01;m=100;td=322;
k0=20.75e6;e=69.71e6;k=k0*exp(-e/tr/8314);
f0=0.03506;p=f0;xp=1-conv;z0=1;alpha=1.5;
%  set vr
    vr=20;
diam=(2*vr/pi)^0.3333;length=2*diam;
z=conv*f0/8.01/k/vr;ca=z*8.01;
ymax=alpha*z/(1+(alpha-1)*z);
xd=z+0.8*(ymax-z);
foverd=(xd-xp)/(z-xp);
d=xp*p/(z*foverd-xd);
if d<=0;return;end
f=d+p;
% Step up to find NT
yp=alpha*xp/(1+(alpha-1)*xp);
x(1)=(d*yp+p*xp)/f;
ycalc=xp;n=1;
while  ycalc<xd
   y(n)=alpha*x(n)/(1+(alpha-1)*x(n));ycalc=y(n);
   x(n+1)=(d*y(n)+p*xp)/f;
   n=n+1;end
nt=n-1,z,ca,xd,d,pause
%  Calculate heat-transfer area
t0=294;tcin=294;u=851;lambda=-69.71e6;roe=801;m=100;cp=3137;cj=4183;roej=1000;
areaj=2*pi*diam^2;q=(1-xp)*f0*(-lambda)-cp*m*f*(tr-td);
tj=tr-q/u/areaj;fj=q/(cj*(tj-tcin));areaj,q,tj,fj
% Calculate column diameter
roe=100*115*144/(1545*580);vel=1/sqrt(roe);vapor=d*100/(.454*0.8*62.3);
areacol=vapor/vel;dcol=sqrt(4*areacol/pi)
```

Figure 2.61 Matlab program for design of reactor–stripper process.

TABLE 2.6 Basis of Economics[a]

Reactor
  Capital cost (\$) = (10)(17,640) $(D)^{1.066}$ $(L)^{0.802}$
Heat exchangers
  Heat duty = (23.24×10^6 J/kmol)(D kmol/s)
  Condenser
    Heat transfer coefficient = 0.852 kW K^{-1} m^{-2}
    Differential temperature = 13.9 K
    Capital cost (\$) = 7296 $(\text{area})^{0.65}$
  Reboiler
    Heat transfer coefficient = 0.568 kW K^{-1} m^{-2}
    Differential temperature = 34.8 K
    Capital cost (\$) = 7296$(\text{area})^{0.65}$
Column
  Length = 1.2 (NT)(0.61)
  Diameter calculated from F factor = 1.22 (in SI units)[b]
  Vessel capital cost = 17,640 $(D)^{1.066}$ $(L)^{0.802}$
Energy
  Annual consumption = (reboiler heat duty in 10^6 J/s)(3600)(24)(365)
  Cost = \$4.7 10^{-6}/kJ

$$\text{TAC} = \frac{\text{capital cost}}{\text{payback period}} + \text{energy cost}$$

$$= \frac{\text{reactor} + \text{column} + \text{heat exchangers}}{\text{payback period}} + \text{energy cost}$$

Payback period = 3 years

[a]dimensions in m, areas in m^2.
[b]F factor = velocity*sqrt(vapor density)
Velocity in m/s.
Vapor density in kg/m^3 using ideal gas $\rho = MP/RT$.
$P = 7.82$ atm, $M = 100$ kg/kmol, $T = 322$ K.

heat exchangers is \$446,000 (area of condenser is 243 m^2 and reboiler is 146 m^2). The capital cost of the reactor is \$1,499,000.

Remember that the capital cost of the 3-CSTR process is \$3,657,000. If a 3-year payback period is established, the TAC of the 3-CSTR process is \$1,219,000 per year. This should be compared with the TAC of the reactor–stripper process of \$1,117,000 per year. So the reactor–stripper process is economically somewhat better than the 3-CSTR process.

How do these two alternative processes compare in terms of dynamic controllability? In Chapter 3 we will compare them quantitatively, but even at this stage in the design, where only steady-state conditions and parameter values are known, we can get some indication of their relative controllability. The first reactor of the 3-CSTR process is expected to be the troublemaker. As shown in Figure 2.21, it has a volume of 14.4 m^3, a heat transfer area of 27.5 m^2, a heat transfer rate of 1.17×10^6 kJ/s, and a jacket temperature of 300 K.

The reactor in the reactor–stripper process has a volume of 20 m^3, a heat transfer area of 34.2 m^2, a heat transfer rate of 0.997×10^6 kJ/s, and a jacket temperature of 316 K. Thus the ΔT driving force is a smaller percentage of the total available ΔT (350 − 294 K) in the

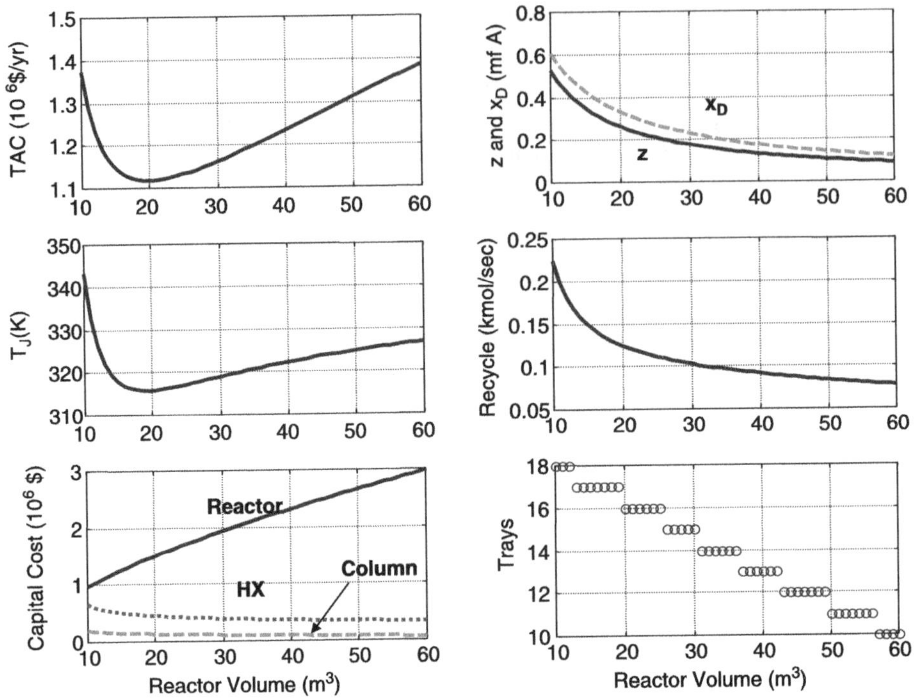

Figure 2.62 Reactor–stripper design.

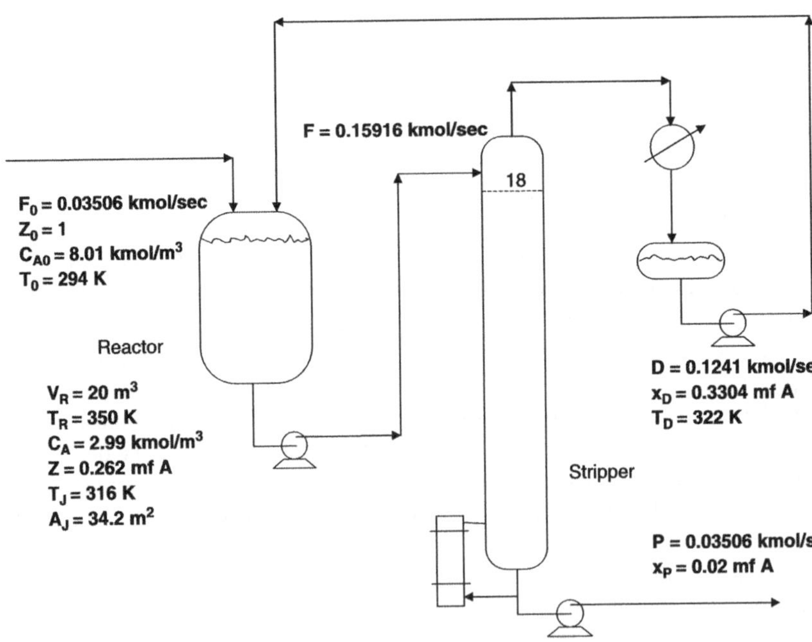

Figure 2.63 Reactor–stripper flowsheet; A $\rightarrow$ B.

reactor–stripper process. We would expect this reactor to be more controllable than the first reactor in the 3-CSTR process, but a 315 K jacket temperature chews up a big percentage of the total available ΔT [(350 − 315)/(350 − 294) = 0.62]. Therefore it is vital to explore the dynamics of this system at the same time as the steady-state economics are considered. Dynamic controllability may dictate that a suboptimal design with a larger reactor should be used. We will revisit this process in Chapter 3 to quantitatively evaluate the dynamics and control of these alternative processes.

The reactor in the reactor–stripper process is larger than the first reactor in the 3-CSTR process and therefore has more heat transfer area. However, all the conversion occurs in this one reactor, so we would expect the heat transfer rate to be high. In fact, it is not high but is even lower than the heat transfer rate in the 1-CSTR process (1.78×10^6 kJ/s, as shown in Fig. 2.21). The reason for this unexpected result is the large recycle stream in the reactor–stripper process. We assume that its temperature is 322 K. The total flow into the reactor is 0.1591 kmol/s (the sum of the recycle 0.1241 kmol/s and the fresh feed is 0.03506 kmol/s), while the flow into the first reactor of the 3-CSTR process is just the fresh feed. Thus the larger sensible heat of the larger stream reduces the heat than must be transferred in the reactor. Note that the total heat of reaction for a 98% conversion is 2.395×10^6 kJ/s. The sensible heat of the large feed stream to the reactor in the reactor–stripper process is 1.398×10^6 kJ/s. In the 3-CSTR process (and in all the multiple CSTR processes) the sensible-heat term in the first reactor is only 0.616×10^6 kJ/s because the flowrate is smaller.

From the standpoint of intensification, the 3-CSTR process has a total reactor volume of $3 \times 14.4 = 43.2$ m^3. The reactor volume in the reactor–stripper process is 20 m^3. In addition, there is liquid holdup in the base (12 m^3) and in the overhead drum (9 m^3), assuming 10 min of holdup in each. So the two processes have essentially the same inventories.

2.9.3 CSTR Processes with Two Reactants

The reaction studied in Section 2.2 has two reactants and therefore offers the possibility of adjusting the compositions of the reactants in the reactor to achieve some economic or control objective. In this section we first find the cost of operating single and multiple CSTR processes to achieve a specified conversion. Then we design an alternative process consisting of a reactor and a distillation column that separates product C from the unreacted A and B in the reactor effluent and recycles them back to the reactor.

Design of Multiple CSTR Systems The reaction and the overall reaction rate are given below:

$$\mathrm{A} + \mathrm{B} \longrightarrow \mathrm{C} \tag{2.91}$$

The reaction rate is first-order in the two reactants.

$$\mathfrak{R} = kC_\mathrm{A}C_\mathrm{B} = C_\mathrm{A}C_\mathrm{B}k_0 \exp\left(\frac{-E}{RT_R}\right) \tag{2.92}$$

Kinetic parameters are given in Table 2.2 with the exception that the preexponential factor k_0 is increased by a factor of 10 so that reasonable reactor volumes are obtained when designing for a high conversion ($k_0 = 2.1045 \times 10^8$ $m^3 s^{-1}$ $kmol^{-1}$). The production rate of product C is 0.03471 kmol/s for all the different designs. The conversion is set at 99% of both components. The compositions of the two pure reactant feedstreams are 8.01 $kmol/m^3$ and their flowrates are 0.004377 m^3/s. The reactor temperature is 333 K, giving a specific reaction rate of 0.002443 $m^3 s^{-1}$ $kmol^{-1}$. For a 99% conversion, the compositions in the product are low ($C_A = C_B = 0.0801$ $kmol/m^3$).

For a single CSTR, the high design conversion produces a very large vessel (2214 m^3) that costs $28,090,000, using the economics given in Table 2.6. This volume is calculated as follows:

$$V_R = \frac{\text{kmol C produced/s}}{kC_A C_B}$$
$$= \frac{0.03471 \text{ kmol C/s}}{(0.002443 \text{ m}^3 \text{ s}^{-1} \text{ kmol}^{-1})(0.0801)(0.0801)} = 2214 \text{ m}^3 \quad (2.93)$$

The heat transfer area in this large reactor is 789 m^2, giving a jacket temperature of 330.7 K for the heat transfer rate of 1.562×10^6 J/s. So this process should give good temperature control.

Figure 2.64 shows how reactor volume and cost vary with the desired conversion.

Using two CSTRs in series drastically reduces the size of the reactors. To calculate the required reactor volume, an iterative procedure is used. The steady-state equations for the first reactor are

$$F_{A0}C_{A0} = F_1 C_{A1} + V_R k C_{A1} C_{B1}$$
$$F_{B0}C_{B0} = F_1 C_{B1} + V_R k C_{A1} C_{B1} \quad (2.94)$$

Since stoichiometric amounts of both A and B are fed to the reactor ($F_{A0} = F_{B0}$ and $C_{A0} = C_{B0}$), the concentrations in the first reactor C_{A1} and C_{B1} are equal. Assuming

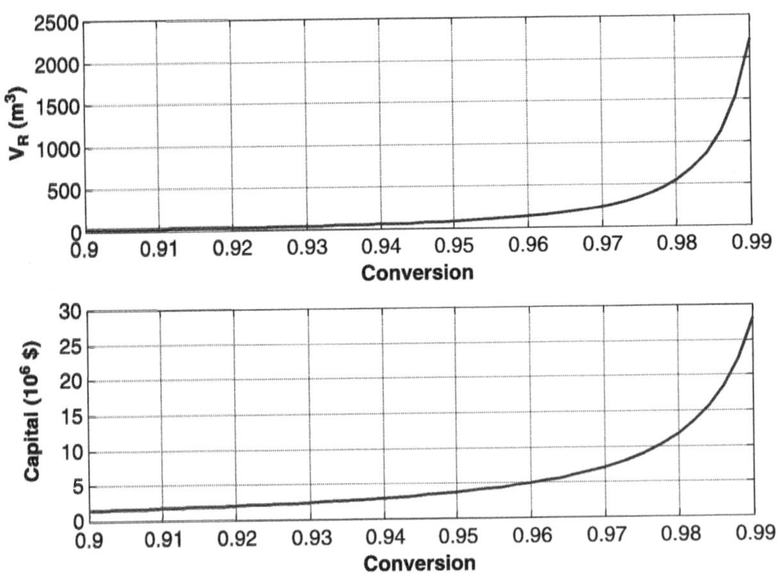

Figure 2.64 1-CSTR process; two reactants.

constant densities, the volumetric flowrate from the first reactor F_1 and from the second reactor are the sum of the two fresh feedstreams:

$$F_1 = F_2 = F_{A0} + F_{B0} \tag{2.95}$$

The component balance for A can be expressed as follows:

$$F_{A0}C_{A0} = F_1 C_{A1} + V_R k(C_{A1})^2$$

$$V_R k(C_{A1})^2 + F_1(C_{A1}) + (-F_{A0}C_{A0}) = 0 \tag{2.96}$$

If the reactor volume were known, this quadratic equation could be solved for C_{A1}. A similar equation can be derived for the second reactor:

$$F_1 C_{A1} = F_2 C_{A2} + V_R k\left(C_{A_2}\right)^2$$

$$V_R k(C_{A2})^2 + F_2(C_{A2}) + (-F_1 C_{A1}) = 0 \tag{2.97}$$

We know that C_{A2} and C_{B2} must be equal to 0.0801 kmol/m^3 since the F_2 stream is the product.

The procedure is to guess a value of V_R and solve Eq. (2.96) for C_{A1}. Then this is substituted into Eq. (2.97), which is solved for C_{A2}. If the calculated value of C_{A2} is larger than the required 0.0801 kmol/m^3, the guessed value of V_R is increased. If it is smaller, the V_R is reduced. Figure 2.65 gives a Matlab program that uses an iterative interval-halving convergence procedure to find the correct volume.

The result is a flowsheet with two 176-m^3 vessels having a total cost of \$11,590,000. This is less than half the cost of the single-vessel process. The reactant

```
%  cstr2multiple.m
%   Irreversible reaction A + B = C; Given conversion and feed streams
%  Calculate size and cost of 1-CSTR, 2-CSTR and 3-CSTR processes
clear
ca0=8.01;e=69.71e6;fa0=0.004377;factor=10;k0=2.1045e+008;fb0=fa0;f1=fa0+fb0;f2=f1;f3=f1;
tr=333;k=k0*exp(-e/tr/8314);conversion=0.99;production=fa0*ca0*conversion;
% 1-CSTR Process  ****************************************************
ca=ca0*(1-conversion),cb=ca;vr1=production/(k*ca*cb)
dr1=(2*vr1/pi)^0.3333;lengthr1=2*dr1;
reactor1=factor*17640*(dr1^1.066)*(lengthr1^0.802)*1e-6
%2-CSTR Process   Guess vr  ***********************************************
vr=200;dvr=5;flagm=-1;flagp=-1;error=10;
while error>.000001
   b1=f1/k/vr;c1=-f1*ca0/k/vr;ca1=(-b1+sqrt(b1^2-4*c1))/2;
   b2=f2/k/vr;c2=-f2*ca1/k/vr;ca2=(-b2+sqrt(b2^2-4*c2))/2;
   error=abs(ca2-ca);
   if (ca2-ca)>0;if flagp>0;dvr=dvr*0.5;end;vr=vr+dvr;flagm=1;end
   if (ca2-ca)<0;if flagm>0;dvr=dvr*0.5;end;vr=vr-dvr;flagp=1;end
end
      vr2=vr,ca1,ca2
      dr2=(2*vr2/pi)^0.3333;lengthr2=2*dr2;
      reactor2=2*factor*17640*(dr2^1.066)*(lengthr2^0.802)*1e-6
%3-CSTR Process   Guess vr  ********************************************************
vr=150;dvr=5;flagm=-1;flagp=-1;error=10;
while error>.000001
   b1=f1/k/vr;c1=-f1*ca0/k/vr;ca1=(-b1+sqrt(b1^2-4*c1))/2;
   b2=f2/k/vr;c2=-f2*ca1/k/vr;ca2=(-b2+sqrt(b2^2-4*c2))/2;
   b3=f3/k/vr;c3=-f3*ca2/k/vr;ca3=(-b3+sqrt(b3^2-4*c3))/2;
   error=abs(ca3-ca);
   if (ca3-ca)>0;if flagp>0;dvr=dvr*0.5;end;vr=vr+dvr;flagm=1;end
   if (ca3-ca)<0;if flagm>0;dvr=dvr*0.5;end;vr=vr-dvr;flagp=1;end
end
      vr3=vr,ca1,ca2,ca3
      dr3=(2*vr3/pi)^0.3333;lengthr3=2*dr3;
      reactor3=3*factor*17640*(dr3^1.066)*(lengthr3^0.802)*1e-6
```

Figure 2.65 Matlab program for CSTR process; two reactants.

concentration in the first reactor is 0.3943 kmol/m^3. The heat transfer area in each of the two reactors is 146 m^2, giving a jacket temperature in the critical first reactor of 321.2 K for the heat transfer rate of 1.466×10^6 J/s. The control of this process should be good.

A similar procedure is used to design a 3-CSTR process. The result is a flowsheet with three reactors that are each 55.8 m^3, with a total cost of $8,523,000. The reactant

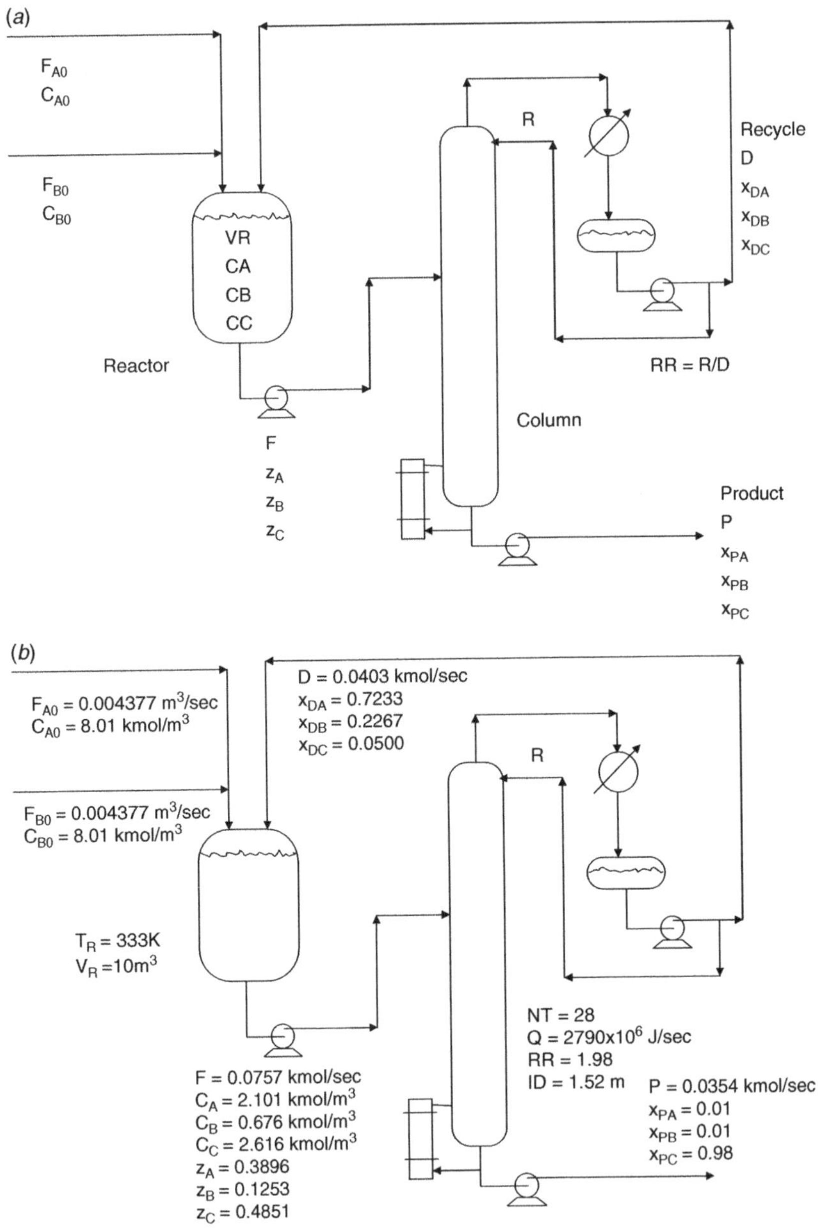

Figure 2.66 Reactor–column flowsheet; two reactants, (*a*) notation; (*b*) results with $V_R = 10$ m^3.

concentrations in the three reactors are 0.6855, 0.1801, and 0.0801 kmol/m^3. The heat transfer area in each of the three reactors is 67.9 m^2, giving a jacket temperature in the critical first reactor of 309.2 K for the heat transfer rate of 1.377×10^6 J/s. Most of the available ΔT driving force is used under steady-state conditions, so the control of this process may be poor.

Design of Reactor–Column Process It is interesting to compare the CSTR systems considered above with an alternative process with exactly the same feedstreams and producing exactly the same product stream. The flowsheet is sketched in Figure 2.66*a* with the nomenclature defined. A small single CSTR is used that has high concentrations of the reactants A and B. These are recovered in a distillation column and recycled back to the reactor. The relative volatilities are assumed to be $\alpha_A > \alpha_B > \alpha_C$, so the distillate D (kmol/s) is the recycle stream, which contains mostly A and B. The composition of C in the distillate D is assumed to be $x_{DC} = 0.05$ mole fraction. The bottom is the product stream P (kmol/s) with compositions $x_{PA} = x_{PB} = 0.01$ and $x_{PC} = 0.98$ mole fraction. The fresh feed streams are fed to the reactor. The reactor effluent F (kmol/s) has a composition C_A, C_B, and C_C (kmol/m^3) and z_A, z_B, and z_C (mole fractions). The reactor temperature is 333 K, so the specific reaction rate k is known (0.002443 m^3 s^{-1} kmol^{-1}). The overall conversion of both reactants is 99%, but the per-pass conversion is only 55% in the optimum economic steady-state design shown in Figure 2.66*b*. The production rate of C is the same as in the alternative processes (0.03471 kmol/s). Density ρ is 801 kg/m^3 of all streams. The molecular weights of the reactants M_A and M_B are 100 kg/kmol and the product M_C is 200 kg/kmol.

All of the product C leaves in the bottoms from the column, which is 0.98 mole fraction C. Therefore the bottoms flowrate is

$$P = \frac{0.03471}{0.98} = 0.0354\,\text{kmol/s} \tag{2.98}$$

The design procedure to find the reactor volume and column that minimize the total annual cost is outlined below; reactor temperature is given, so k and all physical properties are known:

1. Pick a value of reactor volume V_R (to be varied over a range of values) and calculate the capital cost using the economic data given in Table 2.6.
2. Select a range of values of the concentration of B in the reactor C_B.
3. Calculate the product of the two reactant concentrations $C_A C_B$.

$$(C_A C_B) = \frac{0.03471}{k V_R} \tag{2.99}$$

4. For each value of C_B selected above, calculate the required C_A.
5. Calculate C_C from the known density ρ and molecular weights:

$$C_C = \frac{\rho - C_A M_A - C_B M_B}{M_C} \tag{2.100}$$

6. Convert reactor concentrations to mole fractions:

$$z_j = \frac{C_j}{C_A + C_B + C_C} \tag{2.101}$$

7. Solve the overall and component molar balances around the column for the unknowns: D, F, x_{DA} and x_{DB}. These balances are given in Eq. (2.102).

$$\begin{aligned} F &= D + P \\ z_A F &= x_{DA} D + x_{PA} P \\ z_B F &= x_{DB} D + x_{PB} P \\ x_{DA} + x_{DB} + x_{DC} &= 1 \end{aligned} \tag{2.102}$$

The known variables in these four equations are P, z_A, z_B, x_{PA}, x_{PB} and x_{DC}. A little algebra leads to an explicit equation for the distillate recycle flowrate D as a function of known variables.

$$D = \left[\frac{z_A + z_B - x_{PA} - x_{PB}}{1 - z_A - z_B - x_{DC}}\right] P \tag{2.103}$$

Then the other unknowns can be found,

$$\begin{aligned} F &= D + P \\ x_{DA} &= \frac{z_A F - x_{PA} P}{D} \\ x_{DB} &= 1 - x_{DA} - x_{DC} \end{aligned} \tag{2.104}$$

8. Calculate the minimum number of trays using the Fenske Equation.

$$N_{\min} + 1 = \frac{\log\left[\left(\frac{x_{DB}}{x_{PB}}\right)\left(\frac{x_{PC}}{x_{DC}}\right)\right]}{\log \alpha_B} \tag{2.105}$$

The relative volatilities are $\alpha_A = 3$, $\alpha_B = 1.5$, and $\alpha_C = 1$. The number of trays is set equal to twice the minimum number, and the height of the column is calculated using the sizing method given in Table 2.6.

9. Calculate the minimum reflux ratio using the Underwood equations. First solve Eq. (2.106) for the value of the parameter θ that lies between α_B and α_C since the B and C are the light- and heavy-key components in the column. The column feed is assumed to be saturated liquid, so the thermal parameter q is equal to 1:

$$\frac{\alpha_A z_A}{\alpha_A - \theta} + \frac{\alpha_B z_B}{\alpha_B - \theta} + \frac{\alpha_C z_C}{\alpha_C - \theta} = 1 - q \tag{2.106}$$

Then substitute this value of θ into Eq. (2.107) and solve for the minimum reflux ratio:

$$\frac{\alpha_A x_{DA}}{\alpha_A - \theta} + \frac{\alpha_B x_{DB}}{\alpha_B - \theta} + \frac{\alpha_C x_{DC}}{\alpha_C - \theta} = 1 + RR_{min} \tag{2.107}$$

10. Set the actual reflux ratio RR equal to 1.1 times the minimum and calculate the vapor boilup in the column:

$$V = D + R = D + (RR)D = D(1 + RR) \tag{2.108}$$

11. Calculate the diameter of the column using the sizing method given in Table 2.6.
12. Calculate the capital cost of the reactor, column, and heat exchangers. Calculate the energy cost. Calculate the total annual cost.

This design procedure is repeated over a range of reactor volumes, and the reactor volume that gives the minimum TAC is found. Results are shown in Figure 2.67 for three values of reactor volume. The optimum is $V_R = 10\ m^3$, which occurs at a value of $C_B = 0.676$ kmol/m^3 ($z_B = 0.1253$ mole fraction). The TAC of this design is $987,000 per year. This should be compared with the TAC of the 3-CSTR process of $2,840,000 ($8,523,000/3 years).

The volume of the reactor (10 m^3) in this flowsheet is much smaller than that of each reactor in the 3-CSTR process (55.8 m^3) because of the higher reactant concentrations. In the 3-CSTR process the reactant concentrations in the first reactor are $C_A = C_B = 0.6855$ kmol/m^3. In the reactor–column process they are $C_A = 2.101$ kmol/m^3 and $C_B = 0.676$ kmol/m^3.

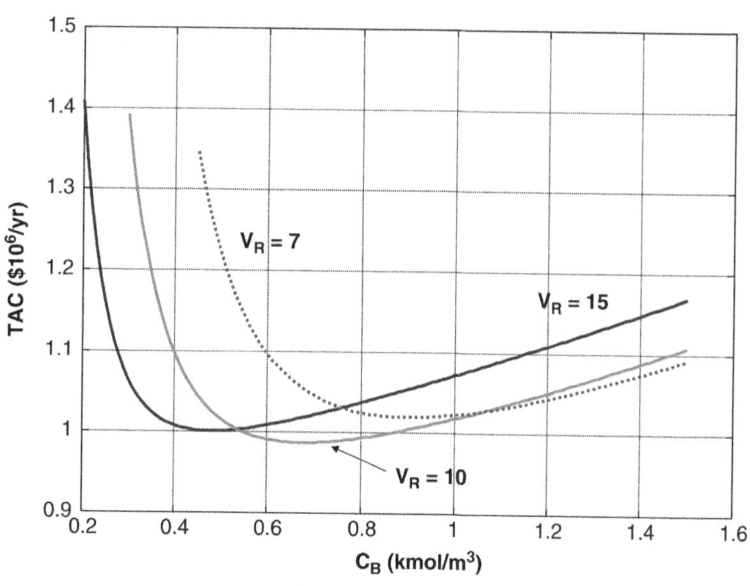

Figure 2.67 Reactor–column process; $\chi = 99$.

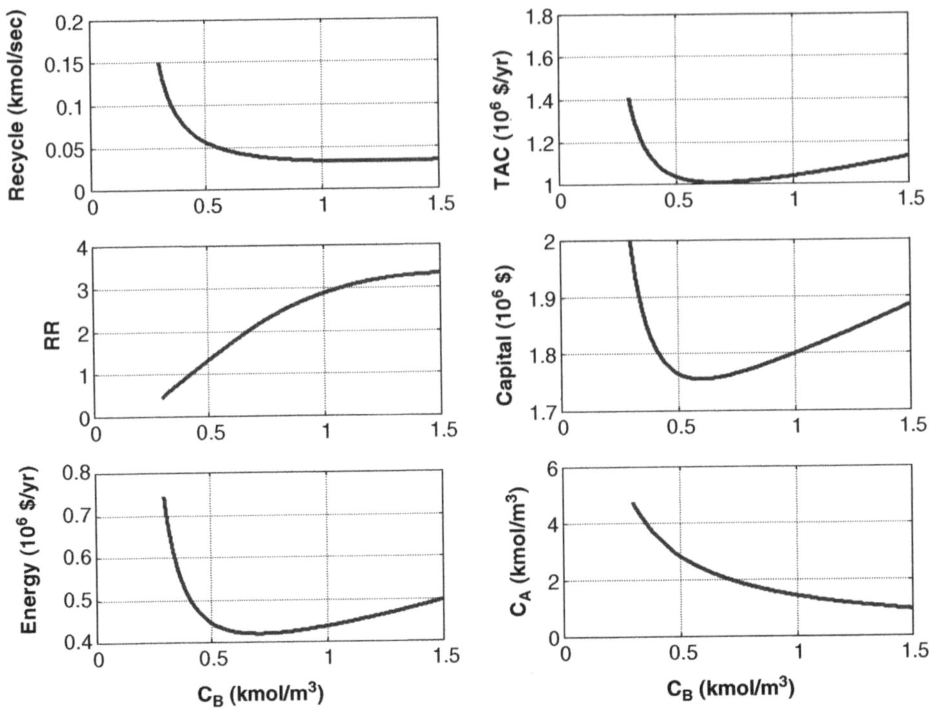

Figure 2.68 Reactor–column process; $\chi = 99$; $V_R = 10\ \text{m}^3$.

Additional information about this process is given in Figure 2.68 for the optimum reactor volume of $10\ \text{m}^3$. At low C_B values, a large recycle is required, which means high energy cost and large column and heat exchanger costs. At higher values of C_B, the feed to the column has more B and less A, so the separation becomes more difficult since B is less volatile. The reflux ratio increases, which increases energy, column and heat exchanger costs.

The important question to ask at this point is:

> Is this economic optimum process controllable?

An energy balance around the reactor at 333 K shows that this small reactor (jacket area $= 21.6\ \text{m}^2$) with a heat removal rate of 1.04×10^6 J/s requires a jacket temperature of 276 K! This is impossible if 294 K cooling water is the cooling medium.

Repeating the design for a much larger reactor ($100\ \text{m}^3$ with $100\ \text{m}^2$ jacket area) gives a more achievable jacket temperature of 318 K. Figure 2.69 gives the condition and equipment sizes for this design. Figure 2.70 give more details of the dependence of a 100-m^3 reactor process on the design optimization parameter C_B. Of course, this design is more expensive. The TAC is $1,729,000 per year compared to the economic optimum of $987,000 per year. However, the $1,729,000 per year is still better than the 2-CSTR process (TAC = $11,590,000/3 = $3,860,000), which has fairly comparable dynamic controllability properties (321 K jacket temperature).

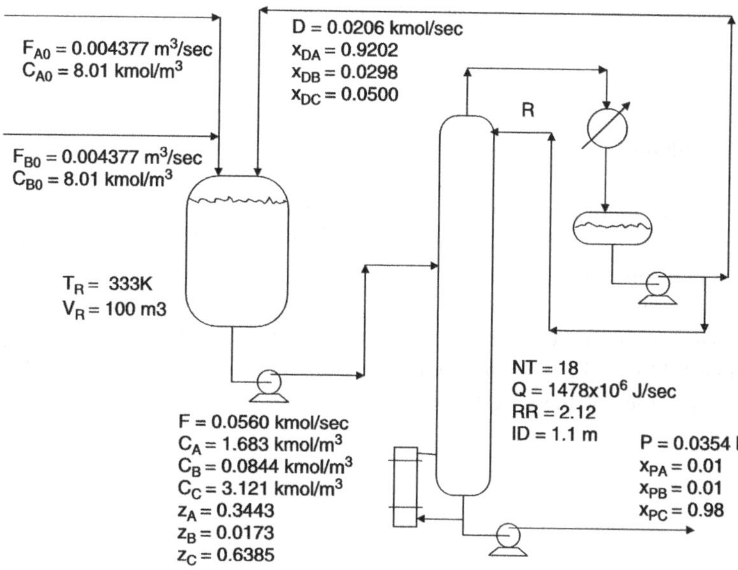

Figure 2.69 Reactor–column flowsheet; two reactants; $V_R = 100\ \text{m}^3$.

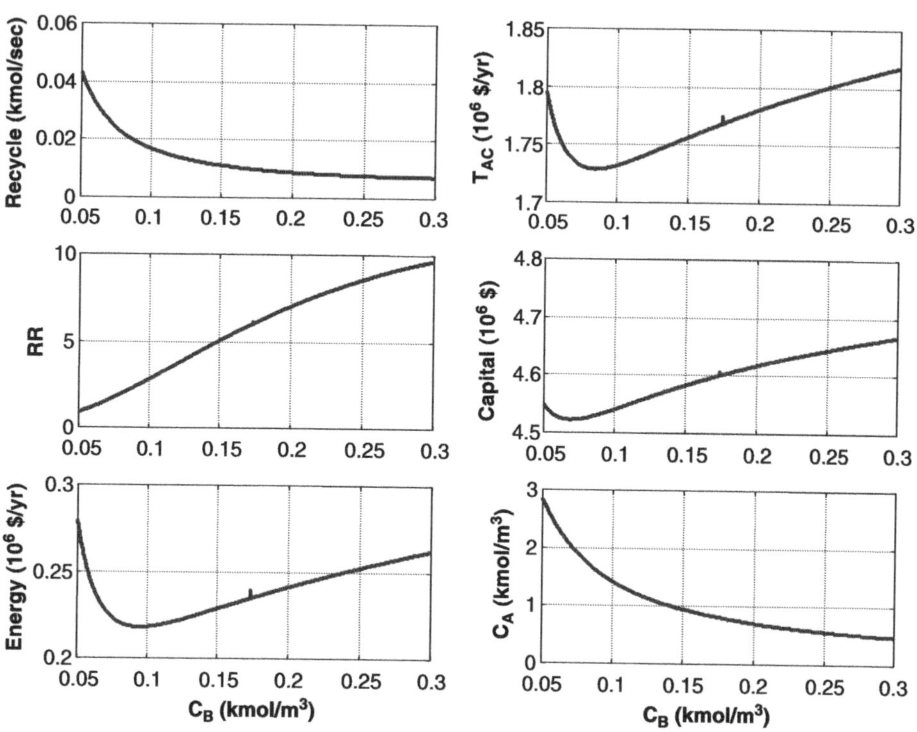

Figure 2.70 Reactor–column process; $\chi = 99$; $V_R = 100\ \text{m}^3$.

The controllability of the suboptimum reactor/column process will be studied quantitatively in Chapter 3.

2.10 CONCLUSION

Several important aspects of designing CSTR systems have been explored in this chapter. Many of the important types of reactions have been studied, and the characteristics that impact dynamic controllability have been discussed.

In the next chapter we take a quantitative look at the dynamics of these CSTR systems using primarily rigorous nonlinear dynamic simulations (time-domain analysis). However, some of the powerful linear Laplace and frequency-domain techniques will be used to gain insight into the dynamics of these systems.

CHAPTER 3

CONTROL OF CSTR SYSTEMS

In this chapter we take the steady-state designs of a variety of CSTR systems discussed in Chapter 2 and study their dynamics and control. The effects of reaction types, kinetics, design parameters, and heat removal schemes on controllability will be quantitatively explored. The important effects of design conversion and heat transfer area on the effectiveness of temperature control will be demonstrated.

Several important types of reactions are considered in the following sections. The dynamic equations describing each of these systems are developed, and the effects of several kinetic and process parameters are explored.

3.1 IRREVERSIBLE, SINGLE REACTANT

The first system studied is a CSTR with jacket cooling in which a first-order irreversible reaction takes place:

$$A \rightarrow B \tag{3.1}$$

The reaction rate is, of course, the same as that used in the steady-state model

$$\Re = kC_A = C_A k_0 \exp\left(\frac{-E}{RT_R}\right) \tag{3.2}$$

where $\Re$ = rate of consumption of reactant A (kmol s^{-1} m^{-3})
k = specific reaction rate (s^{-1})
C_A = concentration of reactant A in reactor (kmol/m^3)

Chemical Reactor Design and Control. By William L. Luyben

k_0 = preexponential factor (s^{-1})
E = activation energy (J/kmol)
$R = 8314$ J kmol^{-1} K^{-1}
T_R = reactor temperature (K)

3.1.1 Nonlinear Dynamic Model

The dynamic model of the reactor and jacket consists of four nonlinear ordinary differential equations:

Total mass balance (kg/s):

$$\frac{d(V_R\rho)}{dt} = \rho_0 F_0 - \rho F \tag{3.3}$$

Component A balance (kmol A/s):

$$\frac{d(V_R C_A)}{dt} = F_0 C_{A0} - F C_A - V_R \Re$$
$$= F_0 C_{A0} - F C_A - V_R k C_A \tag{3.4}$$

Reactor energy balance (J/s):

$$\frac{d(V_R \rho c_p T_R)}{dt} = \rho_0 c_{p0} F_0 T_0 - \rho c_p F T_R - \lambda V_R \Re - Q \tag{3.5}$$

where ρ_0 = density of feedstream (kg/m^3)
F_0 = flowrate of feed (m^3/s)
ρ = density of product stream (kg/m^3)
F = flowrate of product (m^3/s)
C_{A0} = concentration of reactant A in feed (kmol/m^3)
V_R = volumetric holdup of liquid in reactor (m^3)
c_{p0} = heat capacity of feed (J kg^{-1} K^{-1})
T_0 = temperature of feed (K)
c_p = heat capacity of product (J kg^{-1} K^{-1})
λ = heat of reaction (J/kmol)
Q = rate of heat removal from liquid in reactor (J/s)

Note that the heat of reaction λ is negative for exothermic reactions, so the third term on the right-hand side of Eq. (3.5) is positive. This means that an increase in the reaction rate tends to increase the reactor temperature.

With a circulating jacket water system with a jacket temperature T_J, the heat transfer rate depends on the jacket area, the overall heat transfer coefficient, and the differential temperature driving force

$$Q = UA_J(T_R - T_J) \tag{3.6}$$

where U = overall heat transfer coefficient (W K^{-1} m^{-2})
A_J = jacket heat transfer area (m^2) = πDL

D = reactor diameter (m)
L = reactor length (m)

The dynamic model of the jacket is

$$\frac{d(V_J\rho_J c_J T_J)}{dt} = F_J\rho_J c_J T_{C,\mathrm{in}} - F_J\rho_J c_J T_J + Q \tag{3.7}$$

where F_J = flowrate of coolant (m^3/s)
ρ_J = density of coolant (kg/m^3)
c_J = heat capacity of coolant ($J\ kg^{-1}\ K^{-1}$)
$T_{C,\mathrm{in}}$ = supply temperature of cooling medium (K)

If physical properties are assumed constant (densities and heat capacities), these terms can be pulled outside the time derivatives in Eqs. (3.3)–(3.5) and (3.7). If reactor volume is held constant (by a level controller) and the jacket volume is constant, the V_R and V_J terms can also be taken out of the derivatives. Equation (3.3) reduces to

$$F_0 = F \tag{3.8}$$

The other three differential equations reduce to the following set:

$$\frac{dC_A}{dt} = \frac{F}{V_R}(C_{A0} - C_A) - V_R C_A k_0 e^{-E/RT_R} \tag{3.9}$$

$$\frac{dT_R}{dt} = \frac{F}{V_R}(T_0 - T_R) - \frac{\lambda C_A k_0 e^{-E/RT_R}}{\rho c_p} - \frac{UA_J(T_R - T_J)}{V_R \rho c_p} \tag{3.10}$$

$$\frac{dT_J}{dt} = \frac{F_J}{V_J}(T_{C,\mathrm{in}} - T_J) + \frac{UA_J(T_R - T_J)}{V_J \rho_J c_J} \tag{3.11}$$

These three nonlinear ordinary differential equations will be used to simulate the dynamic performance of the CSTR. The "openloop" behavior applies when no controllers are used. In this case the flowrate of the cooling water is held constant. With "closedloop" behavior, a temperature controller is installed that manipulates cooling water flow to maintain reactor temperature.

3.1.2 Linear Model

Considerable insight into the dynamic behavior of the system can be gained by exploring the effects of various parameters on a linearized version of the system equations. Dynamic features such as damping, speed of response, and stability are clearly revealed using a linear model.

The nonlinearity in Eqs. (3.9)–(3.11) occurs in the product of variables and in the exponential temperature term. Expanding these nonlinear terms in a Taylor series and truncating after the first term give three linear ordinary differential equations:

$$\frac{dC_A}{dt} = a_{11}C_A + a_{12}T_R + a_{13}T_J + b_{11}F + b_{12}F_J \tag{3.12}$$

$$\frac{dT_R}{dt} = a_{21}C_A + a_{22}T_R + a_{23}T_J + b_{21}F + b_{22}F_J \tag{3.13}$$

$$\frac{dT_J}{dt} = a_{31}C_A + a_{32}T_R + a_{33}T_J + b_{31}F + b_{32}F_J \tag{3.14}$$

The constant coefficients a_{ij} and b_{ij} are given in Eqs. (3.15) and (3.16). The overscored variables are the steady-state values around which the equations are linearized:

$$\begin{aligned}
a_{11} &= -\frac{\overline{F}}{V_R} - \overline{k} \\
a_{12} &= -\frac{\overline{C_A} E \overline{k}}{R(\overline{T}_R)^2} \\
a_{13} &= a_{31} = 0 \\
a_{21} &= -\frac{\lambda \overline{k}}{M c_P} \\
a_{22} &= -\frac{\overline{C_A} E \overline{k} \lambda}{\rho c_P R(\overline{T}_R)^2} - \frac{\overline{F}}{V_R} - \frac{U A_J}{V_R \rho c_P} \\
a_{23} &= \frac{U A_J}{V_R \rho c_P} \\
a_{32} &= \frac{U A_J}{V_J \rho_J c_J} \\
a_{33} &= -\frac{\overline{F}_J}{V_J} - \frac{U A_J}{V_J \rho_J c_J}
\end{aligned} \tag{3.15}$$

$$\begin{aligned}
b_{11} &= \frac{C_{A0} - \overline{C_A}}{V_R} \\
b_{12} &= b_{22} = b_{31} = 0 \\
b_{21} &= \frac{T_0 - \overline{T}_R}{V_R} \\
b_{32} &= \frac{T_{J0} - \overline{T}_J}{V_J}
\end{aligned} \tag{3.16}$$

Rearranging to find the openloop transfer function between reactor temperature and cooling water flowrate and including two first-order temperature measurement lags (τ_M) give

$$G_{FJ(s)} = \frac{T_{R(s)}}{F_{J(s)}} = \frac{c_1 s + c_0}{(s^3 + b_2 s^2 + b_1 s + b_0)(\tau_M s + 1)^2} \tag{3.17}$$

where $c_1 = a_{23} b_{32}$
$c_0 = -a_{11} a_{23} b_{32}$
$b_2 = -a_{11} - a_{22} - a_{33}$
$b_1 = a_{11} a_{22} + a_{11} a_{33} + a_{22} a_{33} - a_{12} a_{21} - a_{23} a_{32}$
$b_0 = a_{12} a_{21} a_{33} - a_{11} a_{22} a_{33} + a_{11} a_{23} a_{32}$

The two measurement lags are included so that reasonable controller tuning constants can be determined. The reactor itself is only net second-order (first-order polynomial in the numerator and third-order polynomial in the denominator), so the theoretical ultimate gain would be infinite if lags were not included. The linear model is used in the following section to explore stability.

3.1.3 Effect of Conversion on Openloop and Closedloop Stability

The linear model permits the use of all the linear analysis tools available to the process control engineer. For example, the poles and zeros of the openloop transfer function reveal the dynamics of the openloop system. A root locus plot shows the range of controller gains over which the system will be closedloop-stable.

To illustrate these methods and to show quantitatively the impact of conversion on stability, we use the numerical case considered in Chapter 2. Table 2.1 gives the kinetic and process parameters and is repeated here as Table 3.1. The feed flowrate is 4.377×10^{-3} m^3/s, and the reactor temperature is 350 K.

Figure 3.1 gives a Matlab program that sizes the reactor given the conversion, reactor temperature, feed conditions, coolant properties, and kinetic parameters. Then the coefficients of the linear model are evaluated, and the poles and zeros of the openloop transfer function are calculated. If any of the poles have positive real parts, the system is openloop-unstable.

Design conversions of 85 and 95% are considered. The corresponding reactor volumes are 26.1 and 102 m^3, jacket heat transfer areas are 40.9 and 101 m^2 and jacket temperatures are 309 and 330 K.

The one zero and the three poles of the openloop transfer function (not including the two poles from the measurement lags) for the two cases are given in Table 3.2. The 85% conversion case has two complex conjugate poles with *positive* real parts, so it is openloop-unstable. The 95% conversion case is openloop-stable.

Figure 3.2 gives root locus plots for the two designs at reactor temperature of 350 K with the measurement lags included. You may remember that a root locus plot is a plot of the roots of the closedloop characteristic equation as a function of the controller gain K_C. The plots start ($K_C = 0$) at the poles of the openloop transfer function and end ($K_C \to \infty$) at its zeros.

TABLE 3.1 Irreversible Exothermic Reaction Parameters

Preexponential factor k_0	s^{-1}	20.75×10^6
Activation energy E	J/kmol	69.71×10^6
Process molecular weight	kg/kmol	100
Process densities ρ_0 and ρ	kg/m^3	801
Coolant density ρ_J	kg/m^3	1000
Process heat capacities c_{p0} and c_p	$J\ kg^{-1}\ K^{-1}$	3137
Coolant heat capacity c_J	$J\ kg^{-1}\ K^{-1}$	4183
Heat of reaction λ	J/kmol	-69.71×10^6
Feed temperature T_0	K	294
Feed composition C_{A0}	$kmol/m^3$	8.01
Inlet coolant temperature $T_{C,\ in}$	K	294

```
% Program "cstrpolezero.m"
%  Given conversion  and reactor temperature, program calculates
%  (1) Steady-state design (volume,jacket area, jacket temperature and coolant flow)
%  (2) Openloop transfer function T/FJ
%  (3) Plots root locus with two 60 sec. lags
clear
ca0=8.01;k0=20.75e6;e=69.71e6;t0=294;tcin=294;u=851;lambda=-69.71e6;roe=801;m=100;cp=3137;cj=4183;roej=1000;
% Set conversion, reactor temperature and feed rate
conversion =0.95,f=4.377e-3;tr=350,ca=ca0*(1-conversion);k=k0*exp(-e/tr/8314);
vr=f*(ca0-ca)/k/ca; d=(2*vr/pi)^0.3333;areaj=2*pi*d^2;q=(ca0-ca)*f*(-lambda)-cp*roe*f*(tr-t0);
tj=tr-q/u/areaj;vj=0.3333*areaj;
% fj is in cu m/sec
fj=q/(cj*(tj-tcin))/roej;tr,vr,d,areaj,q,tj,fj,pause;vj=0.1*areaj;tjss=tj;fjss=fj;cass=ca;trss=tr;kss=k;
% Linear Model
a(1,1)=-(f/vr)-kss;a(1,2)=-kss*e*cass/8314/(trss^2);a(1,3)=0;
a(2,1)=-lambda*kss/m/cp;a(2,2)=-lambda*kss*e*cass/(roe*cp*8314*trss^2)-f/vr-u*areaj/(vr*roe*cp);
a(2,3)=u*areaj/(vr*roe*cp);
a(3,1)=0;a(3,2)=u*areaj/(vj*roej*cj);a(3,3)=-fj/vj-u*areaj/(vj*roej*cj);
b(1,1)=(ca0-cass)/vr;b(2,1)=(t0-trss)/vr;b(3,2)=(tcin-tjss)/vj;
% Openloop Transfer function
taum=60;b2=-a(1,1)-a(2,2)-a(3,3);
b1=a(1,1)*a(2,2)+a(1,1)*a(3,3)+a(2,2)*a(3,3)-a(1,2)*a(2,1)-a(2,3)*a(3,2);
b0=a(1,2)*a(2,1)*a(3,3)-a(1,1)*a(2,2)*a(3,3)+a(1,1)*a(2,3)*a(3,2);
c1=a(2,3)*b(3,2);c0=-a(1,1)*a(2,3)*b(3,2);
% Use negative since process gain and controller gain must be negative
num=-[c1 c0];den=[1 b2 b1 b0];gm=tf(num,den),poles=roots(den),zeros=roots(num)
% Generate a root locus plot with lags included in the denominator
kc=(0:0.00125:.2);
denlag=conv(den,[taum 1]);denlag=conv(denlag,[taum 1]);gmlag=tf(num,denlag);
[r95lag,kc]=rlocus(gmlag,kc);
clf
plot(real(r95lag)*100,imag(r95lag)*100,'+')
axis([-.25 .05 -.6 .6]);axis square;title('Conversion = 95; TR=350; with lags');
grid;ylabel('Imag*100');xlabel('Real*100')
```

Figure 3.1 Matlab program for linear CSTR.

The loci for the 85% conversion case start in the right half of the s plane (the unstable region) and move into the closedloop-stable left half of the s plane for only a very limited range of controller gains. Thus the linear analysis predicts that a proportional-only controller would have great difficulty in stabilizing this system. The loci for the 95% conversion case start in the stable region, but eventually move across the imaginary axis into the unstable region at the ultimate gain, which can be calculated using either Laplace-domain or frequency-domain methods.

The reactor temperature used in the results discussed above is 350 K. If a lower reactor temperature is used, the reactor will be larger for the same conversion, which will give larger heat transfer area. We would expect the controllability of the reactor to improve. This is quantitatively confirmed by the location of the openloop poles, as shown in Table 3.2, and also in the root locus plots shown in Figure 3.3. Now the 85% conversion case is still openloop-unstable but has a range of controller gains that will stabilize the closedloop system.

The ultimate gains of these four different designs are calculated using a frequency-domain method. The Matlab program given in Figure 3.4 performs this calculation using the linear model. The numerator and the denominator of the transfer function are determined, and the Matlab functions "bode" and "nyquist" are used to calculate the frequency response of the system. The ultimate frequency is found by starting at a high frequency and reducing frequency until the imaginary part becomes negative. This corresponds to the point where the phase angle is $-180°$; it is the ultimate frequency. Then the ultimate gain is calculated from the reciprocal of the magnitude at this frequency. The gain is made dimensionless by using a 50 K temperature transmitter span and

TABLE 3.2 Effect of Conversion for 350 and 330 K Reactor Temperatures

	Conversion (%)	
	85	95
350 K Reactor Temperatures		
Reactor volume (m^3)	30.3	102
Jacket area (m^2)	45.2	101
Jacket temperature (K)	312	330
Poles	-0.00644	-0.00328
	$+0.000179$	-0.000244
	$\pm 0.00327i$	$\pm 0.00186i$
Zero	-9.63×10^{-4}	-8.62×10^{-4}
K_U (dimensionless)	8.54	21.8
P_U (s)	1231	1350
K_C (dimensionless)	2.67	6.80
τ_I (s)	2709	2970
330 K Reactor Temperatures		
Reactor volume (m^3)	129	434
Jacket area (m^2)	119	266
Jacket temperature (K)	313	321
Poles	-0.00392	-0.00281
	$+0.000035$	-0.000054
	$\pm 0.000811i$	$\pm 0.00046i$
Zero	-2.25×10^{-4}	-0.0231
K_U (dimensionless)	41.6	89.7
P_U (s)	1231	1383
K_C (dimensionless)	13.0	28.0
τ_I (s)	2709	3040

assuming that the maximum coolant flowrate is 4 times the design flowrate. The controller tuning parameters are calculate using the Tyreus–Luyben tuning rules:

$$K_C = \frac{K_U}{3.2}, \qquad \tau_I = 2.2P_U \tag{3.18}$$

Table 3.2 shows the results of these calculations. Lower reactor temperature and higher conversion increase the ultimate gain. This indicates better dynamic controllability.

The Nyquist plots for the two conversion are quite interesting and reveal the potential for "conditional stability" that can occur in an openloop unstable process. Figures 3.5 and 3.6 give these plots for reactor temperatures of 330 and 350 K, respectively, for two different conversion cases. The 95% conversion cases are openloop-stable and have a normal polar plot, starting on the positive real axis, moving clockwise, and ending up with a phase angle of $-360°$. The ultimate frequency occurs where the curve crosses the negative real axis. The ultimate gain is the reciprocal of the magnitude of the transfer function at this frequency.

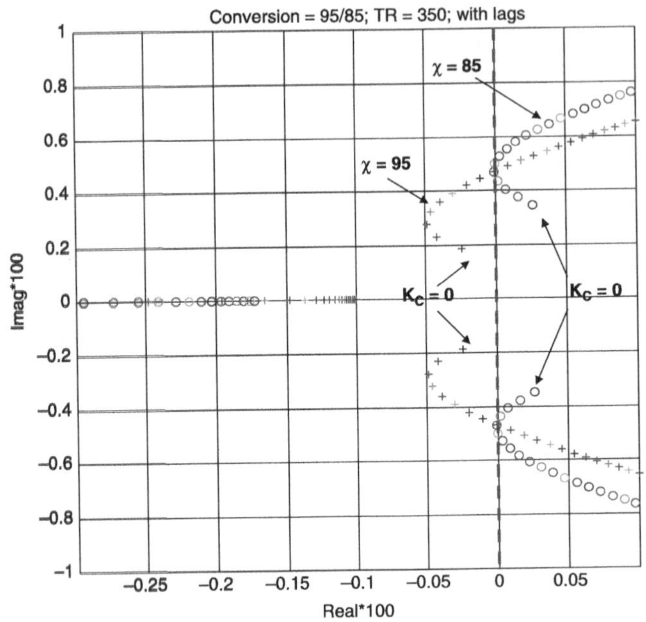

Figure 3.2 Root locus plots; 85 and 95% conversions, $T_R = 350$ K.

However, the openloop unstable 85% conversion cases are very different. The curves start on the positive real axis but move *counterclockwise*, going from the first to the second to the third and eventually back to the second and then the first quadrants before finally ending with a phase angle of $-360°$. Figure 3.7 gives an enlarged view

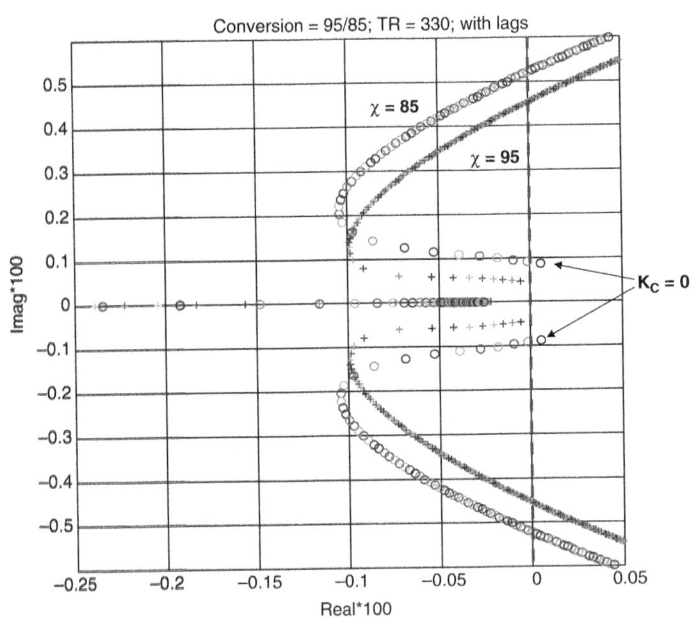

Figure 3.3 Root locus plots; 85 and 95% conversions, $T_R = 330$ K.

```
% Program "cstrku.m"
%  Given conversion  and reactor temperature, program calculates Ku and Pu
clear
ca0=8.01;k0=20.75e6;e=69.71e6;t0=294;tcin=294;u=851;lambda=-69.71e6;roe=801;m=100;cp=3137;cj=4183;roej=1000
% Set conversion, reactor temperature and feed rate
conversion =0.95,f=4.377e-3;tr=330,ca=ca0*(1-conversion);k=k0*exp(-e/tr/8314);
vr=f*(ca0-ca)/k/ca; d=(2*vr/pi)^0.3333;areaj=2*pi*d^2;q=(ca0-ca)*f*(-lambda)-cp*roe*f*(tr-t0);
tj=tr-q/u/areaj;vj=0.3333*areaj;
% fj is in cu m/sec
fj=q/(cj*(tj-tcin))/roej;tr,vr,d,areaj,q,tj,fj,pause;vj=0.1*areaj;tjss=tj;fjss=fj;cass=ca;trss=tr;kss=k;
% Linear Model
a(1,1)=-(f/vr)-kss;a(1,2)=-kss*e*cass/8314/(trss^2);a(1,3)=0;
a(2,1)=-lambda*kss/m/cp;a(2,2)=-lambda*kss*e*cass/(roe*cp*8314*trss^2)-f/vr-u*areaj/(vr*roe*cp);
a(2,3)=u*areaj/(vr*roe*cp);
a(3,1)=0;a(3,2)=u*areaj/(vj*roej*cj);a(3,3)=-fj/vj-u*areaj/(vj*roej*cj);
b(1,1)=(ca0-cass)/vr;b(2,1)=(t0-trss)/vr;b(3,2)=(tcin-tjss)/vj;
% Openloop transfer function
taum=60;b2=-a(1,1)-a(2,2)-a(3,3);
b1=a(1,1)*a(2,2)+a(1,1)*a(3,3)+a(2,2)*a(3,3)-a(1,2)*a(2,1)-a(2,3)*a(3,2);
b0=a(1,2)*a(2,1)*a(3,3)-a(1,1)*a(2,2)*a(3,3)+a(1,1)*a(2,3)*a(3,2);
c1=a(2,3)*b(3,2);c0=-a(1,1)*a(2,3)*b(3,2);
% Use negative since process gain and controller gain must be negative
num=-[c1 c0];den=[1 b2 b1 b0];denlag=conv(den,[taum^2 2*taum 1]);
w=logspace(-4,1,900);
[mag,phase,w]=bode(num,denlag,w);[re,im,w]=nyquist(num,denlag,w);
% Calculate ultimate gain and frequency; start search at high frequency and back up
nw=length(w);wmax=w(nw);kk=nw;
while im(kk)>0;kk=kk-1;end
% dimensionless Kc with TT = 50 and FT =4FJ
wu=w(kk);ku=50/4/fj/mag(kk),pu=2*pi/wu,reset=2.2*pu,kc=ku/3.2
```

Figure 3.4 Calculating ultimate gain and period.

of the high-frequency region around the ultimate frequency for the 85 and 95% conversion cases with a 350 K reactor temperature. The 85% conversion curve must encircle the critical $(-1,0)$ point in a *counterclockwise* direction for the system to be closed-loop-stable. Since there is one pole in the right half of the s plane, $P = +1$ in the Nyquist stability relationship $N = Z + P$, where Z is the number of zeros of the closedloop characteristic equation in the right half of the s plane and P is the number of poles. The

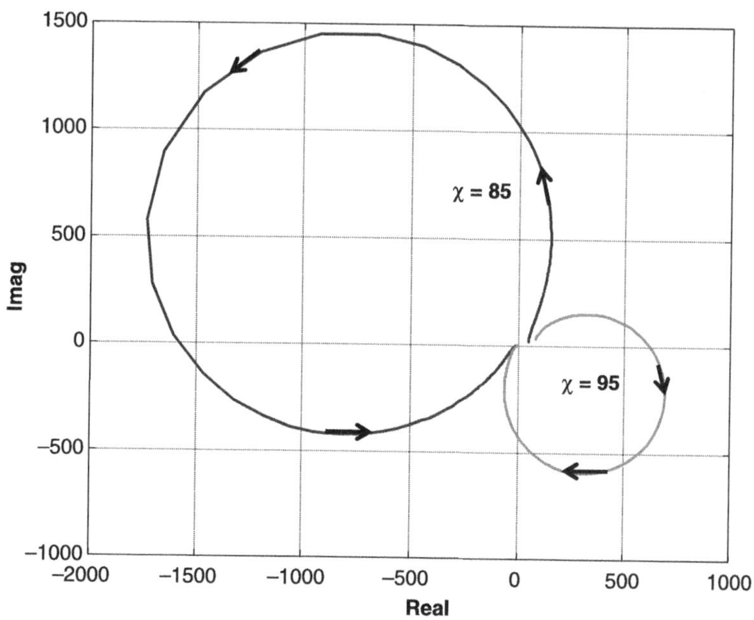

Figure 3.5 Reactor temperature = 330 K.

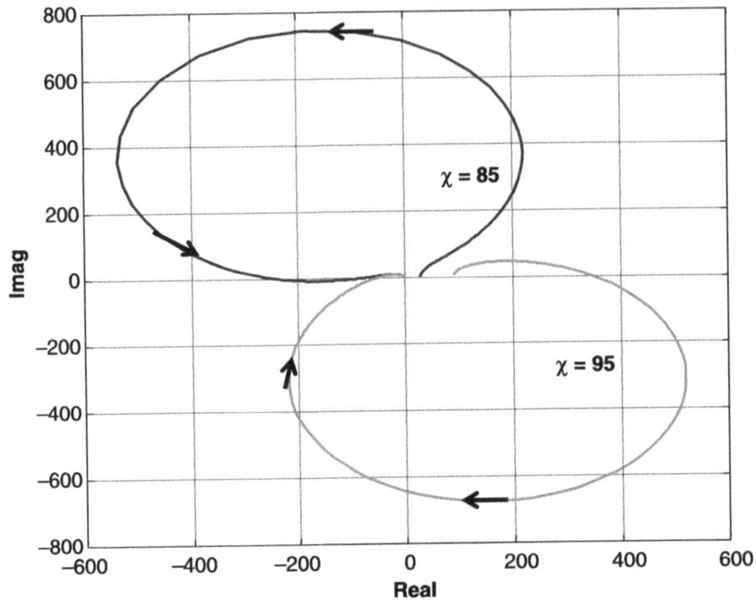

Figure 3.6 Reactor temperature = 350 K.

number of clockwise encirclements is N. For stability, the zeros must be zero, so N must be equal to -1 (one counterclockwise encirclement).

Therefore, the Nyquist plot must dip down into the third quadrant. If is does, there are two values of gain that represent the limits of closedloop stability. The maximum

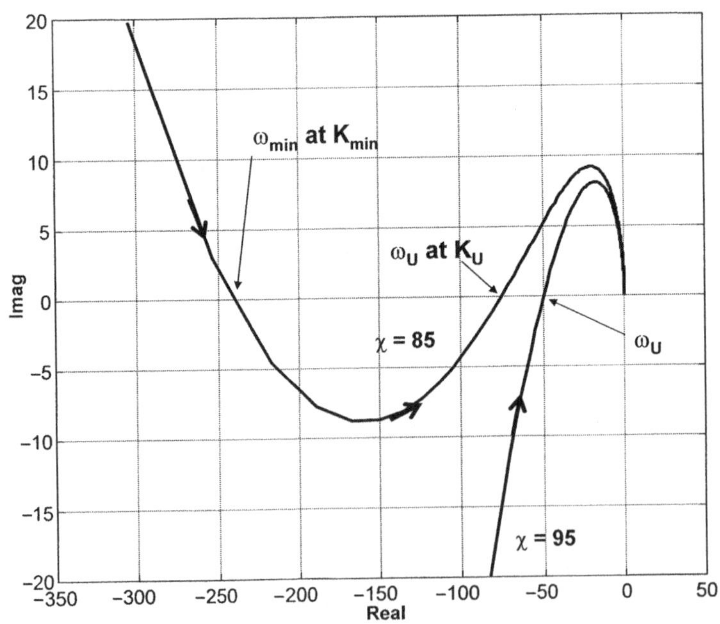

Figure 3.7 High-frequency part; reactor temperature = 350 K.

controller gain K_U occurs at a high frequency ω_U. The minimum controller gain K_{min} occurs at a low frequency ω_{min}. Note that the Nyquist plot for the 85% conversion case at 330 K (Fig. 3.5) dips far down into the third quadrant, which indicates a large difference between K_{min} and K_U. However, for the 85% conversion case at 350 K (Fig. 3.6), the curve barely makes it into the third quadrant, which indicates a small difference between K_{min} and K_U and more difficult controller tuning.

We will illustrate this conditional stability condition in the dynamic simulations developed in the next section.

3.1.4 Nonlinear Dynamic Simulation

To demonstrate the superior dynamic controllability of high-conversion and low-temperature designs, the nonlinear differential equations are numerically integrated for the four different design cases. Disturbances in feed flowrate, temperature controller setpoint, and overall heat-transfer coefficient are made, and the peak deviations in reactor temperature are compared.

The Matlab program given in Figure 3.8 performs the dynamic simulation using the simple Euler integration algorithm. An integration step size "delta" of 5 s is used. The temperature controller tuning constants given in Table 3.2 are used for each case (Tyreus–Luyben tuning rules). Note that the controller gains are dimensionless, using a 50 K temperature transmitter span and assuming the maximum cooling water flowrate is four times the design value. Two 60-s measurement lags are included in the temperature loop. Note the "antireset windup" provision near the end of the integration loop. The integral of the error ("erint") is only evaluated if the controller output signal ("op") is not at a limit, that is, the "op" signal is between zero and one.

```
% Program "simcstr.m"
%  Nonlinear dyanmic simulation
clear
ca0=8.01;k0=20.75e6;e=69.71e6;t0=294;tcin=294;u=851;taum=60;
lambda=-69.71e6;roe=801;m=100;cp=3137;cj=4183;roej=1000;
% Set conversion, reactor temperature and feed rate
conversion =0.85;f=4.377e-3;tr=350;ca=ca0*(1-conversion);k=k0*exp(-e/tr/8314);
vr=f*(ca0-ca)/k/ca; d=(2*vr/pi)^0.3333;areaj=2*pi*d^2;
q=(ca0-ca)*f*(-lambda)-cp*roe*f*(tr-t0);tj=tr-q/u/areaj;vj=0.3333*areaj;
fj=q/(cj*(tj-tcin))/roej;vj=0.1*areaj;tjss=tj;fjss=fj;cass=ca;trss=tr;kss=k;fss=f;
% Controller settings for 350 K 85% case
kc=2.67;taui=2709; fnew=fss*1.2;
% Initial Conditions and Parameters
tstop=4*3600;np=0;erint=0;tplot=0;ca=cass;tj=tjss;tr=trss;
time=0;delta=5;sp=trss;trlag1=trss;trlag2=trss;
% Integration loop
while time<tstop
% Temperature Controller
error=(sp-trlag2)/50;op=0.25-kc*error-erint;if op>1; op=1;end; if op<0;op=0;end;fj=fjss*op*4;
if time>900;f=fnew;end;   % Disturbance
if time>=tplot;   np=np+1;   % store data for plotting
   timep(np)=time/60;trp(np)=tr;tjp(np)=tj;fjp(np)=fj;cap(np)=ca;tplot=tplot+30;end
q=u*areaj*(tr-tj);k=k0*exp(-e/tr/8314);
% Derivative evaluations
dca=f*(ca0-ca)/vr -k*ca;dtj=fj*(tcin-tj)/vj +q/(cj*roej*vj);
dtr=f*(t0-tr)/vr -lambda*k*ca/roe/cp - q/(cp*roe*vr);
dtrlag1=(tr-trlag1)/taum;dtrlag2=(trlag1-trlag2)/taum;
%  Integration
time=time+delta;ca=ca+dca*delta;tr=tr+dtr*delta;tj=tj+dtj*delta;
trlag1=trlag1+dtrlag1*delta;trlag2=trlag2+dtrlag2*delta;
% Anti-reset windup (only integrate error if op signal is between 0 and 1
if op<1;if op>0;erint=erint+error*kc*delta/taui;end;end
end
clf
subplot(2,2,1);plot(timep,trp);grid;ylabel('TR (K)');title('350K; 95 Conversion; +20 Feed')
subplot(2,2,3);plot(timep,tjp);grid;ylabel('TJ (K)');xlabel('Time (min)');
subplot(2,2,2);plot(timep,fjp);grid;ylabel('FJ (cu m/sec)');
subplot(2,2,4);plot(timep,cap);grid;ylabel('CA (kmol/cu m)');xlabel('Time (min)');
```

Figure 3.8 Matlab program for simulating nonlinear CSTR.

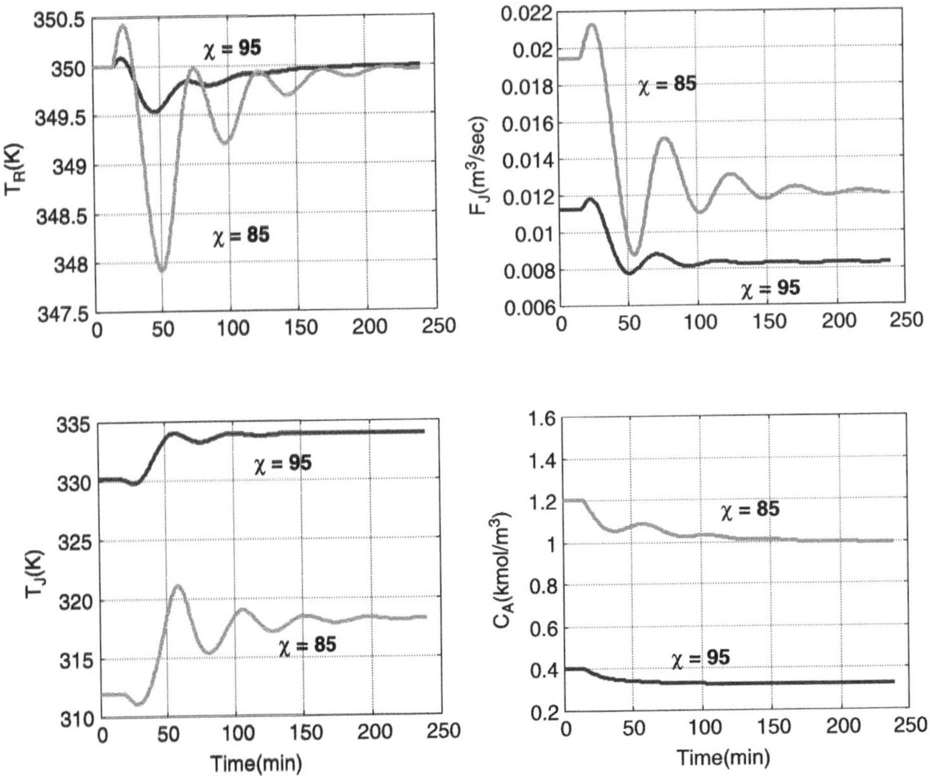

Figure 3.9 Step 20% decrease in feed; reactor temperature = 350 K.

Figure 3.9 gives results for a 20% step decrease in feed flowrate at time equal 20 min for the 350 K reactor design. The maximum deviation in reactor temperature is much smaller in the 95% conversion process than in the 85% case. Note the initial *increase* in the reactor temperature when the feed is decreased. This is caused by the feed being colder than the reactor liquid. This "inverse response" will be discussed in more detail later in this chapter. Figure 3.10 shows the responses for a 20% step increase in feed flowrate. The 85% case is unstable.

The Tyreus–Luyben tuning rules have been used in these simulations. To see the effect of controller tuning on the dynamics of the 85% case, a range of controller gains is explored with the reset time kept constant at 2709 s. Figure 3.11 gives results for three values of K_C (5, 10, and 20). A high gain gives very oscillatory response with a small period. This means that we are close to the K_U limit. A low gain also gives very oscillatory response, but with a large period. This means we are close to the $K_{\min}$ limit.

Linear analysis predicted an ultimate gain of 8.54 (Table 3.2), but the nonlinear dynamic simulations indicate that a higher gain can be used. If a gain of 10 is used, the response is stable, but performance is still worse in this 85% conversion case than in the 95% conversion case as shown in Figure 3.12.

Figures 3.13 and 3.14 give results for the 330 K reactor case with feedrate disturbances using the tuning given in Table 3.2. The 95% conversion process gives better temperature,

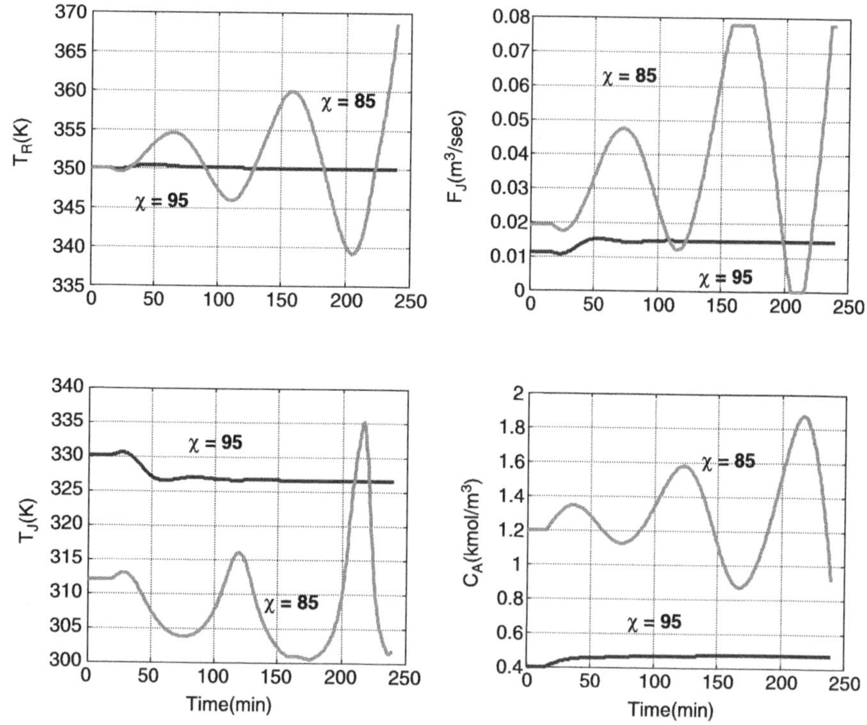

Figure 3.10 Step 20% increase in feed; reactor temperature = 350 K.

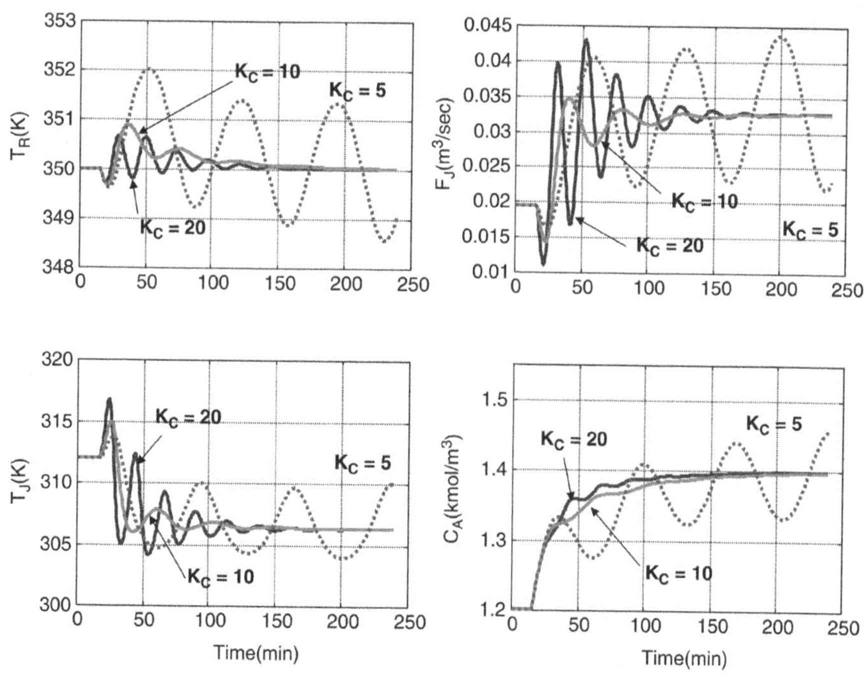

Figure 3.11 $K_C = 20/10/5$; step 20% increase in feed; 350 K.

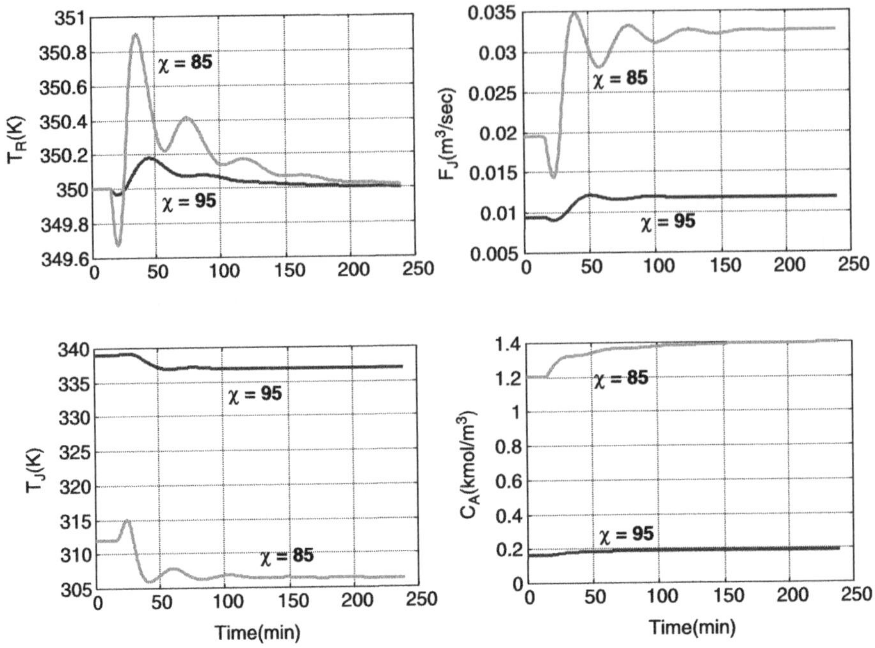

Figure 3.12 $K_C = 10$ for 85% case; step 20% increase in feed; 350 K.

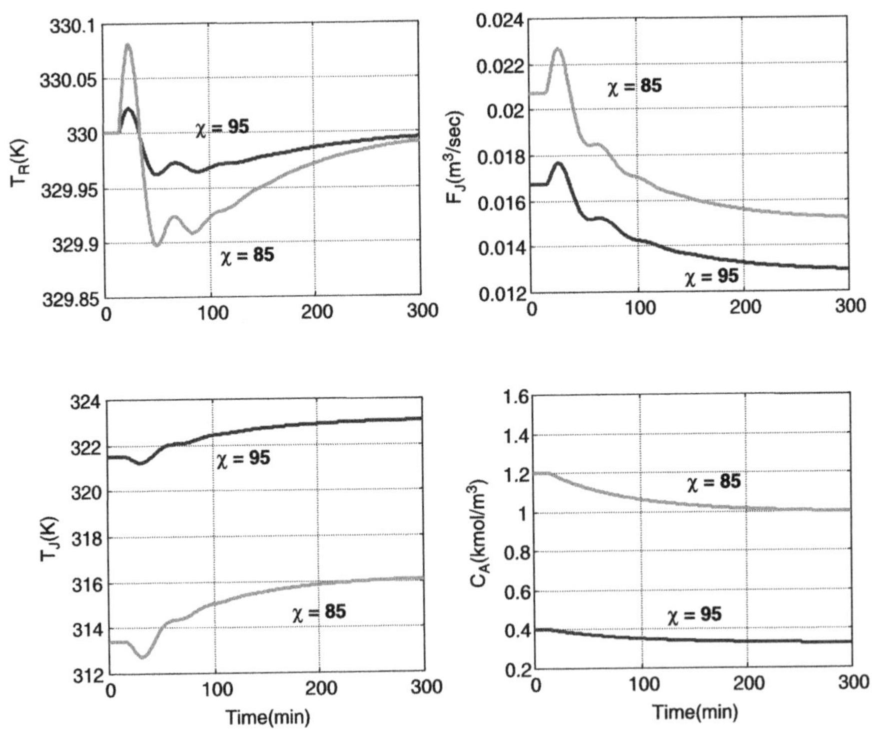

Figure 3.13 Step 20% decrease in feed; reactor temperature = 330 K.

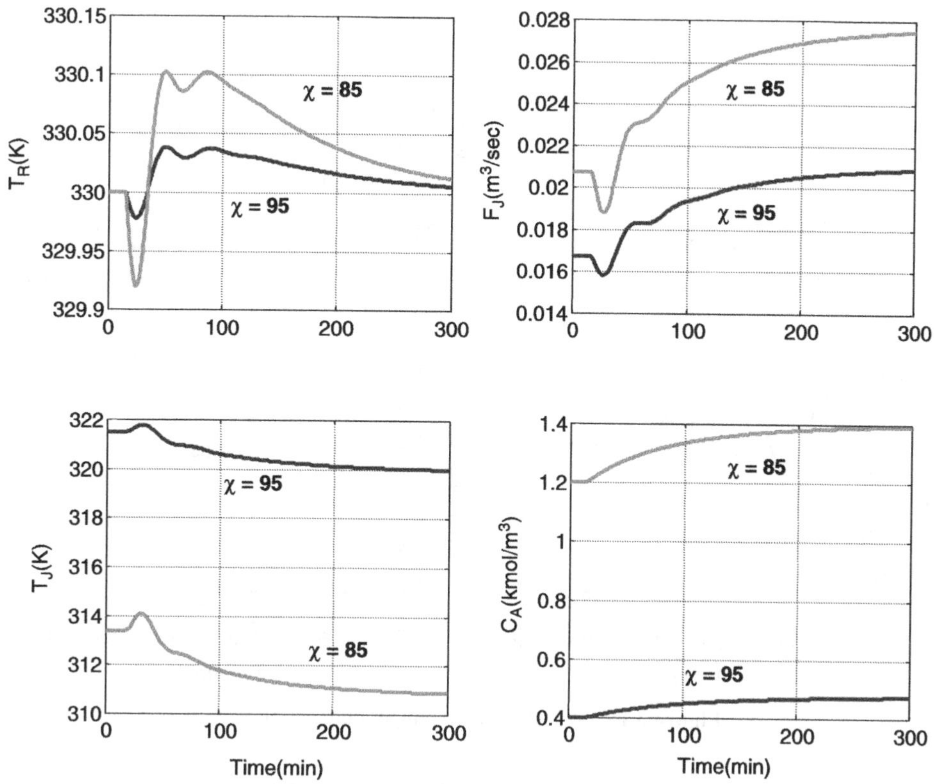

Figure 3.14 Step 20% increase in feed; reactor temperature = 330 K.

but the 85% conversion process is at least stable because the reactors have more heat transfer area in the lower temperature designs.

The responses of the four cases to changes in the setpoint of the temperature controller are given in Figures 3.15–3.18. The superiority of the high-conversion and/or the low-reactor-temperature designs is clearly shown. Note that the controller completely closes the cooling water valve for several minutes (up to 50 min) for the +5 K increase in setpoint. During this period the antireset windup feature prevents the controller from winding up.

The final disturbance tested is a 10% decrease in the overall heat transfer coefficient U. Figure 3.19 shows that the 350 K case with 85% conversion goes unstable. Figure 3.20 shows that the 330 K cases are both stable, but the peak deviation in reactor temperature is over 3 times greater for the 85% conversion process than for the 95% conversion process.

These dynamic results provide graphic and quantitative proof that CSTRs designed for high conversions and/or low temperature are easier to control and provide tighter temperature control than are CSTRs that are designed for low conversion and/or high temperature.

3.1.5 Effect of Jacket Volume

In all the simulations up to now the jacket volume has been calculated by using the jacket heat transfer area and assuming a jacket thickness of 0.1 m. The jacket volume has no

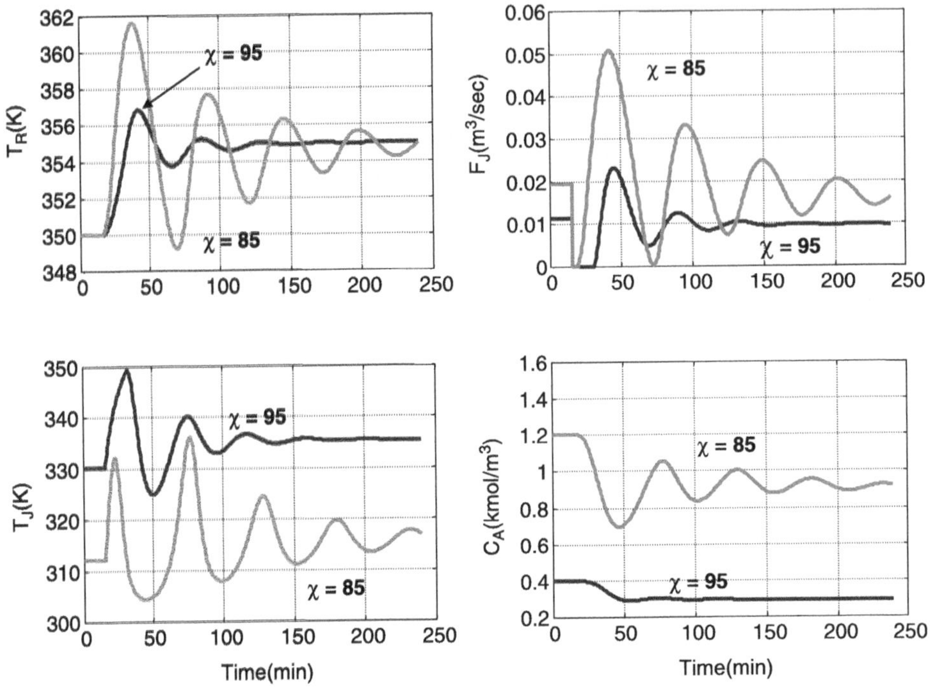

Figure 3.15 +5 SP; reactor temperature = 350 K.

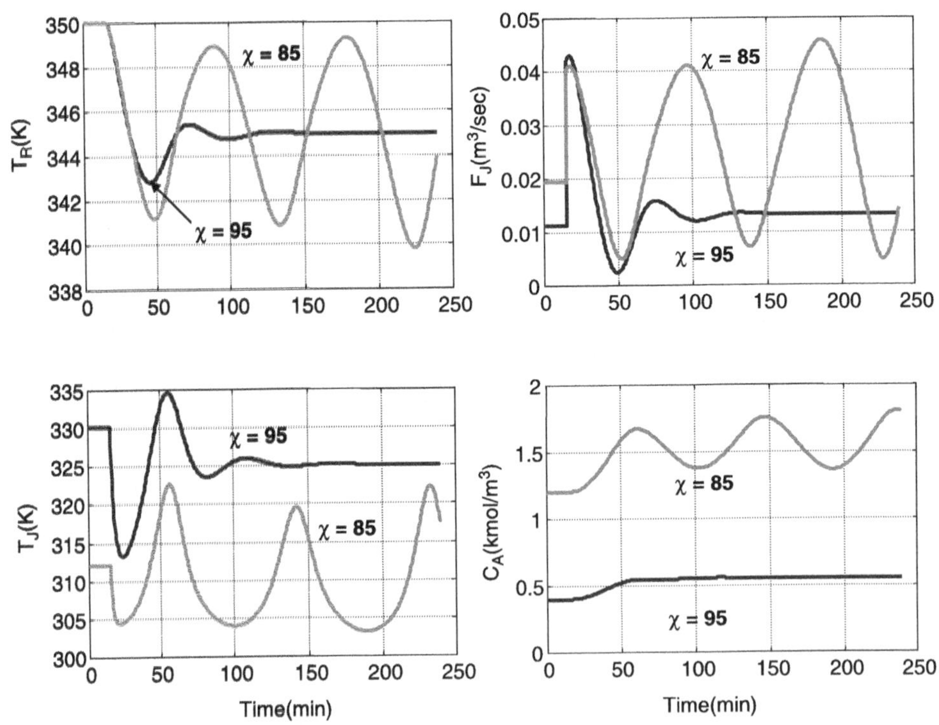

Figure 3.16 −5 K SP; reactor temperature = 350 K.

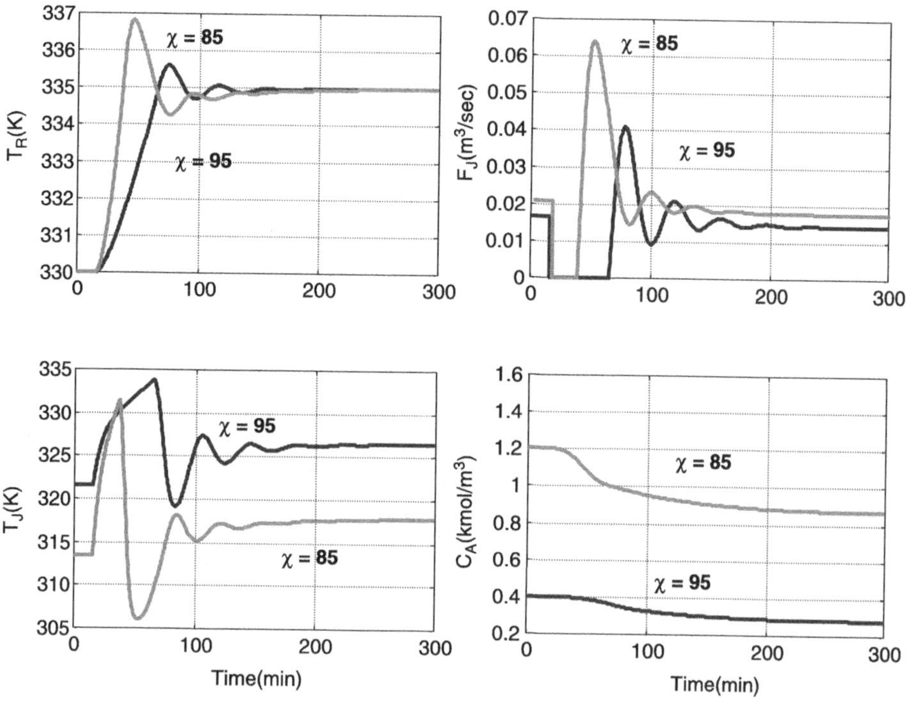

Figure 3.17 +5 SP; reactor temperature = 330 K.

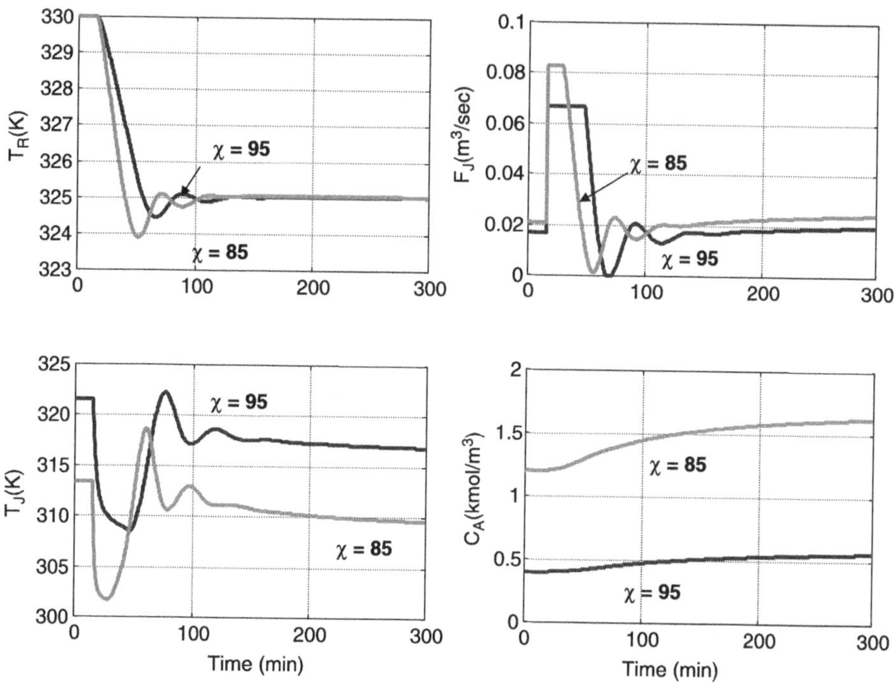

Figure 3.18 −5 K SP; reactor temperature = 330 K.

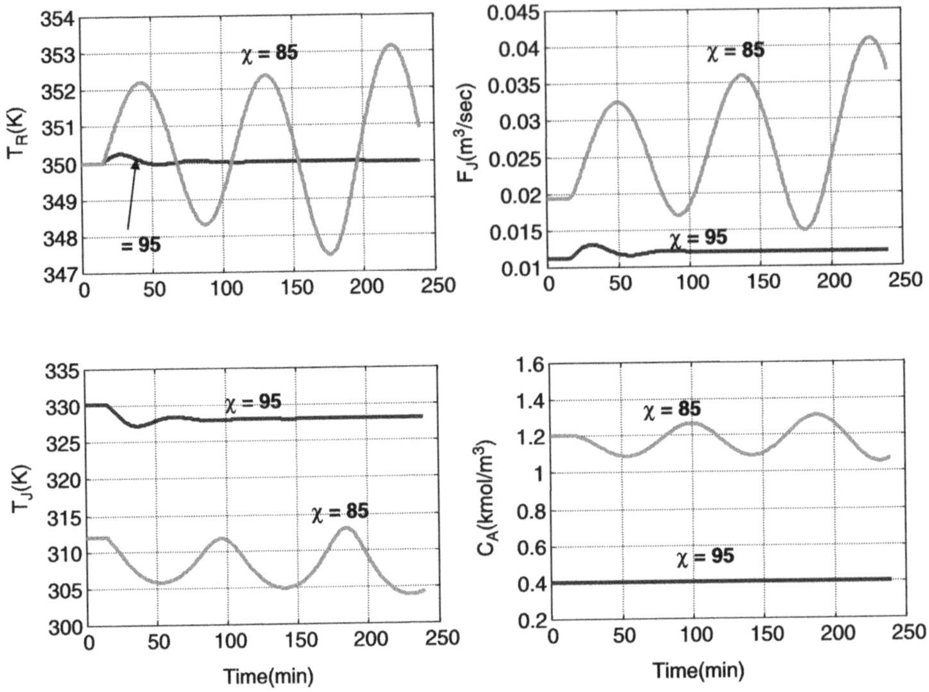

Figure 3.19 0.9 U; reactor temperature = 350 K.

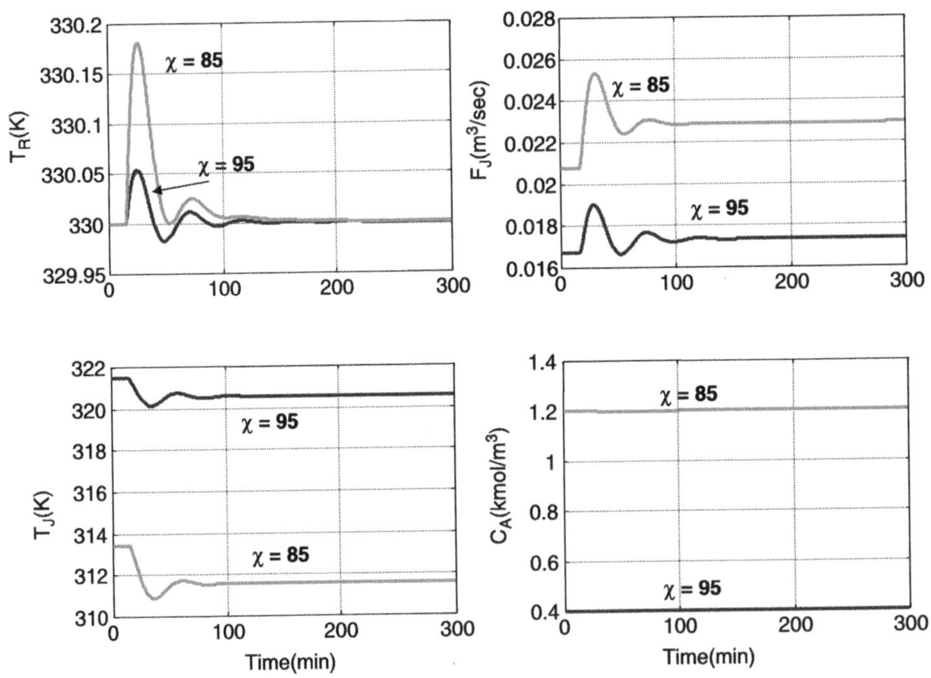

Figure 3.20 0.9 U; reactor temperature = 330 K.

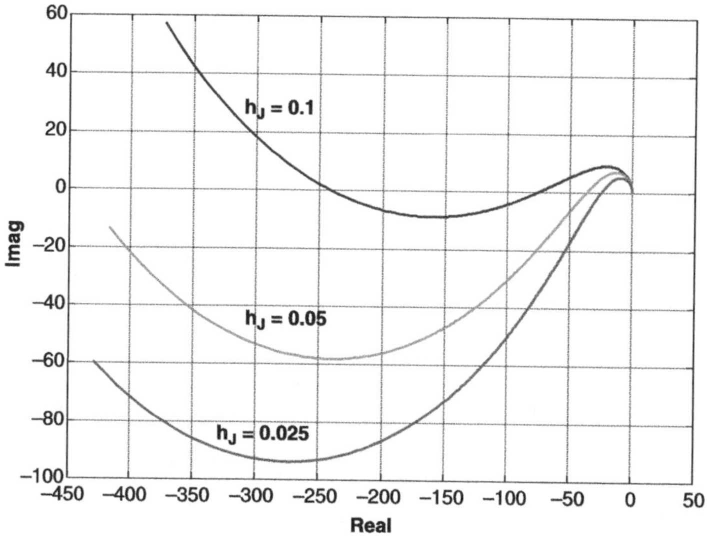

Figure 3.21 Effect of jacket volume; 85%; 350 K; Nyquist plot.

effect on the steady-state design. However, we would expect it to impact the dynamics. Intuition says that the larger the volume, the worse the control.

To quantify this effect, jacket thickness h_J is changed from 0.1 down to 0.025 m. The 350 K reactor temperature with 85% conversion is the case studied. Figure 3.21 shows the effect on the Nyquist plot. The improvement in controllability is indicated by the curves dropping more into the third quadrant as h_J is decreased. The ultimate gain increases from 8.51 to 17.1 to 24.8. The ultimate period decreases from 2711 to 1846 to 1429 s. These indicate improved closedloop performance. Figure 3.22 shows this to be true. The Tyreus–Luyben settings are used in the three cases.

These results show that jacket volume should be kept as small as possible.

3.1.6 Cooling Coil

We have frequently emphasized in this book that the key issue in reactor control is heat transfer area. All cases considered so far in this chapter have assumed jacket cooling, so heat transfer area is limited by the geometry of the vessel. As discussed in Chapter 2, using a cooling coil provides 25% more area, so some improvement in control is expected when a coil is used.

To quantify this expectation, a dynamic simulation of a CSTR with an internal cooling coil is developed. As discussed in Chapter 2, the driving force under steady-state conditions is a log-mean temperature difference, using the differences between the reactor temperature and the temperatures of the inlet and outlet cooling water in and out of the coil. The design for 85% conversion with a 350 K reactor temperature has a coil heat transfer area of 58.1 m^2 compared to the 45.2 m^2 of the jacket-cooled reactor. The volume of the coil is 1.11 m^3 (35.6 loops, 242 m length, 0.0763 m coil diameter) in a 2.71-m-diameter reactor. Note that the coil volume is much smaller than the volume of a jacket with the typical jacket thickness of 0.1 m (4.53 m^3).The cooling water flowrate is 0.00814 m^3/s with an inlet temperature of 294 K and an exit temperature of 336.9 K.

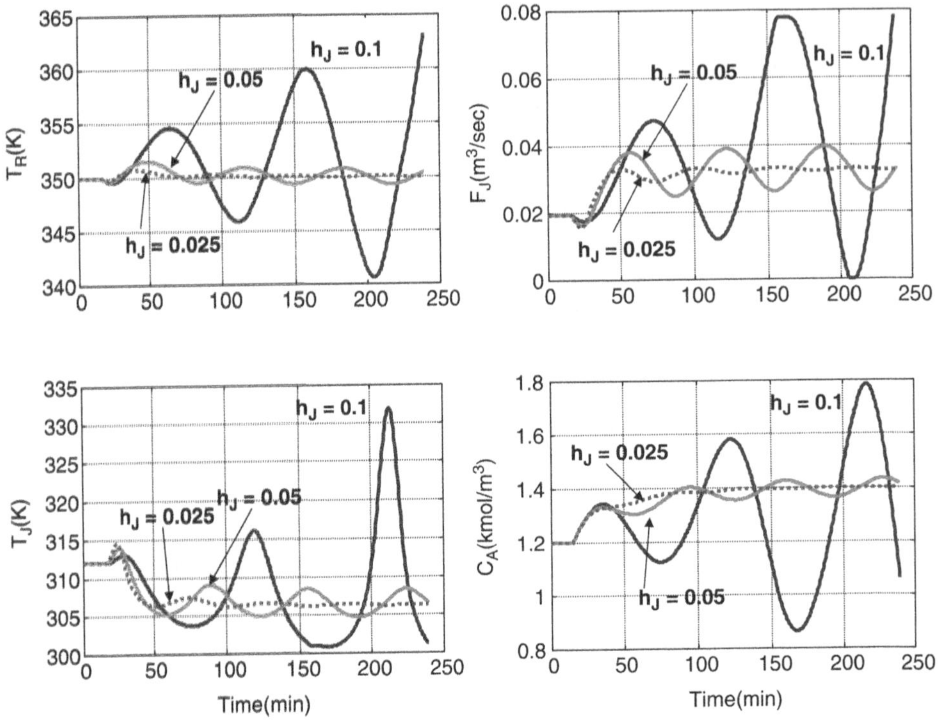

Figure 3.22 Effect of jacket volume; 85%; 350 K; +20% feed.

The dynamics of the coil are described by a partial differential equation for the coil energy balance because temperature in the coil changes with both length down the coil and time. An approximate "lumped" model of the coil is developed so that ordinary differential equations can be used. A 20-lumped model gives an exit temperature 335.7 K, which is close to the rigorous steady-state value of 336.9 K from the rigorous log-mean temperature driving force. The cooling water flowrate from the lumped model is 0.00839 m^3/s, compared to 0.00815 m^3/s from the rigorous model. Figure 3.23 gives a Matlab program used for the dynamic simulation of the CSTR with a cooling coil.

A relay–feedback test on the reactor temperature controller is used to obtain the ultimate gain and frequency ($K_U = 64$ and $P_U = 10$ min), using a 50 K temperature transmitter span and assuming the maximum cooling water flow is twice the steady-state value. The Tyreus–Luyben settings give oscillatory response, so the controller gain is reduced by factor of 2 ($K_C = 10$, $\tau_I = 1320$ s).

Figure 3.24 gives the response to a 20% increase in feed flowrate. Figure 3.25 gives a direct comparison of the coil-cooled CSTR and the jacket-cooled CSTR with a jacket thickness of 0.025 m. Even with this very small jacket holdup, the coil-cooled system has tighter temperature control.

3.1.7 External Heat Exchanger

Another method for increasing heat transfer area is to circulate, usually by pumping, the process liquid from the reactor through an external heat exchanger. The heat exchanger

```
% Program "simcoil.m"
%    Simulations of cooling coil;  A = B Conversion = 85%
clear
ca0=8.01;k0=20.75e6;e=69.71e6;t0=294;tcin=294;u=851;lambda=-69.71e6;
roe=801;m=100;cp=3137;cj=4183;roej=1000;dcoil=0.0763;
conversion =0.85;fa0=4.377e-3;ca=ca0*(1-conversion);tr=350;
k=k0*exp(-e/tr/8314);vr=fa0*(ca0-ca)/k/ca;
% Iterative loop to find diameter using direct substitution
 dguess=(2*vr/pi)^0.3333; error=10;
while error>0.011;lvessel=2*dguess;
nloops=lvessel/2/dcoil;lcoil=0.8*dguess*pi*nloops;vcoil=lcoil*pi*(dcoil^2)/4;
vtot=vr+vcoil;dvessel=(2*vtot/pi)^0.3333;error=abs(dvessel-dguess);dguess=dvessel;end
tr,ca,fa0,vr,nloops,lcoil,vcoil,dvessel,areacoil=lcoil*dcoil*pi
q=(ca0-ca)*fa0*(-lambda)-cp*roe*fa0*(tr-t0);
% Iterative calculation for tcout using log mean temperature difference
tcout=tr-1;flagm=-1;flagp=-1;dtcout=2;loop=0;errorq=1000;
while errorq>10;
   dt1=tr-tcin;dt2=tr-tcout;dtlm=(dt1-dt2)/log(dt1/dt2);qcalc=u*areacoil*dtlm;
   errorq=abs(q-qcalc);  loop=loop+1;if loop>50;q,qcalc,tcout;error('LMTD loop'); end;
   if qcalc>q;if flagp>0;dtcout=dtcout/2;end;tcout=tcout+dtcout;flagm=1;
       if tcout>tr;tcout=tr-0.01;end;end
   if qcalc<q;if flagm>0;dtcout=dtcout/2;end;tcout=tcout-dtcout;flagp=1;end;end
fcoil=q/(cj*roej*(tcout-tcin)),tcout,q,dtlm
% Nonlinear  response for case 85% conversion
% Initial Conditions and Parameters
tstop=5*3600;np=0;erint=0;tplot=0;cass=ca;fa0ss=fa0;trlag1=350;trlag2=350;taum=1*60;
time=0;delta=1;ca0=8.01;kc=20/2;reset=1320;nlump=20;
load coilic tc fcoil;fcoilss=fcoil;tcin=294;
% Integration loop
while time<tstop
% Temperature Controller
error=(350-trlag2)/50;op=0.5-kc*error-erint;if op>1; op=1;end; if op<0;op=0;end
fcoil=fcoilss*op*2;
% Disturbance
fa0new=fa0ss*1.2;if time>900;fa0=fa0new;end
% store data for plotting
if time>=tplot;np=np+1;trp(np)=tr;tcoutp(np)=tc(nlump);fcoilp(np)=fcoil;
   cap(np)=ca;timep(np)=time/60; tplot=tplot+2;end
% Calculate heat transfer to each lump of coil and sum of all q's
 qtot=0;for n=1:nlump;qcoil(n)=u*areacoil*(tr-tc(n))/nlump;qtot=qtot+qcoil(n);end
% Derivative evaluations
k=k0*exp(-e/tr/8314);dca=fa0*(ca0-ca)/vr -k*ca;
dtc(1)=fcoil*(tcin-tc(1))/(vcoil/nlump) + qcoil(1)/(cj*roej*(vcoil/nlump));
for n=2:nlump
dtc(n)=fcoil*(tc(n-1)-tc(n))/(vcoil/nlump) +qcoil(n)/(cj*roej*(vcoil/nlump));end
dtr=fa0*(t0-tr)/vr -lambda*k*ca/roe/cp - qtot/(cp*roe*vr);
dtrlag1=(tr-trlag1)/taum;dtrlag2=(trlag1-trlag2)/taum;
%  Integration
time=time+delta;ca=ca+dca*delta;tr=tr+dtr*delta;
for n=1:nlump;tc(n)=tc(n)+dtc(n)*delta;end;if tr>450;break;end
erint=erint+error*kc*delta/reset;trlag1=trlag1+dtrlag1*delta;trlag2=trlag2+dtrlag2*delta;end
subplot(2,2,1);plot(timep,trp);grid;ylabel('TR (K)');title('Coil Cooling; 85% Conversion; +20 Feed')
subplot(2,2,3);plot(timep,tcoutp);grid;ylabel('Tcoil Out (K)');xlabel('Time (min)');
subplot(2,2,2);plot(timep,fcoilp);grid;ylabel('Fcoil (cu m/sec)');
subplot(2,2,4);plot(timep,cap);grid;ylabel('CA ');xlabel('Time (min)');
```

Figure 3.23 Matlab program for coil cooling; 85%; 350 K; +20% feed.

can be designed for more area than can be achieved using jacket or coil cooling because it is not limited by the geometry of the reactor vessel. From a steady-state perspective, there is no limit on how much area can be installed using an external heat exchanger.

However, there is an important dynamic effect as the size of the heat exchanger is increased. The larger holdup in the heat exchanger introduces more dynamic lag in the heat transfer process, which could degrade dynamic performance. We observed this in the jacket-cooled system discussed in Section 3.1.5. The smaller the thickness of the jacket, the better the temperature control.

The external heat exchanger is assumed to be tube-in-shell with 0.0254-m-diameter tubes, 5 m in length. Given a desired total heat transfer area, the number of tubes and

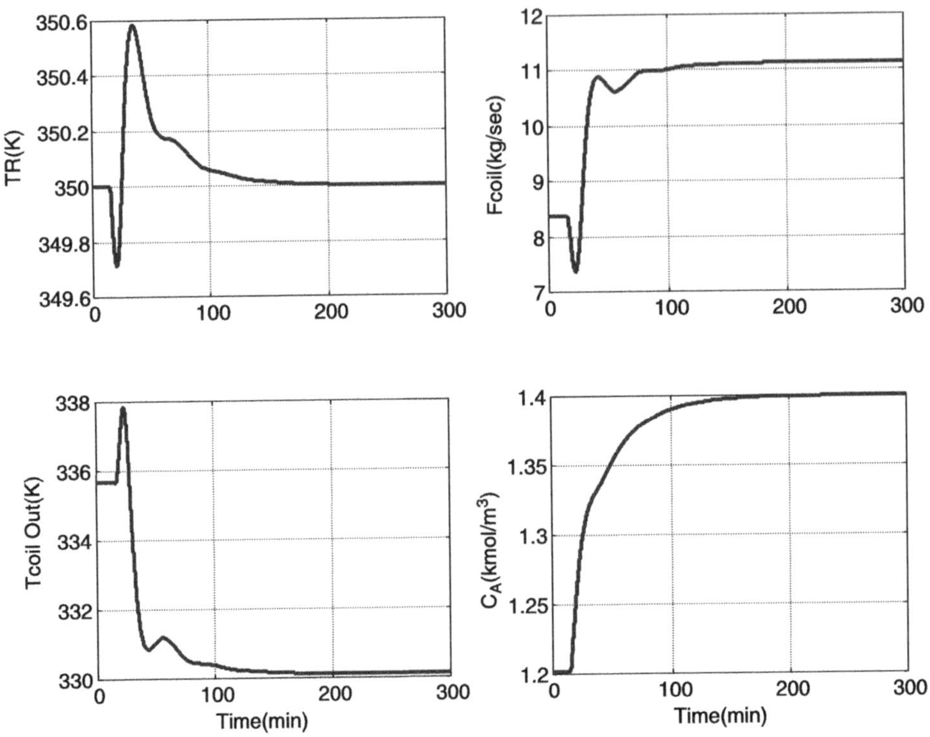

Figure 3.24 Coil cooling; 350 K; +20% feed.

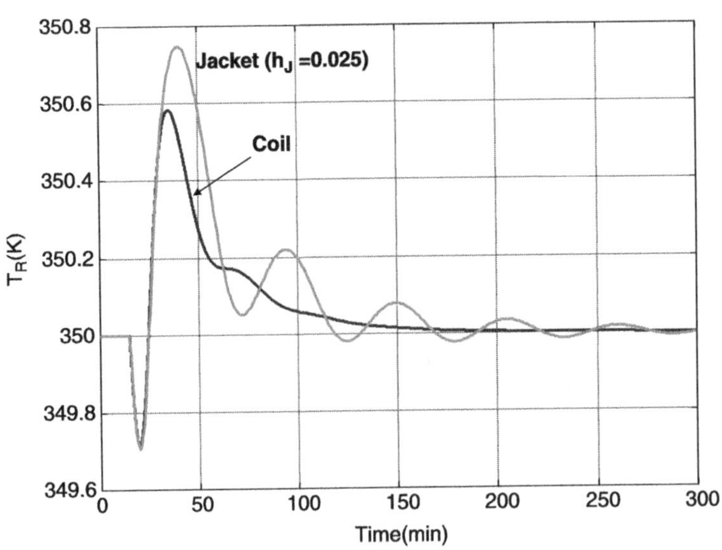

Figure 3.25 Comparison jacket and coil cooling; 350 K; +20% feed.

the volume of the tubes can be calculated:

$$\text{Volume per tube} = \frac{\pi(D_T)^2}{4} L_T = \frac{\pi(0.0254)^2(5)}{4} = 0.00258 \text{ m}^3$$

$$\text{Heat-transfer area per tube} = \pi(D_T)L_T = 0.4026 \text{ m}^2$$

$$\text{Number of tubes} = \frac{A_{\text{HX}}}{0.4026} \tag{3.19}$$

$$\text{Total volume of tubes} = \frac{A_{\text{HX}}}{0.4026}(0.00258)$$

For an area of 58.1 m^2 (the area of a coil), the volume of the tubes is 0.641 m^3. This should be compared with the volume of the coil (1.1 m^3). So the external heat exchanger has less volume (because of the smaller diameter tubes) and should introduce less dynamic lag.

The volume of the liquid in the shell (total shell volume minus tube volume) is typically equal to the tube volume. A circulating cooling water system is assumed, and a high circulation rate of the process liquid is assumed. So the temperature in the shell is T_C, and the temperature in the tubes is T_R. The linear and nonlinear models are the same as for the jacket-cooled CSTR except the volume and area of the heat exchanger are used instead of the jacket volume and area.

Figure 3.26 compares the Nyquist plots for four different cases. The jacket-cooled Nyquist plot is much closer to the critical $(-1,0)$ point. As more area is used in the external heat exchanger, the curves move deeper into the third quadrant, indicating the potential for improved closedloop control. However, the point where they cross the negative real axis moves further to the left. The ultimate gains for the three areas with the external heat exchanger are $K_U = 39.8/24.9/15.1$ (dimensionless) for areas of 45.2/58.1/100 m^2. These results are counter-intuitive since we would expect the controllability to improve with increasing area.

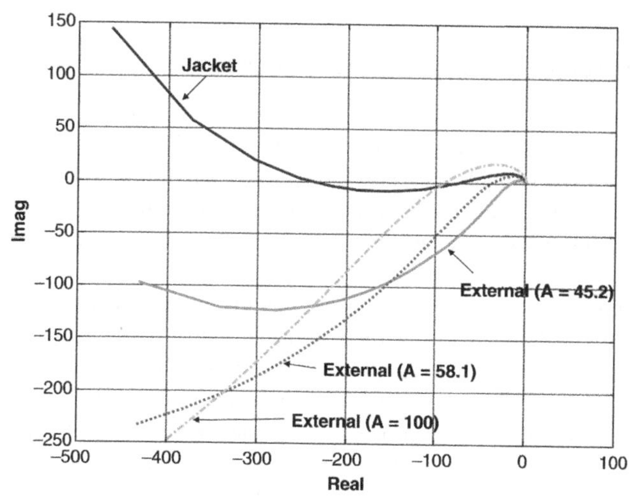

Figure 3.26 Jacket and external HX Nyquist plots.

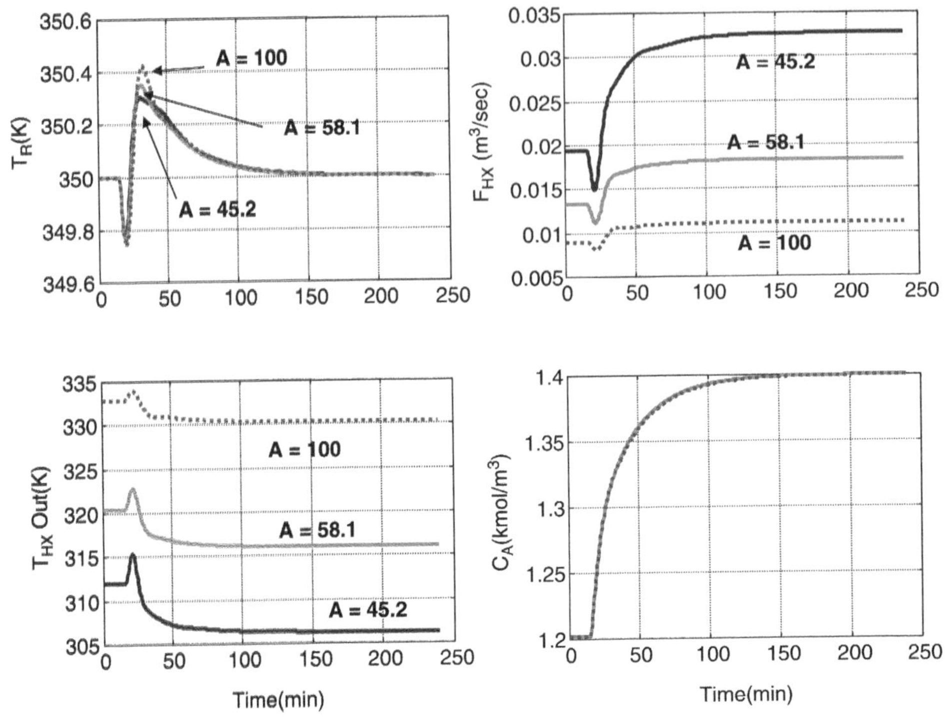

Figure 3.27 External HX; 350 K; +20% feed.

Results of simulations of the external heat exchanger system for a 20% step increase in feed flowrate are shown in Figure 3.27 using the Tyreus–Luyben settings. The higher gain for the 45.2-m^2 case gives a somewhat smaller peak error in temperature. However, if a 40% feed increase is made, as shown in Figure 3.28, the small area process goes unstable. Thus the large area design provides more rangeability.

3.1.8 Comparison of CSTR-in-Series Processes

Let us return to the comparison we made in Chapter 2 of multiple CSTRs in series. In Section 2.6.1 the steady-state designs of 1-, 2-, and 3-CSTR processes were developed, and in Section 2.9.1 the capital costs of the three alternative processes were determined. The steady-state calculations indicate control problems in the first reactor of the 2-CSTR and 3-CSTR processes because the jacket temperatures are getting closer to the supply cooling water temperature.

The 1-CSTR process has a conversion of 98% in the single reactor with a reactant concentration of 0.16 kmol/m^3. The reactor volume is high (262 m^3), and the jacket heat transfer area is large (190 m^2). The resulting jacket temperature is 339 K. Linear analysis gives an ultimate gain of 52.6 (dimensionless) and an ultimate period of 1419 s.

The first reactor in the 2-CSTR process has a conversion of 85.9%, and the reactant concentration in this first reactor is 1.12 kmol/m^3. The reactor volume is much lower (32.6 m^3), and the jacket heat transfer area is reduced to 47.4 m^2. The resulting jacket temperature is much lower (313 K) and approaches the inlet cooling water temperature

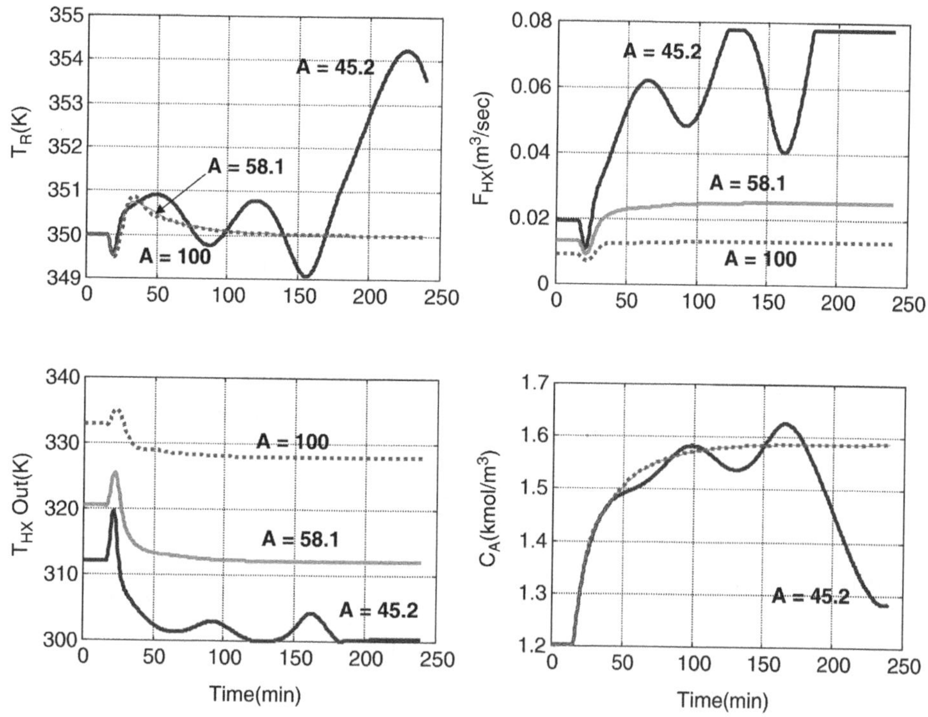

Figure 3.28 External HX; 350 K; +40% feed.

of 294 K. Linear analysis gives an ultimate gain of 9.01 (dimensionless) and an ultimate period of 1232 s.

The first reactor in the 3-CSTR process has a conversion rate of 72.8%, and the reactant concentration in this first reactor is 2.18 kmol/m^3. The reactor volume is low (14.3 m^3), and the jacket heat transfer area is only 24.5 m^2. The resulting jacket temperature (300 K) is almost down to the inlet cooling water temperature of 294 K. Linear analysis gives a Nyquist plot that never drops into the third quadrant, so the critical $(-1,0)$ point cannot be encircled in a counterclockwise direction. This is required for closedloop stability because the openloop system is unstable and has a positive pole. Thus a proportional controller cannot stabilize this first reactor.

Dynamic simulations of these systems verified the expected dynamics. Figure 3.29 gives the responses of the 1-CSTR reactor and the first reactor in the 2-CSTR process to a 20% step increase in feed flowrate. The 2-CSTR reactor is unstable when the Tyreus–Luyben settings are used. For the 1-CSTR process these settings are $K_C = 16.1$ and $\tau_I = 3120$ s. For the first reactor in the 2-CSTR process these settings are $K_C = 2.8$ and $\tau_I = 2710$ s.

To see if the problem in the 2-CSTR process is in the tuning of the controller, several controller settings are explored. Results using gains of 5, 10, and 20 are shown in Figure 3.30 with integral time constant at 2710 s. The conditional stability of this openloop unstable system is clearly shown. For a high gain, the response is very oscillatory with a high frequency. For a low gain, the response again oscillatory but with a low frequency. The Tyreus–Luyben gain is only 2.8, and the system is unstable (Fig. 3.29).

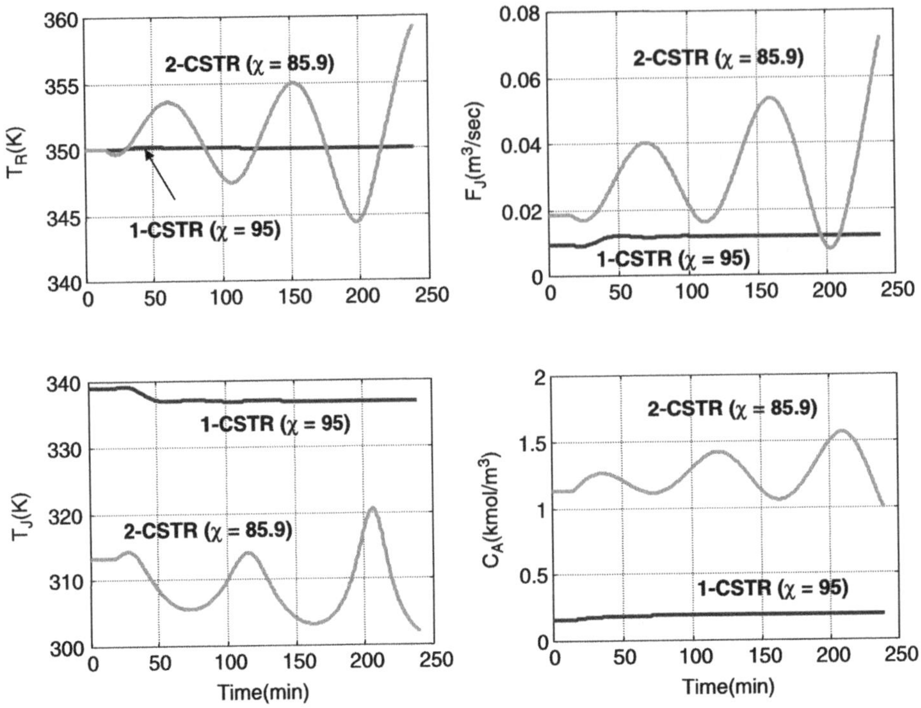

Figure 3.29 1-CSTR and 2-CSTR processes; step 20% increase in feed; 350 K.

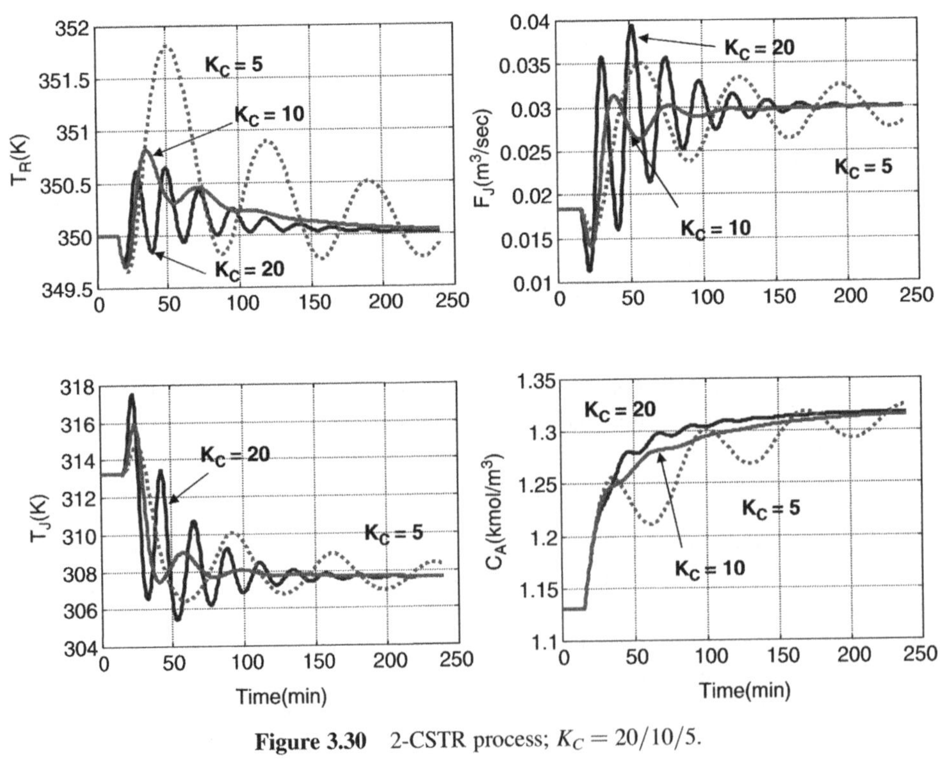

Figure 3.30 2-CSTR process; $K_C = 20/10/5$.

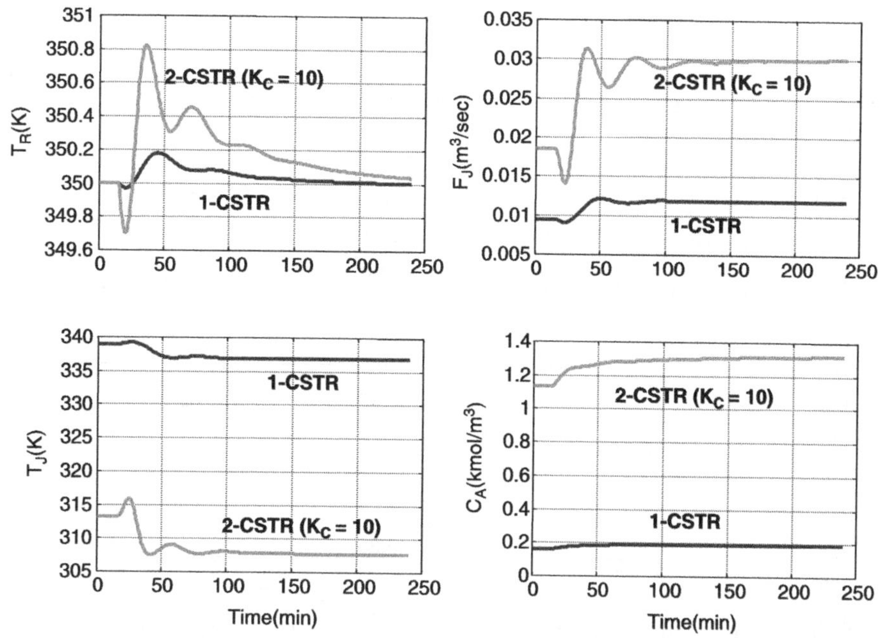

Figure 3.31 1-CSTR and 2-CSTR retuned.

Figure 3.31 compares the 1-CSTR and 2-CSTR processes with the temperature controller in the latter retuned for a gain of 10. Now the 2-CSTR process is stable, but it gives a maximum deviation in temperature that is significantly greater than with the 1-CSTR process. So despite its higher capital investment, the 1-CSTR process is the "best" design. It should provide safe, stable operation in the face of large disturbances.

A 95% conversion design and a 98% conversion design for a single CSTR are compared in Figure 3.32. The improved dynamic controllability of the higher conversion design in clearly demonstrated.

3.1.9 Dynamics of Reactor–Stripper Process

In Chapter 2 (Section 2.9.2) the steady-state design of a reactor–stripper process was studied. Now we investigate the dynamic controllability of this process. The dynamic model of the reactor is the same as Eqs. (3.9)–(3.11) except there is a second stream entering the reactor, the recycle stream D (kmol/s) from the column with composition x_D (mole fraction A). The reactor effluent is F (kmol/s) with composition z (mole fraction A). The reactor component and energy balances are:

$$\frac{dC_A}{dt} = \frac{F_0 C_{A0}}{V_R} + \frac{D x_D}{V_R} - \frac{F z_A}{V_R} - C_A k_0 e^{-E/RT_R} \tag{3.20}$$

$$\frac{dT_R}{dt} = \frac{F_0 T_0}{V_R} + \frac{DMT_D}{\rho V_R} - \frac{FMT_R}{\rho V_R} - \frac{\lambda C_A k_0 e^{-E/RT_R}}{\rho c_p} - \frac{UA_J(T_R - T_J)}{V_R \rho c_p} \tag{3.21}$$

The dynamic model of the stripping column consists of one ordinary differential equation per tray if equimolal overflow, constant liquid holdups on the trays, and instantaneous

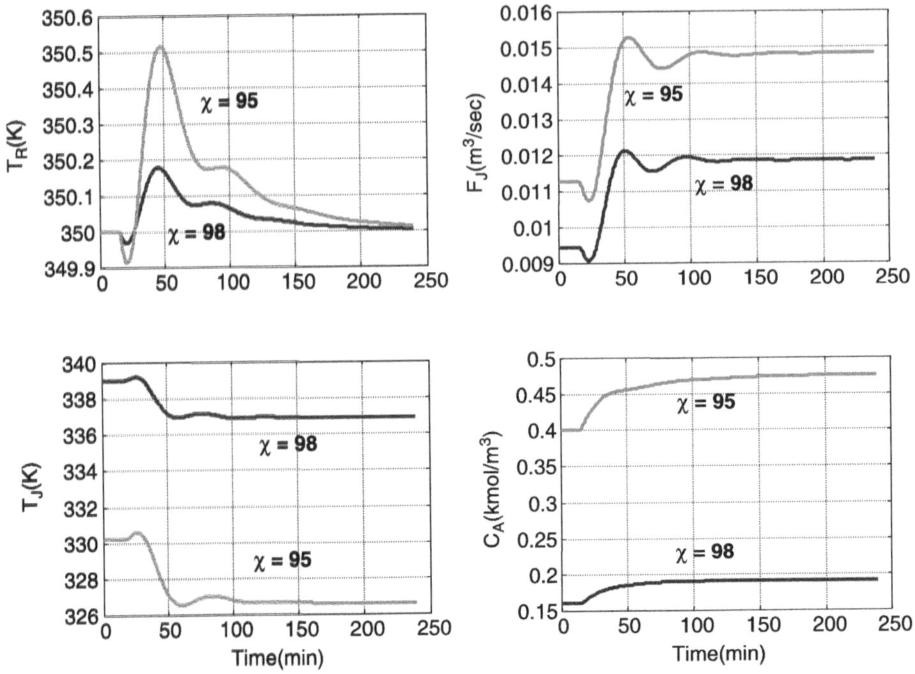

Figure 3.32 $\chi = 95\%$; and 98%; step 20% increase in feed; 330 K.

liquid hydraulics are assumed. Molar flowrates and concentrations in mole fractions are used

$$M_n \frac{dx_n}{dt} = x_{n+1}L_{n+1} + y_{n-1}V - x_nL_n - y_nV \tag{3.22}$$

where the molar liquid holdup on the tray is M_n, the liquid flowrate is L, the vapor flowrate is V, the liquid composition is x, and the vapor composition is y.

Vapor–liquid phase equilibrium is assumed on each tray, so the vapor composition can be calculated from the known relative volatility $\alpha = 1.5$ and liquid composition:

$$y_n = \frac{\alpha x_n}{1 + (\alpha - 1)x_n} \tag{3.23}$$

The molar holdup in the base of the column is M_P and in the overhead accumulator is M_D. These holdups are not constant but vary with time. The total molar and component balances for the base and reflux drum are given below:

Column base:

$$\frac{dM_P}{dt} = L_1 - V - P$$

$$\frac{d(M_P x_P)}{dt} = x_1 L_1 - V y_P - P x_P \tag{3.24}$$

Overhead accumulator:

$$\frac{dM_D}{dt} = V - D$$

$$\frac{d(M_D x_D)}{dt} = V y_{NT} - D x_D \tag{3.25}$$

Remember that the stripping column has no reflux. These levels are controlled by proportional level controllers that manipulate P and D. Tray holdup is 0.3 kmol, and the steady-state holdups in the base and overhead accumulator are each 75 kmol.

```
% Program "simcstrstripper.m"
clear
ca0=8.01;fa0=.004377;k0=20.75e6;e=69.71e6;t0=294;tcin=294;u=851;taum=60;uss=u;
lambda=-69.71e6;roe=801;m=100;cp=3137;cj=4183;roej=1000;td=322;
% Set reactor temperature, volume and overall conversion of A
tr=350;trss=350;vr=60;conversion=.98;k=k0*exp(-e/tr/8314);f0=fa0*ca0;f0ss=f0;z0=1;alpha=1.5;
dr=(2*vr/pi)^0.3333;areaj=2*pi*dr^2;mpss=75;mdss=75;mn=0.5;nt=10;vj=0.1*areaj;
%  Intial conditions
load icreactorstripper60 xp xd x z d p v tr tj fj
dss=d;pss=p;f=p+d;v=d;vss=v;mpxp=mpss*xp;mdxd=xd*mdss;md=mdss;mp=mpss;
trlag1=trss;trlag2=trss;xplag1=.02;xplag2=.02;erint=0;erintxp=0;tjss=tj;fjss=fj;
% Controller settings
kc=22/4;taui=3500;kcxp=5.7;tauixp=20*60;
 tstop=5*3600;np=0;erint=0;tplot=0;time=0;delta=1;sp=350;f0new=f0ss*.8;
% Integration loop
while time<tstop
   k=k0*exp(-e/tr/8314);   f=d+f0;   ca=8.01*z;rate=vr*ca*k;  qj=u*areaj*(tr-tj);
   yp=alpha*xp/(1+(alpha-1)*xp);for n=1:nt;y(n)=alpha*x(n)/(1+(alpha-1)*x(n));end
   % Level Controllers
   f=d+f0;d=dss*md/mdss;p=pss*mp/mpss;
% Temperature Controller
error=(sp-trlag2)/50;op=0.25-kc*error-erint;if op>1; op=1;end; if op<0;op=0;end;fj=fjss*op*4;
% xp composition controller
errorxp=(0.02-xplag2)/0.1;opxp=0.5-kcxp*errorxp-erintxp;if opxp>1; opxp=1;end; if opxp<0;opxp=0;end
v=vss*2*opxp;
% Disturbance
if time>900;f0=f0new;end
% store data for plotting
if time>=tplot;   np=np+1;
   timep(np)=time/60;trp(np)=tr;tjp(np)=tj;fjp(np)=fj;zp(np)=z;xplag2p(np)=xplag2;
   vp(np)=v;pp(np)=p;vp(np)=v;xpp(np)=xp;xdp(np)=xd;trlag2p(np)=trlag2; tplot=tplot+30;end
% Derivative evaluations
dca=fa0*ca0/vr +d*xd/vr -f*z/vr -k*ca;dtj=fj*(tcin-tj)/vj +qj/(cj*roej*vj);
dtr=f0*m*t0/(roe*vr)+d*td*m/(roe*vr)-f*m*tr/(roe*vr) -lambda*k*ca/roe/cp - qj/(cp*roe*vr);
dtrlag1=(tr-trlag1)/taum;dtrlag2=(trlag1-trlag2)/taum;
dxplag1=(xp-xplag1)/taum;dxplag2=(xplag1-xplag2)/taum;
dmpxp=f*x(1)-v*yp-p*xp; dmdxd=v*y(nt)-d*xd;dmp=f-v-p;dmd=v-d;
dx(1)=(f*x(2)+v*yp-f*x(1)-v*y(1))/mn;
for n=2:nt-1;dx(n)=(f*x(n+1)+v*y(n-1)-f*x(n)-v*y(n))/mn;end
dx(nt)=f*z+v*y(nt-1)-f*x(nt)-v*y(nt);
%  Integration
time=time+delta;ca=ca+dca*delta;z=ca/8.01;
tr=tr+dtr*delta;tj=tj+dtj*delta;if tr>500;break;end
trlag1=trlag1+dtrlag1*delta;trlag2=trlag2+dtrlag2*delta;
xplag1=xplag1+dxplag1*delta;xplag2=xplag2+dxplag2*delta;
mpxp=mpxp+dmpxp*delta;mdxd=mdxd+dmdxd*delta;
mp=mp+dmp*delta;md=md+dmd*delta;  xp=mpxp/mp;xd=mdxd/md;
for n=1:nt;x(n)=x(n)+dx(n)*delta;end
% Anti-reset windup
if op<1;if op>0;erint=erint+error*kc*delta/taui;end;end
if opxp<1;if opxp>0;erintxp=erintxp+errorxp*kcxp*delta/tauixp;end;end
end
%   save icreactorstripper60 xp xd x z d p v tr tj fj
% plotting
clf
subplot(4,2,1);plot(timep,trp);grid;ylabel('TR (K)');
title('Reactor/Stripper; VR=60; 350K; 98 Conversion; -20% Feed')
subplot(4,2,3);plot(timep,tjp);grid;ylabel('TJ (K)');
subplot(4,2,2);plot(timep,fjp*roej);grid;ylabel('FJ (kg/sec)');
subplot(4,2,4);plot(timep,zp);grid;ylabel('z (mf)');
subplot(4,2,5);plot(timep,xdp);grid;ylabel('xD (mf)');
subplot(4,2,6);plot(timep,pp);grid;ylabel('P (kmol/sec)');
subplot(4,2,7);plot(timep,vp);grid;ylabel('V (kmol/sec)');xlabel('Time (min)');
subplot(4,2,8);plot(timep,xpp);grid;ylabel('xP (mf)');xlabel('Time (min)');
```

Figure 3.33 Matlab program for CSTR–stripper.

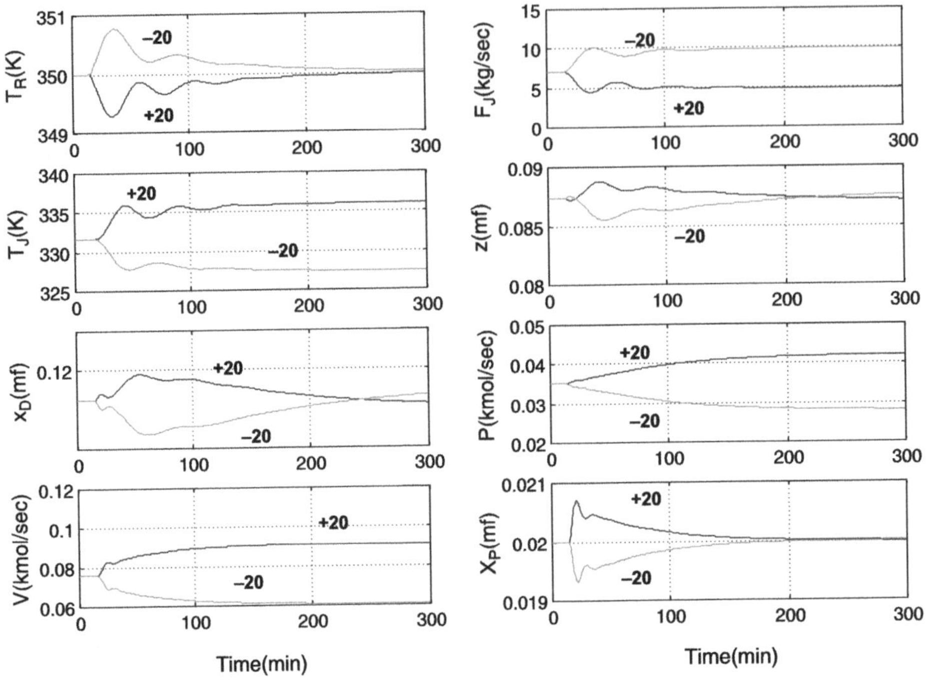

Figure 3.34 CSTR–stripper; feedrate disturbances.

Figure 3.33 gives a Matlab program that simulates this reactor–stripper process. Relay–feedback tests are run on the reactor temperature controller, which manipulates jacket cooling water flow, and on the bottoms composition controller, which manipulates vapor boilup. Two 60-s lags are used in both loops. Tyreus–Luyben tuning is used except the controller gain is reduced by a factor of 2 to give reasonable damping in the reactor

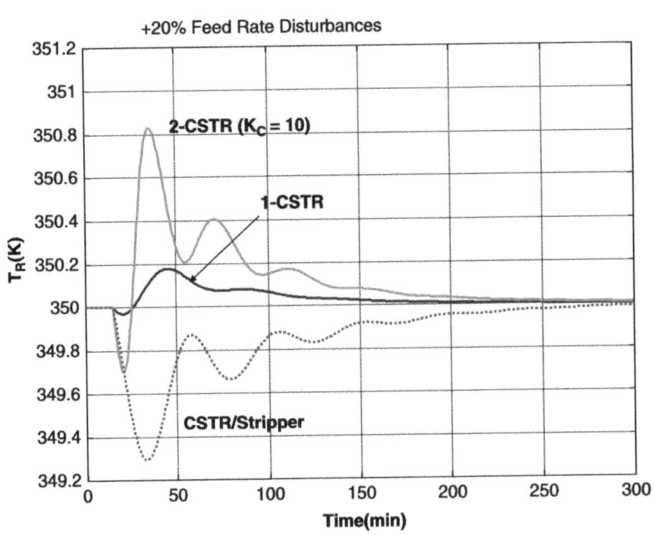

Figure 3.35 Comparison of CSTR–stripper with 1-CSTR and 2-CSTR; +20% feedrate disturbances.

temperature loop. The reactor temperature controller settings are $K_C = 11$ (using a 50 K temperature transmitter span and with the maximum cooling water flow 4 times the design) and $\tau_I = 3500$ s. The product composition controller settings are $K_C = 5.7$ (using a 0.1 mole fraction transmitter span and with the maximum vapor boilup twice the design) and $\tau_I = 1200$ s.

Figure 3.34 gives the responses of the process to 20% step increases and decreases in fresh feed flowrate. Good regulatory control is obtained. Figure 3.35 compares the reactor temperatures for the three alternative processes. The 1-CSTR process gives the best temperature control, but the less expensive reactor–stripper temperature control appears to be quite adequate.

In the next section, we will study a two-reactant case in which the adjustment of the fresh feeds to balance the stoichiometry of the reaction is crucial from a plantwide perspective. In addition, the "limiting reactant" concept can be used to improve dynamic controllability.

3.2 REACTOR–COLUMN PROCESS WITH TWO REACTANTS

The dynamics and control of the reactor–column system studied in Chapter 2 (Section 2.9.3) are investigated in this section. Mathematical models of both the reactor and the column are developed, and a plantwide control structure is evaluated.

The reactor is a CSTR with jacket cooling in which a first-order irreversible reaction takes place:

$$\mathrm{A} + \mathrm{B} \rightarrow \mathrm{C} \tag{3.26}$$

The reaction rate is, of course, the same as that used in the steady-state model:

$$\mathcal{R} = kC_\mathrm{A}C_\mathrm{B} = C_\mathrm{A}C_\mathrm{B}k_0 e^{-E/RT_R} \tag{3.27}$$

3.2.1 Nonlinear Dynamic Model of Reactor and Column

We assume constant densities and constant reactor and jacket volumes. The dynamic model of the reactor and jacket consists of four nonlinear ordinary differential equations:

$$\frac{dC_\mathrm{A}}{dt} = \frac{F_{\mathrm{A}0}C_{\mathrm{A}0}}{V_R} + \frac{Dx_{D\mathrm{A}}}{V_R} - \frac{Fz_\mathrm{A}}{V_R} - C_\mathrm{A}C_\mathrm{B}k_0 e^{-E/RT_R} \tag{3.28}$$

$$\frac{dC_\mathrm{B}}{dt} = \frac{F_{\mathrm{B}0}C_{\mathrm{B}0}}{V_R} + \frac{Dx_{D\mathrm{B}}}{V_R} - \frac{Fz_\mathrm{B}}{V_R} - C_\mathrm{A}C_\mathrm{B}k_0 e^{-E/RT_R} \tag{3.29}$$

$$\frac{dT_R}{dt} = \frac{F_{\mathrm{A}0}T_0}{V_R} + \frac{D_\mathrm{vol}T_0}{V_R} - \frac{F_\mathrm{vol}T_R}{V_R} - \frac{\lambda C_\mathrm{A}C_\mathrm{B}k_0 e^{-E/RT_R}}{\rho c_p} - \frac{UA_J(T_R - T_J)}{V_R\rho c_p} \tag{3.30}$$

$$\frac{dT_J}{dt} = \frac{F_J}{V_J}(T_\mathrm{Cin} - T_J) + \frac{UA_J(T_R - T_J)}{V_J\rho_J c_J} \tag{3.31}$$

where D_vol and F_vol are the volumetric flowrates ($\mathrm{m^3/s}$) of the recycle distillate stream and the reactor effluent, respectively. These are calculated from the molar flowrates and the average molecular weights of the streams. Density is constant at 801 $\mathrm{kg/m^3}$. These four nonlinear ordinary differential equations will be used to simulate the dynamic performance of the CSTR portion of the plant.

The dynamic model of the column consists of two ordinary differential equations per tray if equimolal overflow, constant tray holdup, and instantaneous liquid hydraulics are assumed. Molar flowrates and concentrations in mole fractions are used. The liquid holdup on each tray is 0.4 kmol:

$$
\begin{aligned}
M_n \frac{dx_{nA}}{dt} &= x_{n+1,\mathrm{A}} L_{n+1} + y_{n-1,\mathrm{A}} V - x_{n\mathrm{A}} L_n - y_{n\mathrm{A}} V \\
M_n \frac{dx_{nB}}{dt} &= x_{n+1,\mathrm{B}} L_{n+1} + y_{n-1,\mathrm{B}} V - x_{n\mathrm{B}} L_n - y_{n\mathrm{B}} V
\end{aligned}
\tag{3.32}
$$

Vapor–liquid phase equilibrium is assumed on each tray, so the vapor composition can be calculated from the known relative volatilities and liquid composition. The relative volatilities are $\alpha_{\mathrm{A}} = 3$, $\alpha_{\mathrm{B}} = 1.5$, and $\alpha_{\mathrm{C}} = 1$:

$$
y_{nj} = \frac{\alpha_j x_{nj}}{\alpha_{\mathrm{A}} x_{n\mathrm{A}} + \alpha_{\mathrm{B}} x_{n\mathrm{B}} + \alpha_{\mathrm{C}} x_{n\mathrm{C}}} \tag{3.33}
$$

where $j =$ A,B,C. Holdups and compositions in the base and in the reflux drum are given by the following equations:

Column base:

$$
\begin{aligned}
\frac{dM_P}{dt} &= L_1 - V - P \\
\frac{d(M_P x_{P\mathrm{A}})}{dt} &= x_{1\mathrm{A}} L_1 - y_{P\mathrm{A}} V - P x_{P\mathrm{A}} \\
\frac{d(M_P x_{P\mathrm{B}})}{dt} &= x_{1\mathrm{B}} L_1 - y_{P\mathrm{B}} V - P x_{P\mathrm{B}}
\end{aligned}
\tag{3.34}
$$

Reflux drum:

$$
\begin{aligned}
\frac{dM_D}{dt} &= V - D - R \\
\frac{d(M_D x_{D\mathrm{A}})}{dt} &= V y_{NT,\mathrm{A}} - (R + D) x_{D\mathrm{A}} \\
\frac{d(M_D x_{D\mathrm{B}})}{dt} &= V y_{NT,\mathrm{B}} - (R + D) x_{D\mathrm{B}}
\end{aligned}
\tag{3.35}
$$

The liquid holdup in the base M_P is 40 kmol and in the reflux drum, 40 kmol. Saturated liquid feed and reflux are assumed, so the liquid flowrate in the rectifying section is the reflux R and in the stripping section is $R + F$.

Figure 3.36 gives a Matlab program for the dynamic simulation of this system. The converged steady-state conditions are given in Figure 3.37. Note that the flows and compositions are slightly different from those used in the approximate design given in Chapter 2 (see Fig. 2.60). The dynamic model is rigorous, while the steady-state design made some assumptions such as twice the minimum number of trays, 1.2 times the minimum reflux ratio, and approximate stream compositions leaving the column.

3.2.2 Control Structure for Reactor–Column Process

The control scheme developed for this system is shown in Figure 3.38. It contains the following loops:

1. The reactor effluent F is flow-controlled. This is the production rate handle.
2. Reactor level is controlled by manipulating the fresh feedstream F_{A0}. This is selected because the concentration of A in the reactor is large ($C_A = 1.706$ kmol/m^3). Perfect level control is assumed, so the flowrate F_{A0} is calculated by subtracting the volumetric flowrates of F_{B0} and D_{vol} from F_{vol}.
3. The concentration of B in the reactor C_B is controlled by manipulating the fresh feedstream F_{B0}. This is selected because the concentration of B in the reactor is small ($C_B = 0.0844$ kmol/m^3).
4. Reactor temperature T_R is controlled by manipulating jacket cooling water flow F_J.

```
% Program "simreactorcolumn.m"
clear
ca0=8.01;e=69.71e6;tcin=294;u=851;lambda=-69.71e6;cb0=8.01;roe=801;ma=100;mb=100;mc=200;
cp=3137;cj=4183;roej=1000;t0=294;fa0=0.004377;fb0=fa0;tr=333;k0=2.1045e+008;vj=33.38;vr=100;
aj=100.15;alphaa=3;alphab=1.5;alphac=1;nt=18;nf=9;taum=60;
% Initial Conditions
load ic1 mp xpa xpb xpc p md xda xdb xdc d r v  f
load ic2 m xa xb xc
load ic3 tr ca cb cc tj fj fa0 fb0
md=40;mp=40;fss=0.056;sum=ca+cb+cc;za=ca/sum;zb=cb/sum;zc=cc/sum;
mdxda=md*xda;mdxdb=md*xdb;mdxdc=md*xdc;mpxpa=mp*xpa;mpxpb=mp*xpb;mpxpc=mp*xpc;
for n=1:nt;mxa(n)=m(n)*xa(n);mxb(n)=m(n)*xb(n);mxc(n)=m(n)*xc(n);end
time=0;delta=1;tstop=5*3600;tplot=0;np=0;
%  Controller parameters
mdss=md;mpss=mp;dss=d;pss=p;
vss=v;spxp=0.98;erintxp=0;kcxp=0.33;tauixp=1410;xpclag1=0.98;xpclag2=0.98;
fb0ss=fb0;spcb=0.0844;erintcb=0;kccb=1.33;tauicb=673;cblag1=spcb;cblag2=cblag1;
fjss=fj;sptr=333;erinttr=0;kctr=40/4;tauitr=4950;trlag1=333;trlag2=333;
% Integration loop
while time<tstop
% Column Level controllers
   d=dss*md/mdss;p=pss*mp/mpss;
% Reactor composition cb held by fb0
   spcb=0.0844;errorcb=(spcb-cblag2)/.2;opcb=0.5+kccb*errorcb+erintcb;
   if opcb>1;opcb=1;end;if opcb<0;opcb=0;end;
   % Relay-feedback test%opcb=0.5*1.1;if errorcb<0;opcb=0.5*0.9;end
   fb0=fb0ss*2*opcb;
% Reactor level held by fa0 with f fixed
% Disturbance
   f=fss;if time>900;f=fss*1.2;end
   mwd=xda*100+xdb*100+xdc*200;dvol=d*mwd/801;mwf=za*100+zb*100+zc*200;fvol=f*mwf/801;fa0=fvol-dvol-fb0;
% Reactor TC
   errortr=(sptr-trlag2)/50;optr=0.25-kctr*errortr-erinttr;if optr>1;optr=1;end;if optr<0;optr=0;end;
% Relay feedback test   %optr=0.25*1.1;if errortr>0;optr=0.25*0.9;end
 fj=fjss*4*optr;
% composition controller for xpc
   errorxp=(spxp-xpclag2)/0.1;opxp=0.5+kcxp*errorxp+erintxp;if opxp>1;opxp=1;end;if opxp<0;opxp=0;end
   % Relay-feedback test   %opxp=0.5*1.1;if errorxp<0;opxp=0.5*0.9;end
   v=vss*2*opxp;
   %  r is ratioed to f
   r=.039*f/fss;   ls=f+r;lr=r;
   % VLE calculations
   sum=xpa*alphaa+xpb*alphab+xpc*alphac;ypa=alphaa*xpa/sum;ypb=alphab*xpb/sum;ypc=alphac*xpc/sum;
   for n=1:nt;      sum=xa(n)*alphaa+xb(n)*alphab+xc(n)*alphac;
      ya(n)=alphaa*xa(n)/sum; yb(n)=alphab*xb(n)/sum; yc(n)=alphac*xc(n)/sum;  end
% Reactor
   k=k0*exp(-e/tr/8314);rate=k*vr*ca*cb;qj=u*aj*(tr-tj);qj=u*aj*(tr-tj);
% Save for plotting
   if time>=tplot;np=np+1;timeplot(np)=time/60;xdaplot(np)=xda;xdbplot(np)=xdb;xdcplot(np)=xdc;
      xpaplot(np)=xpa;xpbplot(np)=xpb;xpcplot(np)=xpc;vplot(np)=v;dplot(np)=d;pplot(np)=p;
      trplot(np)=tr;zaplot(np)=za;zbplot(np)=zb;tjplot(np)=tj;fjplot(np)=fj;
      fa0plot(np)=fa0;fb0plot(np)=fb0;cbplot(np)=cb;caplot(np)=ca;trlag2plot(np)=trlag2;
      errorxpplot(np)=errorxp;errorcbplot(np)=errorcb;    tplot=tplot+10;end
```

Figure 3.36 Matlab program for reactor–column simulation.

```
% Evaluate derivatives
  dca=(fa0*ca0+d*xda-f*za-rate)/vr;dcb=(fb0*cb0+d*xdb-f*zb-rate)/vr;
  dcc=(d*xdc-f*zc+rate)/vr; dtj=fj*(tcin-tj)/vj + qj/(roej*cj*vj);
  dtr=(fa0+fb0+dvol)*t0/vr - fvol*tr/vr -lambda*rate/(vr*roe*cp) -qj/(vr*roe*cp);
  dmpxpa=ls*xa(1)-p*xpa-v*ypa;   dmpxpb=ls*xb(1)-p*xpb-v*ypb;   dmpxpc=ls*xc(1)-p*xpc-v*ypc;
  dmxa(1)=ls*xa(2)+v*ypa-ls*xa(1)-v*ya(1);
  dmxb(1)=ls*xb(2)+v*ypb-ls*xb(1)-v*yb(1);
  dmxc(1)=ls*xc(2)+v*ypc-ls*xc(1)-v*yc(1);
  for n=2:nf-1;dmxa(n)=ls*xa(n+1)+v*ya(n-1)-ls*xa(n)-v*ya(n);
               dmxb(n)=ls*xb(n+1)+v*yb(n-1)-ls*xb(n)-v*yb(n);
               dmxc(n)=ls*xc(n+1)+v*yc(n-1)-ls*xc(n)-v*yc(n); end
   dmxa(nf)=lr*xa(nf+1)+v*ya(nf-1)-ls*xa(nf)-v*ya(nf)+f*za;
   dmxb(nf)=lr*xb(nf+1)+v*yb(nf-1)-ls*xb(nf)-v*yb(nf)+f*zb;
   dmxc(nf)=lr*xc(nf+1)+v*yc(nf-1)-ls*xc(nf)-v*yc(nf)+f*zc;
  for n=nf+1:nt-1;dmxa(n)=lr*xa(n+1)+v*ya(n-1)-lr*xa(n)-v*ya(n);
                  dmxb(n)=lr*xb(n+1)+v*yb(n-1)-lr*xb(n)-v*yb(n);
                  dmxc(n)=lr*xc(n+1)+v*yc(n-1)-lr*xc(n)-v*yc(n);end
      dmxa(nt)=r*xda+v*ya(nt-1)-lr*xa(nt)-v*ya(nt);
      dmxb(nt)=r*xdb+v*yb(nt-1)-lr*xb(nt)-v*yb(nt);
      dmxc(nt)=r*xdc+v*yc(nt-1)-lr*xc(nt)-v*yc(nt);
     dmdxda=v*ya(nt)-(r+d)*xda; dmdxdb=v*yb(nt)-(r+d)*xdb; dmdxdc=v*yc(nt)-(r+d)*xdc;
dtrlag1=(tr-trlag1)/taum;dtrlag2=(trlag1-trlag2)/taum;
dxpclag1=(xpc-xpclag1)/taum;dxpclag2=(xpclag1-xpclag2)/taum;
dcblag1=(cb-cblag1)/taum;dcblag2=(cblag1-cblag2)/taum;
|
% Integration
 time=time+delta;tr=tr+dtr*delta;tj=tj+dtj*delta;ca=ca+dca*delta;cb=cb+dcb*delta;cc=cc+dcc*delta;
 sum=ca+cb+cc;za=ca/sum;zb=cb/sum;zc=cc/sum;
 mpxpa=mpxpa+dmpxpa*delta; mpxpb=mpxpb+dmpxpb*delta; mpxpc=mpxpc+dmpxpc*delta;
 mp=mpxpa+mpxpb+mpxpc; xpa=mpxpa/mp; xpb=mpxpb/mp; xpc=mpxpc/mp;
for n=1:nt;   mxa(n)=mxa(n)+dmxa(n)*delta;   mxb(n)=mxb(n)+dmxb(n)*delta;
   mxc(n)=mxc(n)+dmxc(n)*delta;   m(n)=mxa(n)+mxb(n)+mxc(n);
   xa(n)=mxa(n)/m(n);   xb(n)=mxb(n)/m(n);   xc(n)=mxc(n)/m(n);end
mdxda=mdxda+dmdxda*delta;mdxdb=mdxdb+dmdxdb*delta;mdxdc=mdxdc+dmdxdc*delta;
md=mdxda+mdxdb+mdxdc;xda=mdxda/md;xdb=mdxdb/md;xdc=mdxdc/md;
erintxp=erintxp+errorxp*delta*kcxp/tauixp; erinttr=erinttr+errortr*delta*kctr/tauitr;
erintcb=erintcb+errorcb*delta*kccb/tauicb;
trlag1=trlag1+dtrlag1*delta;trlag2=trlag2+dtrlag2*delta;
xpclag1=xpclag1+dxpclag1*delta;xpclag2=xpclag2+dxpclag2*delta;
cblag1=cblag1+dcblag1*delta;cblag2=cblag2+dcblag2*delta;
end
flagsaveic=-1;if flagsaveic>0
save ic1 mp xpa xpb xpc p md xda xdb xdc d  r v f
save ic2 m xa xb xc
save ic3 tr ca cb cc tj fj foa fob
end
clf
subplot(3,2,1);plot(timeplot,trplot);grid;ylabel('TR');title('Reactor/Column; VR=100; +20 F; R/F');
subplot(3,2,2);plot(timeplot,fjplot);grid;ylabel('FJ');
subplot(3,2,3);plot(timeplot,cbplot);grid;ylabel('CB');
subplot(3,2,4);plot(timeplot,fb0plot*100);grid;ylabel('FB0');
subplot(3,2,5);plot(timeplot,caplot);grid;ylabel('CA');xlabel('Time (min)');
subplot(3,2,6);plot(timeplot,fa0plot*100);grid;ylabel('FA0');xlabel('Time (min)');
pause
clf
subplot(3,2,1);plot(timeplot,xpcplot);grid;ylabel('xPC');
subplot(3,2,2);plot(timeplot,vplot);grid;ylabel('V');
subplot(3,2,3);plot(timeplot,dplot);grid;ylabel('D');
subplot(3,2,4);plot(timeplot,xdaplot);grid;ylabel('xDA');
subplot(3,2,5);plot(timeplot,xdcplot);grid;ylabel('xDC');xlabel('Time (min)');
subplot(3,2,6);plot(timeplot,pplot);grid;ylabel('P');xlabel('Time (min)');
```

Figure 3.36 *Continued.*

5. The purity of C in the bottoms product stream x_{PC} is controlled by manipulating vapor boilup in the column V.
6. Reflux flowrate R is ratioed to the column feed F.
7. Column base level is controlled by manipulating bottoms P.
8. Reflux drum level is controlled by manipulating distillate D.

The level controllers on the column base and reflux drum are proportional. The PI controllers for reactor temperature T_R, reactor C_B composition, and bottoms product

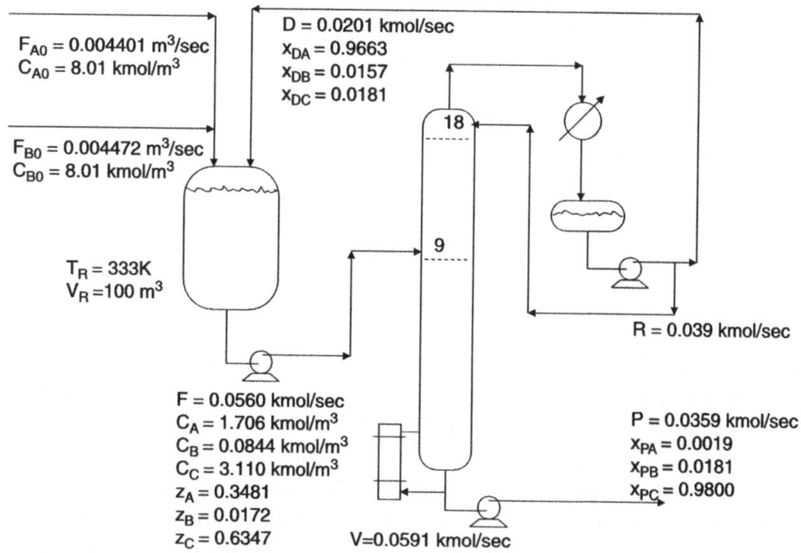

Figure 3.37 Reactor–column steady-state conditions; two reactants, $V_R = 100\ m^3$.

composition x_{PC} are tuned by running relay–feedback tests to get the ultimate gains and ultimate frequencies. Two 1-min lags are included in each of these three loops. Tyreus–Luyben tuning is used in all loops, except the reactor temperature controller gain is reduced by a factor of 4 to give reasonable closedloop damping.

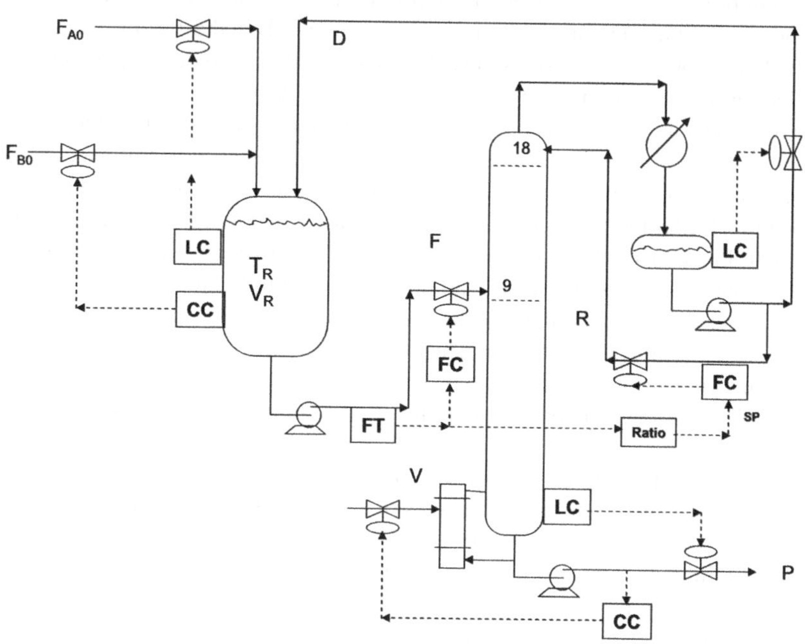

Figure 3.38 Reactor–column plantwide control structure.

TABLE 3.3 Controller Tuning for Reactor–Column Process

Design value	Reactor temperature 333 K	Reactor composition 0.0844 kmol B/m^3	Product composition 0.98 mole fraction C
Transmitter span	50 K	0.20 kmol/m^3	0.1 mole fraction
Manipulated variable	F_J	F_{B0}	V
Design value	0.0136 m^3/s	0.004472 m^3/s	0.0591 kmol/s
Maximum flow	4 times design	Twice design	Twice design
K_C	10	1.3	0.33
τ_I (s)	4950	670	1410

Figure 3.39 gives the response of the system to a step 20% increase in the flowrate of the reactor effluent. The control structure provides good base-level regulatory control. The maximum deviation in reactor temperature is 0.6 K. Three cases are shown. In the first, the reflux flowrate is held constant. In the second, the reflux is ratioed to the feed. There is little difference in the responses of the reactor. But with a fixed reflux flowrate, the impurity of C in the distillate x_{DC} increases from 0.0164 to about 0.025 mole fraction C. With the reflux-to-feed ratio, the impurity remains about the same. The change in the vapor boilup is larger with the reflux-to-feed structure.

The purity of the product x_{PC} undergoes a sharp drop when the ratio structure is used because of the step change in the reflux. This performance is improved by inserting a dynamic lag in the ratio loop, as the dotted curves in Figure 3.39 illustrate.

3.2.3 Reactor–Column Process with Hot Reaction

The process with two reactants offers a design and control degree of freedom that can be utilized to improve dynamic controllability if required. This degree of freedom is the concentration of the "limiting reactant." The concentration of B in the 100-m^3 reactor process considered in the previous section is only 0.0844 kmol/m^3 compared to the concentration of A, which is 1.706 kmol/m^3, so B is the limiting reactant.

With the moderate activation energy (69.71 $\times$ 10^6 J/kmol) used in the study, the temperature dependence of the reaction is not excessive. As we observed above, the reactor could be controlled by manipulating jacket cooling water. However, if the activation energy is high, the temperature dependence is increased, and it may be difficult or impossible to control reactor temperature by manipulating the flowrate of jacket cooling water.

To illustrate this problem, the activation energy is doubled to 139.42 $\times$ 10^6 J/kmol. The preexponential factor is adjusted to give the same specific reaction rate at 333 K ($k_0 = 1.8126 \times 10^{19}$ m^3 s^{-1} kmol^{-1}). The steady-state design is, therefore, not changed. However, the dynamics are changed.

Figure 3.40 shows what happens when the control structure used in previous section is used. The disturbance is a step 20% change in reactor effluent flow. The temperature response is very oscillatory with a reactor controller gain of 10. Figure 3.41 shows that the system goes unstable if the gain is reduced to 5. Figure 3.42 shows that the oscillations become larger if the gain is increased to 20. These results indicate that the process cannot be stabilized using this control structure. Remember that the fresh feed of B is manipulated in this control scheme to control the C_B reactor concentration.

Let us change the control structure to one in which the fresh feedstream F_{B0} is simply flow-controlled. This means that if an increase in temperature produces a large increase in

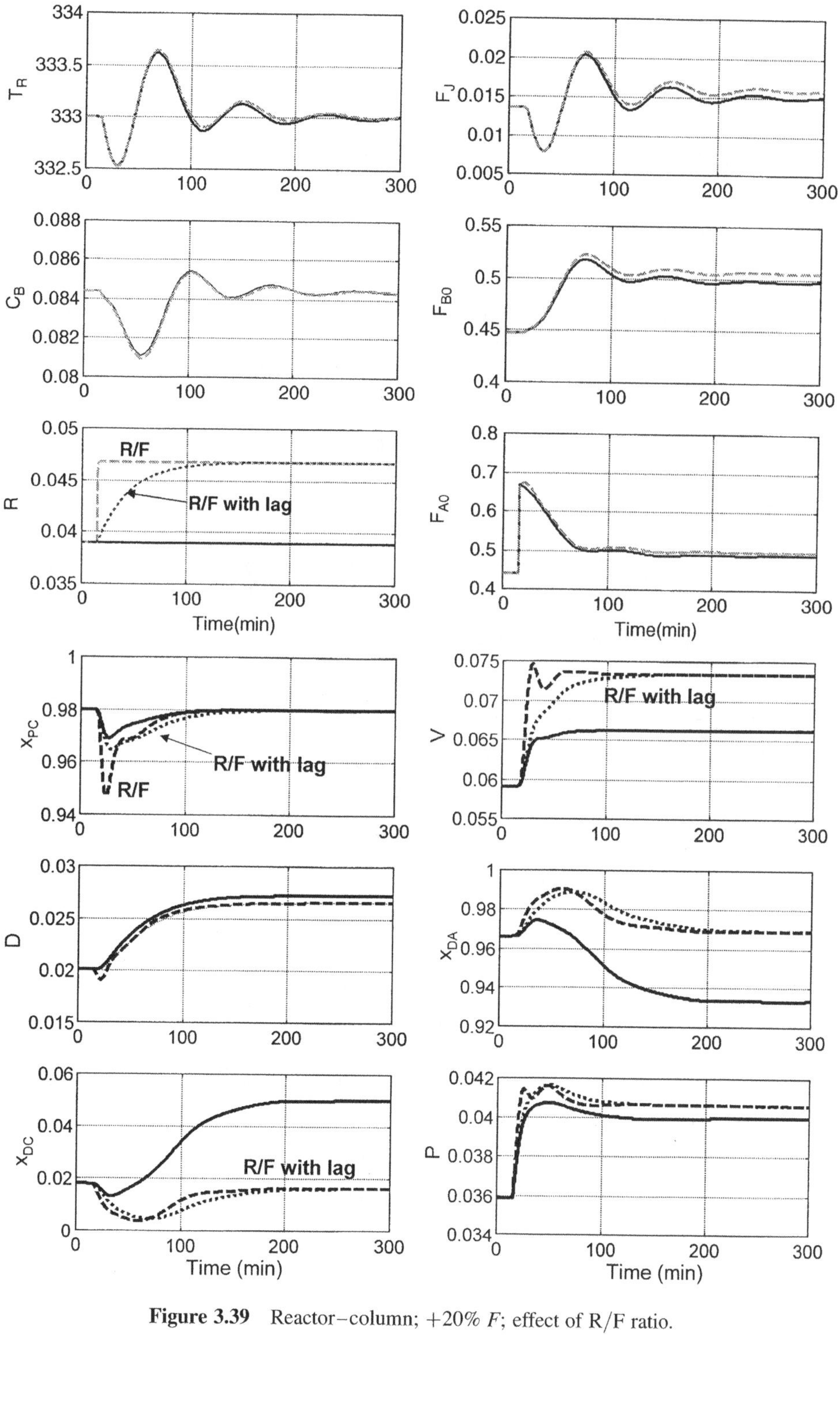

Figure 3.39 Reactor–column; +20% F; effect of R/F ratio.

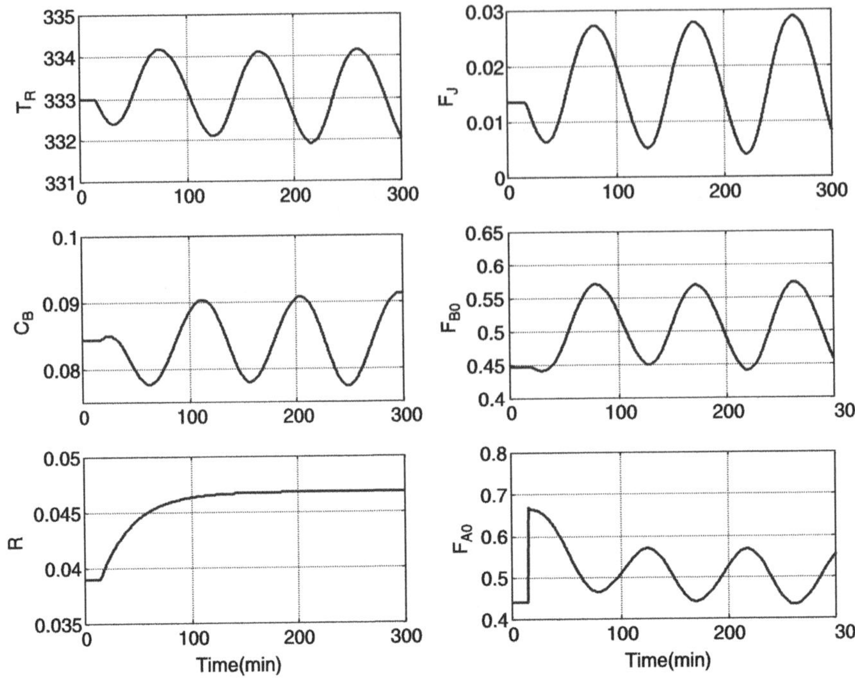

Figure 3.40 Hot reaction; T_R ($K_C = 10$) and C_B on automatic.

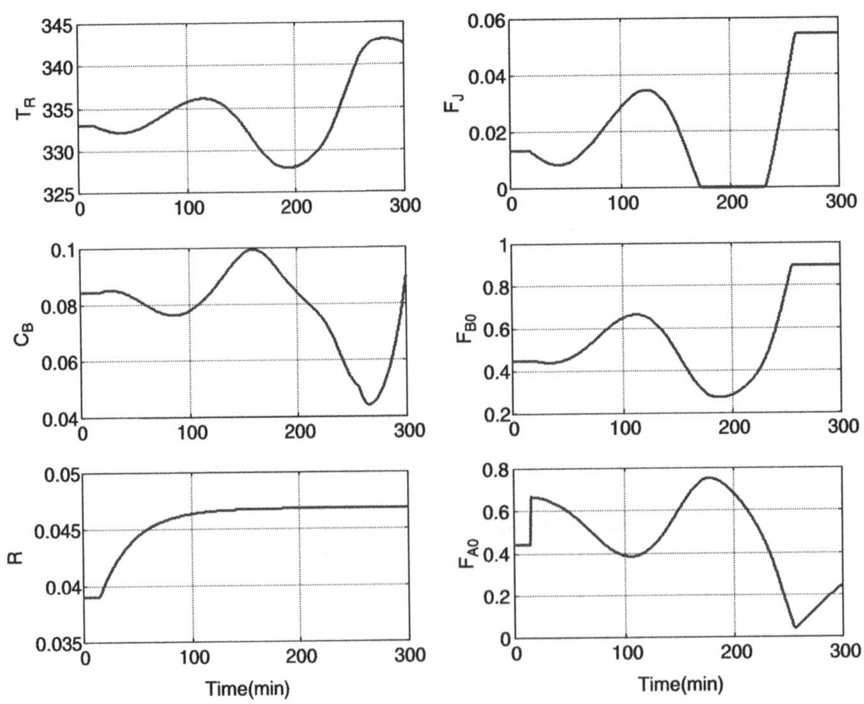

Figure 3.41 Hot reaction; T_R ($K_C = 5$) and C_B on automatic.

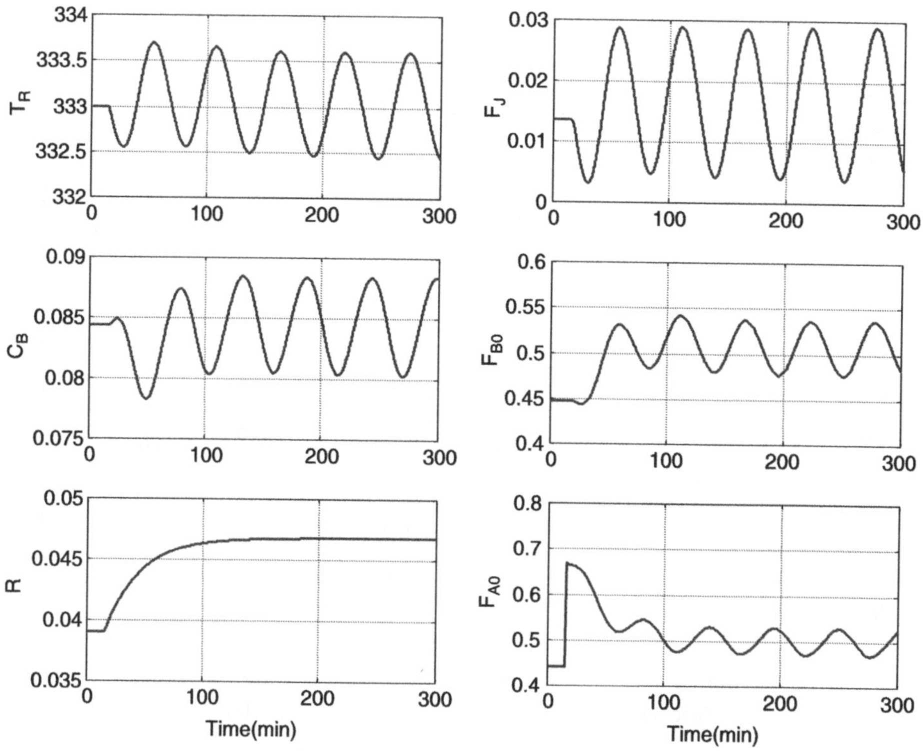

Figure 3.42 Hot reaction; T_R ($K_C = 20$) and C_B on automatic.

the specific reaction rate k because of the high activation energy, the B reactant will be consumed, the C_B concentration will decrease and the overall reaction rate will not change as much as if the C_B concentration were maintained. This "limiting reactant" structure provides some self-regulation, which helps compensate for a "hot" specific reaction rate.

The improvement in system stability is clearly shown in Figure 3.43. Reactor temperature is well controlled in the face of the disturbance in F. The C_B concentration drops to about 0.06 kmol/m^3. The F_{B0} flowrate is constant. The F_{A0} flowrate increases initially because the reactor level drops, but it gradually returns to the initial value. Reflux, vapor boilup and recycle flows all increase because of the increase in F, but the product stream P returns to its initial level.

To increase the production rate, the flowrates of both F_{B0} and F are increased in Figure 3.44. This revised control structure provides stable base-level regulatory control for this system with a very temperature-sensitive specific reaction rate.

This system is an excellent example of the interaction, conflicts, and tradeoffs that are required in designing a chemical plant for safe, stable operation. From a purely steady-state economic perspective, the optimum plant has a very small (10-m^3) reactor. However, this reactor is uncontrollable. Revising the reactor to provide much more heat transfer area produces an operable plant, provided the temperature sensitivity of the reaction is not too high. If the activation energy of the reaction is high, even the large reactor process is uncontrollable unless a "limiting reactant" control structure is used to provide some inherent self-regulation of the overall reaction rate. This example illustrates

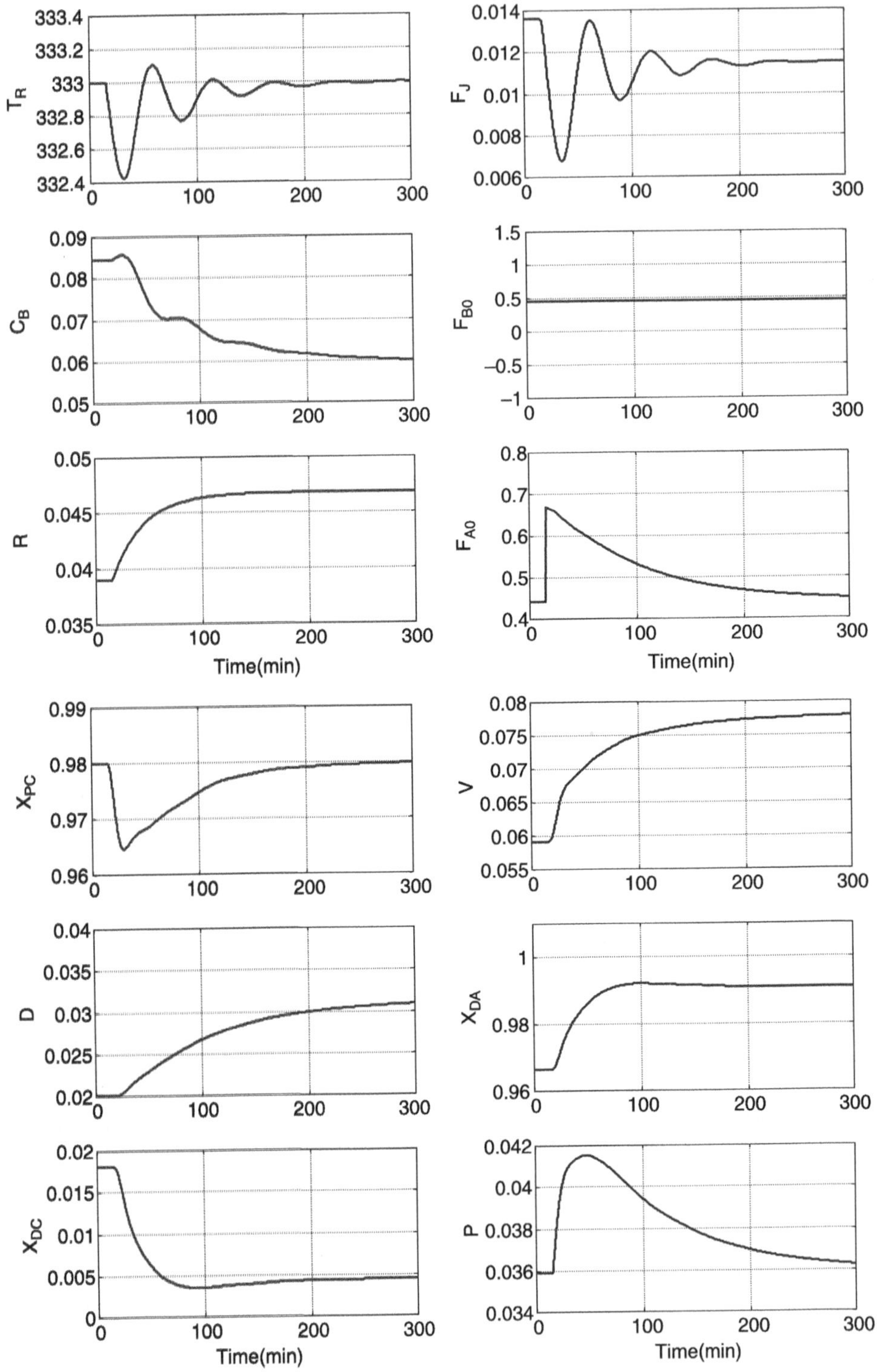

Figure 3.43 Hot reaction; T_R ($K_C = 10$) on automatic and fixed F_{B0}.

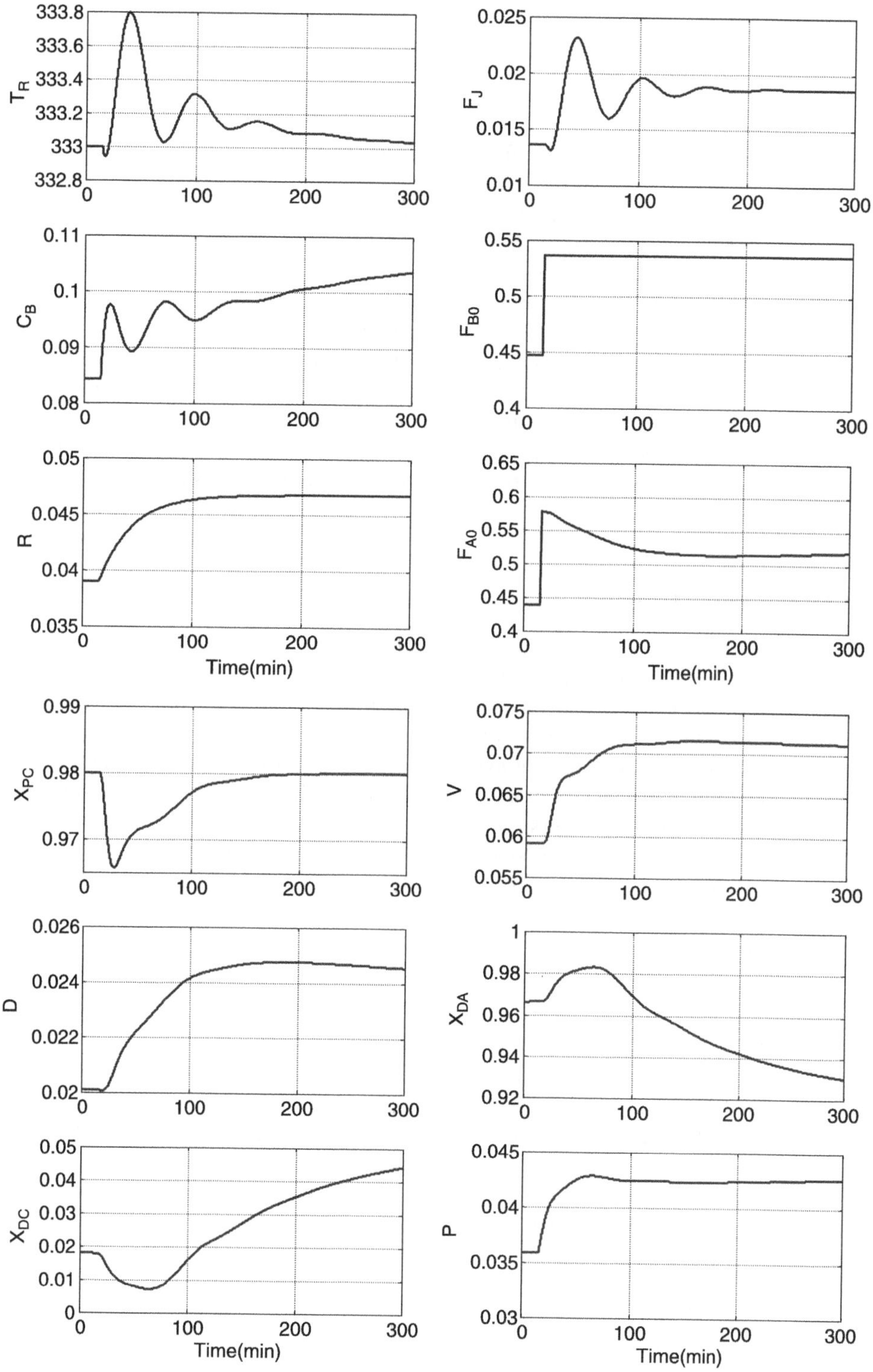

Figure 3.44 Hot reaction; T_R ($K_C = 10$) on automatic; $+20\%$ F and F_{B0}.

very graphically the need for "simultaneous design" (consideration of both steady-state and dynamics aspects) during all phases of the development of a chemical process.

3.3 AUTOREFRIGERATED REACTOR CONTROL

The steady-state design of an autorefrigerated reactor system was studied in Chapter 2 (Section 2.7). The dynamics and control of this process are considered in this section. Parameter values are given in Table 2.5. Two cases were investigated in Chapter 2 with conversions of 80 and 90%. For each of these designs, two values for the design condenser cooling water temperature were studied. The larger the conversion, the larger the required condenser heat transfer area because more heat of reaction must be removed. The higher the design value for condenser cooling water temperature, the larger the condenser heat transfer area because the temperature differential driving force between the process temperature and the cooling water temperature is smaller. We will demonstrate quantitatively in this section that reactor controllability is improved by increasing condenser heat transfer area.

3.3.1 Dynamic Model

The mathematical model describing an autorefrigerated reactor includes mass, component, and energy balances for both the reactor and the condenser. It also includes total mass and component balances on the liquid in the liquid return piping. The liquid height h in the return pipe changes as the pressure differential between the reactor and the condenser changes, assuming gravity flow. The "ρh" hydraulic head permits the liquid to flow from the condenser at a lower pressure into the reactor at a higher pressure.

We assume that reaction occurs only in the reactor. In addition to the feedstream F_0 entering the reactor and the product stream F_{vol} leaving the reactor (both with steady-state volumetric flowrates of 7.812×10^{-4} m^3/s), there is a vapor stream V (kmol/s) leaving the vessel and a liquid stream L_P (kmol/s) returning to the vessel from the return pipe:

Reactor total mass balance:

$$\frac{d(\rho V_R)}{dt} = F_0\rho + L_{P,\text{vol}}\rho - F_{\text{vol}}\rho - VM_V \tag{3.36}$$

where F_{vol} and $L_{P,\text{vol}}$ are the volumetric flowrates (m^3/s) of the reactor effluent and the liquid entering the reactor from the return pipe, respectively. These are calculated from the molar flowrates F and L_P and the molecular weights of the streams. Liquid density ρ is constant at 801 kg/m^3. The molecular weight of the vapor V leaving the reactor M_V (kg/kmol) is calculated from the vapor composition y_j:

Reactor component balances:

$$\frac{d(V_R C_A)}{dt} = F_0 C_{A0} + L_P x_{PA} - F x_A - V y_A - C_A C_B V_R k \tag{3.37}$$

$$\frac{d(V_R C_B)}{dt} = F_0 C_{B0} + L_P x_{PB} - F x_B - V y_B - C_A C_B V_R k \tag{3.38}$$

where x_{Pj} is the mole fraction of component j in the liquid from the return pipe and y_j is the mole fraction of component j in the vapor leaving the reactor.

Reactor energy balance:

$$\frac{d(V_R T_R)}{dt} = F_0 T_0 + L_{P,\text{vol}} T_C - F_{\text{vol}} T_R - \frac{V(c_P M_V T_R + \Delta H_V)}{\rho c_p} - \frac{\lambda C_A C_B V_R k}{\rho c_p} \tag{3.39}$$

Reactor volume V_R can vary with time and is controlled by a proportional controller that manipulates the reactor effluent stream F_{vol}. Heat capacity c_P is assumed constant (3137 J kg^{-1} K^{-1}), as is the heat of vaporization ΔH_V (23.24×10^6 J/kmol). The heat of reaction is removed by vaporizing the liquid. Note that there is no heat transfer in the reactor:

Effective vapor volume total molar balance:

$$\frac{V_{\text{eff}}}{8.314\, T_C^{\text{av}}} \frac{dP_C}{dt} = V - L_C \tag{3.40}$$

Equation (3.40) gives the dynamic changes in the condenser pressure P_C as changes occur in the flowrate of the vapor V entering the condenser from the reactor and the flowrate of liquid L_C condensing as a result of heat transfer in the condenser. The effective vapor volume in the whole system V_{eff} is assumed to be 10 times the volume of the process side of the condenser. It includes all the vapor volume in the top of the reactor, piping, and process side of the condenser. The process side is calculated from the heat transfer area assuming tubes that are 0.0254 m in diameter and 4 m in length. An average constant temperature T_C^{av} of 333 K is used in the total mass balance.

The holdup in the condenser is assumed to be negligible, so the thermal and composition dynamics are neglected. This means that x_{Cj} is equal to y_j at each point in time. The condenser process temperature T_C is determined from a bubblepoint calculation from the known pressure P_C and liquid composition x_{Cj}. Instantaneous thermal effects mean that the energy balance can be used to calculate the rate of condensation:

Rate of condensation:

$$L_C = \frac{Q_C}{\Delta H_V + c_P M_V (T_R - T_C)} \tag{3.41}$$

Heat transfer:

$$Q_C = U A_C (T_C - T_{CC}) \tag{3.42}$$

Condenser cooling water energy balance:

$$\frac{dT_{CC}}{dt} = \frac{F_{CW}(T_{C,\text{in}} - T_{CC})}{\rho_{CW} V_{CW}} + \frac{Q_C}{\rho_{CW} c_{CW} V_{CW}} \tag{3.43}$$

where the flowrate of cooling water is in kg/s.

Total molar balance on liquid in return pipe:

$$\rho A_P \frac{d(h/M_P)}{dt} = L_C - L_P \tag{3.44}$$

where A_P (m^2) is the cross-sectional area of the liquid return pipe and M_P is the molecular weight of the liquid with composition x_{Pj}. The diameter of the pipe is sized to give a liquid velocity of 0.05 m/s at design conditions. For example, the 90% conversion case has a liquid molar flowrate of 0.00975 kmol/s and a composition $x_{PA} = 0.5837$, $x_{PB} = 0.0905$, and $x_{PC} = 0.3258$. The molecular weight is 66.3 kg/kmol, the pipe diameter is 0.143 m, and the cross-sectional area is 0.143 m^2.

Component balances for liquid in return pipe:

$$A_P \rho \frac{d(hx_{PA}/M_P)}{dt} = L_C x_{CA} - L_P x_{PA}$$

$$A_P \rho \frac{d(hx_{PB}/M_P)}{dt} = L_C x_{CB} - L_P x_{PB} \tag{3.45}$$

Note that both the liquid height h and the composition of the liquid in the pipe vary with time. The latter means the molecular weight of the liquid M_P also is time-varying.

Hydraulic equations are used to determine the dynamic changes in the flowrates of the liquid and vapor. The flowrates of vapor V from the reactor to the condenser and the liquid from the condenser to the reactor vary with the square root of the appropriate pressure differentials

$$V = k_V \sqrt{P_R - P_C}$$

$$L_P = k_L \sqrt{(9.8\rho h/1000) - (P_R - P_C)} \tag{3.46}$$

where k_V and k_L are hydraulic constants that satisfy the steady-state flow and pressure conditions: 10 kPa pressure differential between the reactor and the condenser and a 1.84 m liquid height with the steady-state flowrates of V and L_P, which are, of course, equal under steady-state conditions.

3.3.2 Simulation Results

Figure 3.45 gives a Matlab program for the nonlinear simulation of the autorefrigerated reactor. The specific case is the 90% conversion with a cooling water temperature in the condenser of 317 K. The reactor volume is 1.68 m^3, and the reactor temperature T_R is 353 K. The condenser area is 19.9 m^2, and the condenser temperature T_C is 331 K. The temperature differential driving force is 331 – 317 = 14 K to transfer 0.237 MW.

There are two controllers. The proportional reactor level control has a gain of 5. The reactor temperature controller is tuned by running a relay–feedback test. The manipulated variable is the cooling water flowrate in the condenser. With a 50-K temperature transmitter span and the cooling water control valve half open at design conditions, the resulting tuning constants are $K_C = 4.23$ and $\tau_I = 25$ min.

Figure 3.46 shows that this process is closedloop-stable for 20% positive and negative step changes in feed flowrate. The increase in feed flowrate causes a slight temporary decrease in the reactor temperature because the feed (294 K) is colder than the reactor

```
% Program "autosim90317.m"
%  Given conversion and cooling water temp (tcc) with tr = 353 K
%       steady-state design and dynamic simulation
clear
a=[14.9287 14.2347 13.3187];b=[-2825.7 -2825.7 -2825.7];
% Sizes reactor given tr, tcc and conversion and feed
f0=.0007812;ca0=8.804;cb0=7.203;tr=353;conv=0.9;tcc=317;tcin=294;
k=37.35e6*exp(-69.71e6/8314/tr);ma=50;mb=50;mc=100;
cb=cb0*(1-conv),ca=ca0-conv*cb0,cc=conv*cb0,vr=f0*cc/(k*ca*cb)
dr=(2*vr/pi)^0.3333,ajacket=2*pi*dr^2;
x(1)=ca/(ca+cb+cc);x(2)=cb/(ca+cb+cc);x(3)=cc/(ca+cb+cc);
% Calculate reactor pressure and vapor composition
 for j=1:3;ps(j)=exp(a(j)+b(j)/tr);end; pr=0;for j=1:3;pr=pr+ps(j)*x(j);end
 for j=1:3;y(j)=ps(j)*x(j)/pr;xc(j)=y(j);xp(j)=y(j);end;  pc=pr-10;
 % Bubblepoint calculation in condenser using Newton-Raphson to get tc
 tc=tr-40;error=10;
 while error>.0001
    for j=1:3;ps(j)=exp(a(j)+b(j)/tc);end; pcalc=0;for j=1:3;pcalc=pcalc+ps(j)*xc(j);end
    error=abs(pcalc-pc); func=pcalc-pc;dfunc=0;for j=1:3;dfunc=dfunc-xc(j)*ps(j)*b(j)/(tc^2);end
    tc=tc-func/dfunc; end
 % Overall energy balance
 roe=800;cp=3137;lambda=69.71e6;dhvap=23.24e6;t0=294;u=851;qc=k*vr*ca*cb*(lambda)+f0*roe*cp*(t0-tr)
 v=qc/((cp*50*(tr-tc)+dhvap)),liqc=v;liqp=v;area=qc/(tc-tcc)/u,fcw=qc/(tcc-tcin)/4183
 dtube=0.0254;ltube=4;areatube=pi*dtube*ltube;numtube=area/areatube;voltube=pi*(dtube^2)*ltube*numtube/4,
 voleff=10*voltube;volcw=voltube; mp=xp(1)*ma+xp(2)*mb+xp(3)*mc;liqpvol=liqp*mp/roe;csareapipe=liqpvol/0.05;
 %***************************************************************
 %  Initial conditions for dynamic simulation
 load autoic tr vr ca cb cc pr pc tc h fcw tcc v xp mp
 time=0;delta=1;tstop=2*3600;tplot=0;np=0;ap=0.01614;trss=353;fcwss=fcw;kc=4.23;erint=0;taui=25.3*60;
 vrca=vr*ca;vrcb=vr*cb;vrtr=vr*tr;mp=xp(1)*ma+xp(2)*mb+xp(3)*mc;hxpam=h*xp(1)/mp;hxpbm=h*xp(2)/mp;hmp=h/mp;
 vrss=vr;fvolss=f0; f0ss=f0; fvol=f0;
 ;sum=ca+cb+cc;x(1)=ca/sum;x(2)=cb/sum;x(3)=cc/sum;mf=x(1)*ma+x(2)*mb+x(3)*mc;
 f=fvolss*roe/mf; kvap=v/sqrt(pr-pc);kliq=v/sqrt(roe*h/102.4-(pr-pc));
 trlag1=353;trlag2=353;taum=60;
% Integration loop
while time<tstop
   % Disturbance
   if time>0.2;f0=f0ss*1.2;end
%Controllers
    errorlevel=(vrss-vr)/vrss;oplevel=0.5-5*errorlevel; fvol=fvolss*oplevel*2;
    error=(trss-trlag2)/50;op=0.5-kc*error-erint;if op>1;op=1;end;if op<0;op=0;end;
    %  relay-feedback test     %op=0.5*1.1;if error>0;op=0.5*0.9;end
    fcw=fcwss*2*op;
        if time>=tplot;np=np+1;timep(np)=time/60;trp(np)=tr;fcwp(np)=fcw;hp(np)=h;vp(np)=v;
       prp(np)=pr;pcp(np)=pc;tcp(np)=tc;tccp(np)=tcc;vrp(np)=vr;fvolp(np)=fvol;
       liqcp(np)=liqc;liqpp(np)=liqp;cap(np)=ca;cbp(np)=cb;qcp(np)=qc/1e6;
       trlag2p(np)=trlag2;tplot=tplot+5;end
% Calculate reactor pressure and vapor composition
    sum=ca+cb+cc;x(1)=ca/sum;x(2)=cb/sum;x(3)=cc/sum;
    for j=1:3;ps(j)=exp(a(j)+b(j)/tr);end; pr=0;for j=1:3;pr=pr+ps(j)*x(j);end
for j=1:3;y(j)=ps(j)*x(j)/pr;xc(j)=y(j);end; if pr<pc;pc=pr;end
%  Calculate TC in condenser
errortc=10; while errortc>.0001
   for j=1:3;ps(j)=exp(a(j)+b(j)/tc);end;  pcalc=0;for j=1:3;pcalc=pcalc+ps(j)*xc(j);end
   errortc=abs(pcalc-pc); func=pcalc-pc;  dfunc=0;for j=1:3;dfunc=dfunc-xc(j)*ps(j)*b(j)/(tc^2);end
   tc=tc-func/dfunc; end
 % Molar Flowrates
    v=0;if pr-pc>0;v=kvap*sqrt(pr-pc);end
    liqp=0;if (roe*h/102.4-(pr-pc))>0;liqp=kliq*sqrt(roe*h/102.4-(pr-pc));end
    mp=xp(1)*ma+xp(2)*mb+xp(3)*mc;mf=x(1)*ma+x(2)*mb+x(3)*mc;mv=y(1)*ma+y(2)*mb+y(3)*mc;
    liqpvol=liqp*mp/roe;f=fvol*roe/mf;
    k=37.35e6*exp(-69.71e6/8314/tr);qc=u*area*(tc-tcc); liqc=qc/(dhvap+cp*mv*(tr-tc));
    dvr=f0+liqpvol-fvol-v*mv/roe;
    dvrca=f0*ca0+liqp*xp(1)-f*x(1)-v*y(1)-ca*cb*vr*k;dvrcb=f0*cb0+liqp*xp(2)-f*x(2)-v*y(2)-ca*cb*vr*k;
    dvrtr=f0*t0+liqpvol*tc-fvol*tr-v*(cp*mv*tr+dhvap)/roe/cp  + lambda*ca*cb*k*vr/roe/cp;
    dpc=(v-liqc)*8.314*333/voleff;dhmp=(liqc-liqp)/roe/csareapipe;
    dtcc=fcw*(tcin-tcc)/(volcw*1000) + qc/(volcw*4183*1000);
    dhxpam=(liqc*xc(1)-liqp*xp(1))/csareapipe/roe;dhxpbm=(liqc*xc(2)-liqp*xp(2))/csareapipe/roe;
    dtrlag1=(tr-trlag1)/taum;dtrlag2=(trlag1-trlag2)/taum;
% Integration
     time=time+delta;vr=vr+dvr*delta;vrca=vrca+dvrca*delta;vrcb=vrcb+dvrcb*delta;ca=vrca/vr;cb=vrcb/vr;
     vrtr=vrtr+dvrtr*delta;tr=vrtr/vr; pc=pc+dpc*delta;tcc=tcc+dtcc*delta;hmp=hmp+dhmp*delta;
     hxpam=hxpam+dhxpam*delta;hxpbm=hxpbm+dhxpbm*delta;
     xp(1)=hxpam/hmp;xp(2)=hxpbm/hmp;xp(3)=1-xp(2)-xp(1);mp=xp(1)*ma+xp(2)*mb+xp(3)*mc;
     h=hmp*mp;if h<0.01;h=0.01;end;
     erint=erint+error*kc*delta/taui; trlag1=trlag1+dtrlag1*delta; trlag2=trlag2+dtrlag2*delta; end
    %save autoic tr vr ca cb cc pr pc tc h fcw tcc v xp mp
clf; subplot(5,2,1); plot(timep,trp); grid;ylabel('TR (K)');title('Autorefrig; 90%; 317 Tcc;  +20% F0');
subplot(5,2,2); plot(timep,fcwp); grid;ylabel('FCW (kg/sec)');
subplot(5,2,3); plot(timep,hp); grid;ylabel('h (m)');
subplot(5,2,4); plot(timep,vp*100); grid;ylabel('100V (kmol/sec)');
subplot(5,2,5); plot(timep,vrp); grid;ylabel('VR (m^3)');
subplot(5,2,6); plot(timep,fvolp*100); grid;ylabel('100F (m^3/sec)');
subplot(5,2,7); plot(timep,cap,timep,cbp,'--'); grid;ylabel('CA/CB (kmol/m^3)');
subplot(5,2,8); plot(timep,qcp); grid;ylabel('QC (MM J/sec)');
subplot(5,2,9); plot(timep,prp,timep,pcp,'--'); grid;ylabel('PR/PC (kPa)');xlabel('Time (min)');
subplot(5,2,10); plot(timep,tcp,timep,tccp,'--'); grid;ylabel('TC/Tcc (K)');xlabel('Time (min)');
```

Figure 3.45 Matlab program for autorefrigerated reactor simulation.

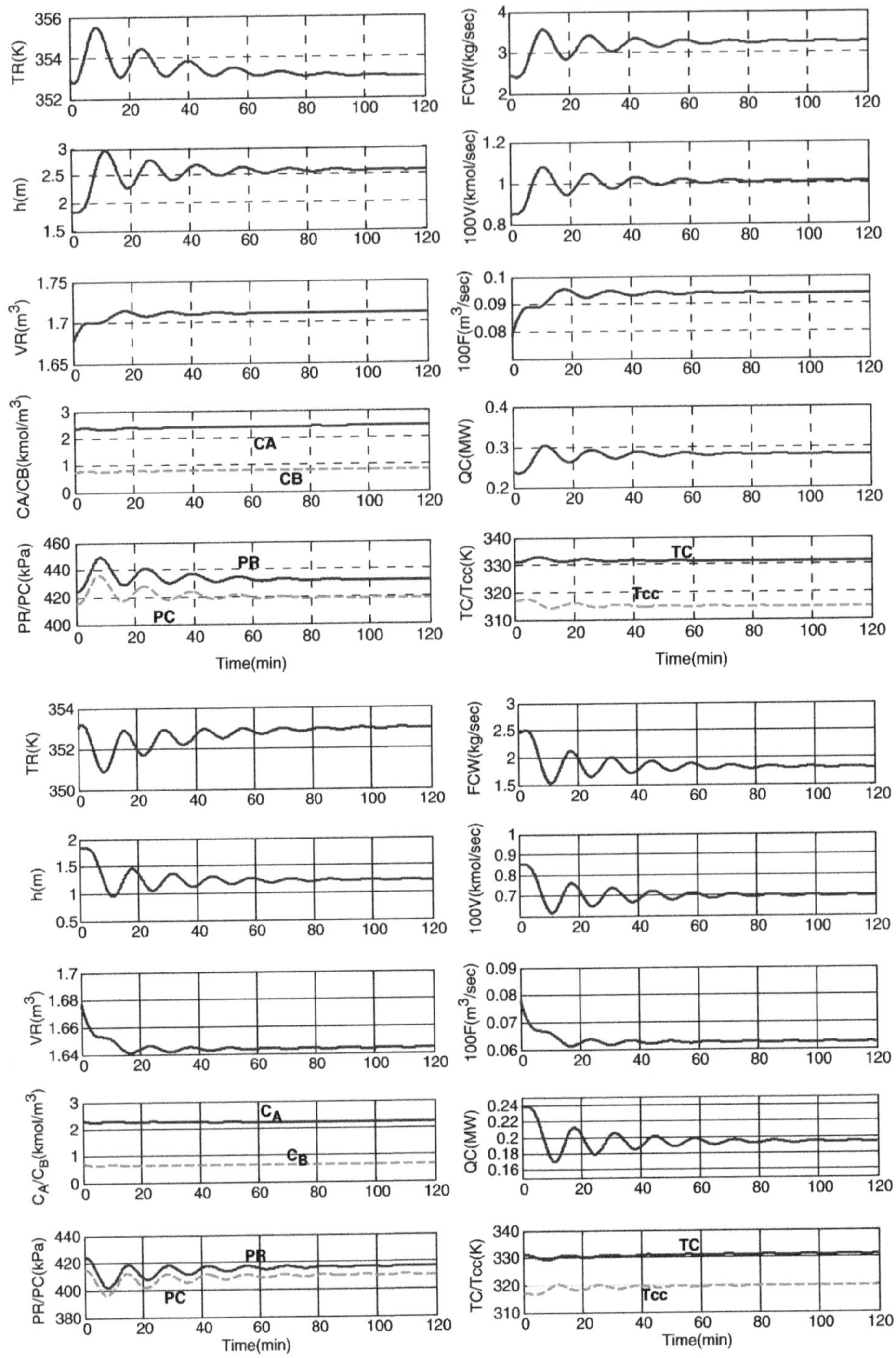

Figure 3.46 Autorefrigerated reactor; 90% conversion; $T_{CC} = 317$ K; +20% F_0.

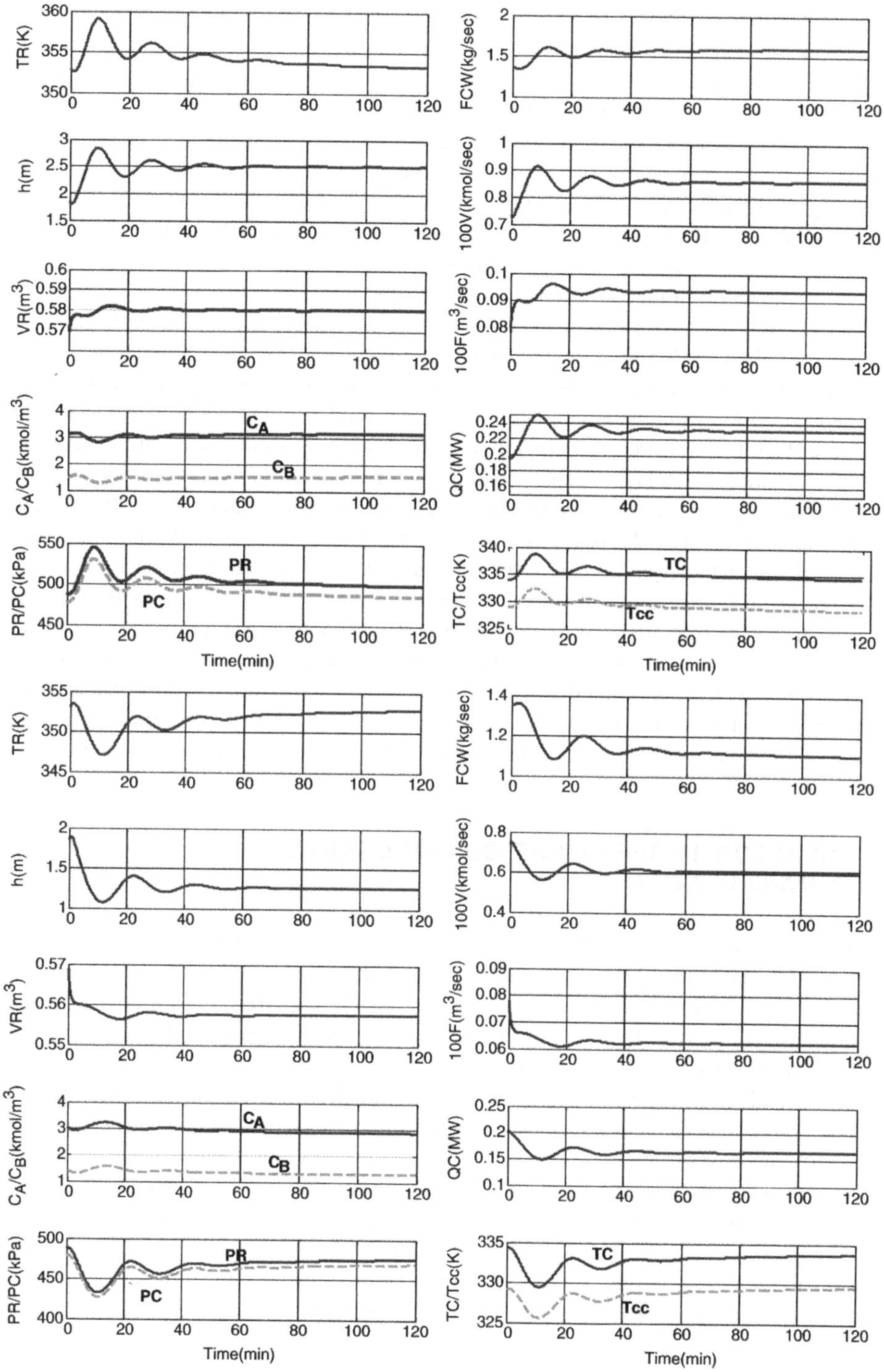

Figure 3.47 Autorefrigerated reactor; 80% conversion; $T_{cc} = 329$ K; $+20\%$ F_0.

(353 K). But as the reactant concentrations of A and B increase, the reaction rate picks up, which generates more heat and raises reactor temperature. The pressure in the reactor also increases. The temperature controller brings in more cooling water. The vapor and liquid flowrates increase as the heat removal increases because of the lower cooling water temperature in the condenser. The final steady-state difference in pressure between the reactor and the condenser increases for an increase in feed. Condenser cooling water temperature decreases, and the height of liquid in the return pipe increases. The reverse is true for a reduction in feed flowrate. Note that reactor pressure is not controlled. It varies with reactor composition and temperature. Pressure changes of about 20 kPa occur. The maximum deviation in reactor temperature is a little over 2 K.

The 80% conversion case with a cooling water temperature in the condenser of 317 K has reactor volume of 0.569 m^3 and a condenser area of only 13.6 m^2. Simulations of this process showed that it could not be stabilized. When the design cooling water temperature is raised to 329 K (which is only 5 K lower than the condenser temperature at 334 K), the condenser heat transfer area increases to 45.8 m^2, and the system is closedloop stable as shown in Figure 3.47. Relay–feedback testing gives reactor temperature controller settings $K_C = 1.4$ and $\tau_I = 26$ min, but the resulting response has a low damping coefficient. The results shown in Figure 3.47 used a controller gain of 0.7. Note that the changes in reactor pressure are much larger in this case (70 kPa). The maximum deviations in reactor temperature (6 K) are also larger in the 80% conversion case even with a very large condenser area.

Thus the autorefrigerated reactor system provides yet another example of the importance of heat transfer area and the increased difficulty of controlling reactors that do not have high conversion rates. Keep in mind that we are considering exothermic reactions that are irreversible. Control problems are much less severe in reactors with endothermic reactions or with reversible reactions because of the inherent self-regulatory nature of the chemistry.

3.4 REACTOR TEMPERATURE CONTROL USING FEED MANIPULATION

When reactor capacity is limited by heat removal, an often-recommended control structure is to run with maximum coolant flow and manipulate feed flowrate to control reactor temperature ($T_R \leftarrow F_0$ control). This control scheme has the potential to achieve the highest possible production rate. However, if the feed temperature is lower than the reactor temperature, the transfer function between temperature and feed flowrate contains a positive zero, which degrades dynamic performance, as we demonstrate quantitatively in this section. The choice of a control structure for this process presents an example of the often encountered conflict between steady-state economics and dynamic controllability.

The reactor is the jacket-cooled CSTR with an irreversible, exothermic, liquid-phase reaction $A \rightarrow B$, which was considered in Section 3.1. In that section the flowrate of the cooling water F_J to the jacket was the manipulated variable for the reactor temperature controller ($T_R \leftarrow F_J$ control). In this section we explore the use of the flowrate of the fresh feed F_0 to control reactor temperature ($T_R \leftarrow F_0$ control).

3.4.1 Introduction

The flowrate of the cooling/heating medium is usually the manipulated variable that is changed by a reactor temperature controller, either directly or through a reactor-to-jacket temperature cascade structure. A less commonly encountered control structure is to use feed flowrate to control reactor temperature. This concept is mentioned in several books, but only qualitative discussions are presented. The reason for using this kind of control structure is to maximize production when reactor capacity is limited by heat transfer. If the flowrate of the cooling/heating medium is set at its maximum value, this maximizes heat transfer. Then the $T_R \leftarrow F_0$ control structure will feed in as much fresh feed as the system can handle and still maintain the desired reactor temperature.

Greg Shinskey[1] presents a practical discussion of some of the advantages and problems of this control structure. The use of a "valve position control" (VPC) structure is recommended. A reactor temperature controller sets the coolant valve position. Then a VPC looks at the position of the coolant valve and adjusts the flowrate of the reactor feed to keep the coolant valve near its wide open position.

The temperature of the feed to a CSTR has a fairly minor impact on the steady-state design. Unlike tubular reactors (to be discussed in Chapters 5 and 6), in which feed temperature is a critictal design parameter (lower inlet temperature requires a larger reactor), the temperature of the feed to a CSTR affects only the sensible-heat component of the total heat removal/addition rate. The heat of reaction is typically considerably larger than this sensible heat.

Since its steady-state economic effect is minor, the temperature of the feed to a CSTR is usually set by an upstream supply temperature. If the feed is at ambient temperature, it is fed to the reactor at this temperature, despite the fact that the reactor may be operating at a significantly higher temperature for kinetic reasons. If the reaction is exothermic and heat must be removed, the colder the feed, the less heat must be transferred. This reduces the amount of coolant required because of two effects: (1) the total heat transfer rate is lower and (2) the exit temperature of the coolant is higher since a smaller differential temperature driving force is required.

However, in terms of dynamics, the temperature of the feed can have a significant impact on controllability. This is particularly true when the feed is colder than the reactor because the immediate effect of increasing the feed flowrate can be a temporary decrease in reactor temperature. This is illustrated in Figure 3.48 for three different design cases with feed temperatures of 294, 310, and 320 K. The feed temperature has no effect on the reactor size since this is set by conversion, feed flowrate, and reactor temperature (330 K). With a low feed temperatures, the sensible heat of the cold feed initially reduces the reactor temperature when the feed flowrate is increased. Eventually the increase in fresh feed raises the concentration of the reactant in the reactor, reaction rate increases, and reactor temperature begins to increase. This "inverse" or "wrongway" response is represented by a *positive* zero in the process openloop transfer function. This "non-minimum-phase feature" degrades feedback control. Thus there may be an inherent conflict between steady-state economics and dynamic controllability because we want to maximize production rate, which is achieved by the $T_R \leftarrow F_0$ control structure. But reactor temperature control may not be as tight as when the $T_R \leftarrow F_J$ control structure is used, which does not maximize production rate.

Figure 3.48 also shows that the design value for the jacket temperature decreases as the feed temperature increases because more heat must be transferred.

[1]F. G. Shinskey, *Process Control Systems*, 3rd edition, McGraw-Hill, 1988, p. 390.

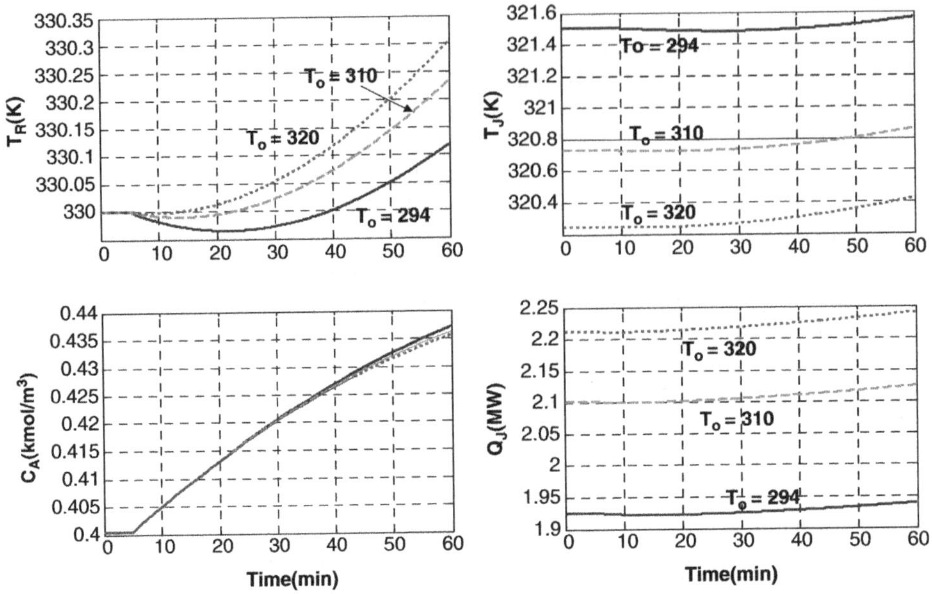

Figure 3.48 Openloop response ($+10\%$ F_0); 95% conversion; 330 K.

3.4.2 Revised Control Structure

The nonlinear model and the linear model are the same as those given in Section 3.1.1. For the $T_R \leftarrow F_0$ control structure, the openloop transfer function required to design the temperature controller is $T_{R(s)}/F_{0(s)}$. Using Eqs. (3.12)–(3.16), solving for $T_{R(s)}/F_{0(s)}$ and including two first-order temperature measurement lags give

$$G_{F0(s)} = \frac{T_{R(s)}}{F_{0(s)}} = \frac{c_2 s^2 + c_1 s + c_0}{(s^3 + b_2 s^2 + b_1 s + b_0)(\tau_M s + 1)^2} \tag{3.47}$$

$$b_2 = -a_{11} - a_{22} - a_{33}$$

$$b_1 = a_{11}a_{22} + a_{11}a_{33} + a_{22}a_{33} - a_{12}a_{21} - a_{23}a_{32}$$

$$b_0 = a_{12}a_{21}a_{33} - a_{11}a_{22}a_{33} + a_{11}a_{23}a_{32}$$

$$c_2 = b_{21} = \frac{T_0 - \bar{T}_R}{V_R} \tag{3.48}$$

$$c_1 = a_{21}b_{11} - a_{11}b_{21} - a_{33}b_{21}$$

$$c_0 = -a_{33}a_{21}b_{11} - a_{11}a_{33}b_{21}$$

A comparison of Eq. (3.47) with Eq. (3.17) shows that the denominators are identical but the numerators are different. The $T_R \leftarrow F_0$ control structure has a second-order numerator (two zeros) while the $T_R \leftarrow F_J$ control structure has a first-order numerator.

Note that the b_{21} coefficient is negative when the feed temperature T_0 is less than the reactor temperature T_R. This produces a positive root of the numerator polynomial given in Eq. (3.47), so the openloop transfer function has a *positive* zero.

3.4.3 Results

The case examined is for a conversion of 95% and a reactor temperature of 330 K. The effect of the feed temperature T_0 is explored. The fresh feed is pure A ($C_{A0} = 8.01$ kmol/m^3). The concentration of reactant A in the reactor is 0.40 kmol/m^3. The reactor volume is 434 m^3, and the heat transfer rate is 1.926×10^6 J/s through a jacket area of 266 m^2. The jacket temperature is 321.5 K. The cooling water flowrate is 16.7 kg/s.

The openloop transfer function between T_R and F_0 for this case has a positive zero equal to +0.00889. Results of a relay–feedback test are shown in Figure 3.49. Note the very large ultimate period ($P_U = 300$ min). This is caused by the inverse response, which can be observed from the "shark tooth" shape of the PV signal (TRlag2) to the temperature controller. The temperature controller settings used are gain $K_C = 3.7$ (dimensionless using a 50-K temperature transmitter span and a feed valve that is 50% open at design) and integral time $\tau_I = 440$ min.

The performance of this loop is shown in Figure 3.50. The disturbance is a step change in the inlet cooling water temperature $T_{C,\,in}$ from 294 to 310 K. The cooling water flowrate is fixed. The controller is successful in driving the reactor temperature back to the setpoint by reducing the fresh feed into the system, but it takes over 20 h because of the large integral time.

If the design value for the feed temperature is increased to 320 K, the jacket temperature is only slightly lower (320.2 K), the heat transfer rate is slightly higher (2.21×10^6 J/s) and the cooling water flowrate increases to 20.1 kg/s. The positive zero becomes +0.0322. Controller settings become $K_C = 5.5$ and $\tau_I = 440$ min. Figure 3.51 shows that the response of this system is also very slow, but the peak deviation in reactor

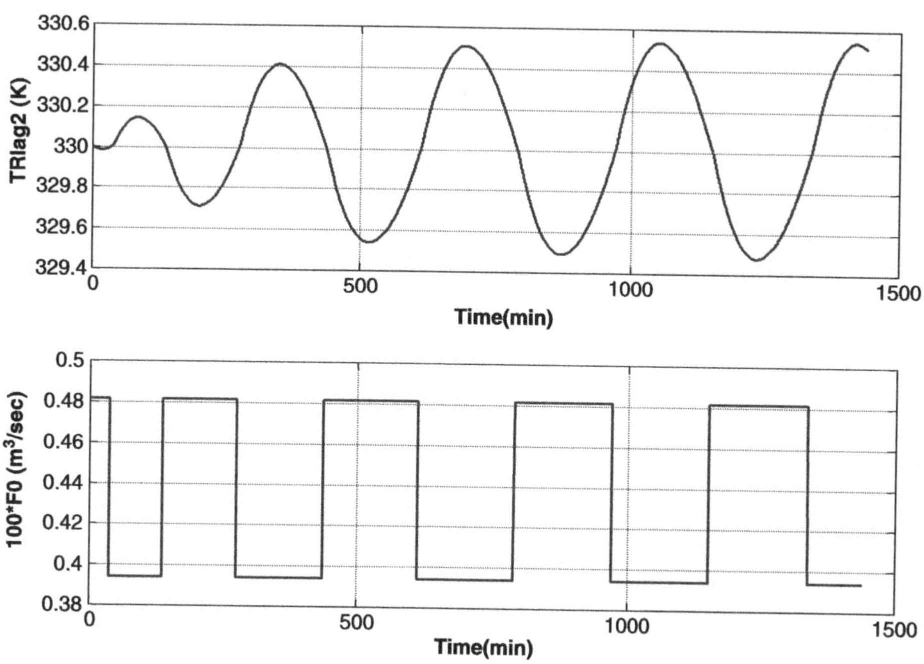

Figure 3.49 Relay–feedback test with feed manipulation; $T_R \leftarrow F_0$.

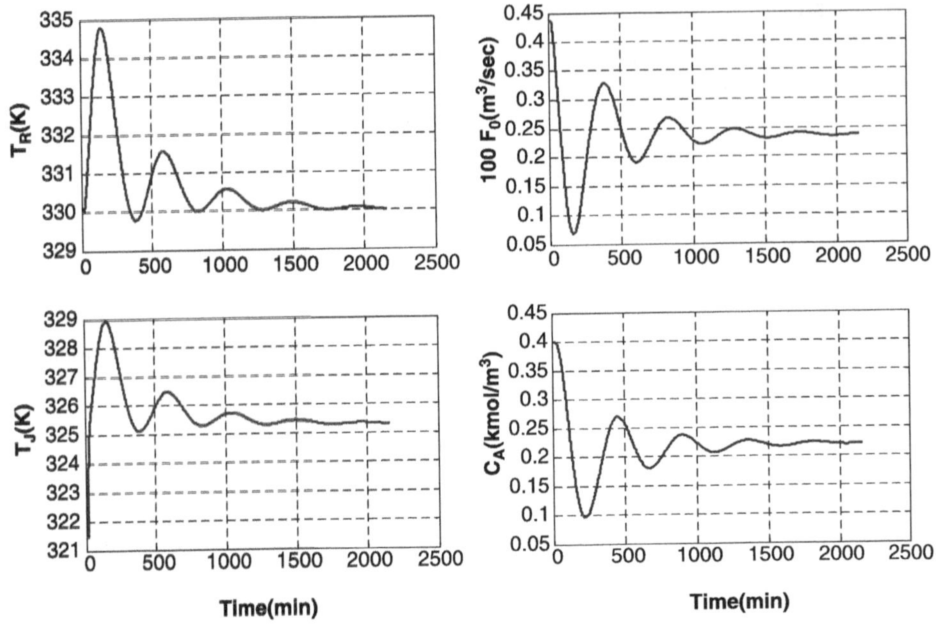

Figure 3.50 Feed manipulation; $T_0 = 294$ K; disturbance $T_{c,\mathrm{in}} = 294 \rightarrow 310$ K.

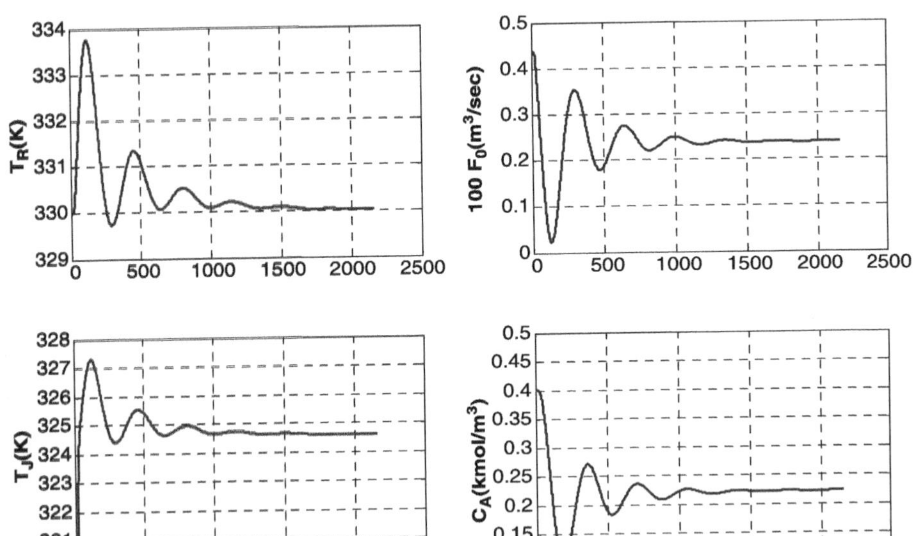

Figure 3.51 Feed manipulation; $T_0 = 320$ K; disturbance $T_{c,\mathrm{in}} = 294 \rightarrow 310$ K.

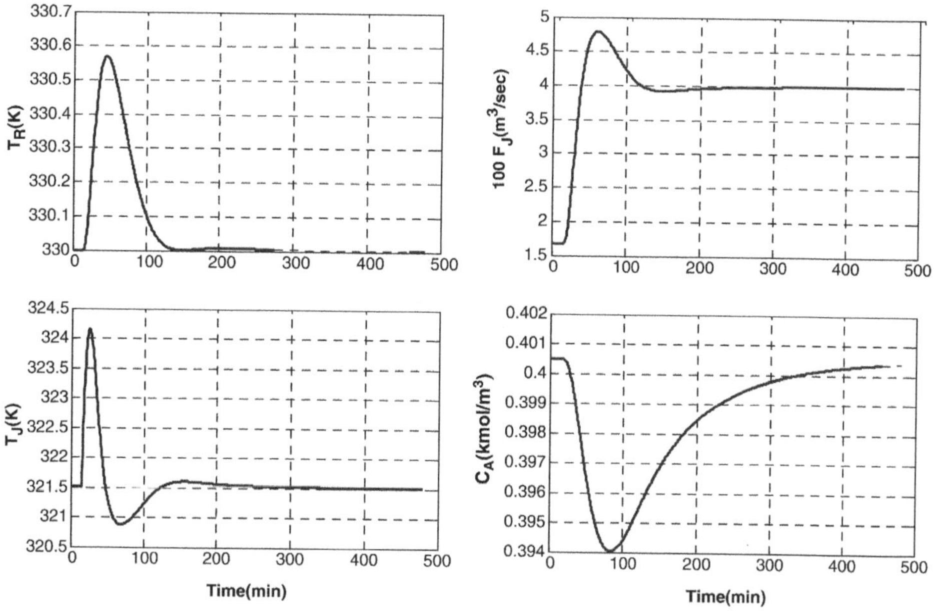

Figure 3.52 Coolant manipulation; $T_0 = 320$ K; disturbance $T_{c,\text{in}} = 294 \rightarrow 310$ K.

temperature is reduced from 4.7 to 3.7 K. Thus the use of feed to control reactor temperature gives very large time constants in this example.

The performance of the convention control structure, in which cooling water flow is manipulated to control temperature, is shown in Figure 3.52. The disturbance is the same increase in cooling water temperature. Feed flowrate is constant. The cooling water flowrate more than doubles to control reactor temperature, but the temperature is returned to the desired value in about 2 h. The peak deviation in temperature is less than 0.6 K. Controller settings are those given in Table 3.2 for the 95% conversion case with a 330 K reactor temperature (the integral time is 50 min).

3.4.4 Valve Position Control

These results demonstrate that the $T_R \leftarrow F_J$ control structure gives much tighter control than does the $T_R \leftarrow F_J$ control structure. Thus the valve position control system suggested by Shinskey should give good temperature control while still pushing the reactor toward maximum feed flowrates. The VPC structure is shown in Figure 3.53. The cooling water valve is air-to-close (AC), so that it will fail in the safe position of wide open. Therefore the desired output signal OP from the temperature control should be as small as possible and still be able to handle disturbances that require an increase in cooling water. We select a setpoint for the VPC of 20%. The VPC has integral-only action with $\tau_I = 100$ min. This is selected so that the VPC action is slower than the temperature controller with a 50 min integral time. The output signal from the VPC sets the fresh feed flowrate.

The effectiveness of this temperature–VPC structure is demonstrated in Figure 3.54. Initially the setpoint of the VPC is 0.5, which matches the initial output signal OP of

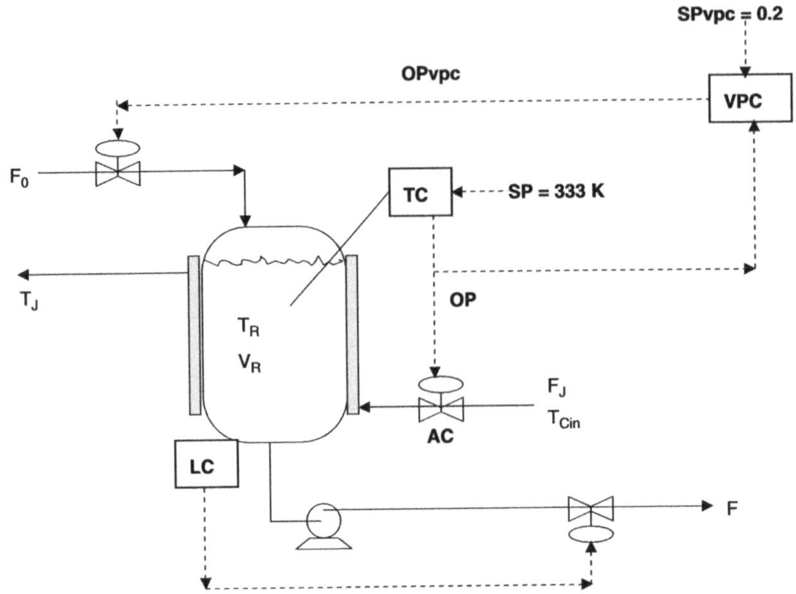

Figure 3.53 Valve position control.

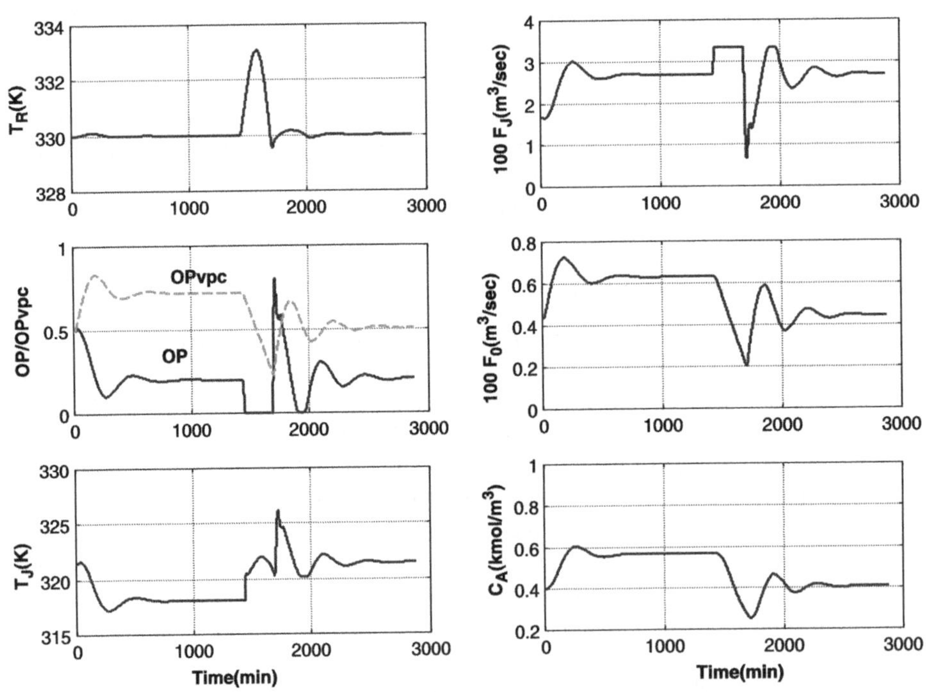

Figure 3.54 VPC; SP 0.5 → 0.2; disturbance $T_{c,\text{in}} = 294 \rightarrow 304$ K.

the temperature controller. The setpoint is then decreased to 0.2. The VPC gradually increases the feed flowrate, which causes reactor temperature to begin to increase. The temperature controllers sees a higher temperature and reduces its output signal to increase cooling water flowrate through the air-to-close valve. The feed eventually increases from 4.377 to about 6.4×10^{-3} m^3/s. Reactor temperature is controlled very close to the desired value.

At time equal 1400 min, the inlet temperature of the cooling water is increased from 294 to 304 K. This cause reactor temperature to increase, and the temperature controller drops its output signal OP to increase the flowrate of cooling water. The VPC controller sees this reduction in OP and reduces the feed flowrate until reactor temperature can be controlled with the cooling water valve 80% open (OP = 0.2). The feed is reduced to about 4.5×10^{-3} m^3/s. The maximum deviation in reactor temperature is 3 K for this 10-K increase in cooling water inlet temperature. Note that the cooling water valve goes wide open (OP = 0) for over 2 h for this large step disturbance.

Figure 3.55 shows the effect of moving the VPC setpoint further from the minimum. Initially the setpoint of the VPC is 0.5, which matches the initial output signal OP of the temperature controller. The setpoint is then decreased to 0.3 instead of the previous 0.2. The feed is increased to only 5.6×10^{-3} m^3/s. However, when the disturbance in cooling water temperature occurs, the deviation in reactor temperature is only 1.2 K, and the time the cooling water valve is saturated is about one hour.

Thus the VPC control structure is able to achieve good temperature control while adjusting feedflow to close to the maximum.

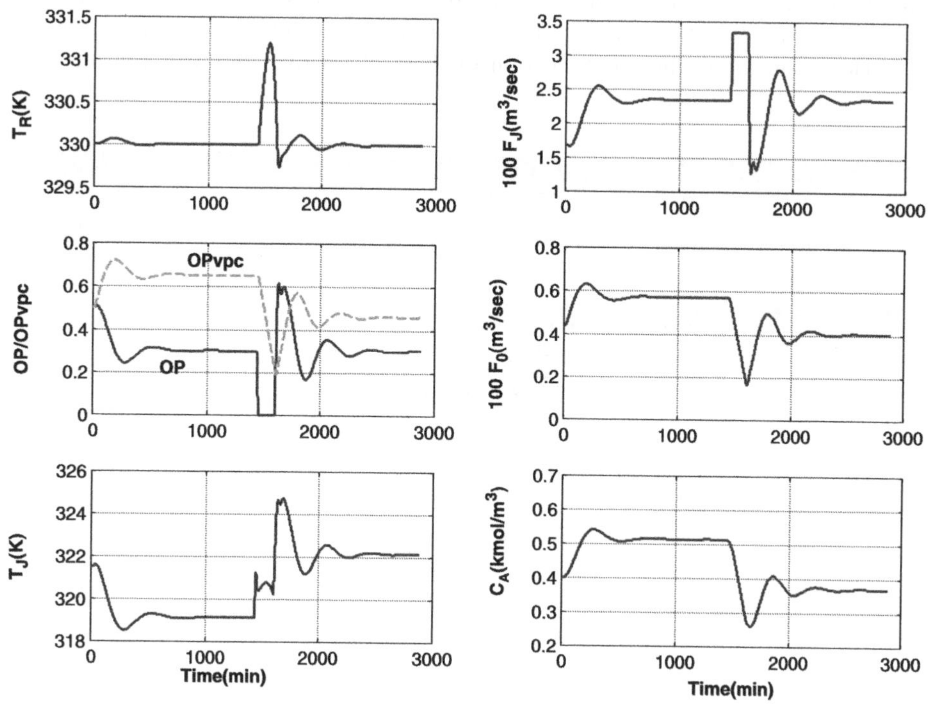

Figure 3.55 VPC; SP 0.5 → 0.3; disturbance $T_{c,in} = 294 \rightarrow 304$ K.

3.5 ASPEN DYNAMICS SIMULATION OF CSTRs

The ethylbenzene CSTR considered in Chapter 2 (Section 2.8) is used in this section as an example to illustrate how dynamic controllability can be studied using Aspen Dynamics. In the numerical example the 100-m^3 reactor operates at 430 K with two feedstreams: 0.2 kmol/s of ethylene and 0.4 kmol/s of benzene. The vessel is jacket-cooled with a jacket heat transfer area of 100.5 m^2 and a heat transfer rate of 13.46×10^6 W. As we will see in the discussion below, the steady-state simulator Aspen Plus does not consider heat transfer area or heat transfer coefficients, but simply calculates a required "UA" given the type of heat removal specified.

All of the dynamic simulations discussed in this book use "pressure-driven" flows. The alternative of using a "flow-driven" simulation is more simple, but not at all realistic of the actual situation in a real physical process. The plumbing in the real process has to be set up so that "water flows downhill." Pumps, compressors, and valves must be used in the appropriate locations to make the hydraulics of the system operate. If valves are not designed with sufficient pressure drop under steady-state conditions, they may not be able to provide the required increase in flow even when wide open. So valve saturation must be included in the rigorous nonlinear dynamic simulation. It is much better to simulate a realistic system by using a pressure-driven simulation.

To convert a steady-state simulation into a dynamic simulation, the *Dynamic* button on the top toolbar shown at the top of Figure 3.56 is clicked. If this button is not showing, go to the top toolbar and click *View* and *Toolbars*. The window shown at the bottom of Figure 3.56 opens, on which the box in front of *Dynamic* is clicked. Then the *Dynamic* item under the reactor block *R1* is clicked, which opens the window shown in Figure 3.57. The *Heat Transfer* page tab lists six possible options that can be used in the dynamic simulation. In our discussion of these alternatives, we will consider the

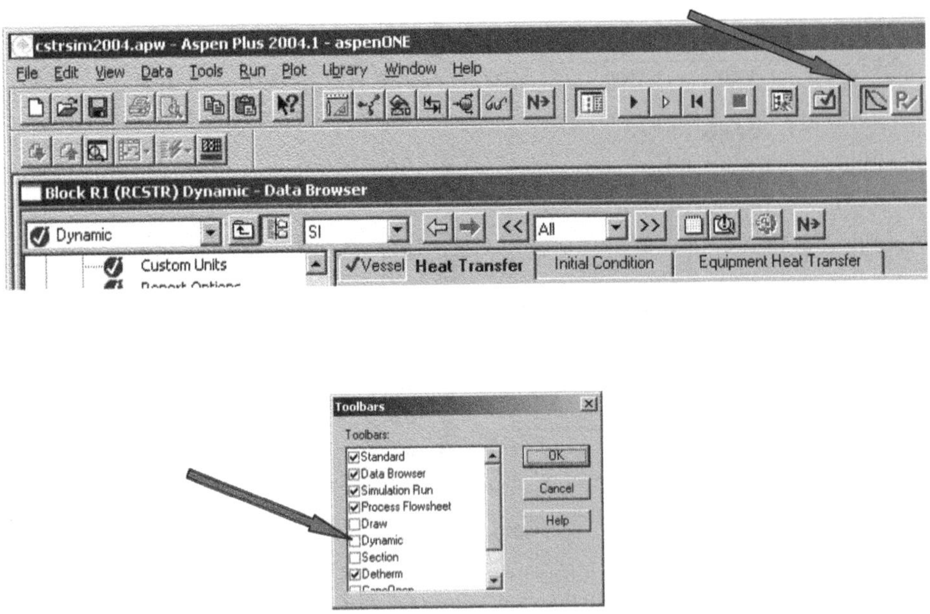

Figure 3.56 Selecting *Dynamic*.

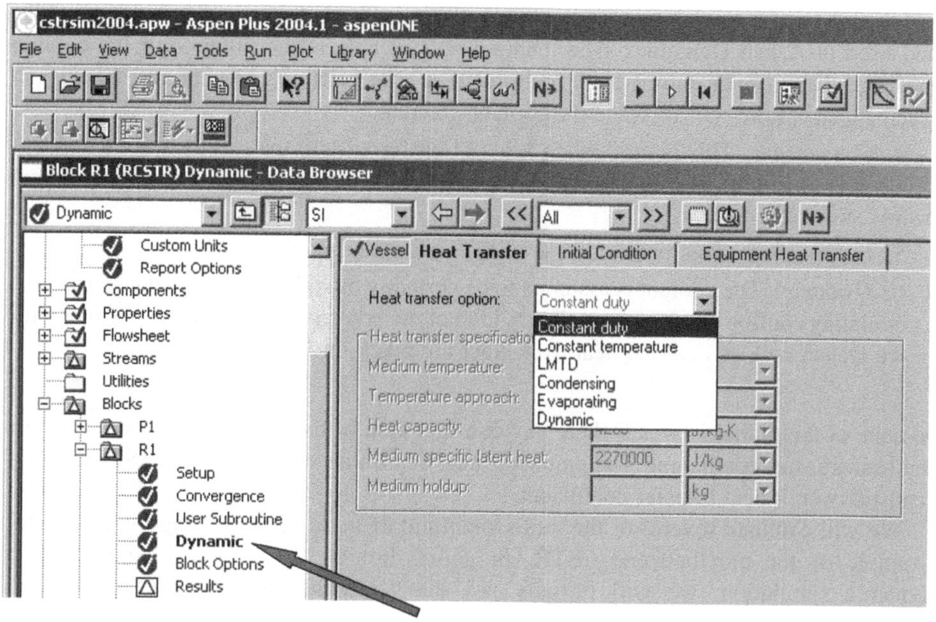

Figure 3.57 Selecting *Heat transfer option.*

case in which the reaction is exothermic and heat must be removed from the reactor. Of course, for endothermic reactions, heat must be added to the reactor, and the actions of the controllers would be opposite that used for cooling:

1. *Constant Duty.* With this mode of heat transfer, which is called "constant duty," the heat transfer rate is the manipulated variable. This is the most simple of the alternatives, but it is the most unrealistic. There is no consideration of the available area or of the temperature of the heat sink. In the dynamic simulation, the output signal of the reactor temperature controller is Q. The sign convention in both Aspen simulators is that heat added to the reactor is positive and heat removed is negative. This means that in a reactor with heat removal, the *minimum* output signal from a controller is a negative number (the maximum heat removal rate). The maximum output signal is zero. The action of the temperature controller is "reverse" because an increase in temperature should reduce the controller output signal to increase heat removal.

2. *Constant Temperature.* With this mode of heat transfer, the temperature of the cooling medium is the manipulated variable. Aspen Plus calculates the required "UA" for the given reactor temperature, coolant temperature, and heat transfer rate. There is no consideration of whether this UA is attainable with the geometry of the reactor vessel. The reactor temperature controller is reverse-acting so that an increase in reactor temperature will decrease the coolant temperature. This mode is often used when the reactor is cooled by generating steam or with a circulating cooling water system. In a real reactor with steam generation, the temperature controller would change the setpoint of a steam pressure controller.

3. *LMTD.* This mode is equivalent to an internal heat transfer coil where the temperature driving force is a log-mean average of the differences between the reactor temperature

and the inlet and outlet coolant temperatures. The required UA is calculated in Aspen Plus and used in Aspen Dynamics. The manipulated variable is the flowrate of the cooling medium. The reactor temperature controller is direct-acting, so that an increase in reactor temperature will increase the coolant flowrate.

4. *Condensing*. This is used when heat is transferred into the reactor from a condensing vapor medium. The manipulated variable is the flowrate of the medium. Latent heat is assumed constant.

5. *Evaporating*. This is the reverse of the condensing mode.

6. *Dynamic*. The coolant is assumed to be perfectly mixed as would be the situation in a circulating cooling water system. The holdup of the coolant is specified, so the dynamics of the jacket, coil, or external heat exchanger are taken into consideration.

In none of these alternatives is there any consideration of the actual heat transfer area or the heat transfer coefficient. The engineer must determine what area is required, using a realistic overall heat transfer coefficient.

We will evaluate several of the more important of these options using the numerical example of the ethylbenzene CSTR. In going through the details of setting up a dynamic simulation, we will initially use the "constant temperature" option. The

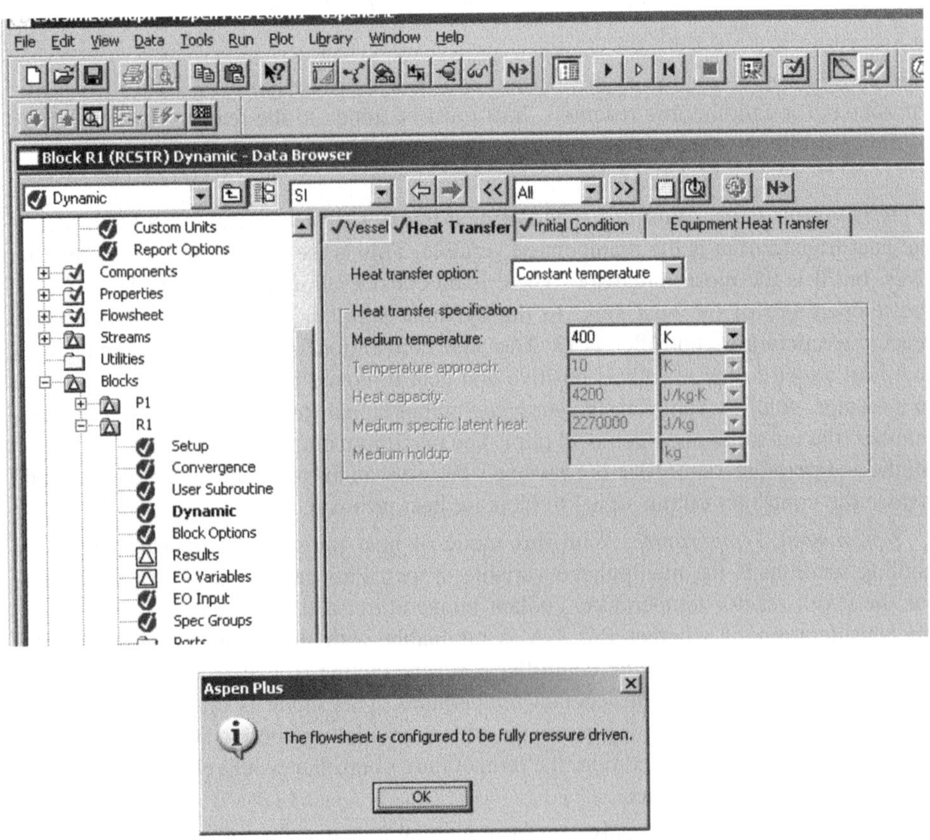

Figure 3.58 Selecting *Constant temperature, Heat transfer option.*

coolant medium temperature is specified to be 400 K, which gives a reasonable 30 K driving force. This means the value of "UA" must be 415,500 J s^{-1} K^{-1}. If a typical value for the overall heat transfer coefficient of 851 W m^{-2} K^{-1} is assumed, the area required to transfer the 12.46×10^6 W is 488 m^2. The jacket area is only 100.5 m^2, so additional area must be used. The 488 m^2 could be achieved in an external heat exchanger by using 518 tubes that are 0.05 m in diameter and 6 m in length.

3.5.1 Setting up the Dynamic Simulation

Figure 3.58 shows the window when *Constant temperature* is selected on the *Heat Transfer* page tab. We specify a *Medium temperature* of 400 K. With the reactor at 430 K, this gives a 30 K differential driving force.

The *Pressure check* button on the top tool bar is clicked, and if the plumbing has been set up correctly, the message shown at the bottom of Figure 3.58 appears. Now go to *File* on the top toolbar and select *Export*. The window shown in Figure 3.59 opens. The arrow to the right of the *Save as type* gives a list of possible types of files. To select a pressure-driven dynamic simulation, click *P Driven Simulation* [**.dynf, *dyn.appdf*]. As indicated in the extensions of the files, two files are generated that will be used by Aspen Dynamics. If the filename is specified to be "cstrsim2004," the "cstrsim2004.dynf" file will contain the details of the dynamic simulation including process units and controllers. The "cstrsim2004dyn.appdf" file contains all the physical property information used by Aspen Dynamics.

The message shown in Figure 3.60 provides some information about the exported file and indicates the exported file is ready to run in Aspen Dynamics. A simple way to

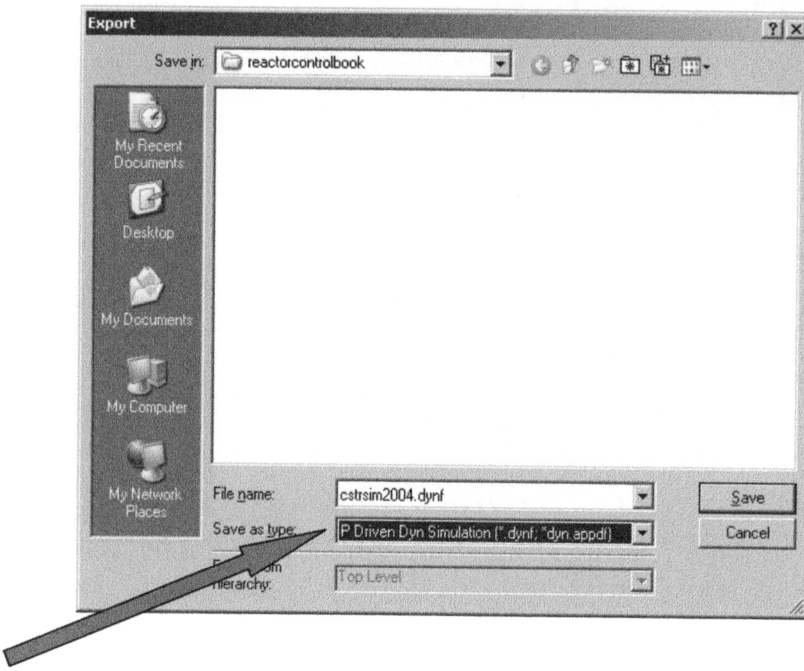

Figure 3.59 Specifying pressure-driven dynamic simulation.

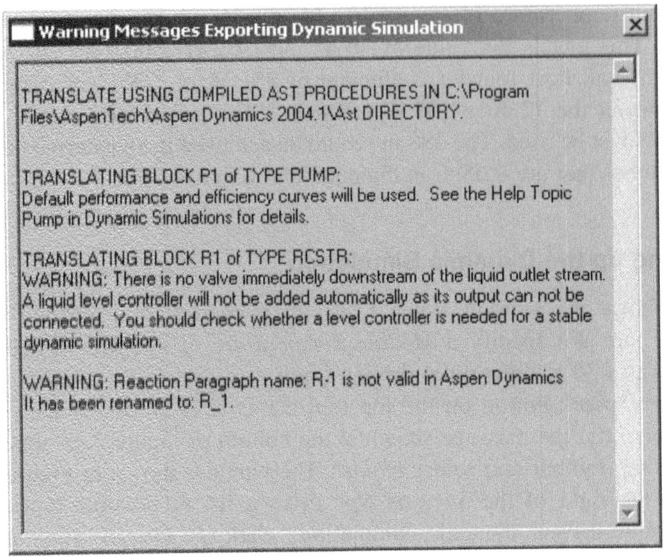

Figure 3.60 Message.

open Aspen Dynamics is the use Windows Explorer and double-click on the "cstrsim2004. dynf" file. The main Aspen Dynamics window opens, as shown in Figure 3.61. There are several windows. The *Process Flowsheet window* shows all reactor, control valve, pump, and connecting streams. The *Simulation Messages* window at the bottom is used to follow

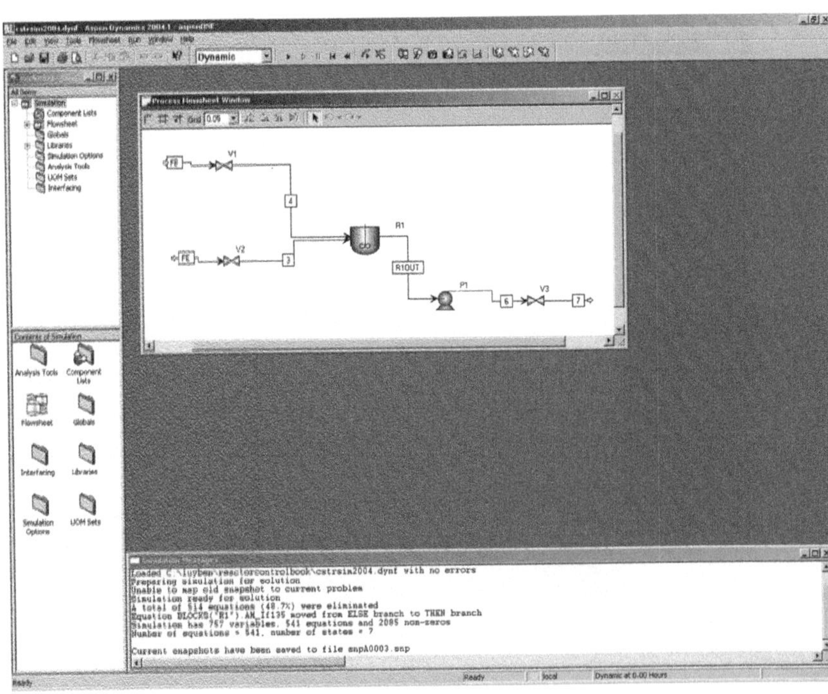

Figure 3.61 Initial Aspen Dynamics screen.

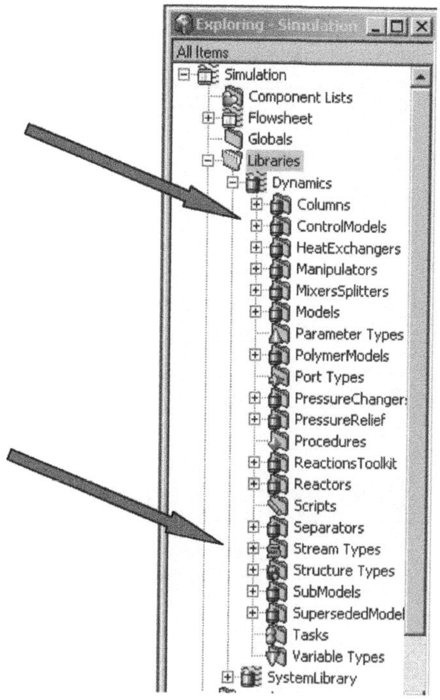

Figure 3.62 Dynamics libraries.

the progress of the dynamic simulation. The *Exploring Sim* window on the left has several subitems. The most important is *Libraries.* Clicking the " + " box to the left of *Libraries* reveals two subitems, one of which is *Dynamics.* When this is expanded by clicking the " + " box, the subitems shown in Figure 3.62 are listed. The ones used to set up the control structure are *ControlModels* and *Stream Types.*

Clicking the *ControlModels* produces a list of controllers, dynamic elements, and other functions shown in Figure 3.63. The procedure for installing a controller is to click the *PIDIncr* item in the list, hold down the left mouse button, drag it to the flowsheet window, and release the mouse button. This "drag and drop" action puts a block called "B1" on the PFD as shown in Figure 3.63. To rename the block, click on the icon, click the right mouse button, and select *Rename block.* We label this controller "LC" as shown at the bottom of Figure 3.63.

Next the input (PV signal) and the output (OP signal) of the controller are specified by expanding *Stream Types* to get a list of various types. Click *Control signal,* hold down the left mouse button, and drag it to the flowsheet, but continue to hold down the mouse button. A number of arrows appear that indicate possible connections that could be made. We want to control the liquid level in the reactor, so we place the cursor on the arrow pointing out from the reactor. The window shown in the upper left corner of Figure 3.64 opens and indicates four possible choices. Clicking the *BLOCKS("R1").level* attaches the control signal to the reactor level transmitter. Then move the cursor to the arrow pointing into the level controller *LC* and click. The window shown in the upper right of Figure 3.64 opens, on which we select *LC.PV* (the PV signal to this controller).

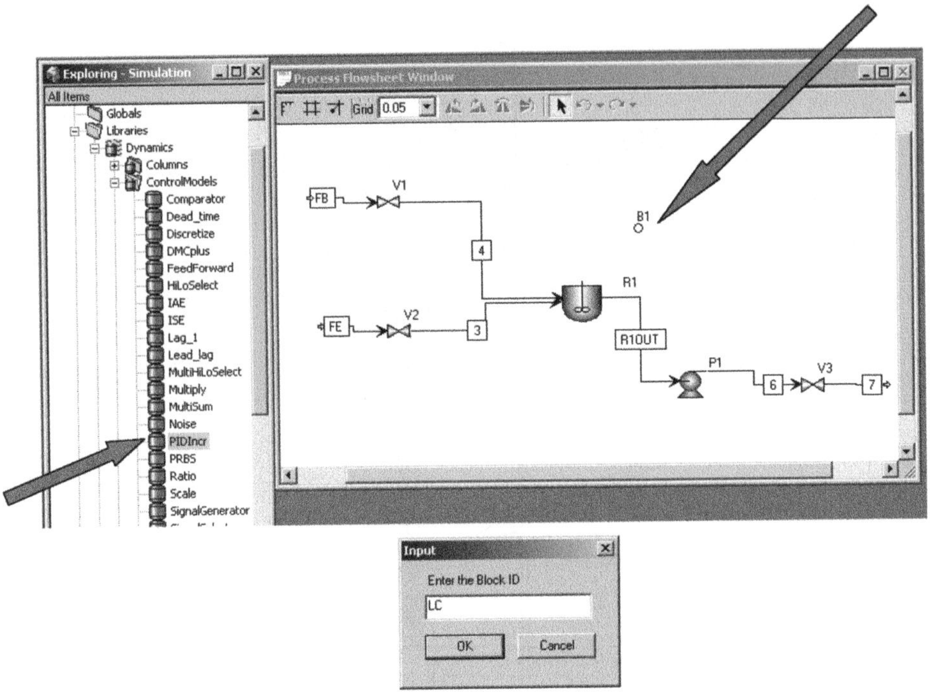

Figure 3.63 Adding controller to flowsheet.

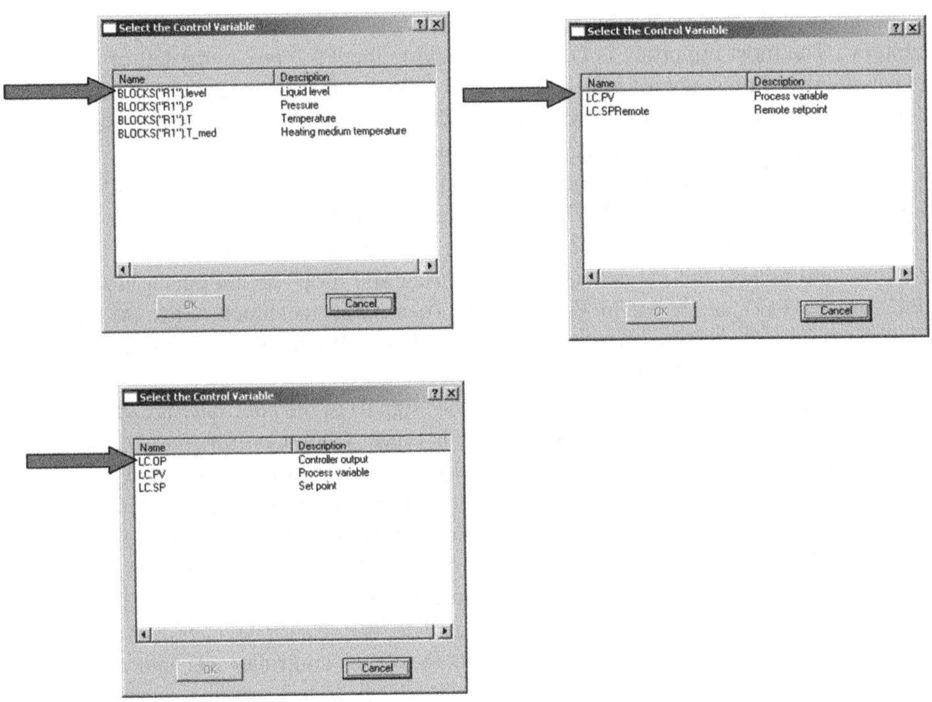

Figure 3.64 Selecting *PV* and *OP* of level controller.

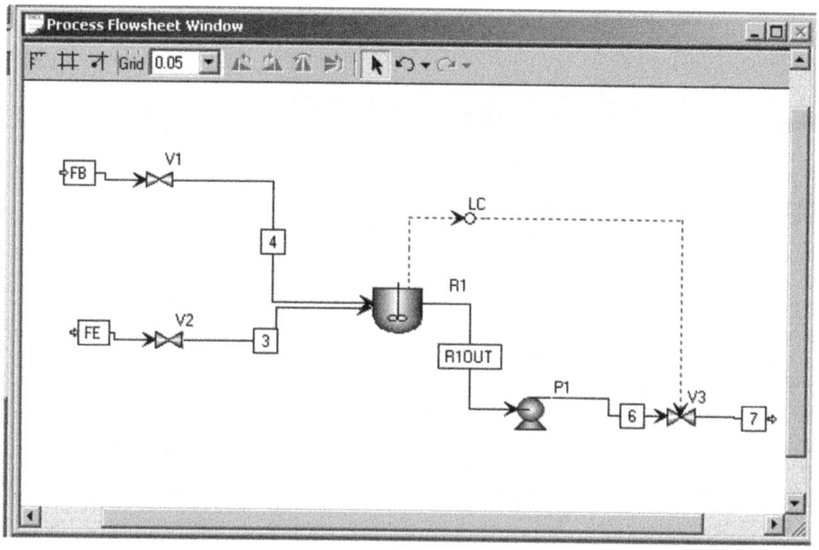

Figure 3.65 Flowsheet with level controller connected.

In a similar way, a control signal is moved to the flowsheet and positioned on the arrow pointing *out* of the LC controller. The window at the bottom left of Figure 3.64 opens on which we select *LC.OP* (the OP signal from this controller). Then the cursor is clicked on the control valve V3 on the flowsheet in the product line. Figure 3.65 shows the level control loop installed on the flowsheet.

Double-clicking the controller icon opens a controller faceplate, as shown in Figure 3.66. The controller faceplate is where all the features of the controller can be examined and adjusted. There are seven buttons at the top. The first one from the left puts the controller in automatic. The second puts the controller on manual. The setpoint (SP), process variable (PV), and controller output (OP) signals are displayed in the bottom three bar charts, with numerical values given in the boxes.

The first thing to do is to click the third button from the right (the "Configure" in Figure 3.66). This opens the window shown on the left in Figure 3.67 below the faceplate. The *Tuning* page tab shows the default values. Clicking the *Initialize Values* button at the

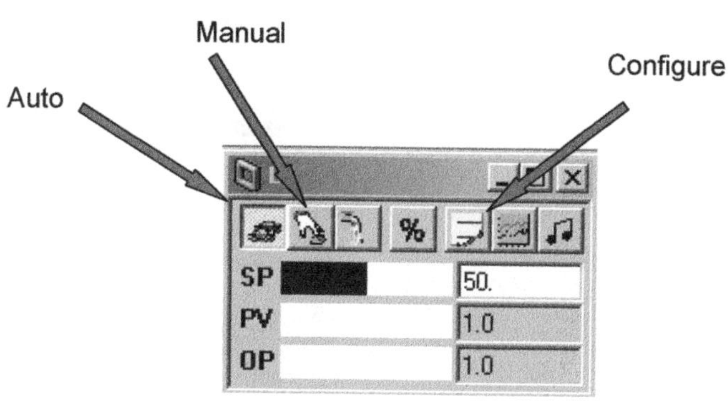

Figure 3.66 Controller faceplate.

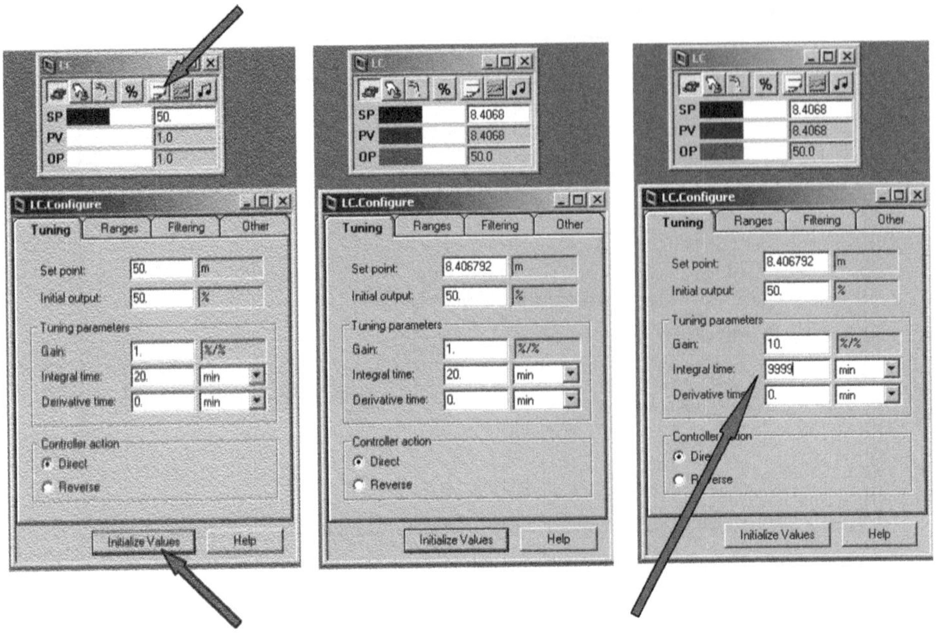

Figure 3.67 Initializing and configuring controller.

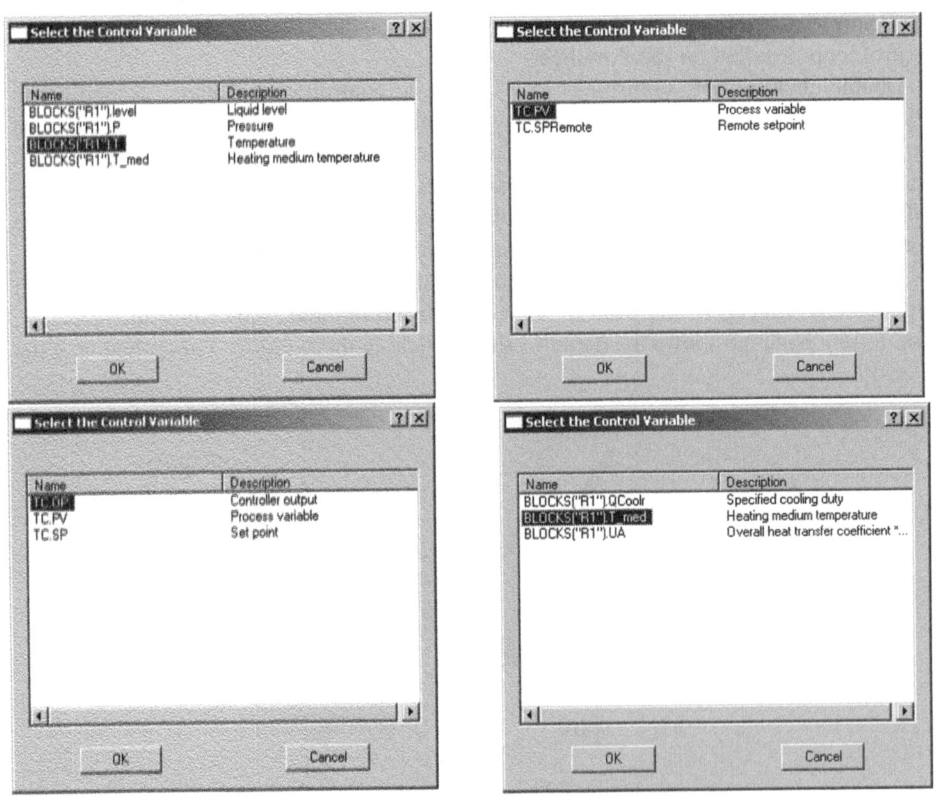

Figure 3.68 Setting up the temperature controller.

bottom changes the PV, OP, and SP signals to their appropriate values, as shown in the middle of Figure 3.67. The controller OP signal is set at 50%, which means that the valve is 50% open at design conditions. Since the level controller manipulates the outflow from the vessel, its action should be *Direct* (in increase in level opens the valve). We make the controller proportional-only by changing the gain to 10 and the integral time to 9999 minutes. The level controller is completely installed and ready to run.

Next the temperature controller is installed. The procedure is the same. Figure 3.68 shows the various windows for selecting the reactor temperature from the R1 block, specifying it as the PV signal, selecting the temperature controller OP signal, and specifying it as the medium temperature in the reactor: *Blocks("R1").T_med*. Figure 3.69 shows the controller faceplate and the tuning page tab before and after initialization. Before initialization, the default values of 323.15 K are inserted for the setpoint and the initial output signals. After initialization, the PV and the SP are set equal to the reactor temperature of 430 K, and the OP signal is the medium temperature of 400 K. It is very important to note that the controller action is set to *reverse*. If reactor temperature goes up, medium temperature should go down. The tuning of this controller will be discussed in the next section.

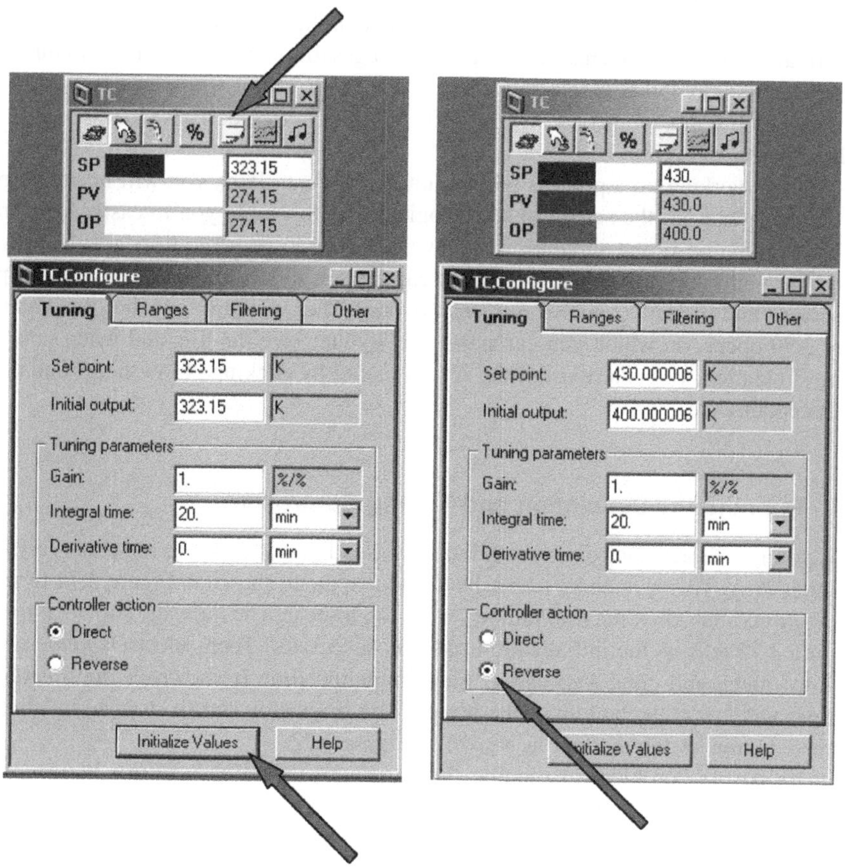

Figure 3.69 Initializing temperature controller.

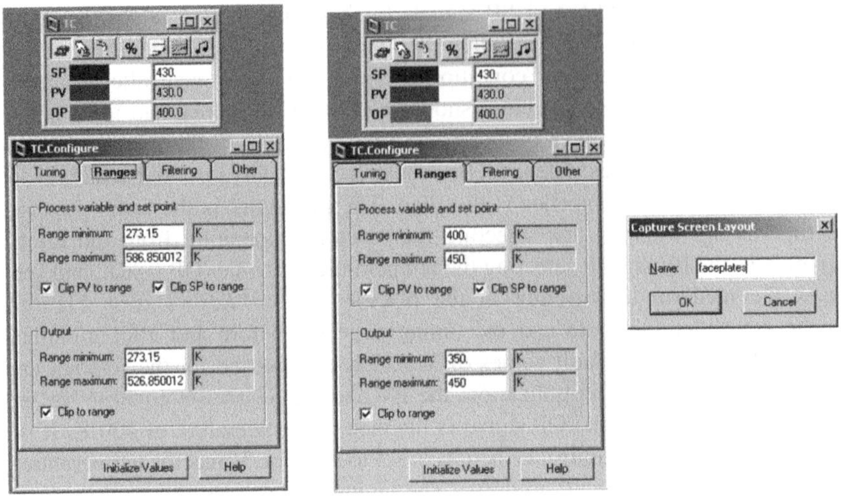

Figure 3.70 Specifying ranges of temperature controller.

Clicking the *Ranges* page tab opens the window shown on the left side of Figure 3.70. Aspen Dynamics inserts default values for the range of the PV signal and of the OP signal. It is normally convenient to change temperature ranges to some even numbers. In this case, a range from 400 to 450 K is selected for the PV signal (reactor temperature). For the OP signal (medium temperature) a range of 350–450 K is selected. These are shown in the middle window in Figure 3.70.

If you are working on a process with a number of controllers, you will normally position the faceplates on the screen in some logical way to enable you to easily keep track of what is going on in the simulation. To avoid having to generate this faceplate layout every time you open the simulation file, it can be saved. Go to the *Tools* at the top of the screen and select *Capture Screen Layout*. The window on the far right in Figure 3.70 opens, on which you can name the layout. Save the file, and when you next run the simulation, you can reinstall the screen layout by clicking "Flowsheet" and clicking "Faceplates."

3.5.2 Running the Simulation and Tuning Controllers

At this point, with two controllers installed, it is usually a good idea to see that everything is running okay. This is done by using the dropdown menu shown in Figure 3.71 to select *Initialization*. Then click the *Run* button on the top toolbar. The message shown at the top of Figure 3.72 tells us the initialization run was successful. Then select *Dynamic* on the dropdown menu and click *Run* again. The simulation time is shown in the *Simulation Messages* window at the bottom of the screen. The simulation can be stopped by clicking the *Pause* button on the toolbar, as shown in Figure 3.72.

If there is some problem in running the simulation, a variety of error messages are displayed. These are seldom very useful in determining what the problem is. The most common problem is an incorrect setup of the plumbing, which causes errors in the

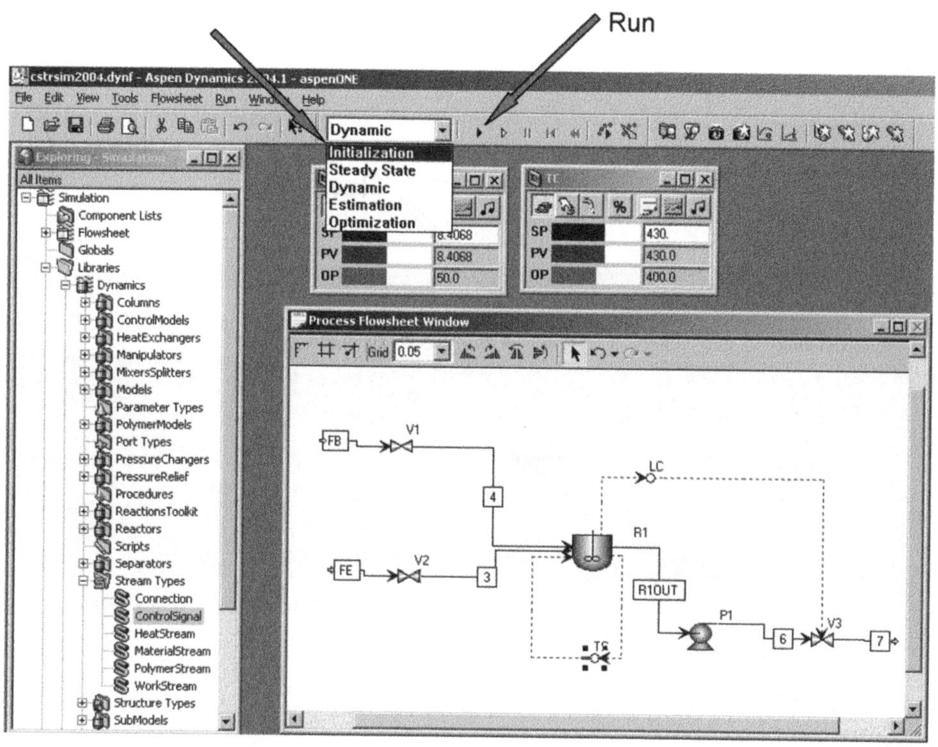

Figure 3.71 *Initialization* and *Dynamic* runs.

flow-solving portion of the simulation. To correct this problem, you usually have to go back to Aspen Plus and correct the plumbing setup.

The tuning of the temperature controller involves inserting a deadtime element in the loop to account for the dynamic lags that are always present in any real controller. Temperature measurement lags are not insignificant (30–60 s). In addition, changing the medium temperature cannot be achieved instantaneously. Therefore a one-minute deadtime is inserted. This is done by clicking on *Dead_time* in the list of *Control Models* and dragging it to the flowsheet, as shown in Figure 3.73. The block is renamed *dead*. We want to position the deadtime element between the reactor and the controller. Click the signal line coming from the reactor, click the right mouse button, and select *Reconnect Destination*. Then position the cursor on the arrow pointing to the deadtime element and click. Figure 3.74 shows this connection. Then another control signal line is connected between the deadtime and the controller (PV signal).

The value of deadtime is set by clicking on the icon, clicking the right mouse button, and selecting *Forms* and *All Variables*. The window shown at the top of Figure 3.75 opens, which shows some default values for input and output. Make an initialization run to get the correct values, as shown in at the bottom of Figure 3.75. The deadtime value has also been inserted. Figure 3.76 shows the flowsheet and faceplate with the temperature and level controllers installed.

The tuning of the temperature controller is achieved by running a relay–feedback test, which the recent versions of Aspen Dynamics has made quite easy to do. The button on the

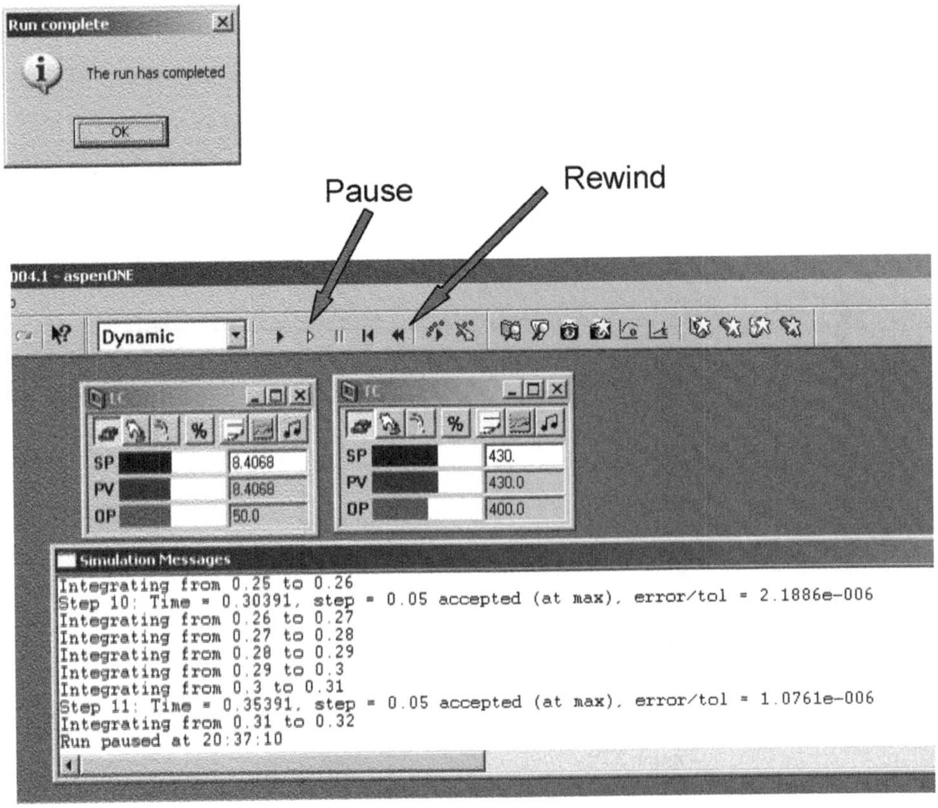

Figure 3.72 Running the simulation.

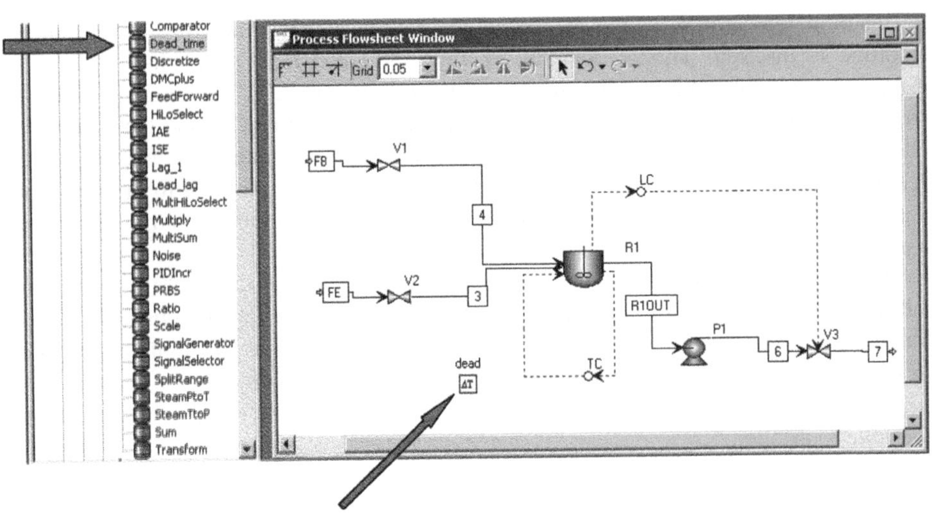

Figure 3.73 Inserting deadtime.

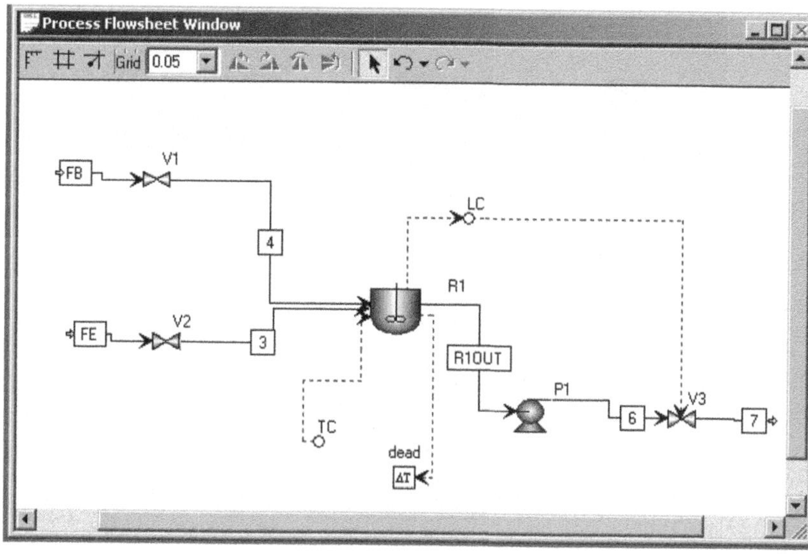

Figure 3.74 Connecting input to deadtime.

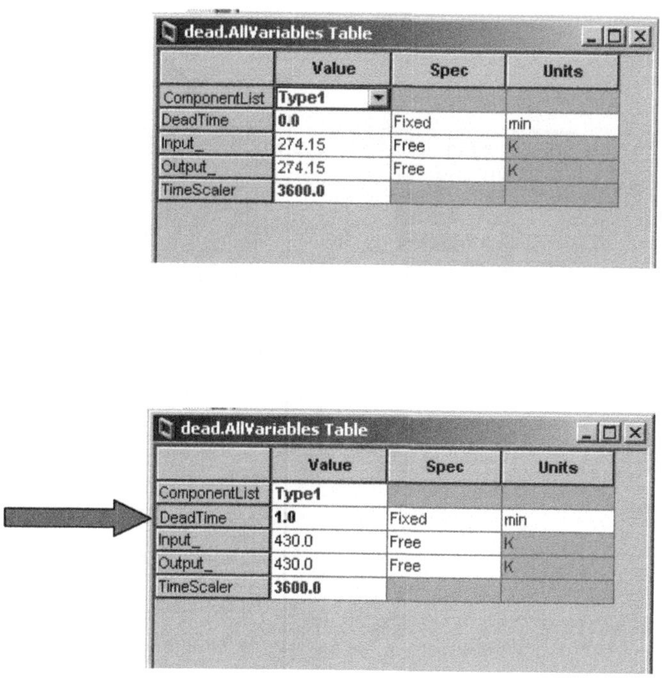
dead.AllVariables Table

	Value	Spec	Units
ComponentList	Type1		
DeadTime	0.0	Fixed	min
Input_	274.15	Free	K
Output_	274.15	Free	K
TimeScaler	3600.0		

dead.AllVariables Table

	Value	Spec	Units
ComponentList	Type1		
DeadTime	1.0	Fixed	min
Input_	430.0	Free	K
Output_	430.0	Free	K
TimeScaler	3600.0		

Figure 3.75 Specifying deadtime.

far right of the toolbar on the controller faceplate is the *Tune* button (see Fig. 3.77). The button next to it opens a stripchart on which the PV, SP, and OP signals of the controller are recorded as functions of time. Clicking the *Tune* button opens the window shown at the lower right in Figure 3.77. Before running the test, it is a good idea to change what

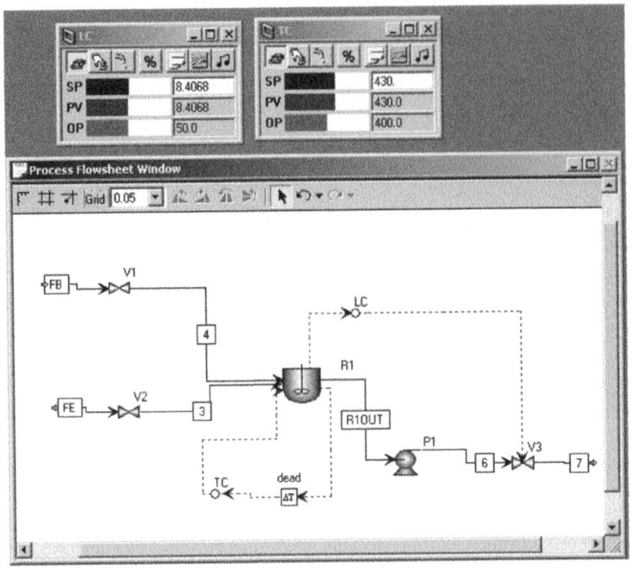

Figure 3.76 Flowsheet and faceplate with LC and TC.

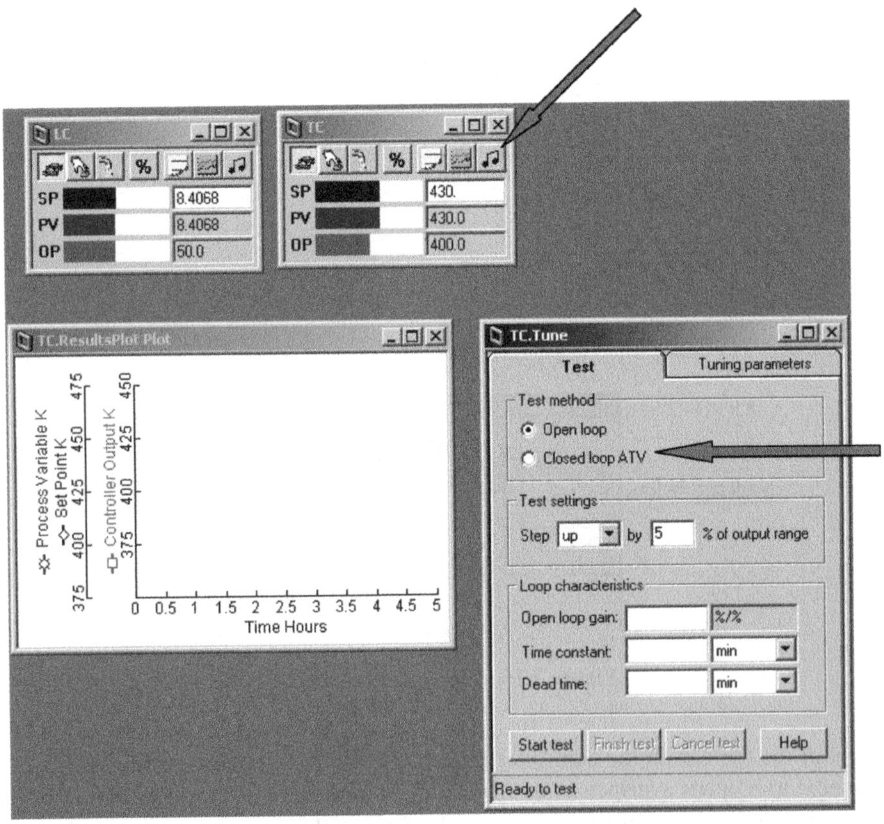

Figure 3.77 Running relay–feedback test on temperature controller.

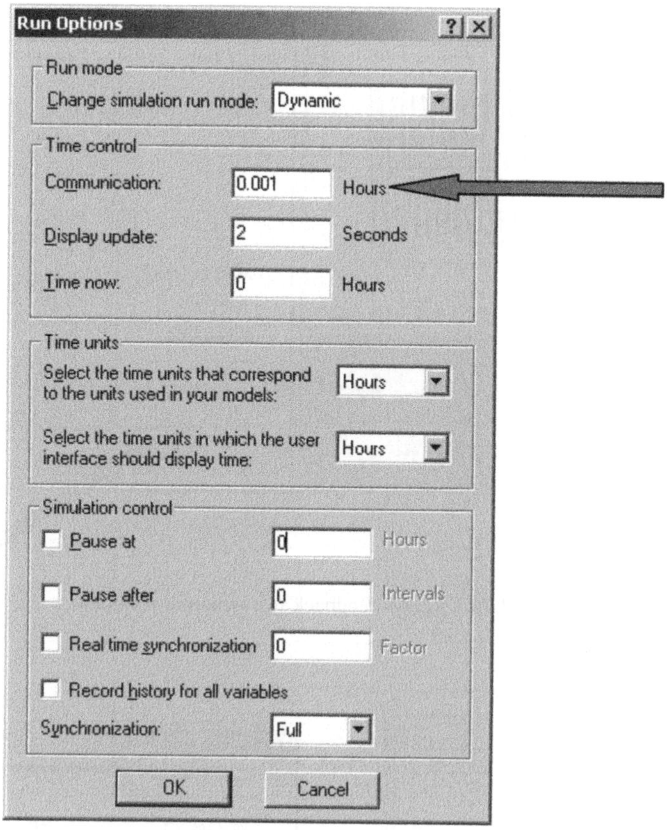

Figure 3.78 Shortening communication time.

Aspen calls the "communication time." This is the sampling period for sending information to the stripchart recorder. The default value (0.01 h) is usually too large to get good-looking plots, so it should be changed to 0.001 h by clicking *Run* on the toolbar at the top of the window and selecting *Run Options*. The window shown in Figure 3.78 opens, on which the *Communication Time* is set.

It is important to remember that a deadtime or several lags must be inserted in most control loops in order to run a relay–feedback test. To have an ultimate gain, the process must have a phase angle that drops below $-180°$. Many of the models in Aspen Dynamics have only a first-order transfer function between the controller variable and the manipulated variable. In the CSTR temperature controller example, the controlled variable is reactor temperature and the manipulated variable is medium temperature. The phase angle of a first-order process goes to only $-90°$, so there is no ultimate gain. The relay–feedback test will fail without the deadtime element inserted in the loop.

Click the *Closed loop ATV* bullet, start the simulation running, and click the *Start test* button. After several cycles, click the *Pause* button to stop the simulation and click the *Finish test* button at the bottom of the *Tune* window (see Fig. 3.79). The ultimate gain (3.73) and the ultimate period (4.8 min) are displayed, as shown in the left side of Figure 3.80. To calculate the controller tuning constants, click the *Tuning parameters* page tab on the *Tune* window and select either Ziegler–Nichols or Tyreus–Luyben.

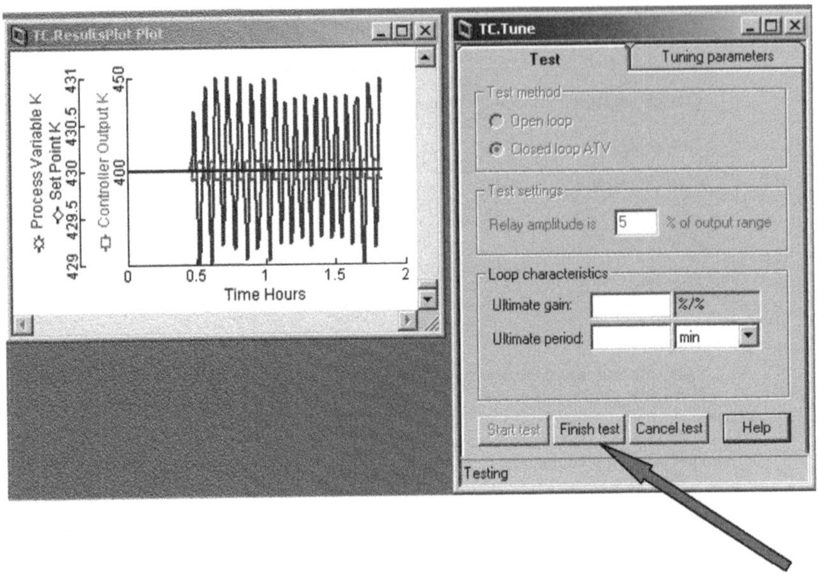

Figure 3.79 Relay–feedback test dynamic results.

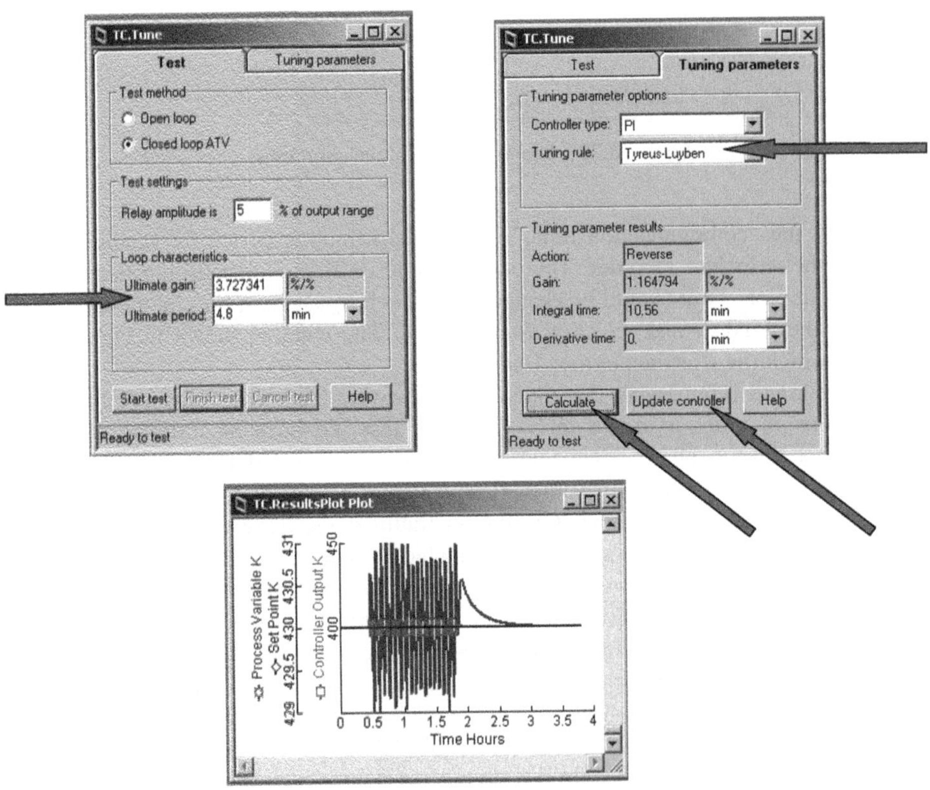

Figure 3.80 Relay–feedback test results.

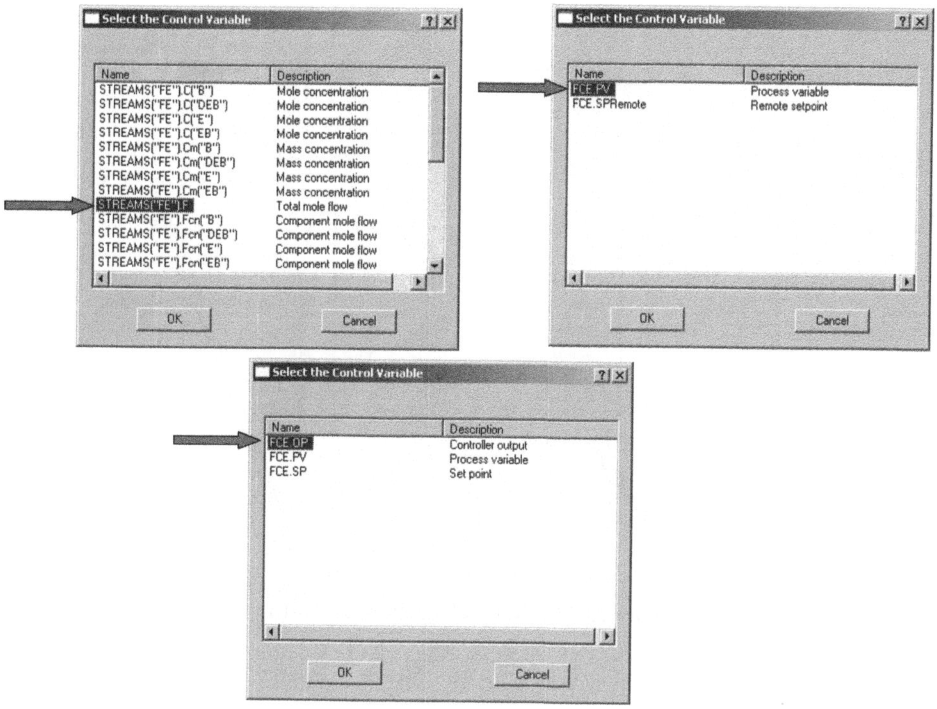

Figure 3.81 Installing the ethylene feedflow controller.

Then click the *Calculate* button shown on the right side of Figure 3.80. The calculated tuning can be used in the controller by clicking the *Update controller* button. The Tyreus–Luyben settings are $K_C = 1.16$ and $\tau_I = 10.6$ min. The plot at the bottom of the figure shows that these settings yield a stable temperature controller. When the simulation is restarted after the test, the temperature is driven to the setpoint value in about 30 min with a smooth trajectory.

Finally we now install a flow controller on each of the feedstreams. Figure 3.81 shows the two windows in which the molar flowrate of the ethylene feedstream (*FE*) is selected and specified as the PV signal to the flow controller *FCE*. The third window at the bottom of the figure shows that the OP signal of the flow controller is selected to send to the valve in the ethylene feedline V2. Double-clicking the icon places a faceplate in the window. Clicking the third button from the right on the faceplate opens the window shown below the faceplate on the left side of Figure 3.82. Clicking the *Initialize Values* button creates the faceplate and window shown in the middle of the figure. It is very important to note that the controller action is set to *reverse*. The tuning of a flow controller requires no dynamic analysis or testing since the loop is quite fast. Simply use a gain of 0.5 (dimensionless) and an integral time of 0.3 min. It also seems to avoid some problems with the integration algorithm if some filtering is used. This is done by clicking the *Filtering* page tab and type in a filter time constant of 0.1 min. Figure 3.83 shows the flowsheet and the controller faceplates with the two flow controllers, the level controller, and the temperature controller installed.

The simulation is run until all variables stop changing and the desired reactor temperature is achieved. These steady-state conditions should be saved as initial

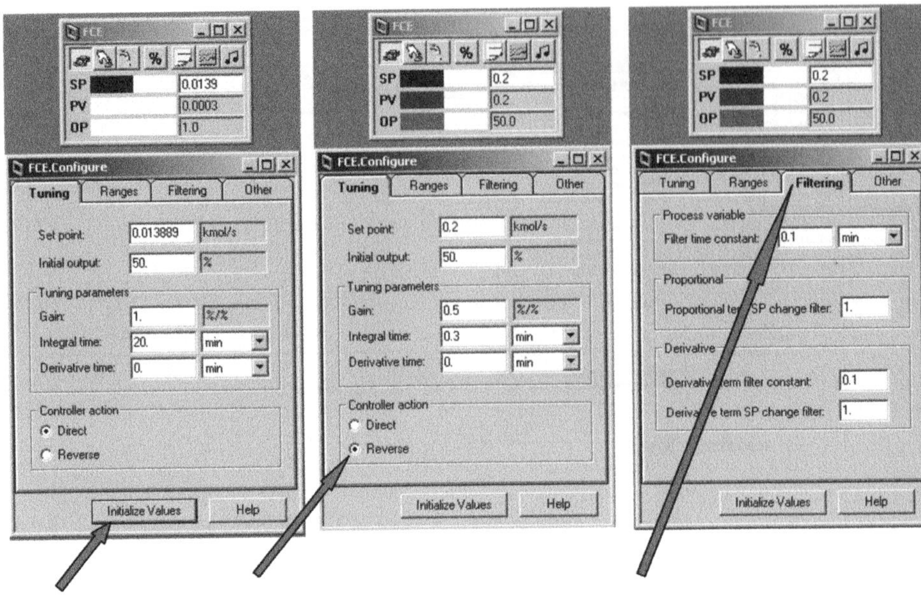

Figure 3.82 Setting up the flow controller.

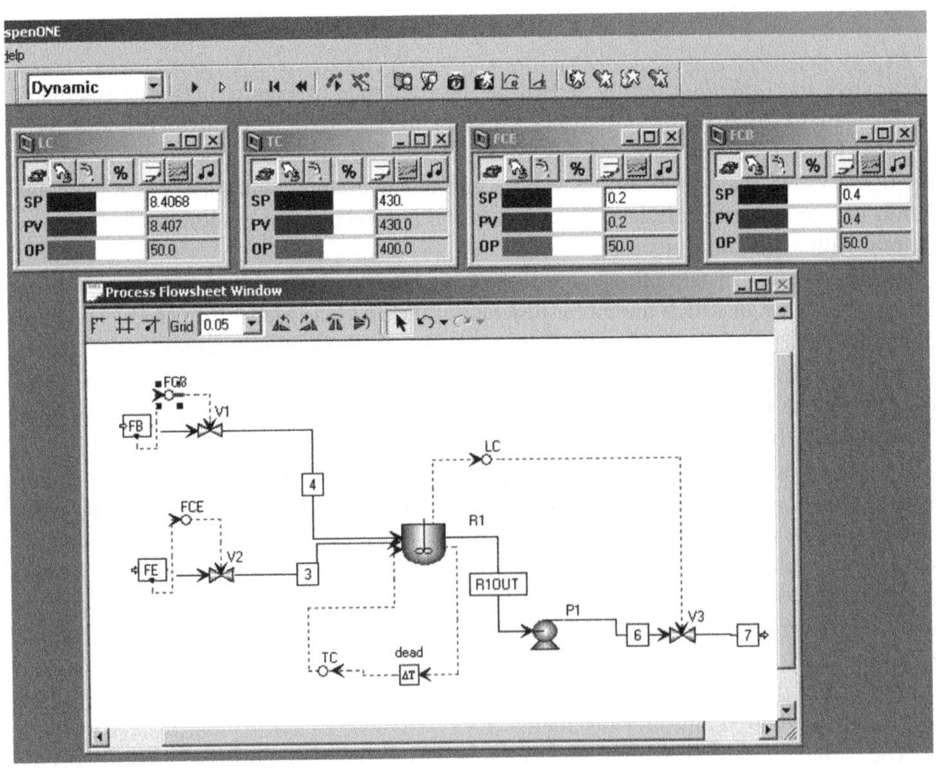

Figure 3.83 Flowsheet and faceplates.

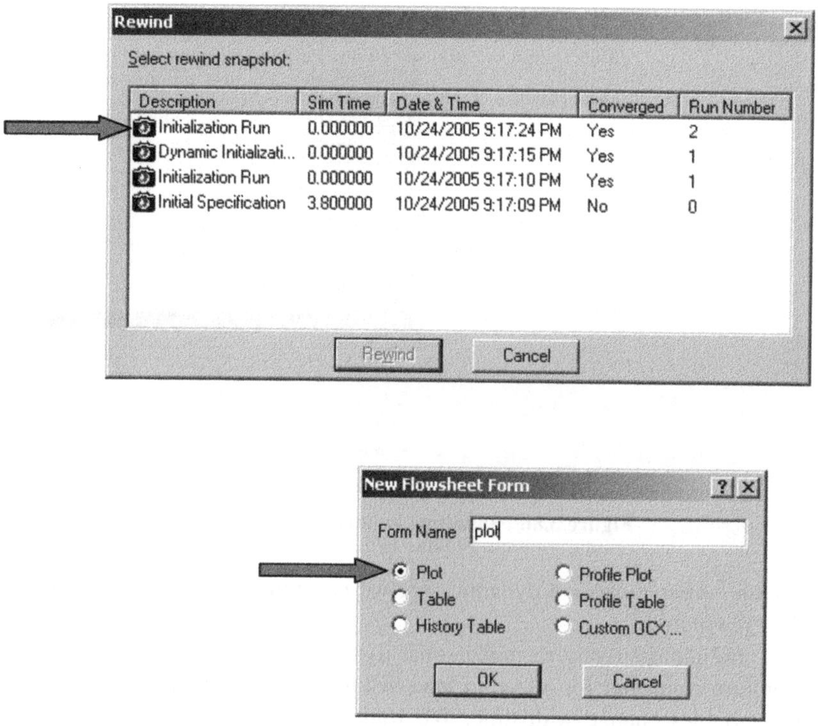

Figure 3.84 Saving steady-state conditions.

conditions for future runs. To store this steady state, select *Initialization* on the drop-down menu and click *Run*. Then select *Dynamic* and click the *Rewind* button (see Fig. 3.72). The window shown at the top of Figure 3.84 opens on which we select *Initialization Run*. This sets all variables to their steady-state values. Save the Aspen

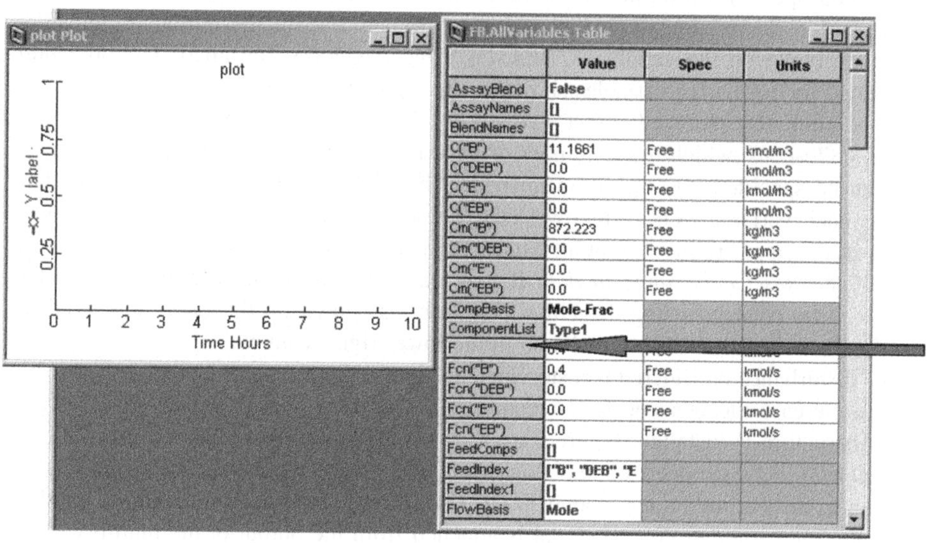

Figure 3.85 Installing a variable on the plot.

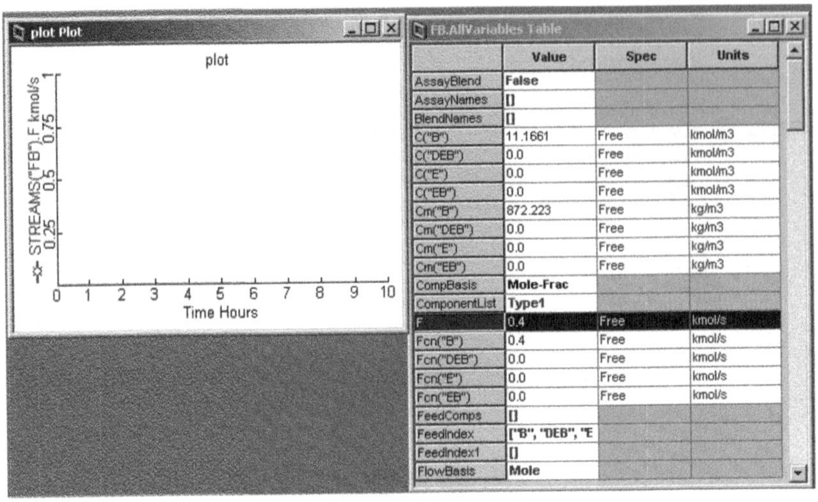

Figure 3.86 Installing a variable on the plot.

Dynamics file. After making a dynamic run, we can return to these stored initial conditions by "rewinding."

We are now almost ready to make some dynamic runs to test the effectiveness of the temperature controller for rejecting load disturbances. We will want to see how the important variables change with time. We want plots of reactor temperature, cooling water temperature, heat transfer rate, and the two feed flowrates versus time. A plot is installed by clicking the *Tools* button at the top of the window and selecting *New form* and then clicking the bullet for *Plot* as shown at the bottom of Figure 3.84. Insert an appropriate name and click *OK*. The plot shown to the left in Figure 3.85 opens. To plot the molar flowrate of the benzene feed *FB*, click on the stream, press the left mouse button, and select *Forms* and *AllVariables*. The window on the right in Figure 3.85 opens. Position the cursor on *F* and click. Then click again and holding the mouse button down, drag it to the plot window and release the button. Figure 3.86 shows the plot with the variable *STREAMS("FB").F kmol/s"* installed. The procedure is repeated for four more variables shown at the top of Figure 3.87.

Right click the plot and select *Properties*. The window shown at the bottom of Figure 3.87 opens, which has a number of page tabs on which various properties of the plot can be specified. On the *AxisMap* page tab, click the *One for Each* button. This permits each variable to have its own axis. These are defined on the *Axis* page tab, as shown in the upper right window of Figure 3.88. Remove the checkmark in the box in front of the *Reset axis range to data* if you want the axis scale to remain fixed. The labels can be changed using the *Label* page tab, as shown on the upper left window. The time axis can also be set, as shown in the lower right window.

The disturbance considered is a step increase in the flowrate of the ethylene feed. We want to increase the benzene feed whenever the ethylene feed is increased, so a ratio control structure is installed. Figure 3.89 shows a *Mulitply* block selected from the list of *ControlModels*, dropped on the flowsheet, and renamed *ratio*. A control signal is attached to the *FE* stream and connected to *ratio.Input1*, as shown in the upper window of Figure 3.90. Another control signal is attached from the output of the multiply block to the setpoint of the benzene flow controller. Clicking the *ratio* icon, clicking the right

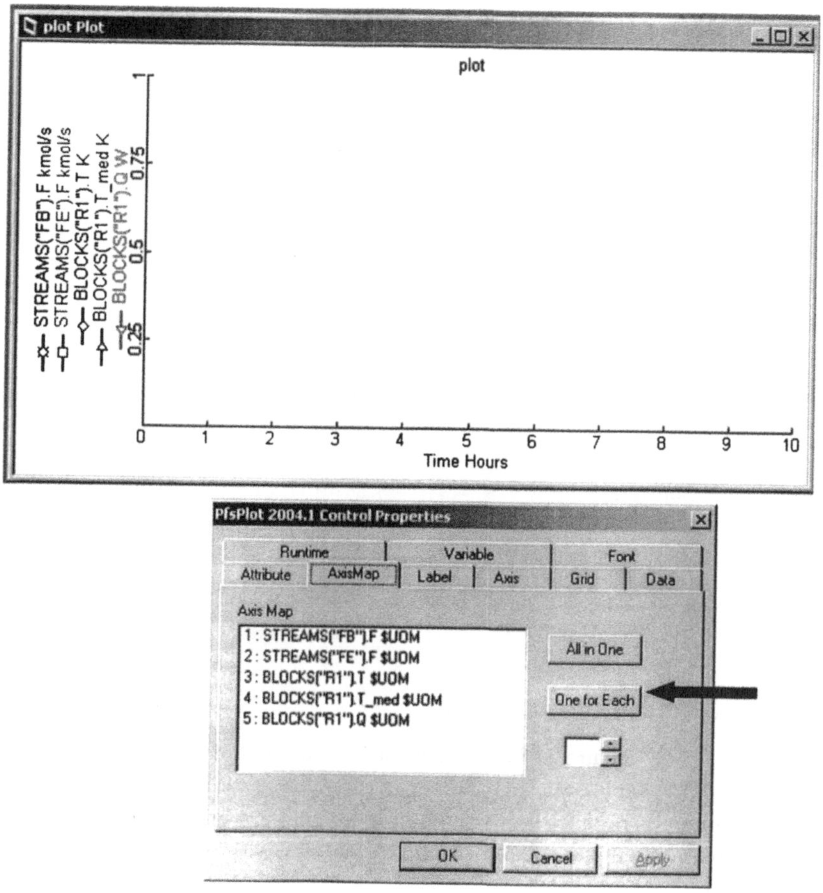

Figure 3.87 Plot with five variables inserted.

mouse button, and selecting *Forms* and *AllVariables* opens the window shown at the bottom left of Figure 3.90. The fixed *Input2* is specified to be 2 so that the ratio of benzene to ethylene feedstreams is 2. The lower right window shows variable values after an *Initialization* run has been made. Figure 3.91 shows the controller faceplates and the plot. Note that the benzene feedflow controller *FBC* is on *Cascade* with its setpoint coming from the *ratio* block.

When the *Run* button is clicked, the simulation starts at time = 0. To pause the run at some point in time, click the *Run Options* button on the top toolbar and select *Pause Time*. We will run with no disturbance for 0.2 h, so the pause time is specified as shown in Figure 3.92. Clicking the *Run* button starts the simulation. When it pauses, change the pause time to 2 h. The setpoint of the ethylene flow controller is changed to 0.24 by highlighting the number in the box on the *SP* line of the faceplate (see Fig. 3.93), typing in the new setpoint signal, and hitting *Enter* on the keyboard.

Figure 3.93 shows the results. The ethylene feed increases, and the ratio increases the benzene feed. The reactor temperature climbs to about 431.6 K in 0.08 h from its desired value of 430 K. The temperature controller decreases the cooling water temperature to about 395 K from its initial value of 400 K to handle the increase in throughput.

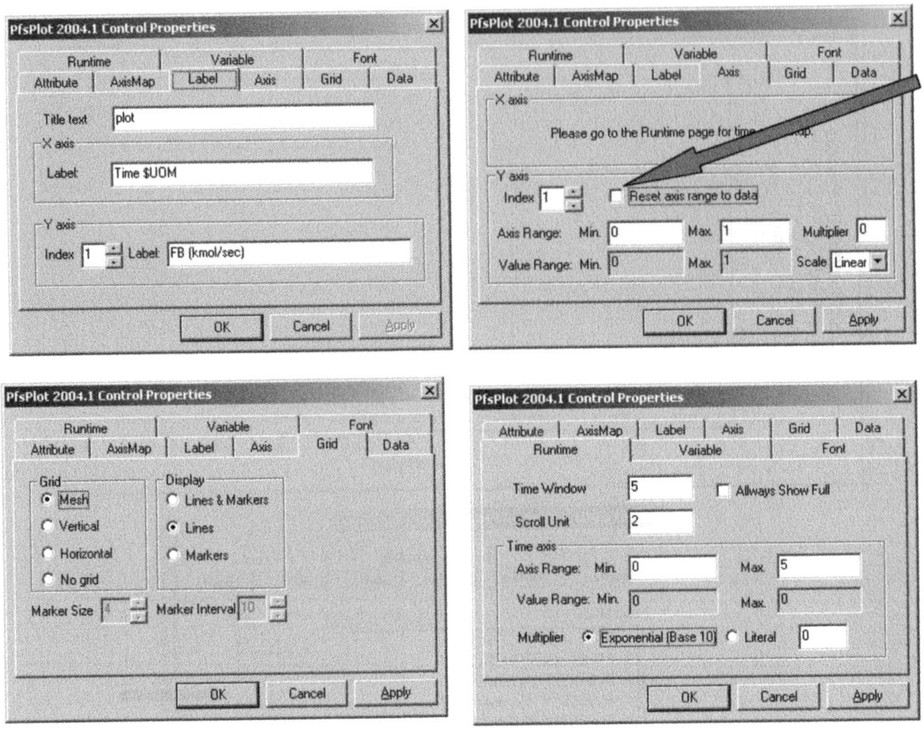

Figure 3.88 Changing labels and axis scales.

Reactor temperature is returned to its setpoint in about 1 h. Figure 3.94 shows responses for a 20% decrease in ethylene feed flowrate.

3.5.3 Results with Several Heat Transfer Options

The results above are for the "constant medium temperature" heat transfer option. Effective temperature control is obtained. In this section we want to compare these results with

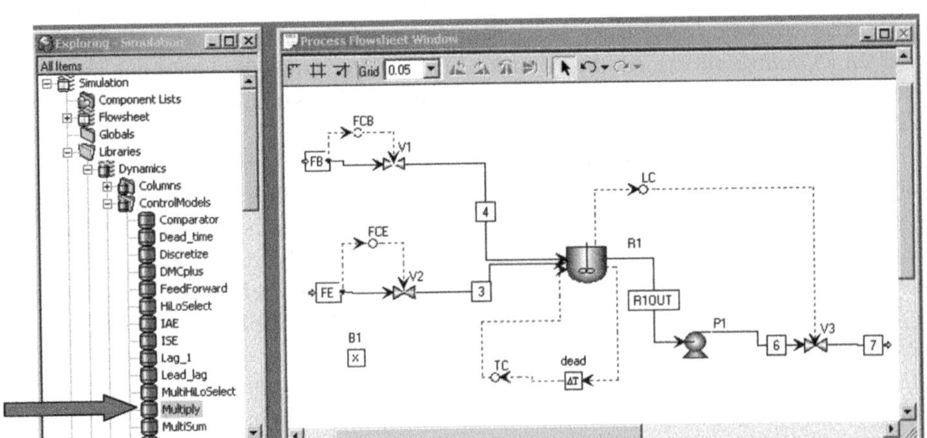

Figure 3.89 Selecting multiplier block.

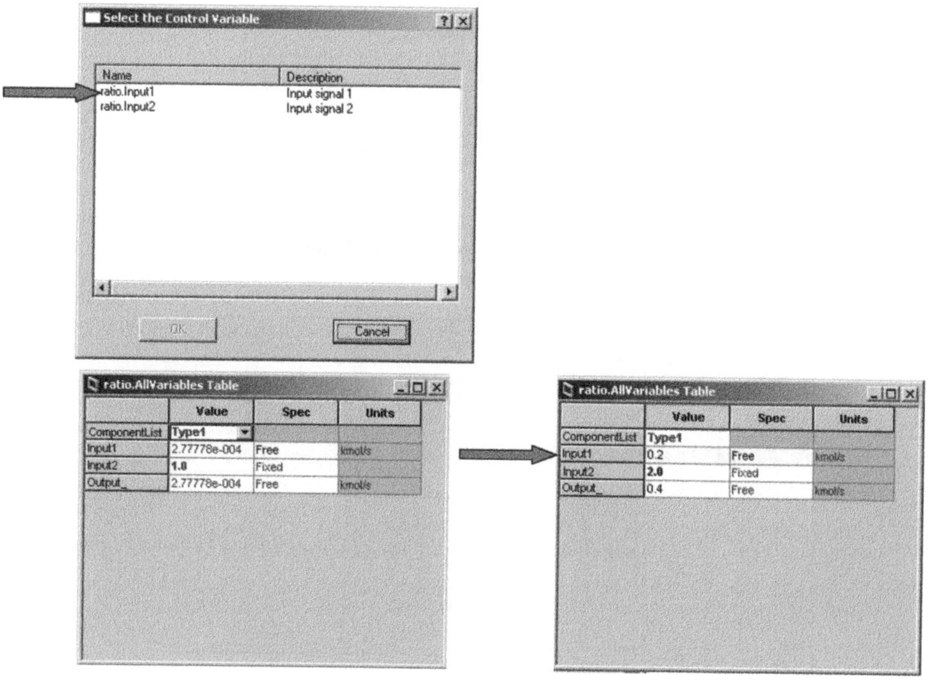

Figure 3.90 Installing ratio configuration.

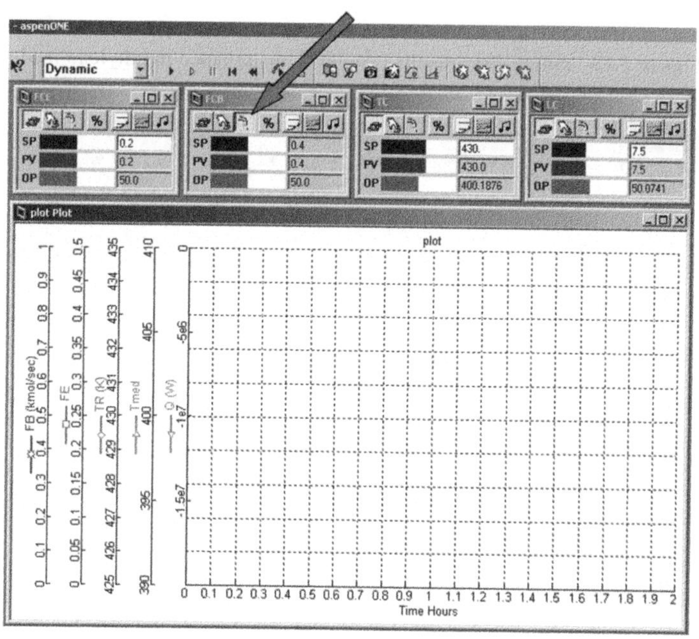

Figure 3.91 Benzene flow controller on cascade.

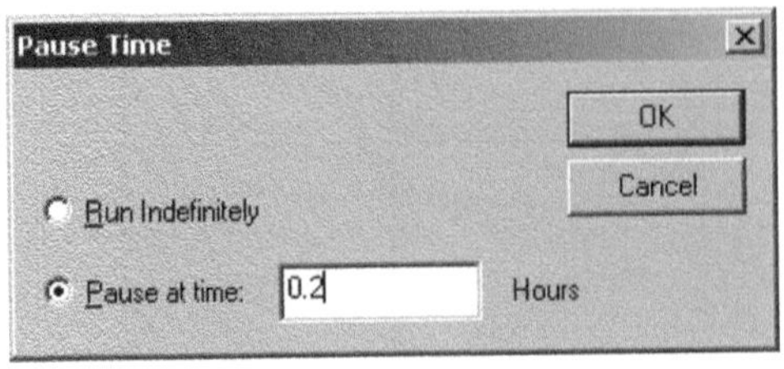

Figure 3.92 Pausing the simulation.

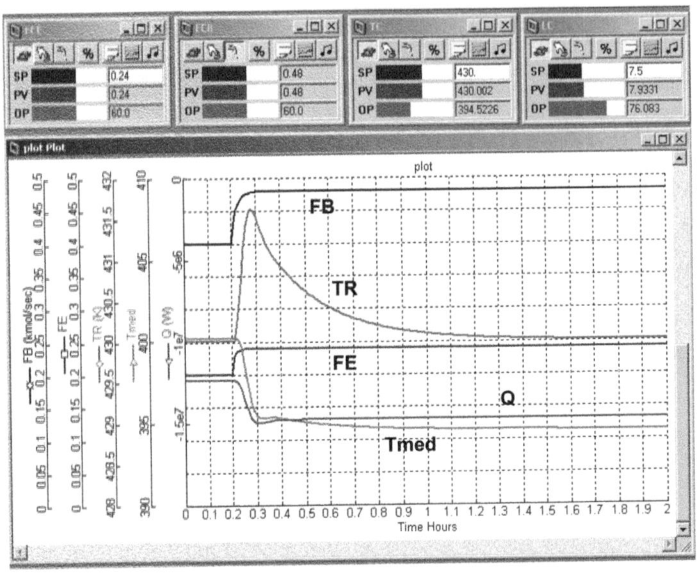

Figure 3.93 Simulation results for +20% disturbance in ethylene feed.

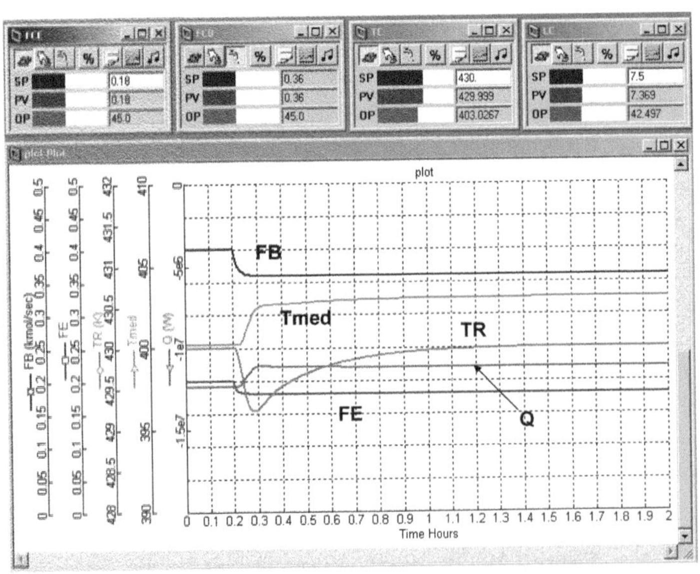

Figure 3.94 Simulation results for −20% disturbance in ethylene feed.

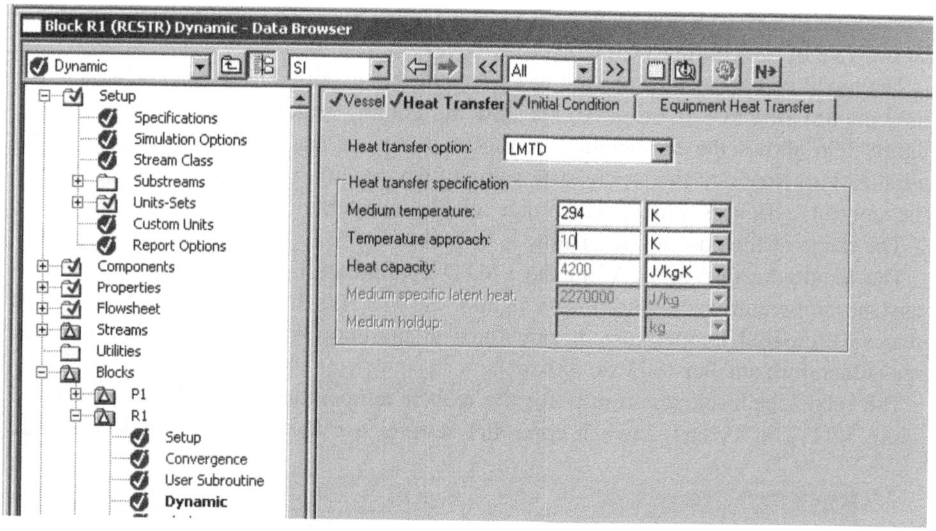

Figure 3.95 Selecting the LMTD heat transfer option.

those for two other heat transfer options. The file in Aspen Plus is modified and exported to Aspen Dynamics.

LMTD Figure 3.95 shows the parameters that must be specified when this heat transfer option is selected. The inlet temperature of the cooing water is set at 294 K. The temperature approach is set at 10 K. This means that the cooling water leaves the system at a

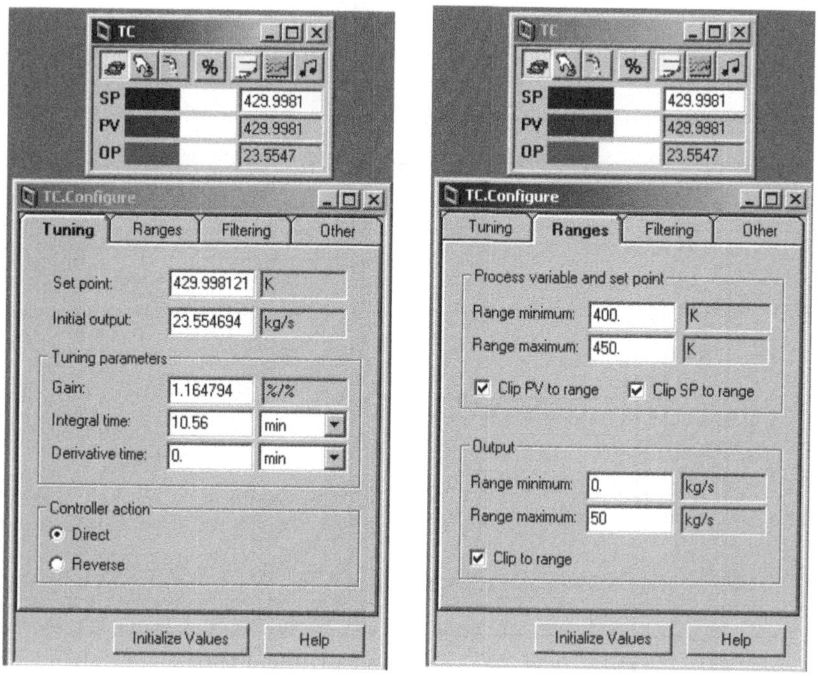

Figure 3.96 Specifying temperature controller parameters with LMTD.

temperature that is 10 K lower than the reactor temperature. The manipulated variable is the flowrate of the cooling water.

The modified file is saved under a new filename and exported to Aspen Dynamics. The same control structure is installed, but now the cooling water flowrate is manipulated. Figure 3.96 shows the temperature controller faceplate and configuration pages after initialization. Note that the steady-state value of the cooling water flowrate is 23.55 kg/s. The control action is direct. The range of the temperature transmitter is 400–450 K. The range of the cooling water flow is 0–50 kg/s.

The required value of "UA" in the LMTD configuration is lower than that for the constant-temperature configuration (258,200 vs. 415,500 J s^{-1} K^{-1}) because of the greater differential temperature driving force at the cooling water inlet. The required area is therefore smaller (303 vs. 488 m^2).

The relay–feedback test results for the reactor temperature controller are shown in Figure 3.97. The Tyreus–Luyben controller settings are $K_C = 2.25$ and $\tau_I = 11.9$ min.

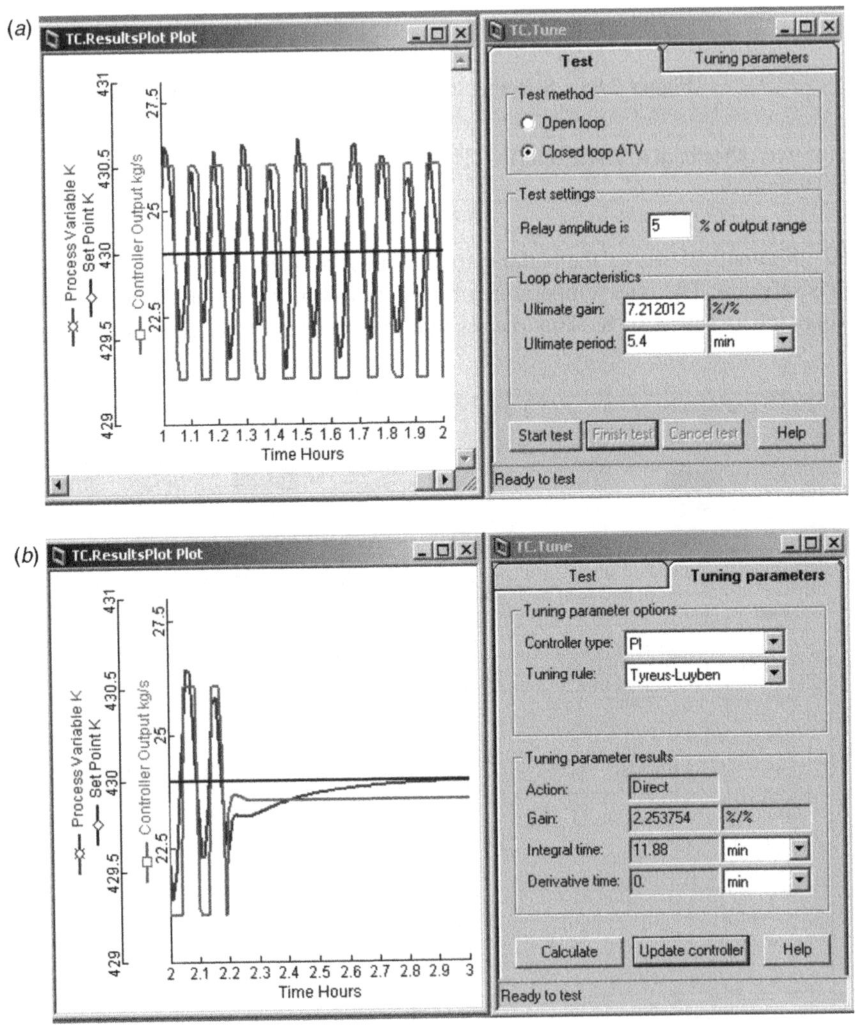

Figure 3.97 Relay–feedback test with LMTD.

Note that this controller gain is higher than that found for the constant-temperature heat transfer option.

Figure 3.98*a* shows results for a 20% increase in setpoint of the ethylene feed flow controller. The ethylene feed increases, and the ratio increases the benzene feed. The reactor temperature climbs to about 432 K in 0.1 h from its desired value of 430 K. The temperature controller increases the flowrate of the cooling water from 23.55 up to about 30 kg/s. Figure 3.98*b* shows results for a 20% decrease in setpoint of the ethylene feedflow controller.

Dynamic Heat Transfer Option Figure 3.99 shows the parameters that must be specified when this heat transfer option is selected. The manipulated variable with this option is the flowrate of the cooling water. The temperature of the inlet cooling water is

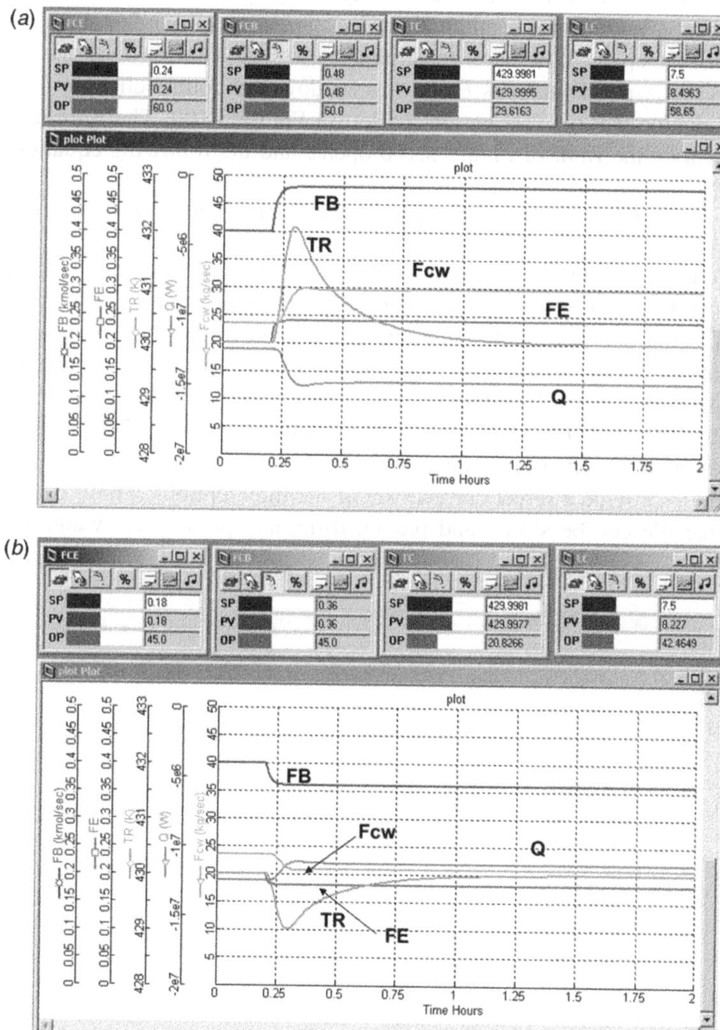

Figure 3.98 Response for +20% disturbance (*a*) and −20% disturbance (*b*) in feed with LMTD.

set at 294 K. The temperature approach is set at 30 K. This means the temperature of the perfectly mixed cooling water is 400 K. This is the same temperature used in the constant-temperature option, so the "UA" is the same, and the required heat transfer area is the same (488 m^2). The tube volume of 0.0254-m-diameter tubes, 6 m in length to provide this much heat transfer area is 6.1 m^3. Therefore, the medium holdup specified is 6100 kg, as shown in Figure 3.99. It should be noted that this heat transfer option is the only one that has some dynamic component, that is, the thermal inertia of the cooling water contained in the coil, jacket, or external heat exchanger is explicitly considered in the "medium holdup" term.

When the Aspen Plus file was exported to Aspen Dynamics, a bug was discovered that caused the simulation to not run properly. The medium temperature ("T_med") was incorrectly specified to be "fixed" (held constant). The T_med can be changed to "free" by clicking on the reactor icon and selecting *Forms* and *All Variables* and making the change. However, an error message was displayed saying that the simulation was underspecified, namely, more variables than equations. The "workaround" provided by Aspen Support is to set up a "flowsheet constraint" equation. This is done by clicking on Flowsheet in the Aspen Dynamics *Exploring* window and double-clicking the flowsheet symbol (the double bars), as shown on the left in Figure 3.100. The *Editing Flowsheet* window shown on the right of Figure 3.100 opens, and the following equation is typed:

$$Blocks(\text{"}R1\text{"}).T\_\text{-}med = Blocks(\text{"}R1\text{"}).Tmed\_out.$$

Then right click the editor window and select *Compile*. This solves the problem.

The temperature controllers is retuned giving controller tuning constants $K_C = 1.59$ and $\tau_I = 18.5$ min. A one-minute deadtime is included in the loop. This integral time is larger than the other two options because the dynamics of the cooling system are incorporated in the model. Results are shown in Figure 3.101 for a +20% change in feed flowrate. The peak reactor temperature deviation is about 3.7 K, which is larger than that seen with the other heat-removal options.

It is useful to provide a direct comparison of the three options. To do this, the results from the three file can be stored and plotted using any plotting software. To save the

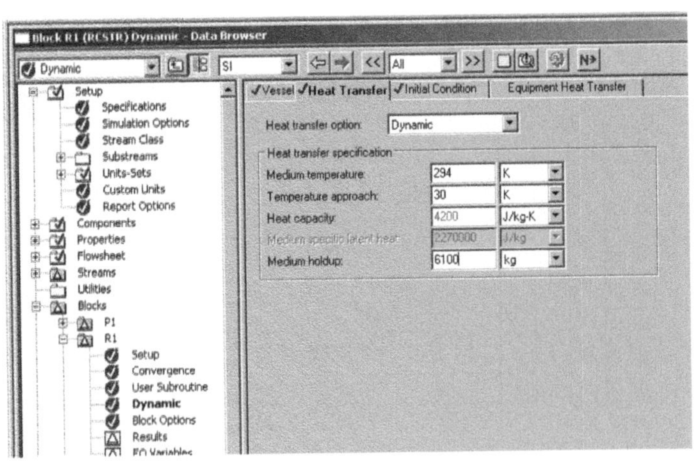

Figure 3.99 Selecting the *Dynamic* heat transfer option.

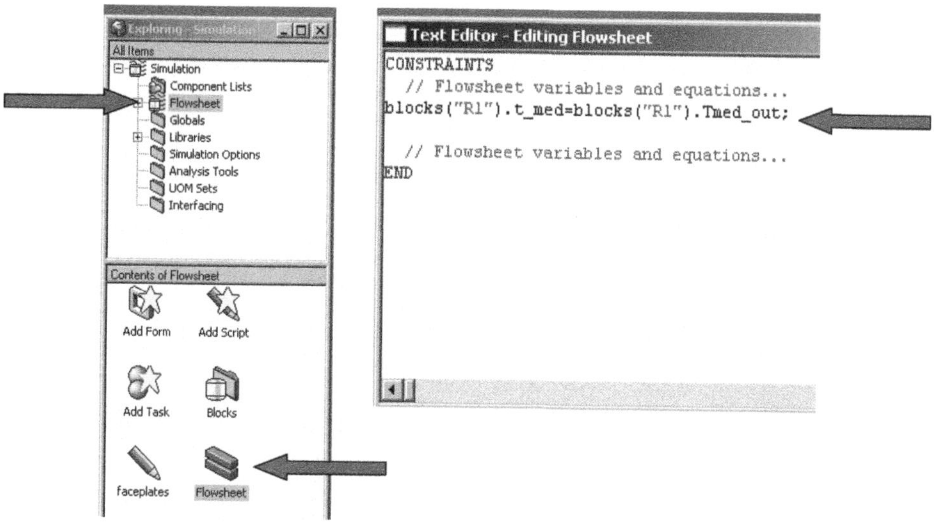

Figure 3.100 Flowsheet constraint to correct bug in *Dynamic* option.

data of a dynamic run, right-click the plot window after the run has stopped. Select *Show as History*, highlight the resulting table (see Fig. 3.102), copy it, and save it as a file. The files were loaded into Matlab and plotted. Results are shown in Figures 3.103 and 3.104. The Dynamic option shows larger deviations, as expected, but it is probably more realistic of the true performance to be expected in an actual plant reactor. It would certainly give a more conservative design.

It is important to remember that with all these options, the heat transfer area required is significantly greater than that attainable with just jacket cooling.

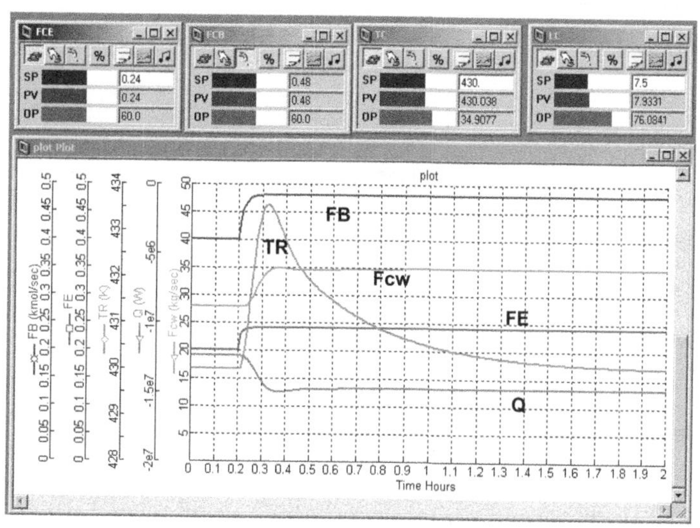

Figure 3.101 Response with *Dynamic* option; +20% feed.

plot History Table

Time	STREAMS("FB").F	STREAMS("FE").F	BLOCKS("R1").T	BLOCKS("R1").Q	BLOCKS("R1").FI_med
Hours	kmol/s	kmol/s	K	W	kg/s
0.0	0.4	0.2	429.993	-1.24626e+007	23.55
0.01	0.4	0.2	429.993	-1.24624e+007	23.54
0.02	0.4	0.2	429.993	-1.24623e+007	23.54
0.03	0.4	0.2	429.993	-1.24623e+007	23.54
0.04	0.4	0.2	429.994	-1.24624e+007	23.54
0.05	0.4	0.2	429.994	-1.24624e+007	23.54
0.06	0.4	0.2	429.994	-1.24624e+007	23.54
0.07	0.4	0.2	429.994	-1.24625e+007	23.54
0.08	0.4	0.2	429.994	-1.24625e+007	23.54
0.09	0.4	0.2	429.995	-1.24626e+007	23.54
0.1	0.4	0.2	429.995	-1.24626e+007	23.54
0.11	0.4	0.2	429.995	-1.24626e+007	23.
0.12	0.4	0.2	429.995	-1.24627e+007	23.
0.13	0.4	0.2	429.995	-1.24627e+007	23.
0.14	0.4	0.2	429.996	-1.24628e+007	23.
0.15	0.4	0.2	429.996	-1.24628e+007	23.
0.16	0.4	0.2	429.996	-1.24628e+007	23.
0.17	0.4	0.2	429.996	-1.24628e+007	23.
0.18	0.4	0.2	429.996	-1.24628e+007	23.
0.19	0.4	0.2	429.996	-1.24628e+007	23.
0.2	0.4	0.2	429.996	-1.24628e+007	23.
0.21	0.382969	0.186076	429.992	-1.24624e+007	23.54
0.22	0.37462	0.183151	429.916	-1.24591e+007	23.55
0.23	0.369109	0.181659	429.771	-1.24228e+007	23.50
0.24	0.365573	0.180876	429.589	-1.23103e+007	23.27
0.25	0.363402	0.180472	429.402	-1.21308e+007	22.8
0.26	0.36206	0.180255	429.24	-1.1918e+007	22.42
0.27	0.361285	0.180146	429.124	-1.17023e+007	21.95
0.28	0.360791	0.180082	429.05	-1.15087e+007	21.52

Figure 3.102 Show data from plot as history.

3.5.4 Use of RGIBBS Reactor

As mentioned in Chapter 2, it is sometimes convenient to use the "RGIBBS" reactor model instead of the "RCSTR" model illustrated above. If the reactions are known to be reversible and the forward and reverse reaction rates are fast, a reasonable residence time in the reactor will produce an exit stream with chemical equilibrium conditions. In

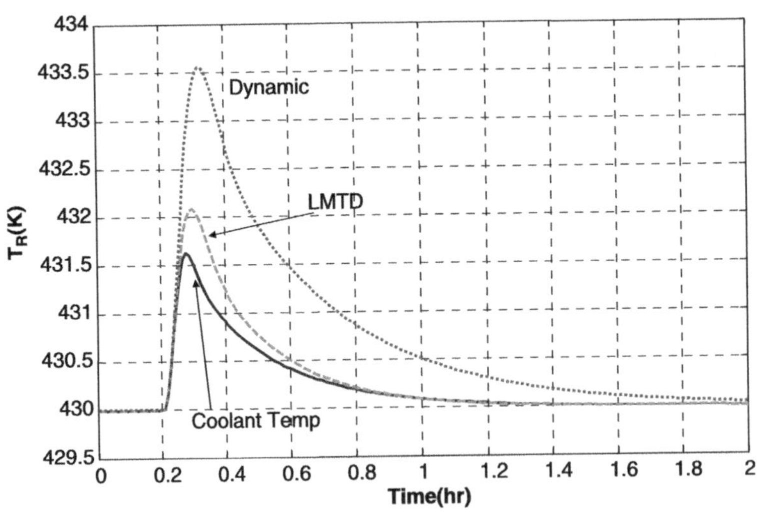

Figure 3.103 Comparison of CSTR heat transfer options; reactor temperatures.

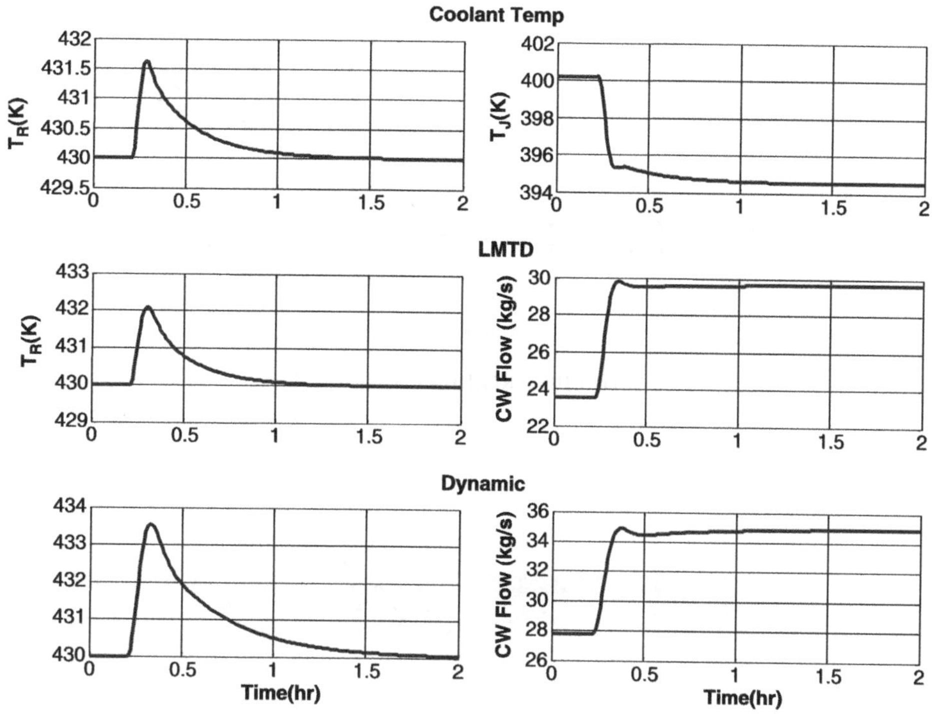

Figure 3.104 Comparison of CSTR heat transfer options; temperature and manipulated variables.

this case, the RGIBBS reactor is much easier to set up and use. No reactions need to be defined, and no kinetic parameters are required. Given the chemical components in the feed, the reactor temperature and pressure are specified, and the composition of the reactor exit stream that minimizes the Gibbs free energy is calculated.

To use this model for dynamics, the size of the reactor vessel must be specified. For example, if a 10 min residence time is known to be adequate to achieve equilibrium conditions, the reactor diameter and length are selected to give the required volume.

There are two options as to what type of approximate dynamics to assume: CSTR or PFR. The former uses a first-order lag with the appropriate time constant based on the vessel residence time. The latter used a deadtime.

To briefly illustrate this option, let us consider the following liquid-phase reversible reaction:

$$\text{Methanol} + \text{acetic acid} \Longleftrightarrow \text{methyl acetate} + \text{water} \tag{3.49}$$

A reactor feed of 50 mol% methanol and 50 mol% acetic acid is fed to a reactor operating at 400 K and 5 atm. The RGIBBS reactor is selected, as shown in Figure 3.105. Figure 3.106 shows that the reactor temperature and pressure are specified, and the option to select all components as products of the reaction is selected.

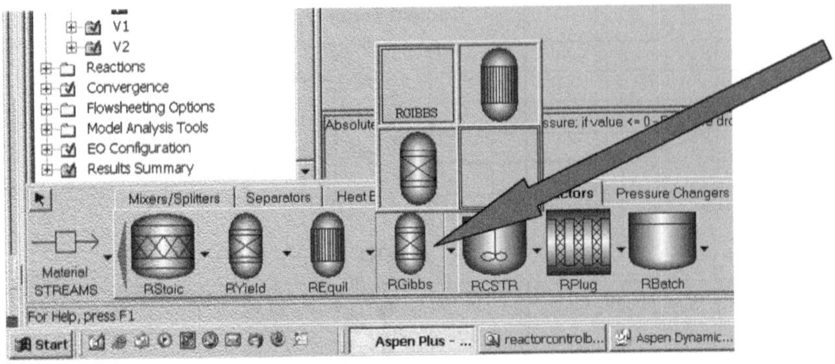

Figure 3.105 RGIBBS reactor model.

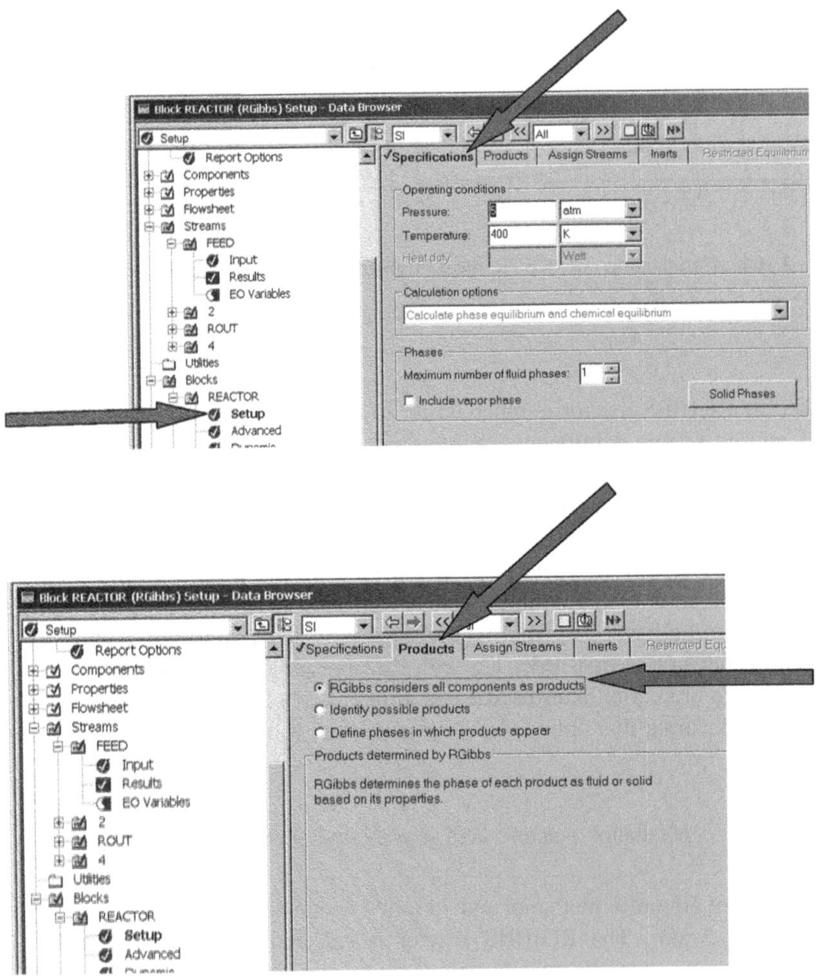

Figure 3.106 Setting up the RGIBBS reactor.

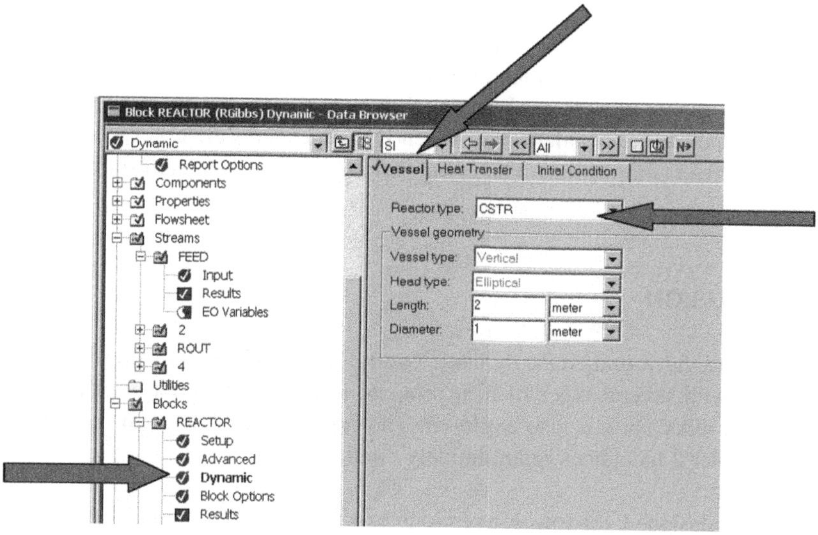

Figure 3.107 Specifying *Dynamic* for RGIBBS reactor.

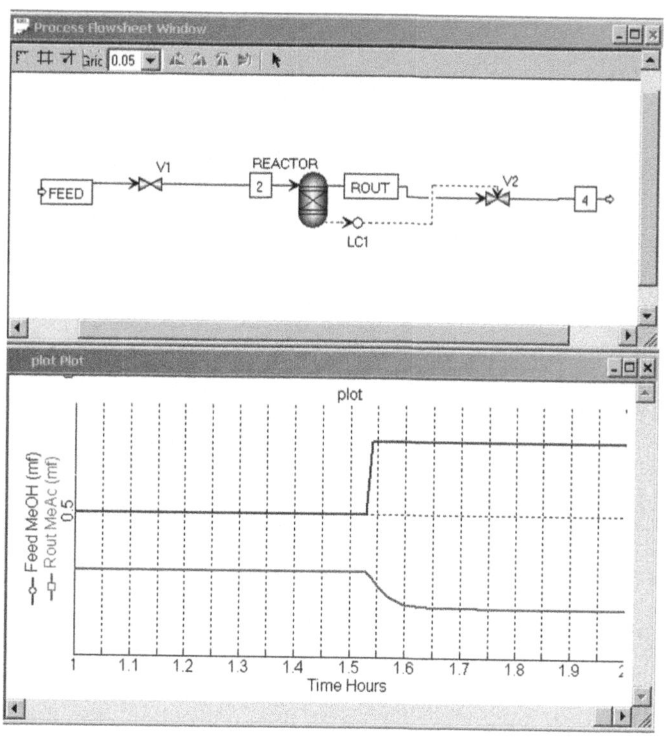

Figure 3.108 Dynamic response of RGIBBS reactor as CSTR.

Under the *Dynamics* item, the vessel is specified to have CSTR-type dynamics with a diameter of 1 m and a length of 2 m, as shown in Figure 3.107. This corresponds to a residence time of 5 min for the given feed flowrate (0.1 kmol/s). The file is pressure-checked and exported as a pressure-driven dynamic simulation.

Results of running the simulation in Aspen Dynamics are shown in Figure 3.108 for a step change in feed composition. The compositions of the exit stream show a first-order lag response.

3.6 CONCLUSION

In this chapter we have studied the dynamic controllability of CSTRs with several types of reactions and with several types of heat removal schemes. The control of a CSTR in a flowsheet with other units is also explored. The predictions made from the steady-state results of Chapter 2 have been quantitatively confirmed:

1. Reactors designed for low conversions can present severe control problems with exothermic irreversible reactions.
2. Large heat transfer area improves reactor controllability.

CHAPTER 4

CONTROL OF BATCH REACTORS

The batch reactor is perhaps the most "natural" of all chemical reactors. Most events in the natural world are time-varying batch events. They start from some initial conditions and proceed to change dynamically with time over some batch time. The whole cycle of animal and plant life is a batch operation. The conception, gestation, birth, growth, and death of an animal occur in a batch cycle. Our first exposure to a chemical reaction is probably in a high school chemistry class in which the reaction takes place batchwise in a test tube or flask.

The batch chemical reactor is widely used by the chemist in laboratory studies of the chemistry of various systems. Reactions are run at different temperatures, with different ratios of reactants, with different catalyts, and for different batch times in order to explore the effects of various parameters on conversion, yield, and reaction rates. The laboratory batch reactor is usually quite small (test tube, flask, or Parr bomb). The surface-to-volume area is very large, so heat transfer is extremely good. Placing the small reactor in a constant-temperature bath pretty well guarantees that the reaction temperature is the bath temperature. Of course, the scaleup of batch reactors to pilot plant and industrial scale can present significant challenges because of the large reduction in area-to-volume ratio and the resulting problems in temperature control. These are the same as discussed for CSTRs in Chapter 3.

At the very beginning of the chemical industry in the seventeenth and eighteenth centuries, most commercial reactors were batch. They were simply large versions of the chemist's laboratory apparatus. The production of gunpowder, iron, and soap are classical examples. During the period of rapid growth of the chemical industry in the twentieth century, continuous reactors were developed for many reactions because they proved to be more economical in some systems. Continuous reactors are most often used for high-throughput plants.

Chemical Reactor Design and Control. By William L. Luyben

However, batch reactors have very important inherent kinetic advantages for some reaction systems, so they continue to be fairly widely used, even when production rates are high. Two important classical examples are polymerization and fermentation. Batch polymerization reactors permit the production of polymer with a more narrow molecular weight distribution. Batch fermentors use the lifecycle of the "bugs" to grow the organisms by feeding them substrate and letting them produce the desired chemical.

Batch reactors are also frequently used in situations where production rates are low, such as specialty chemicals. The batch reactor is quite flexible and can be used to produce a number of different products under a variety of conditions in the same vessel. For small production rates, capital investment for a batch reactor is low.

The safety issues in batch reactors are more pronounced than in continuous reactors, and this consideration should be kept in mind when making a decision about what type of system to design. Once a continuous reactor is up and running, it can usually be operated in a steady and safe mode by have a good control system and eliminating disturbances. A batch reactor, however, must follow a sometimes complex recipe of operating steps for each batch run. Any equipment or instrumentation malfunction or any human error can lead to unsafe conditions. For example, if the operator mistakenly adds a double charge of catalyst, the resulting rapid reaction rate could cause safety valves or rupture disks to open, which could result in atmospheric pollution. In extreme cases the vessel may overpressure and come apart, sending destructive and deadly missiles through the plant that can cause massive damage to other equipment and inflict numerous casualties.

Several types of batch reactors are considered in this chapter. In a "pure" batch reactor all the reactants are charged initially to the vessel. The reactor is heated to the desired reaction temperature and products are formed. At the termination of the batch cycle, the products are removed. The inherent problem with this type of batch reactor is that all the "fuel" is sitting in the vessel. If temperature is increased too quickly or if adequate heat transfer area is not available, an exothermic reaction can easily cause a runaway.

A "semi"batch reactor comes in two flavors. The first is when some material is *fed* to the reactor during the batch cycle. There is an initial charge of some of the reactants, but the rest of the reactants or catalysts are continuously fed into the reactor during the cycle. This is called a "fed-batch reactor." Many batch polymerization reactors operate in the "fed-batch mode," with monomer fed into the reactor during the batch cycle. The fed-batch reactor has the inherent advantage that the concentration of the limiting reactant (or catalyst) can be kept low enough to prevent runaway reactions.

The second type of semibatch reactor is when some material is *removed* from the reactor during the batch. The material is typically one of the products of the reaction. A common example is in fermentors producing ethanol in which the byproduct carbon dioxide is vented off during the batch cycle.

The design and control problems for batch reactors are inherently more difficult than for CSTRs because of the time-varying nature of the batch process. The typical design problem is to be given the desired annual production rate. The myriad of variables that define a batch reactor and its operation makes finding the "best" design (equipment and operating conditions) quite complex. Some of the many design choices include the size of the vessel, the number of vessels, the heat transfer area required, the time-varying temperature trajectory, and the time to stop the batch.

One of the major control issues with batch reactors is the tuning of the controllers. Process time constants and gains vary over the course of the batch because of the large changes in the values of variables. The system is inherently nonlinear, and the control system must be capable of handling this nonlinearity. "Gain scheduling" (changing controller tuning constants) is frequently used in batch reactors. But determining how these tuning constants should be changed with time is not a simple job.

4.1 IRREVERSIBLE, SINGLE REACTANT

We begin by considering the simple chemistry of a liquid-phase, first-order, irreversible, exothermic reaction occurring in a batch reactor:

$$\mathrm{A} \longrightarrow \mathrm{B} \tag{4.1}$$

The chemical kinetics are the same for any type of reactor. Most of the kinetic information used to design CSTRs is obtained from batch laboratory reactors:

$$\Re = kC_\mathrm{A} = C_\mathrm{A} k_0 e^{-E/RT_R} \tag{4.2}$$

We use the same reaction kinetic parameters considered in Chapters 2 and 3 for this reaction (given in Table 2.1).

The optimum operation of a batch reactor with this type of reaction is intuitively obvious. The reactor temperature should be increased to its maximum value as quickly as possible, without violating heat removal constraints, and held there until the desired conversion is attained. The high temperature will give the maximum reaction rate and therefore the shortest batch time. The maximum temperature constraint may occur because of side reactions, catalyst degradation, or materials of construction issues.

4.1.1 Pure Batch Reactor

All of reactant A is charged to the vessel at the beginning of the batch at a temperature $T_0 = 294$ K. The amount of the initial charge fills the vessel. In the discussion below, different heat transfer areas are considered, starting with the jacket heat transfer area and increasing the area if necessary by using an external heat exchanger.

To illustrate some of the design and control issues, a vessel size ($D_R = 2$ m, $V_R = 12.57$ m^3, jacket heat transfer area $A_J = 25.13$ m^2) and a maximum reactor temperature ($T_R^{\max} = 340$ K) are selected. The vessel is initially heated with a hot fluid until the reaction begins to generate heat. Then a cold fluid is used. A split-range-heating/cooling system is used that adds hot or cold water to a circulating-water system, which is assumed to be perfectly mixed at temperature T_J. The setpoint of a reactor temperature controller is ramped up from 300 K to the maximum temperature over some time period.

Batch Reactor Equations The dynamic equations describing the batch reactor are similar to those of a CSTR except that the feed and product streams are missing:

Total mass balance (kg/s):

$$\frac{d(V_R\rho)}{dt} = 0 \tag{4.3}$$

Component A balance (kmol A/s):

$$\frac{d(V_R C_A)}{dt} = -V_R \Re = -V_R k C_A \tag{4.4}$$

Reactor energy balance (J/s):

$$\frac{d(V_R \rho c_p T_R)}{dt} = -\lambda V_R \Re - Q = -\lambda V_R k C_A - UA_J(T_R - T_J) \tag{4.5}$$

Coolant energy balance (J/s):

$$\begin{aligned}\frac{d(V_J \rho_J c_J T_J)}{dt} &= F_{hot}\rho_J c_J T_{hot} + F_{cold}\rho_J c_J T_{cold} \\ &\quad - (F_{hot} + F_{cold})\rho_J c_J T_J + UA_J(T_R - T_J)\end{aligned} \tag{4.6}$$

where F_{hot} and F_{cold} are the flowrates of the hot and cold streams fed to the circulating water system going through the jacket or the external heat exchanger with volume V_J.

Constant liquid density and physical properties are assumed, so the volume of liquid in the reactor is constant with time, as is the volume in the cooling system (jacket or external heat exchanger). Equations (4.4)–(4.6) become

$$\frac{dC_A}{dt} = -kC_A \tag{4.7}$$

$$\frac{dT_R}{dt} = \frac{-\lambda k C_A}{\rho c_p} - \frac{UA_J(T_R - T_J)}{V_R \rho c_p} \tag{4.8}$$

$$\frac{dT_J}{dt} = \frac{F_{hot}T_{hot} + F_{cold}T_{cold}}{V_J} - \frac{(F_{hot} + F_{cold})T_J}{V_J} + \frac{UA_J(T_R - T_J)}{V_J \rho_J c_J} \tag{4.9}$$

Batch Temperature Control Structure The output signal from the reactor temperature controller positions two control valves, using a split-range setup. Figure 4.1 shows how the flowrates of the hot and cold streams change with the OP signal and gives the batch reactor and temperature controller setup when just jacket cooling is used. Note that the cold-stream control valve is "air-to-close" (AC) so it will fail wide open, which is the safe position. Using the same logic, the hot-stream control valve is "air-to-open" (AO). The hot-stream temperature is 373 K. The cold stream temperature is 294 K. The maximum flowrate of each of these streams is 1 kg/s.

The setpoint of the temperature control SP is ramped from 300 to 340 K over a period of time. The effect of this ramp rate is investigated below. If the ramp rate is too fast, the reactor temperature may run away because the heat removal system may not be able to remove the heat generated with the high initial concentrations of reactant A. If the ramp rate is too slow, the batch time will be long and therefore reduce productivity.

The tuning of the temperature controller is another important issue, which is explored below. We will see that a low controller gain leads to reactor runaways and a high

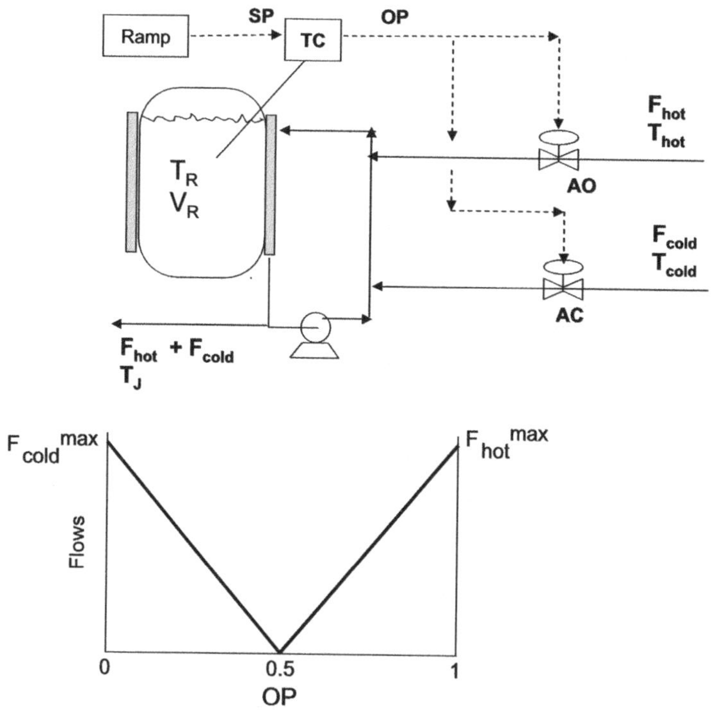

Figure 4.1 Batch reactor split-range temperature control.

controller gain leads to excessive oscillatory response. Two 1-min first-order lags are used in the temperature control loop.

Nonlinear Dynamic Simulation The nonlinear ordinary differential equations are numerically integrated in the Matlab program given in Figure 4.2. A simple Euler integration algorithm is used with a step size of 2 s. The effects of several equipment and operating parameters are explored below.

The effect of heat transfer area is illustrated in Figure 4.3. Three different areas are used. The temperature controller is proportional with a gain of 0.1 (dimensionless using a 50-K temperature transmitter span and split-range flows shown in Figure 2.1). The setpoint is ramped to 340 K in 60 min. Clearly in the numerical example, a jacket-cooled batch reactor of the size selected (2 m diameter) and with the given heat of reaction would produce runaway reactions. An external heat exchanger with 4 times the jacket area would be required to catch the reaction.

The effect of controller gain is shown in Figure 4.4. The heat transfer area is 4 times the jacket area, and the ramp time is 60 min. The importance of controller tuning is clearly illustrated. If the gain is too small, temperature control is poor (for $K_C = 0.05$, there is a 20 K overshoot of the desired temperature; for smaller gains, the reactor runs away). If the controller gain is too large, the response is very oscillatory. The $K_C = 0.1$ response shows a 6 K overshoot.

Figure 4.5 shows what happens when the setpoint ramp time is reduced from 60 to 45 min, using the same heat transfer area and the same three values of controller gain.

```
% Program "simbatch3.m"
%  Nonlinear dyanmic simulation of batch reactor A = B;  Tr=340 K
%  Run for controller gain = 0.05 with 45 minute ramp in setpoint and area = 4*areaj; P only
clear
ca0=8.01;k0=20.75e6;e=69.71e6;t0=294;tcin=294;u=851;taum=60;lambda=-69.71e6;
roe=801;m=100;cp=3137;cj=4183;roej=1000;
dr=2; vr=pi*(dr^2)*dr/2;areaj=2*pi*dr^2;areahx=areaj*4;vj=.00258*areahx/.4026;
% Initial conditions
tj=t0;tr=t0;ca=ca0;time=0;trsp0=300;trss=340;
% Controller settings
kc=0.1;taui=60*10; tramp=45*60;conversionstop=.999;castop=ca0*(1-conversionstop);
np=0;erint=0;tplot=0;thot=373;tcold=294;fhotmax=1;fcoldmax=1;
delta=2;trlag1=t0;trlag2=t0;
% Integration loop
while ca>castop
% Temperature Controller P only
   trsp=trsp0+time*40/tramp;if trsp>trss;trsp=trss;end; if tr>400;break;end
if time>tramp+10*60;kc=0.05;end % Use for gain scheduling
error=(trsp-trlag2)/50;op=0.5+kc*error;if op>1; op=1;end; if op<0;op=0;end;
% Calculate hot and cold flowrates
fcold=0;fhot=0;if op<0.5;fcold=fcoldmax*(0.5-op)*2;end;
if op>0.5;fhot=fhotmax*(op-0.5)*2;end;fjout=fhot+fcold;
% store data for plotting
if time>=tplot;   np=np+1;timep1(np)=time/60;trp1(np)=tr;tjp1(np)=tj;fcoldp1(np)=fcold*1000;
      fhotp1(np)=fhot*1000;   cap1(np)=ca;trspp1(np)=trsp;    tplot=tplot+10;end
q=u*areahx*(tr-tj);k=k0*exp(-e/tr/8314);
% Derivative evaluations
dca= -k*ca;dtj=(fcold*tcold+fhot*thot)/vj -fjout*tj/vj+q/(cj*roej*vj);
dtr= -lambda*k*ca/roe/cp - q/(cp*roe*vr);dtrlag1=(tr-trlag1)/taum;dtrlag2=(trlag1-trlag2)/taum;
%  Integration
time=time+delta;ca=ca+dca*delta;tr=tr+dtr*delta;tj=tj+dtj*delta;
trlag1=trlag1+dtrlag1*delta;trlag2=trlag2+dtrlag2*delta;
end
subplot(2,2,1);plot(timep1,trp1,timep1,trspp1,'--');grid;ylabel('TR (K)');
title('45 ramp; 4*areaj; kc=0.1 for 55 min, then 0.05 ')
subplot(2,2,3);plot(timep1,tjp1);grid;ylabel('TJ (K)');xlabel('Time (min)');
subplot(2,2,2);plot(timep1,fhotp1,timep1,fcoldp1,'--');grid;ylabel('Fcold/Fhot (kg/sec)');
subplot(2,2,4);plot(timep1,cap1);grid;ylabel('CA (kmol/cu m)');xlabel('Time (min)');
```

Figure 4.2 Matlab program for batch reactor simulation.

Now the reactor with $K_C = 0.05$ runs away. The high gain is still oscillatory. The $K_C = 0.1$ response shows a 10 K overshoot.

As mentioned earlier, the inherent problem with the "pure" batch reactor is that at the beginning of the batch all the reactant is available for reaction at a high concentration. If the initial reactant concentration is lower, there is less likelihood of a runaway. This is

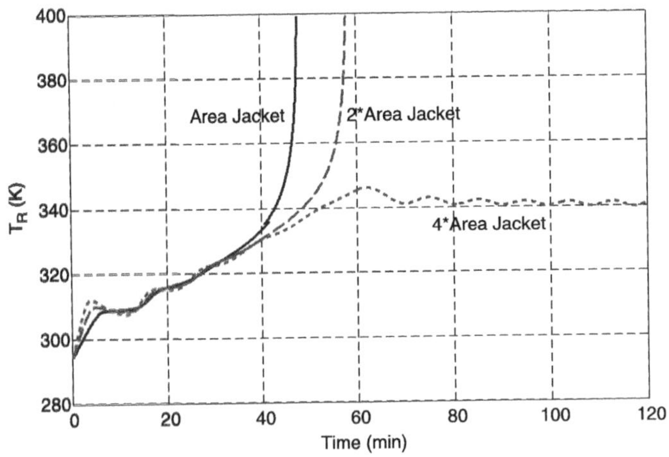

Figure 4.3 Effect of heat transfer area; $K_C = 0.1$; 60 min ramp.

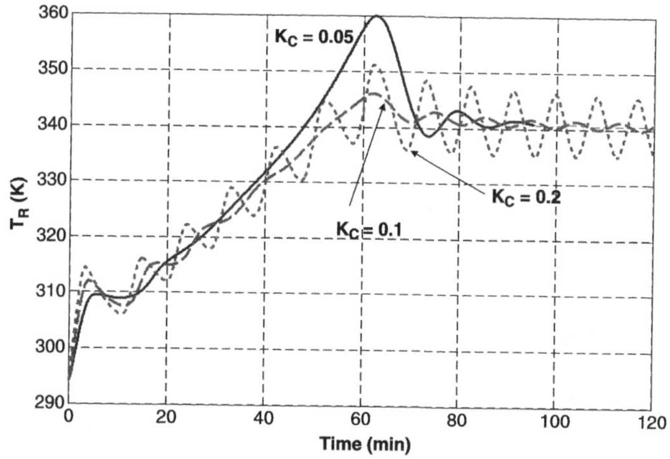

Figure 4.4 Effect of controller gain; 4^*A_J; 60 min ramp.

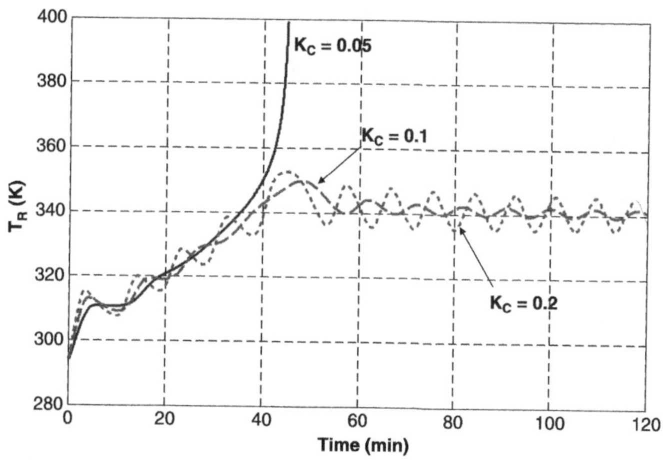

Figure 4.5 Effect of controller gain; 4^*A_J; 45 min ramp.

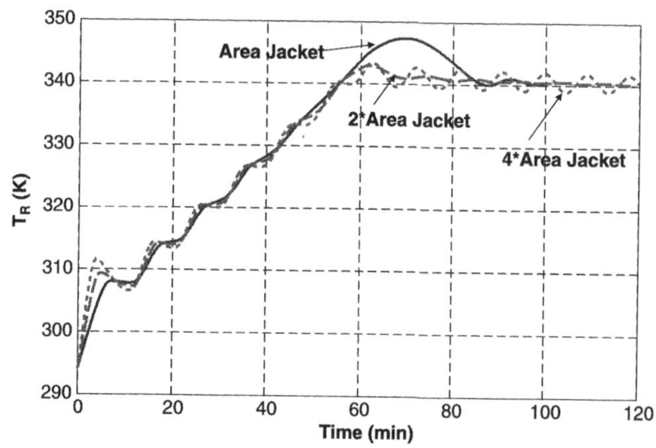

Figure 4.6 Effect of area with lower reactant concentration; $K_C = 0.1$; 60 min ramp.

illustrated in Figure 4.6, where the initial reactant concentration is reduced from 8.01 to 4.005 kmol/m^3. Now the reactor with just jacket cooling does not run away.

Figure 4.7 shows the effect of the heat of reaction on reactor controllability. The heat of reaction is reduced from $\lambda = -69.71 \times 10^6$ to -34.85×10^6 J/kmol. The jacket-cooled reactor does not run away.

The final parameter explored in this section is the size of the reactor. A diameter of 2 m has been used in the previous cases. The aspect ratio is 2, so the vessel length is 4 m. Making the reactor smaller increases the area-to-volume ratio and should improve controllability. Figure 4.8 gives results for three different reactor sizes ($D_R = 1$, 0.5, and 0.25 m) when only jacket cooling is used. The smaller reactors do not run away, but their responses are quite oscillatory. These results used the same maximum flowrates of the hot and cold streams that are used in the 2-m reactor case. If these maximum flowrates are scaled on the basis of reactor volume, the results in Figure 4.9 are obtained. The reactor size must be reduced to 0.25 m diameter to prevent a runaway reaction.

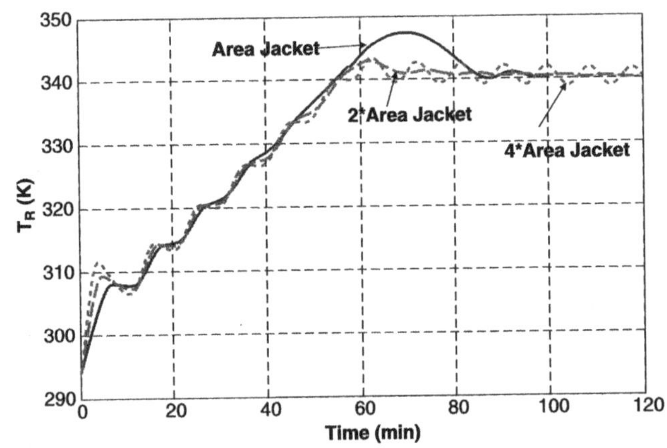

Figure 4.7 Effect of area with lower heat of reaction; $K_C = 0.1$; 60 min ramp.

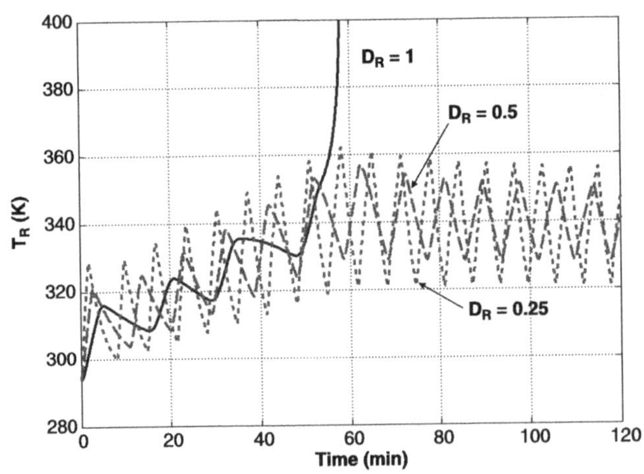

Figure 4.8 Effect of reactor size; jacket area; $K_C = 0.1$; 60 min ramp.

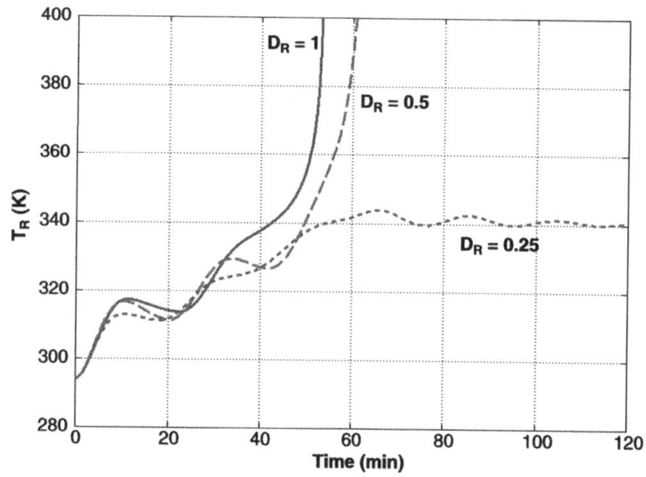

Figure 4.9 Effect of reactor size; jacket area; flows scaled with reactor size; $K_C = 0.1$; 60 min ramp.

The results presented above illustrate the effects of various parameters on a "pure" batch reactor. High initial reactant concentrations, large heats of reaction, large vessels, small controller gains, and high ramp rates can lead to reaction runaways.

Gain Scheduling The nonlinear nature of the batch reactor often requires that the tuning of the temperature controller change with time. The results shown in Figures 4.4 and 4.5 illustrate that a high gain is useful early in the batch to prevent runaways, but a lower gain is useful later to prevent oscillatory response. To demonstrate the use of gain scheduling, we consider the base case reactor system: 4 times jacket area, 45 min setpoint ramp, and normal kinetic parameters. Figure 4.5 shows that a controller gain of 0.1 works well during the period of setpoint ramping, but thereafter the temperature response is somewhat oscillatory.

There are several approaches to gain scheduling. One simple procedure is to change the controller gain at some point in time. This is illustrated in Figure 4.10. The controller gain is changed from $K_C = 0.1$ to $K_C = 0.05$ at time = 55 min. The oscillatory response later in the batch cycle is eliminated. The figure also shows how other important variables change during the batch. The hot stream is used for a very short time. The maximum flowrate of the cold stream occurs at the end of the ramp when the temperature has exceeded 340 K. Note that the reactant concentration C_A has dropped to about half its initial value at this point in time. The conversion increases gradually until it reaches the desired value of 99.9% at time = 291 min. So this is the batch time.

The capacity of this 2-m-diameter reactor (with an external heat exchanger) is 12.57 m^3 per 291 min of batch time. Assuming a 30 min turnaround time to cool down the reactor, discharge the products, and recharge the initial material, the daily production rate (assuming 24-h operation) is

$$\left(\frac{12.57}{291 + 30}\right)(60)(24) = 56.4 \text{ m}^3/\text{day}$$

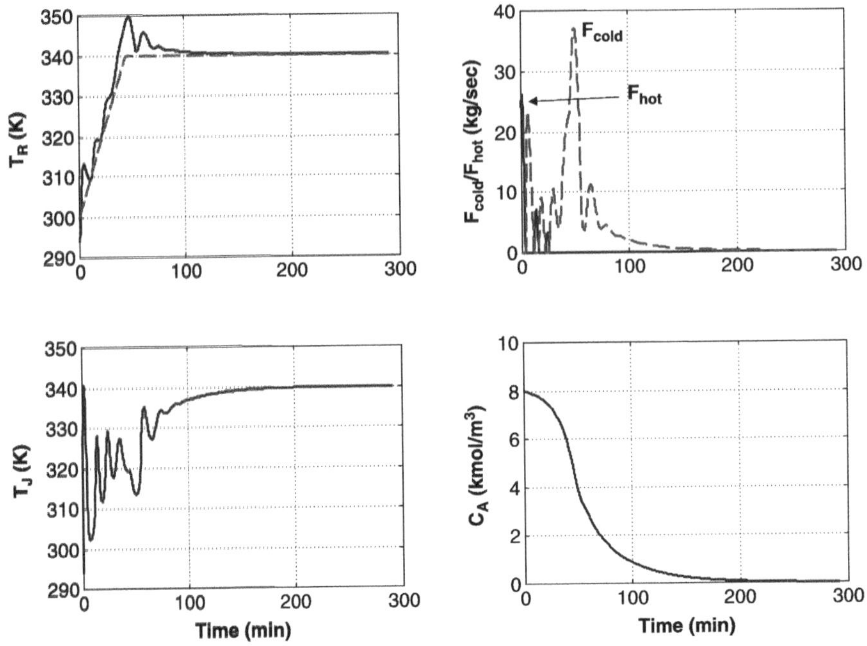

Figure 4.10 Gain scheduling; K_C changed from 0.1 to 0.05 at 55 min; 45 min ramp; 4* jacket area.

4.1.2 Fed-Batch Reactor

For batch reactors in which two reactants are involved, the problems encountered with a "pure" batch reactor can sometimes be reduced if only one of the reactants is initially charged to the vessel and the other reactant is fed gradually at a rate such that the heat removal capacity is not exceeded. We will study this mode of operation in the next section.

Two reactor temperature controllers are now used. The first manipulates the hot and cold streams used to heat or cool the reactor. The second controller manipulates the feed flowrate. More details about this control structure are presented in Section 4.2.

You might think that the fed-batch technique would work in the single-reactant case considered above. A number of simulations were run, and results showed that there was little improvement when the base case kinetic parameters were used. The problem is illustrated in Figure 4.11. The material initially charged to the reactor is assumed to contain no reactant A ($C_A = 0$ at time = 0). The concentration of reactant has to buildup to quite high levels ($C_A = 6\ \text{kmol/m}^3$) because the specific reaction rate k is relatively small ($4.045 \times 10^{-4}\ \text{s}^{-1}$ at 340 K). Eventually there is enough fuel to cause a runaway.

One important issue of the fed-batch operation is the variable volume of material in the reactor and its effect on heat transfer area. If jacket cooling is used, the heat transfer area covered by the liquid in the reactor will be proportional to the volume of the liquid at any point in time. However, if the reaction liquid is circulated through an external heat exchanger, the full heat transfer area is available throughout the batch cycle.

In the results shown in Figure 4.11, the heat transfer area changes with reactor volume, and its maximum value is the jacket area. Figure 4.12 give results when the

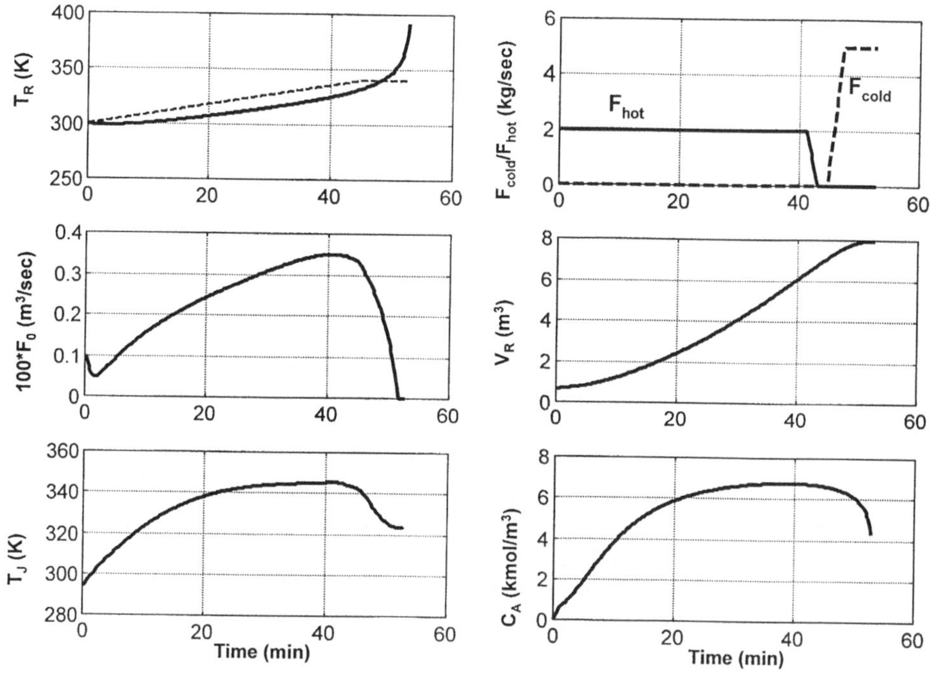

Figure 4.11 Fed batch; base case k_0; 45 min ramp; 4* jacket area.

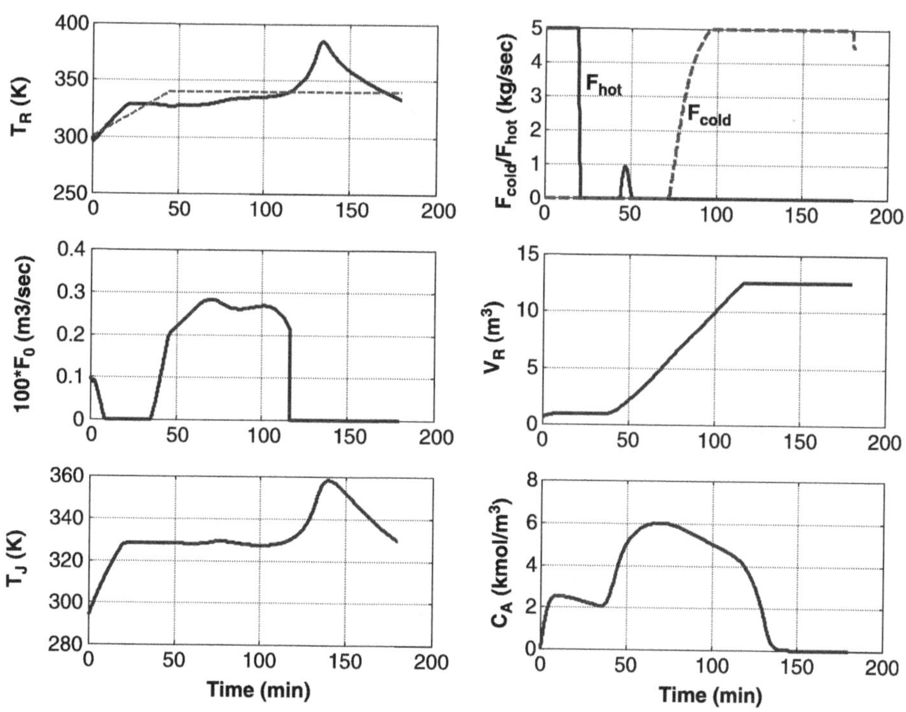

Figure 4.12 Fed batch; base case k_0; full area; 4* jacket area.

total area is 4 times the jacket area and is available throughout the batch. A partial runaway up to about 380 K still occurs.

If a much higher specific reaction rate is used, the reactant concentration is lower and the tendency for a runaway is reduced. This is illustrated in Figure 4.13, where the preexponential factor k_0 is increased by a factor of 10. In this "hot" reaction case, the reactant concentration C_A builds to only ~1 kmol/m^3 and no runaway occurs. Note that the fresh feed is cut off when the reactor is full ($V_R = 12.57$ m^3). The batch is terminated when the desired conversion (99.99%) is attained at 460 min.

The response is fairly oscillatory during the first 2 h when a controller gain of $K_C = 0.2$ and a 40 min integral time are used. Note also the "droop" or "offset" in the reactor temperature, that is, the temperature does not come up to the setpoint of 340 K. This is caused by the constantly changing load on the system. The feed flowrate is ramping up as the reactor fills and more heat transfer area permits more feed. The cooling water is wide open until the feed is cut off.

If the controller gain is reduced to $K_C = 0.1$, the response is less oscillatory, as shown in Figure 4.14. But the droop is greater, which lengthens the batch time to 510 min. These results indicate that gain scheduling should be used. At the beginning of the batch there is little material in the vessel. Adding a kilomole of reactant liberates a certain amount of heat that tends to significantly increase the temperature of the small amount of liquid. So the initial process gain between temperature and feed is large. As the amount of liquid in the reactor increases, the process gain decreases. Therefore a controller whose gain increases with reactor volume should provide improved control. This is demonstrated

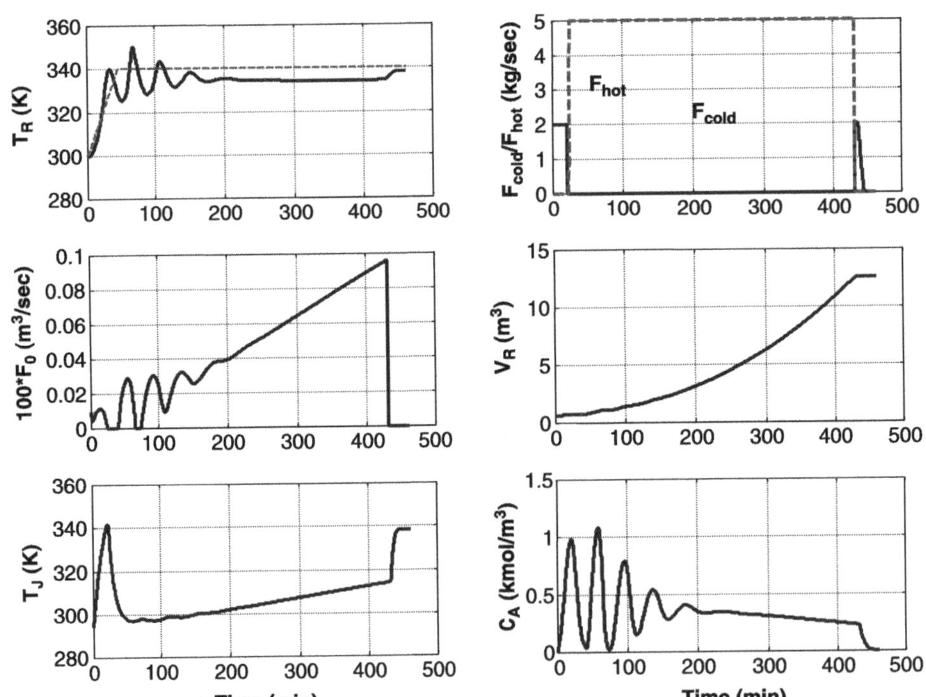

Figure 4.13 Fed batch; hot reaction; jacket area; $K_{C2} = 0.2$.

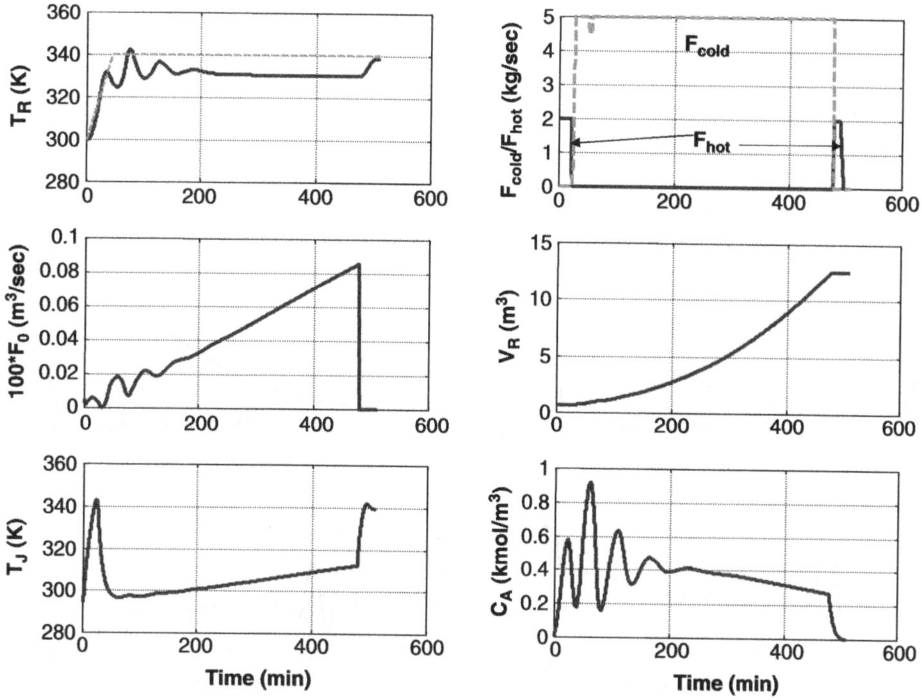

Figure 4.14 Fed batch; hot reaction; jacket area; $K_{C2} = 0.1$.

in Figure 4.15. The controller gain is changed with reactor volume:

$$K_{C2} = 0.05 + \frac{(2 - 0.05)}{(V_{R\text{total}} - V_{R\min})}(V_R - V_{R\min}) \tag{4.10}$$

The initial oscillatory response is reduced somewhat, and the droop toward the end of the batch is greatly reduced. The batch time decreases to 440 min.

The split-range hot-water valve is set up in this example to be wide open when the controller output is 100% and closed when the output is 75%. This gives a "gap" between the two control actions that seems to work better in this example.

The equations used for these simulations of fed-batch reactors are similar to Eqs. (4.3)–(4.6), but the reactor volume is time-varying and a feed term is present:

Total mass balance (m^3/s), assuming constant density:

$$\frac{dV_R}{dt} = F_0 \tag{4.11}$$

Component A balance (kmol A/s):

$$\frac{d(V_R C_A)}{dt} = F_0 C_{A0} - V_R k C_A \tag{4.12}$$

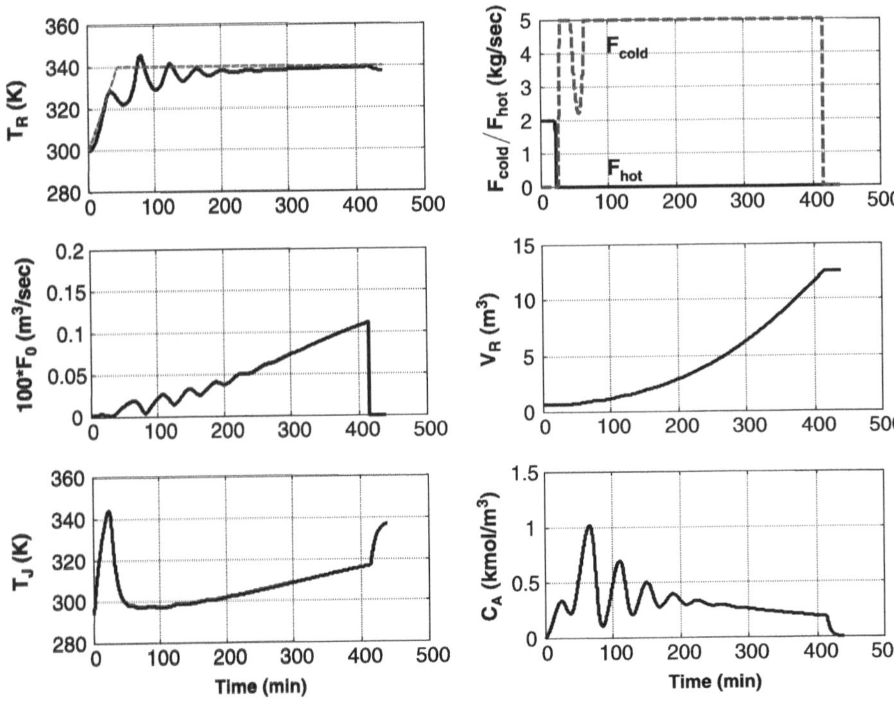

Figure 4.15 Fed batch; hot reaction; jacket area; gain scheduling.

Reactor energy balance (J/s):

$$\frac{d(V_R \rho c_p T_R)}{dt} = F_0 \rho c_p T_0 - \lambda V_R k C_A - U A_J (T_R - T_J) \tag{4.13}$$

For the case where the heat transfer area changes with the liquid holdup in the reactor, we obtain

$$A_{J(t)} = V_{R(t)} \frac{A_{J\max}}{V_{R\max}} \tag{4.14}$$

4.2 BATCH REACTOR WITH TWO REACTANTS

The results presented in the previous section are for the simple reaction A → B. In this section we consider the reaction A + B → C. For the pure batch reactor in which the two reactants are initially charged to the vessel, the results are quite similar to those found in the simple A → B reaction. The effects of various design and kinetic parameters are essentially the same because both reactants are charged in their stoichiometric amounts.

For fed-batch operation, however, there is a significant difference between the reaction A → B and the reaction A + B → C. One of the reactants can be charged to the reactor. Then the other can be fed at a rate consistent with the heat transfer capacity of the system.

The following numerical example illustrates the dynamics of a fed-batch reactor with two reactants.

The kinetics used are those given in Chapter 2 (Table 2.2). The desired operating temperature is 340 K. The diameter of the reactor is 2 m, giving a total volume of 12.57 m^3 and jacket heat transfer area of 25.13 m^2. The reactor is initially charged with 6.285 m^3 of pure B with a composition $C_B = 8.01$ kmol/m^3. The initial reactor temperature is 300 K.

Two temperature controllers are used. The first manipulates the flowrate of the A feed. A 45 min ramp in this reactor temperature controller is used with $K_{C2} = 0.5$ and $\tau_{I2} = 20$ min. Two 30-s lags are included in the loop. The span of the temperature transmitter is 50 K, and the maximum flowrate F_{A0} of the reactant A is 0.004 m^3/s. The second temperature controller setting the flowrates of the hot and cold streams to the jacket is proportional-only with a 330 K setpoint and a gain of 0.05. The maximum cold water and hot water flowrates are 0.005 and 0.002 m^3/s, respectively.

Figure 4.16 gives the results. Note that the reactor is initially half full, so half of the heat transfer area is available at the very beginning of the batch cycle. This permits a fairly rapid feedrate. The concentration of A in the reactor is low throughout the batch cycle. The concentration of B declines steadily as it reacts with the incoming A.

Once the vessel is full, the fresh feed is cut off. The setpoint of the second controller is raised to 340 K at this point in time, which is about 130 min into batch. The batch cycle is stopped when the conversion reaches 98%. This takes about 353 min. Figure 4.17 gives a Matlab program for this fed-batch system.

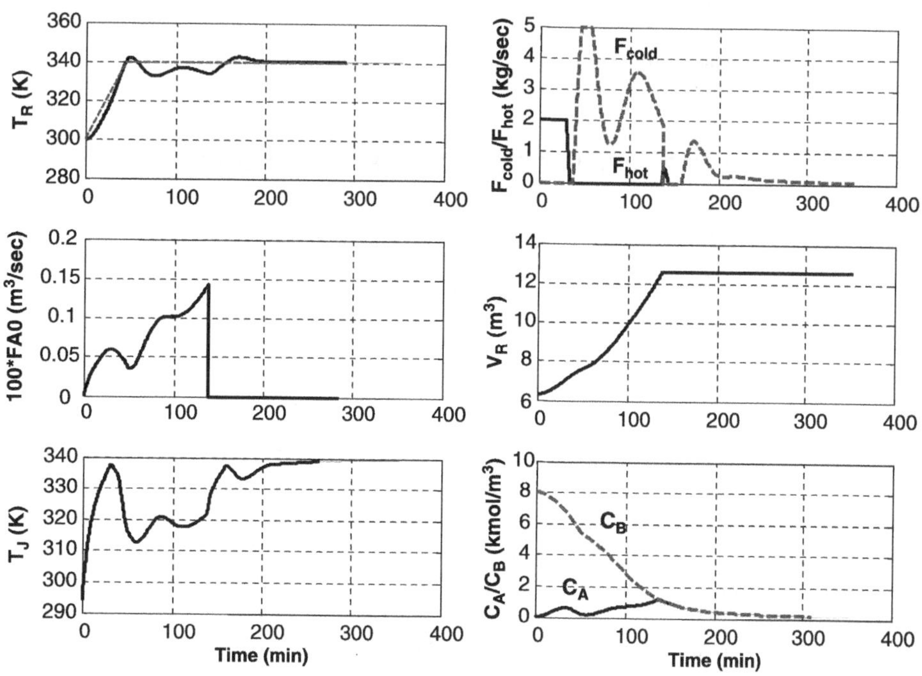

Figure 4.16 Fed batch; A + B → C; jacket area.

```
% Program "fedbatchtworeactants.m"
%  Nonlinear dyanmic simulation of batch reactor A+B=C;  Tr=340 K;
%  Fed-batch operation; 50% full at t=0 with reactant B;  two temp controllers
% %   with 45 minute ramp in setpoint and area = jacket only;
clear
ca0=8.01;k0=21.045e6;e=69.71e6;t0=294;tcin=294;u=851;lambda=-69.71e6;
roe=801;ma=100;mb=100;mc=200;cp=3137;cj=4183;roej=1000;
dr=2; vrtotal=pi*(dr^2)*dr/2;areaj=2*pi*dr^2;areahx=areaj;vj=.1*areahx;areatotal=areahx;
% Initial conditions
vr=0.5*vrtotal;  fa0max=0.004; trss=340;
tj=t0;tr=300;ca=0;cb=8.01;time=0;trsp0=300;vrtr=vr*tr;vrca=vr*ca;vrcb=vr*cb;
% Controller settings
kc1=0.05;kc2=0.5;taui2=60*20; tramp=45*60;conversionstop=.98;cbstop=ca0*(1-conversionstop);
np=0;erint2=0;tplot=0;thot=373;tcold=294;fhotmax=0.002;fcoldmax=0.005;%  flows in cu m/sec
delta=2;trlag1=300;trlag2=300;tstop=6*3600;taum=30;trsp1=330;
% Integration loop
while  cb>cbstop;
   if time>tstop;break;end
   trsp2=trsp0+time*40/tramp;if trsp2>trss;trsp2=trss;end; if tr>400;break;end
% Temperature Controller No. 1 sets hot and cold flows
   error1=(trsp1-trlag2); op1=0.5+kc1*error1;if op1>1; op1=1;end; if op1<0;op1=0;end;
   fcold=0;fhot=0;if op1<0.5;fcold=fcoldmax*(0.5-op1)*2;end;if op1>0.75;fhot=fhotmax*(op1-0.75)*4;end;fjout=fhot+fcold;
% Temperature controller No. 2 sets feed of A
   error2=(trsp2-trlag2)/50;op2=kc2*error2+erint2;if op2>1; op2=1;end; if op2<0;op2=0;end;
   fa0=fa0max*op2; if vr>vrtotal;fa0=0;trsp1=trss;end
% store data for plotting
if time>=tplot;   np=np+1;timep1(np)=time/60;trp1(np)=tr;tjp1(np)=tj;fcoldp1(np)=fcold*1000;
   fhotp1(np)=fhot*1000;cap1(np)=ca;cbp1(np)=cb;trspp1(np)=trsp2;vrp1(np)=vr;fa0p1(np)=fa0;tplot=tplot+10;end
areahx=vr*areatotal/vrtotal;q=u*areahx*(tr-tj);k=k0*exp(-e/tr/8314);
% Derivative evaluations
dvr=fa0;dvrca= fa0*ca0-vr*k*ca*cb;dvrcb= -vr*k*ca*cb;dtj=(fcold*tcold+fhot*thot)/vj -fjout*tj/vj+q/(cj*roej*vj);
dvrtr= fa0*t0-lambda*vr*k*ca*cb/roe/cp - q/(cp*roe);dtrlag1=(tr-trlag1)/taum;dtrlag2=(trlag1-trlag2)/taum;
%  Integration
time=time+delta;vr=vr+dvr*delta;vrca=vrca+dvrca*delta;vrcb=vrcb+dvrcb*delta;vrtr=vrtr+dvrtr*delta;
tj=tj+dtj*delta;trlag1=trlag1+dtrlag1*delta;trlag2=trlag2+dtrlag2*delta;tr=vrtr/vr;ca=vrca/vr;cb=vrcb/vr;
if op2<1;if op2>0;erint2=erint2+error2*kc2*delta/taui2;end;end
end
subplot(3,2,1);plot(timep1,trp1,timep1,trspp1,'--');grid;ylabel('TR (K)');
title('A+B=C;340 K; areaj;  ')
subplot(3,2,5);plot(timep1,tjp1);grid;ylabel('TJ (K)');xlabel('Time (min)');
subplot(3,2,2);plot(timep1,fhotp1,timep1,fcoldp1,'--');grid;ylabel('Fcold/Fhot (kg/sec)');
subplot(3,2,6);plot(timep1,cap1,timep1,cbp1,'--');grid;ylabel('CA/CB (kmol/cu m)');xlabel('Time (min)');
subplot(3,2,3);plot(timep1,100*fa0p1);grid;ylabel('100*FAO (cu m/sec)');
subplot(3,2,4);plot(timep1,vrp1);grid;ylabel('VR (cu m)');
```

Figure 4.17 Matlab program for fed batch; A + B → C.

4.3 BATCH REACTOR WITH CONSECUTIVE REACTIONS

If consecutive reactions are conducted in a batch reactor, the optimization of the process includes finding the optimum time to stop the batch and determining the optimum temperature. To illustrate the issues, we take the simple reactions

$$A \xrightarrow{k_1} B \xrightarrow{k_2} C \tag{4.15}$$

We assume the desired product to be B. If the reaction is allowed to run for too long, the amount of undesirable C produced may be too high and the yield of B may be lower than if the batch were stopped earlier. If the activation energy of the second reaction is larger than the first, increasing reactor temperature lowers the yield of B but reduces the batch time. Therefore both the reactor temperature and the batch time must be optimized.

For an illustrative numerical example, the preexponential factor and activation energy given in Table 2.1 are used for the first specific reaction rate k_1 ($k_{01} = 20.75 \times 10^6\ s^{-1}$ and $E_1 = 69.71 \times 10^6$ J/kmol). The activation energy of the second reaction is assumed to be twice that of the first. The preexponential factor for the second reaction is calculated to give a ratio of k_1 to k_2 of 10 when the temperature is 340 K ($k_{02} = 10.642 \times 10^{16}\ s^{-1}$).

Pure A is charged to the reactor $[C_{A(t=0)} = 8.01\ \text{kmol/m}^3]$. The operation runs isothermally at a given temperature. Results for three values of temperature are selected: 335, 340, and 345 K. Figure 4.18 give the results. The maximum concentration of B occurs sometime during the batch. The time when this maximum concentration occurs decreases as temperature increases, which means a shorter batch time (higher productivity). However, the value of the maximum concentration decreases as reactor temperature increases, which means a lower yield of the desired product.

Figure 4.19 gives results for the case in which the preexponential factor for the second reaction is calculated to give a ratio of k_1 to k_2 of 5 when the temperature is 340 K ($k_{02} = 21.285 \times 10^{16}\ \text{s}^{-1}$). As expected, the maximum attainable yield of B is lower.

Determining the optimum time and optimum temperature involves an economic balance between shorter batch times and less concentrated reaction liquid at the end of the batch, which implies higher capital and energy costs to purify the product and recover byproducts.

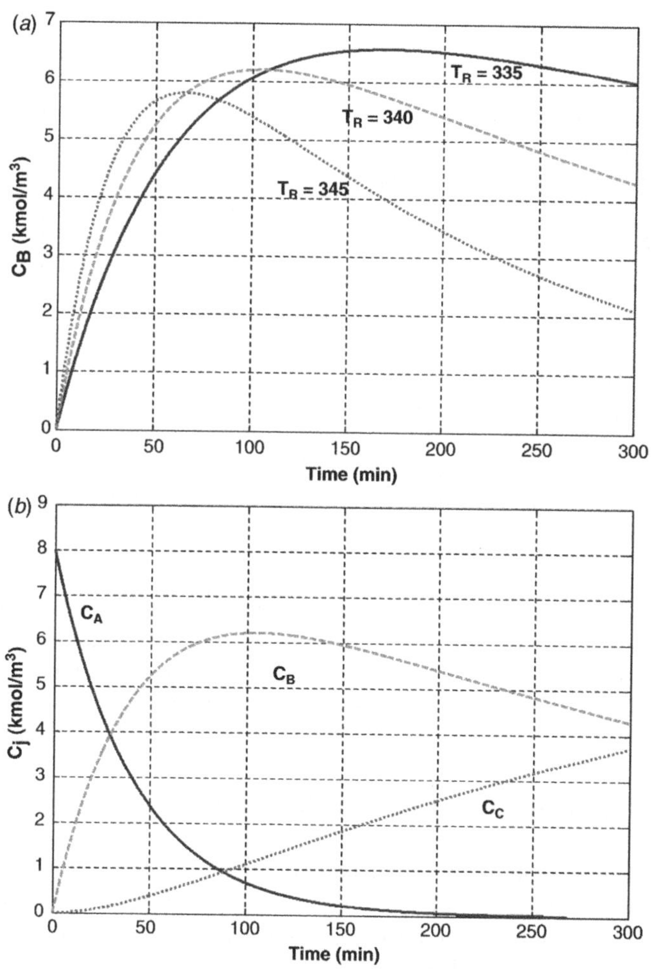

Figure 4.18 Batch consecutive reactions, A → B → C. (a) effect of temperature, $K_1/K_2 = 10$; (b) 340 K, $K_1/K_2 = 10$.

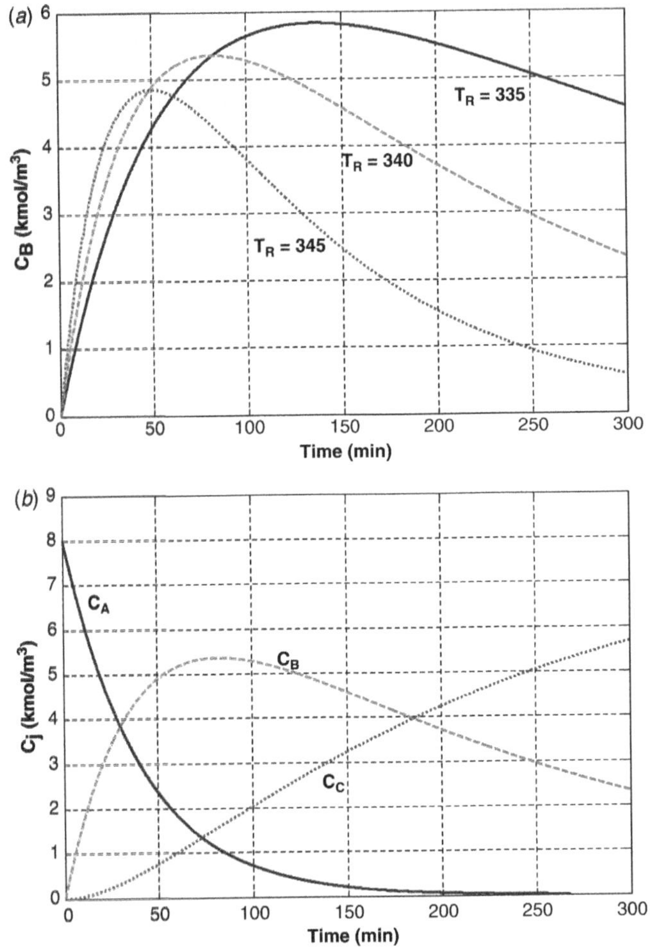

Figure 4.19 Batch consecutive reactions; A → B → C; (*a*) effect of temperature, $K_1/K_2 = 5$; (*b*) 340 K, $K_1/K_2 = 5$.

Note that we have said nothing about the size of the reactor vessel. If the reactor operates isothermally, the composition profiles shown in Figures 4.18 and 4.19 are independent of reactor size. Of course, attaining a constant-temperature trajectory becomes more difficult as the vessel size increases because of the reduction in area-to-volume ratio.

4.4 ASPEN PLUS SIMULATION USING RBatch

Aspen Plus has the capability of simulating a batch reactor. The *RBatch* reactor is selected from the possible types of reactors that appear at the bottom of the Aspen Plus window, as shown in Figure 4.20. Several streams can be attached to the reactor. Two of these are "pseudostreams." The pseudostream named "CHARGE" is connected to the upper red arrow called *Batch Charge (required)* shown in Figure 4.20. As we will see shortly, the hourly flowrate specified for this stream is used as the total amount of material charged to the reactor at the beginning of the batch.

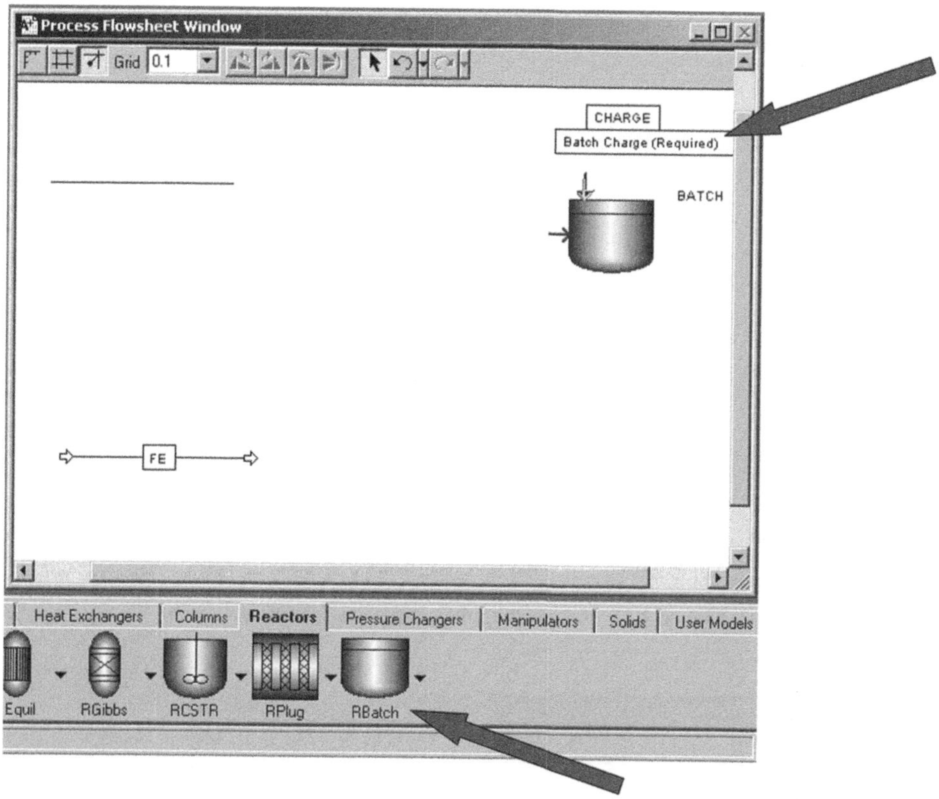

Figure 4.20 Installing RBatch and attaching the pseudostream batch charge.

A second stream can also be used in *RBatch* by attaching it to the blue arrow on the side of the reactor called *Continuous Feed (optional)* (see Fig. 4.21). As the name implies, this stream is used if fed-batch operation is desired. This stream is a real flowing stream, whose flow-versus-time profile can be specified, as we will illustrate later.

A third stream is another pseudostream, which is attached on the red arrow at the bottom of the vessel called *Reactor Product (required)* (see Fig. 4.22). This pseudostream represents the contents of the reactor at the end of the batch cycle. Note the blue arrow pointing out of the vessel in Figure 4.22. This can be used as a product stream that is withdrawn from the vessel during the batch cycle. For example, a pressure controller can be used to withdraw a vapor stream to hold the pressure in the reactor. Figure 4.23 shows the flowsheet with the three "streams" attached to the RBatch reactor. Remember that the "charge" and "product" streams are required connections for the reactor.

The CHARGE pseudostream is specified by clicking on *Data*, *Setup*, and *Streams* in the same way used in Chapter 2. Figure 4.24 shows the *Input* for this stream. Specifying the total flow to be 100 kmol/h represents putting 100 kmol into the reactor at the beginning of the batch cycle. This 100 kmol charge is pure benzene at 300 K. The density is displayed by Aspen Plus after the program has been run by clicking *Results* under the CHARGE stream. The molar density is 11.14 kmol/m^3, so the volume of the initial charge is 9 m^3. Assuming that the vessel is initially 70% full and using an aspect ratio of 2, the reactor diameter is 2 m and the jacket heat transfer area is 25.5 m^2.

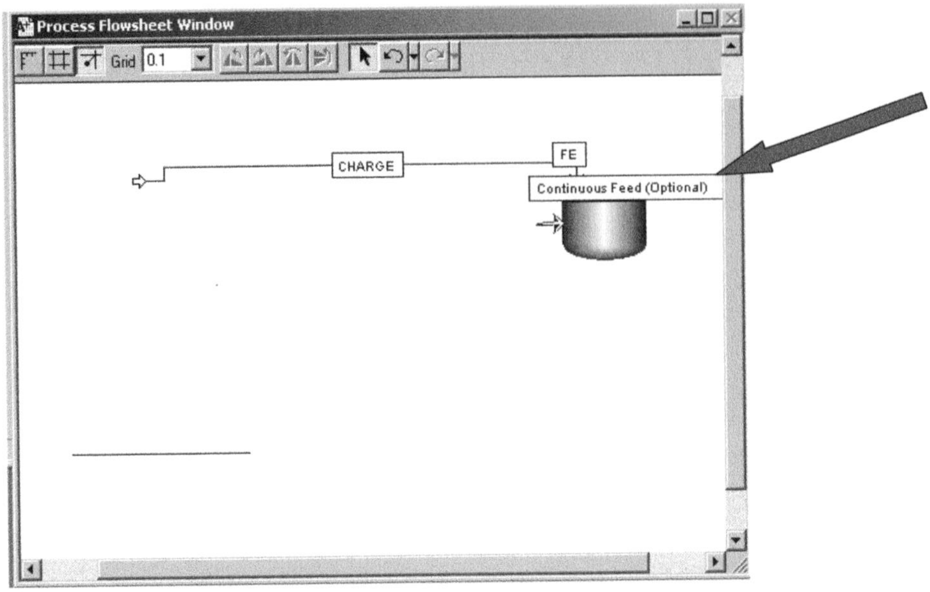

Figure 4.21 Connecting the continuous feedstream.

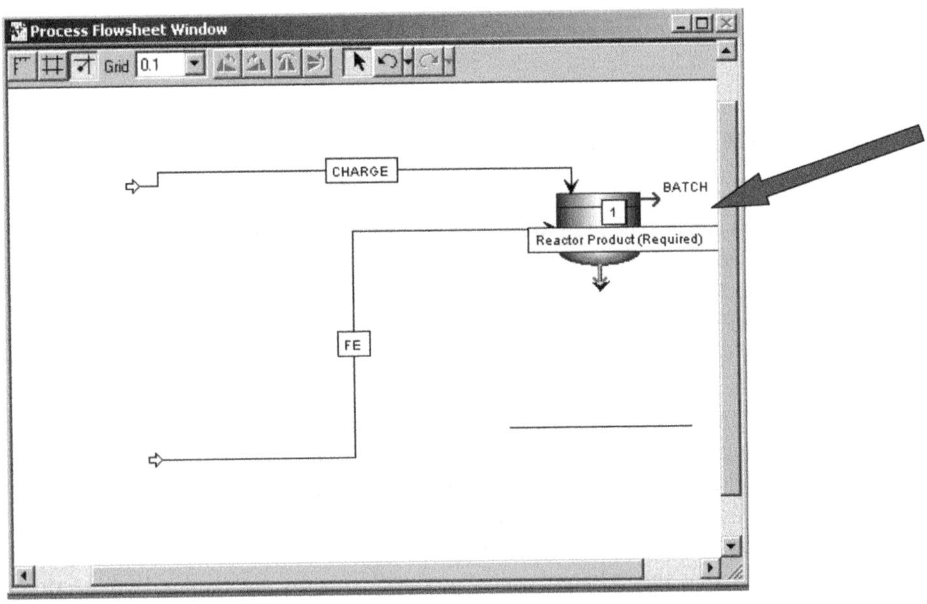

Figure 4.22 Connecting the product pseudostream.

The continuously fed stream is called "FE" and is specified to be fed at a rate of 50 kmol/h as shown in Figure 4.25. It is pure ethylene at 296 K. As mentioned earlier, the flow-versus-time profile of this stream can be specified.

Now we are ready to set up the reactor. Figure 4.26 shows the *Setup* window for the reactor block with the *Specifications* page tab open. There are several alternative

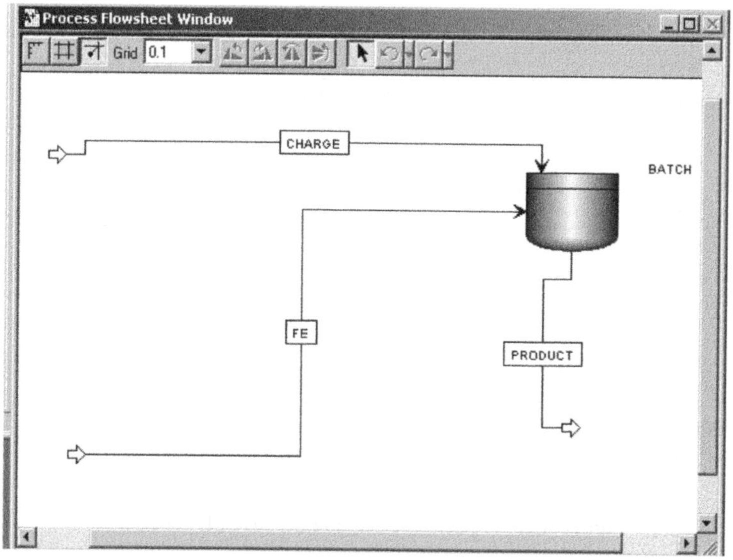

Figure 4.23 Completed flowsheet.

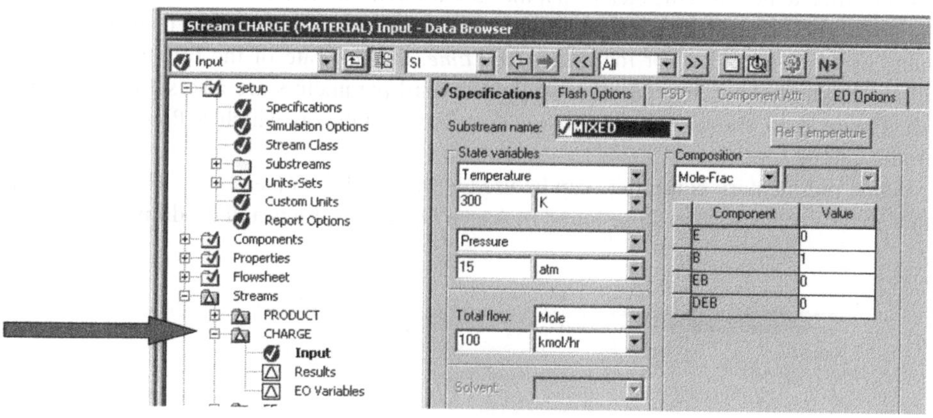

Figure 4.24 Specifying the pseudostream "CHARGE."

Reactor operating specifications that can be used. We will demonstrate the use of two: *Temperature profile* and *Coolant temperature.*

In Figure 4.26 a temperature profile has been specified. The reactor starts at 300 K at time equal zero, and the temperature is ramped to 400 K at 10 min. Then it is ramped to 430 K at 20 min and remains at this temperature for the rest of the 120-min batch. Figure 4.27 shows that the *Reactions* page tab permits the installation of reactions in the normal way, as discussed in Chapter 2. The primary reaction is ethylene and benzene forming ethylbenzene.

The *Stop Criteria* page tab opens the window shown in Figure 4.28, on which one or more criteria for ending the batch cycle can be specified. We select a *Time* variable and set the *Stop value* to 7200 s (2 h). Other criteria could also be specified, such as a maximum reactor temperature and a desired concentration.

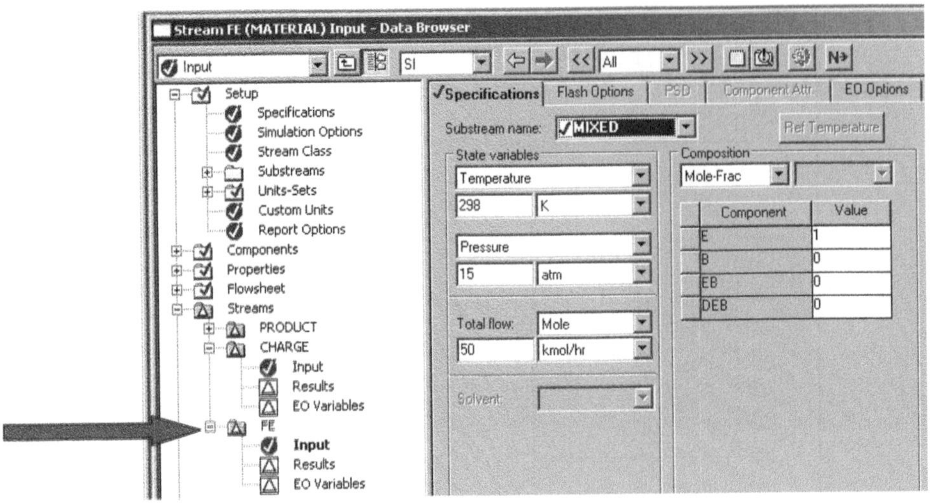

Figure 4.25 Specifying the continuous "FE" stream.

Figure 4.29 shows the *Operating Times* page tab. The *Batch feed time* bullet is clicked and specified to be 1 h. This means that the "50 kmol/h" specified for the FE feedstream is on an hourly basis. To help clarify this none-too-intuitive aspect of the feedstream, if a value of 2 h is specified for the *Batch feed time*, the flowrate of the FE stream will be set at "50 kmol/2 h." Several other straightforward parameters are also set on this page tab. Figure 4.30 shows the *Continuous Feeds* page tab, on which a flow-versus-time profile can be specified.

Some rudimentary controllers can be used with the RBatch (see Fig. 4.31) reactor, but they are less realistic than those found in Aspen Dynamics. Lags and deadtimes cannot be

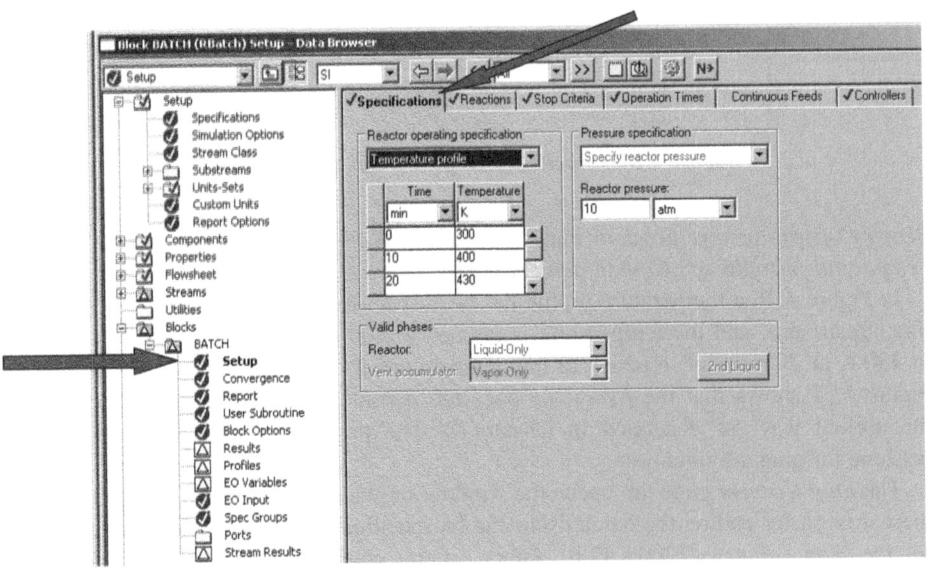

Figure 4.26 Specifying the reactor.

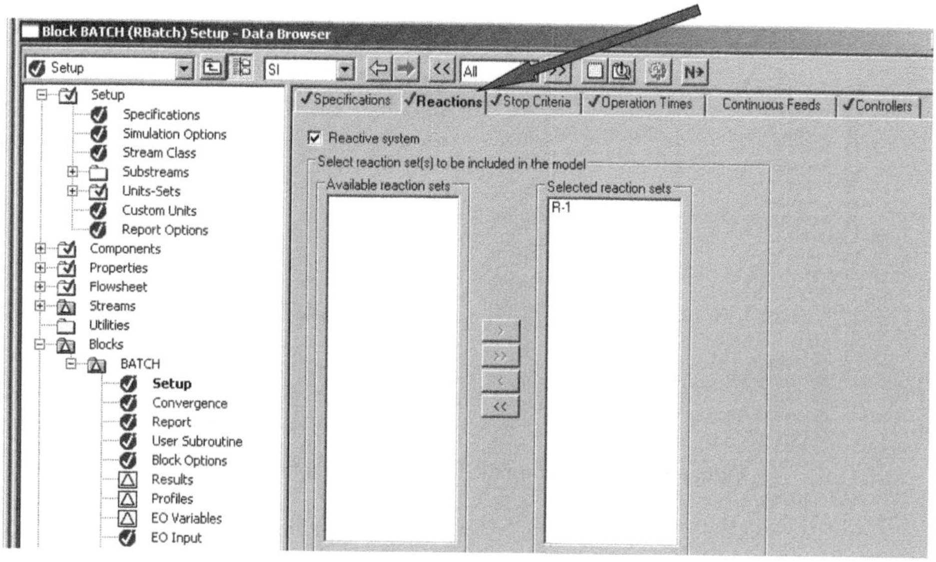

Figure 4.27 Installing reactions.

installed in the loop, and automatic relay–feedback testing for controller tuning is not available. The specified temperature (or temperature profile) is interpreted as the setpoint of a PID controller:

$$Q = M_C\left[K(T - T^S) + (K/I)\int_0^t (T - T^S)dt + KD\frac{d(T - T^S)}{dt}\right] \quad (4.16)$$

where Q = reactor heat duty (J/s)
M_C = reactor charge (kg)
K = controller gain (J kg^{-1} K^{-1})

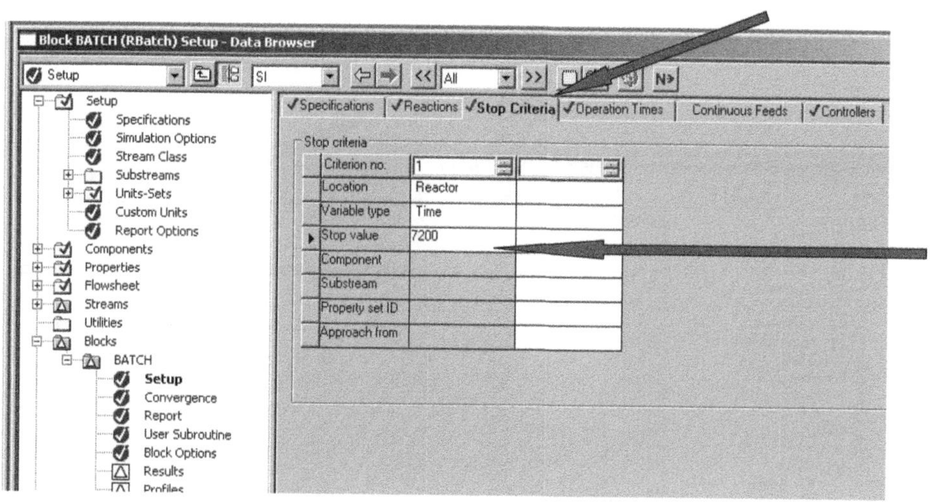

Figure 4.28 Specifying when to stop the batch cycle.

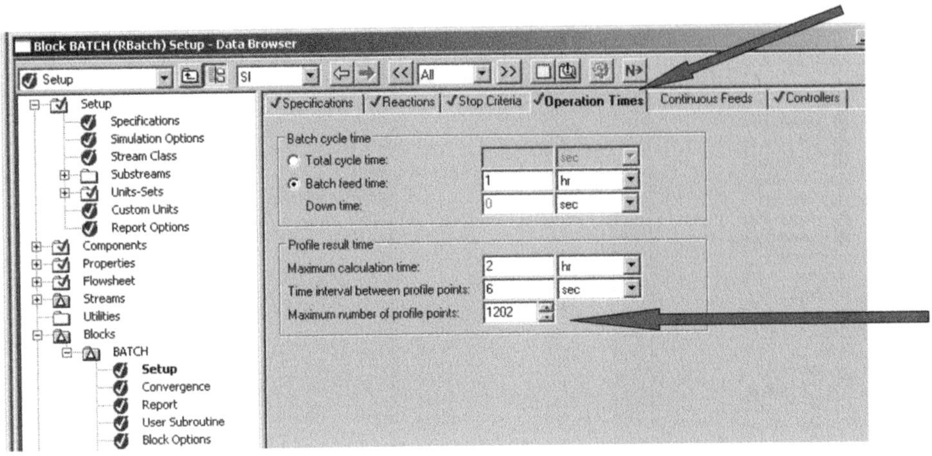

Figure 4.29 Specifying the feed time and profile points.

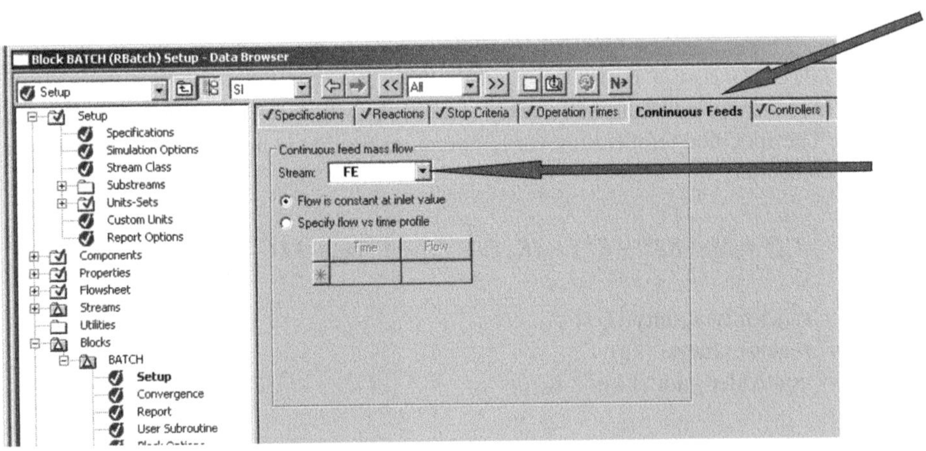

Figure 4.30 Specifying a constant ethylene feedflow.

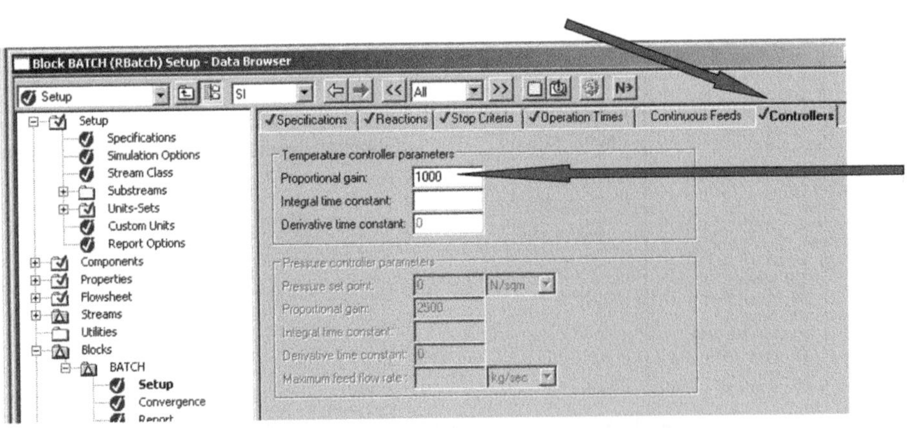

Figure 4.31 Specifying the temperature controller.

T = reactor temperature (K)
T^S = setpoint (K)
I = integral time (s)
D = derivative time (s)
t = time (sec)

Note that the controller gain is a specific gain per unit mass.

The simulation is run by clicking the blue *N* button. The time profiles can be seen by clicking *Profiles* under the reactor block in the *Data Browser* window. Figure 4.32 shows the tabular results. Plots can be easily made by clicking the *Plot Wizard* button at the top of the window and selecting what variables are to be plotted. In Figure 4.33 *Temperature* is selected to be plotted versus time in minutes. The top graph in Figure 4.34 shows the resulting temperature profile. The second graph shows the composition profile over

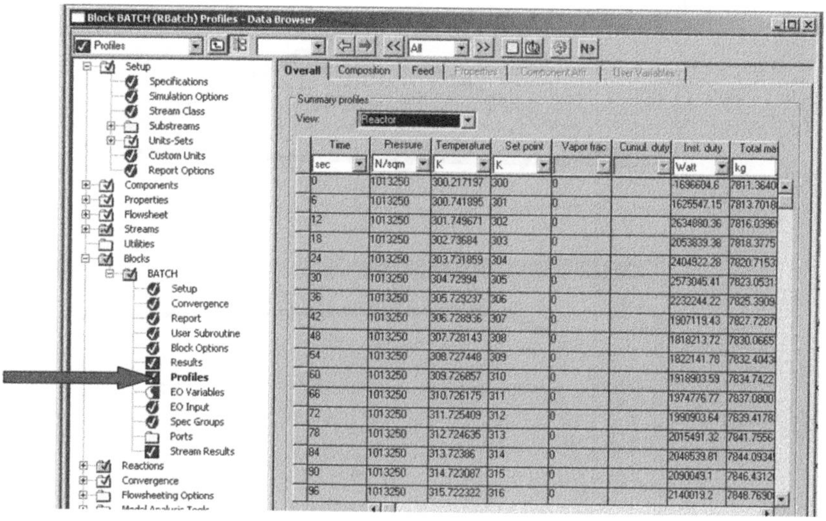

Figure 4.32 Time profile results.

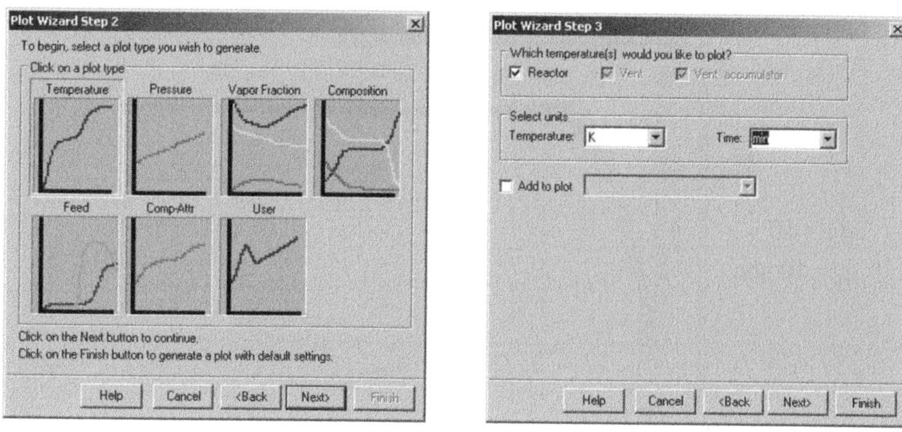

Figure 4.33 Plotting profile results; selecting temperature.

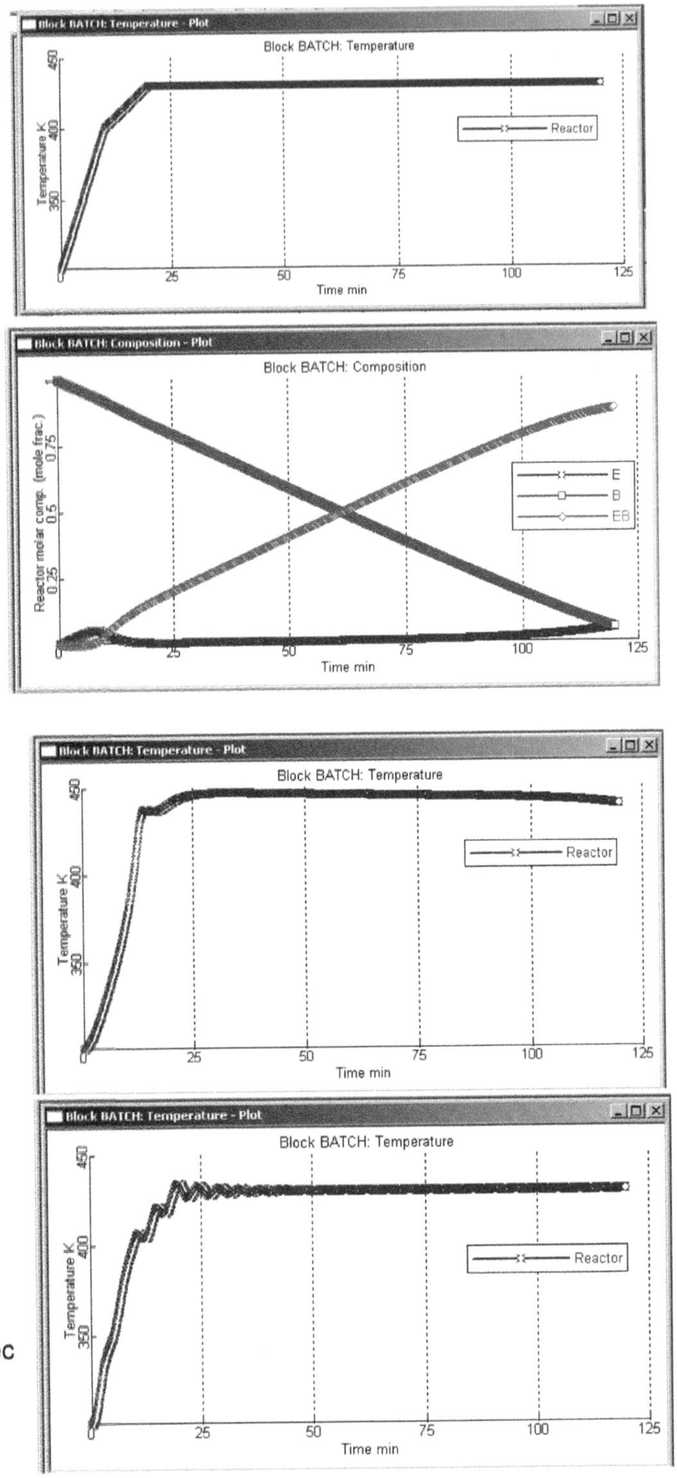

Figure 4.34 Temperature and composition profiles.

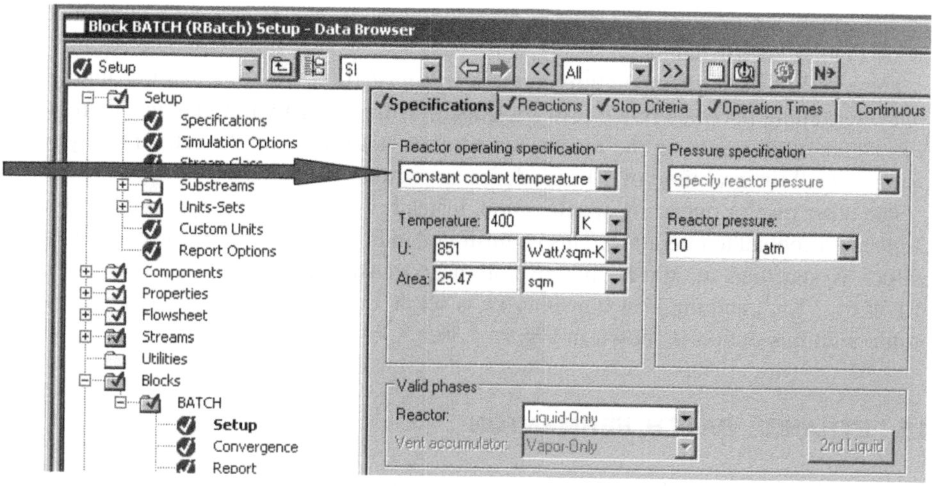

Figure 4.35 Selecting a coolant temperature.

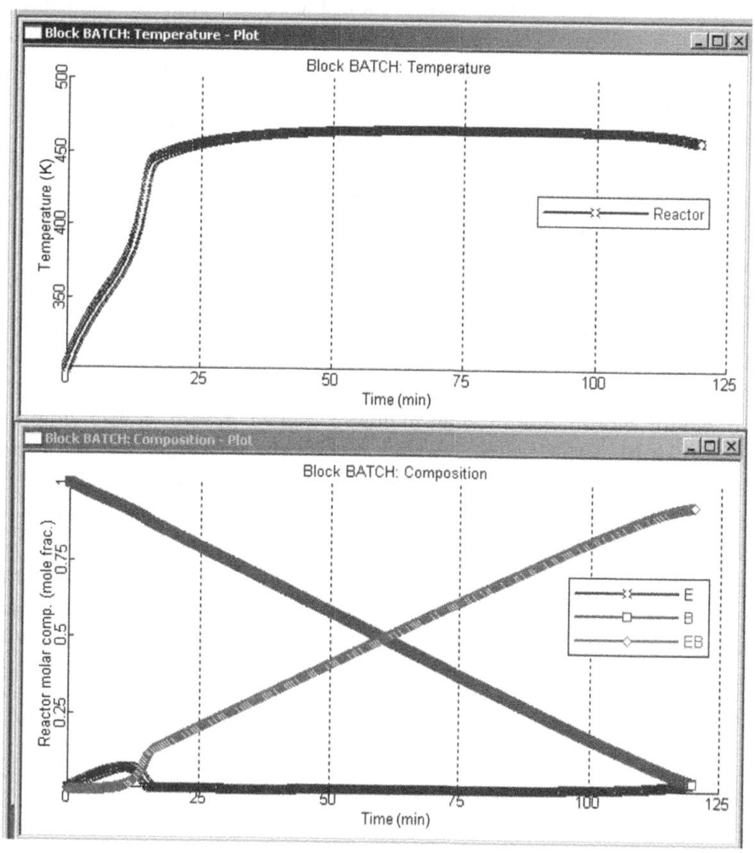

Figure 4.36 Profiles with constant coolant temperature.

the 2-h batch. The original 100 kmol of benzene in the reactor is gradually consumed by the 50 kmol/h of ethylene feed.

These results used a temperature controller gain of 1000. The effects of changing the controller tuning parameters are also illustrated in the lower two graphs in Figure 4.34. A low gain of 10 results in the reactor temperature not following the setpoint very well. Using a small integral time of 10 s produces an oscillatory response.

The other reactor operating specification, instead of selecting a temperature profile, is to specify a coolant temperature. This is illustrated in Figure 4.35, where the coolant temperature is specified at 400 K, the overall heat transfer coefficient U is specified at 851 W m^{-2} K^{-1}, and the heat transfer area is set at 25.46 m^2. The resulting temperature profile with this option is shown in Figure 4.36.

4.5 ETHANOL BATCH FERMENTOR

There has been considerable interest in recent years in the concept of producing ethanol from renewable biological raw materials for use as a supplement for fossil-fuel-based gasoline. New laws (as of 2000) even require its addition. A very small fraction of the world's transportation fuel is currently supplied by ethanol.

Corn is one source of glucose, which yeast can convert into ethanol and carbon dioxide. While it is claimed that corn is a renewable resource, there is considerable disagreement on this subject. The raising of corn is one of the most energy-intensive farming operations, requiring significant energy for tillage, fertilizer production, irrigation, pesticide production, harvesting, and drying. Most of this energy comes from fossil fuels (gasoline, diesel, natural gas, and propane). Corn also has serious environmental problems in fertilizer runoff causing pollution of streams and in soil erosion causing depletion of topsoil. Beyond the technical and economic questions about using corn for fuel, there looms the important ethical and societal issue of using a human food source for fuel when 20% of the world's population is living with inadequate food supplies.

Putting these important issues aside, the production of ethanol by batch fermentation is an important example of a batch reactor. The basic regulatory control of a batch ethanol fermentor is not a difficult problem because the heat removal requirements are modest and there is no need for very intense mixing. In this section we develop a very simple dynamic model and present the predicted time trajectories of the important variables such as the concentrations of the cells, ethanol, and glucose. The expert advice of Bjorn Tyreus of DuPont is gratefully acknowledged. Sources of models and parameter values are taken from three publications.[1–3]

The simple fermentor model has three "state variables" that change with time during the batch. The first is the concentration of cells X (grams of cells per liter of reactor liquid) that grow during the batch from some initial small value. This is provided by a "seed fermentor" that is itself a small batch fermentor in which a small number of cells are grown.

[1]A. Dourado, G. Goma, U. Albuquerque, and Y. Sevely, Modeling and the static optimization of the ethanol production in a cascade reactor—1. Modeling, *Biotechnol. Bioeng.*, **29**, 187–194 (1987).

[2]G. R. Cysewski and C. R. Wilke, Process design and economic studies of alternative fermentation methods for the production of ethanol, *Biotechnol. Bioeng.*, **20**, 1421–1444 (1978).

[3]J. Nielsen, J. Villadsen, and G. Lindén, *Bioreaction Engineering Principles*, 2nd ed., Springer, 2005.

The second state variable is the concentration of glucose or substrate S (grams of glucose per liter of reactor liquid). This starts at some large value whose value depends on the glucose source, and is depleted as it is consumed by the cells. The third state variable is the concentration of ethanol product P (grams of ethanol per liter of reactor liquid). This starts at zero and increases during the batch.

As the ethanol concentration increases, it begins to inhibit the growth rate of the cells. This can be seen in the expression used to describe the specific growth rate of the cells μ (grams of cells produced per hour per gram of cells in the fermentor).

$$\mu = \mu_0 \left(\frac{S}{S + K_S} \right) \left(\frac{K_P}{P + K_P} \right) \left(1 - \frac{P}{P_L} \right) \tag{4.17}$$

The values of the parameters used are $K_S = 5.0$ g/L, $K_P = 4.5$ g/L, and $P_L = 85$ g/L. The inhibition of cell growth is quantified by the last term. As the ethanol concentration gets closer to 85 g/L, the specific growth rate of the cells becomes smaller and smaller.

There are two other kinetic parameters that are used in the model. The specific production rate of ethanol $\Re_P$ (grams of ethanol produced per hour per gram of cells in the fermentor) depends on the specific growth rate μ. The most simple relationship is a linear one:

$$\Re_P = \alpha\mu + \beta \tag{4.18}$$

The values of the parameters used are $\alpha = 4.4$ and $\beta = 0.09$.

The specific rate of consumption of glucose $\Re_S$ (grams of glucose consumed per hour per gram of cells in the fermentor) also depends on the specific growth rate μ. The most simple relation ship is a linear one:

$$\Re_P = \delta\mu + \gamma \tag{4.19}$$

The values of the parameters used are $\delta = 10.2$ and $\gamma = 0.18$.

The dynamic model consists of three ordinary differential equations:

$$\begin{aligned} \frac{dX}{dt} &= \mu X \\ \frac{dP}{dt} &= \Re_P X \\ \frac{dS}{dt} &= -\Re_S X \end{aligned} \tag{4.20}$$

These equations are integrated from the initial conditions $X_{(t=0)} = 1$ g (gram) of cells per liter, $P_{(t=0)} = 0$ g of ethanol per liter, and $S_{(t=0)} = 100$ g of glucose per liter. Figure 4.37 gives a Matlab program that performs the integration and produces the time trajectories. The program is stopped when the glucose is completely consumed ($S = 0$).

Figure 4.38 gives results for three cases with different initial conditions for the glucose and cell concentrations at the beginning of the batch fermentation. The base case is 100 g/L of glucose and 1 g/L of cells. For these initial conditions, the batch time is about 40 h and the final ethanol concentration is 44 g/L. If the initial glucose

```
% Program "etohfermentor.m"
clear
% Initial conditions
s=100;x=1;p=0;time=0;tplot=0;delta=0.005;tstop=50;np=0;
% Integration loop
while time<tstop
    mu=0.3*(s/(5+s))*(4.5/(4.5+p))*(1-p/85);
    rp=4.4*mu+0.09;
    rs=10.1*mu+0.18;
    % save for plots
    if time<=tplot;np=np+1;timep(np)=time;xp(np)=x;pp(np)=p;sp(np)=s;tplot=tplot+.005;end
    % derivative evaluation
    dx=mu*x;dp=rp*x;ds=-rs*x;
    %Integration
    x=x+dx*delta;p=p+dp*delta;s=s+ds*delta;time=time+delta;
    if s<=0;break;end
end
clf
subplot(3,1,1)
plot(timep,xp);grid;ylabel('X (g cells/L)');title('BDT model');

subplot(3,1,2)
plot(timep,sp);grid;ylabel('S (g sugar/L)');
subplot(3,1,3)
plot(timep,pp);grid;ylabel('EtOH (g/L)');xlabel('Time (hr)');
```

Figure 4.37 Matlab program for batch ethanol fermentor.

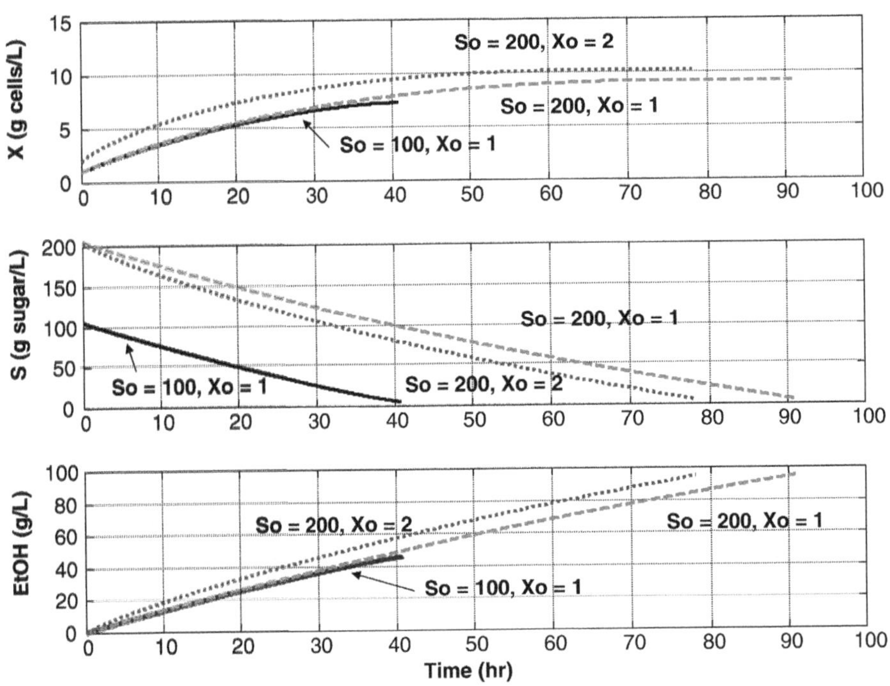

Figure 4.38 Batch ethanol fermentor.

concentration is increased to 200 g/L, the batch time increases to about 90 h and the final ethanol concentration increases to 94 g/L. If the initial glucose and cell concentrations are both increased, the batch time is reduced to about 78 h with the final ethanol of 94 g/L. The optimum operation of a batch fermentor involves tradeoffs between raw-material costs and productivity.

As the trajectories in Figure 4.38 show, the higher concentration of ethanol toward the end of the batch inhibits the growth rate of the cells, so the cell concentration levels off.

The subject of biochemical reactors is discussed in more detail in a textbook by Bequette.[4]

4.6 FED-BATCH HYDROGENATION REACTOR

Hydrogenation reactions are frequently run in fed-batch reactors. The chemical component to be hydrogenated is charged to the reactor vessel. The hydrogen is then fed into the vessel on pressure control. The temperature of the reactor is controlled by manipulating the flowrate of coolant to the jacket, coil, or external heat exchanger. Thus this system has two manipulated variables (the flowrate of hydrogen and the flowrate of coolant) and two controlled variables (pressure and temperature).

Since hydrogenation reactions are very exothermic, the situation often arises where the heat removal capacity cannot maintain the desired temperature with the normal operating hydrogen pressure. This usually occurs early in the batch when the concentration of the other reactant is high because it has not yet been diluted by the formation of the product compound. This situation requires that the flowrate of hydrogen be restricted so that temperature control is maintained. Thus pressure control should be temporarily abandoned.

There are several ways to achieve this "variable structure" control strategy. The elegant approach is to use model predictive control (MPC). The simple approach is to use override control. The latter technique is demonstrated in this section.

As a specific numerical example, we use the reaction of aniline with hydrogen to form cyclohexylamine (CHA):

$$C_6H_7N + 3H_2 \rightarrow C_6H_{13}N \tag{4.21}$$

We assume that the kinetics of this reaction are given by the following equation.

$$\mathcal{R}_{CHA} = kC_A C_{H_2} = C_A C_{H_2} k_0 e^{-E/RT_R} \tag{4.22}$$

where $\mathcal{R}_{CHA}$ = rate of production of CHA (kmol s^{-1} m^{-3})
C_A = concentration of aniline in reactor liquid (kmol/m^3)
C_{H_2} = concentration of hydrogen in reactor liquid (kmol/m^3)
k_0 = preexponential factor = 4×10^4 (m^3 s^{-1} $kmol^{-1}$)
E = activation energy = 46.49×10^6 J/kmol

The reaction has a large heat of reaction ($\lambda = -190 \times 10^6$ J/kmol).

[4]B. W. Bequette, *Process Dynamics: Modeling, Analysis and Simulation*, Module 8, Prentice-Hall, 1998, p. 529.

The dynamic mathematical model describing the system consists of a total mass balance, two component balances, an energy balance on the reactor liquid, and a jacket energy balance:

Total mass balance (m^3/s), assuming constant density:

$$\frac{dV_R}{dt} = F_{H_2} M_{H_2}/\rho \tag{4.23}$$

where F_{H_2} is the molar flowrate of hydrogen feed (kmol/s) and M_{H_2} is the molecular weight of hydrogen.

Component balance for aniline (kmol/s):

$$\frac{d(V_R C_A)}{dt} = -V_R k C_A C_{H_2} \tag{4.24}$$

Component balance for hydrogen (kmol/s):

$$\frac{d(V_R C_{H_2})}{dt} = F_{H_2} - 3V_R k C_A C_{H_2} \tag{4.25}$$

Reactor energy balance (J/s):

$$\frac{d(V_R T_R)}{dt} = \frac{F_{H_2} M_{H_2} T_0}{\rho} - \frac{\lambda V_R k C_A C_{H_2}}{\rho c_p} - \frac{UA_J(T_R - T_J)}{\rho c_p} \tag{4.26}$$

Coolant energy balance (J/s):

$$\frac{dT_J}{dt} = \frac{F_J(T_{c,\text{in}} - T_J)}{V_J} + \frac{UA_J(T_R - T_J)}{V_J \rho_J c_J} \tag{4.27}$$

The reactor pressure is calculated from the temperature and the liquid composition. Vapor pressure constants for aniline and CHA and a Henry's law constant for hydrogen were calculated from data obtained from Aspen Plus using the Chao–Seader physical property package

$$\begin{aligned} \ln\left(P_A^S\right) &= 11.6606 - 5329/T \\ \ln\left(P_{CHA}^S\right) &= 11.9125 - 4959/T \\ H_{H_2(450\,K)} &= 7300 \end{aligned} \tag{4.28}$$

where P_j^S = vapor pressure of component j (atm)
T = temperature (K)
H_{H_2} = Henry's law constant (atm)

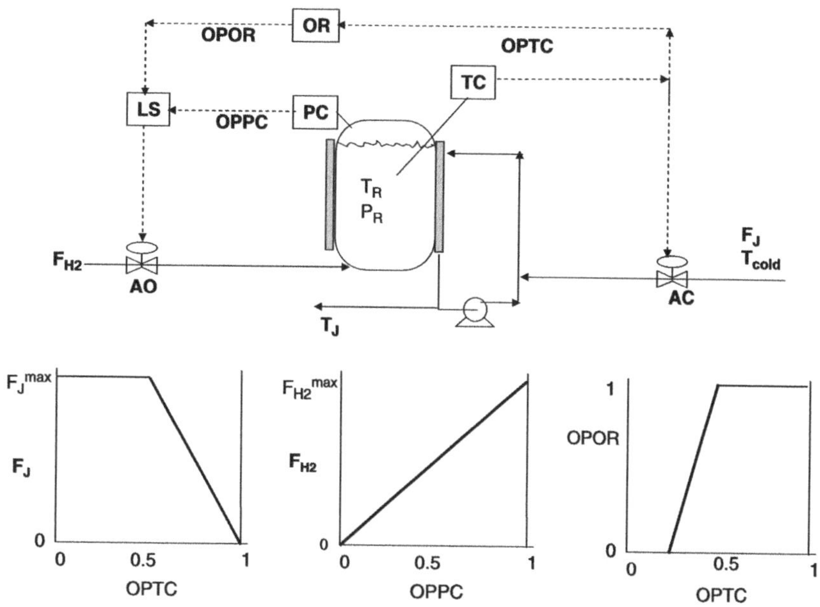

Figure 4.39 Fed-batch hydrogenation reactor.

The reactor is 2 m in diameter and 4 m in length. The charge of pure aniline fills the vessel to 90% of its total volume. The temperatures of the reactor liquid and the jacket are both initially 300 K. Cooling water at 300 K is used.

Figure 4.39 gives a sketch of the system and its control scheme. The temperature controller loop has a temperature transmitter span of 50 K, two 30-s lags, and tuning constants of $K_C = 0.05$ and $\tau_I = 10$ min. The pressure controller loop has a pressure transmitter span of 5 atm and tuning constants of $K_C = 0.5$ and $\tau_I = 10$ min. The maximum flowrate of cooling water (valve wide open) is 0.005 m^3/s, and the maximum flowrate of hydrogen is 0.035 kmol/s. The cooling water valve is "AC," so that it fails wide open. The output signal from the temperature controller "OPTC" positions the cooling water valve. When the signal is 50% of scale or less, the valve is wide open (see the left input output plot at the bottom of Fig. 4.39). The pressure controller output signal "OPPC" positions the "AO" hydrogen feed valve (see the middle plot at the bottom of Fig. 4.39).

The high-temperature override controller looks at the output signal from the temperature controller "OPTC" and generates an output signal "OPOR" that decreases from 100 to 0% of scale as the input signal decreases from 50 to 25% of scale (see the plot at the right at the bottom of Fig. 4.39). Figure 4.40 gives a Matlab program that simulates this system.

Figure 4.41 gives time trajectories during a batch cycle. At time equal zero, hydrogen feed is begun and this rapidly brings the pressure up to the setpoint value of 10 atm. The setpoint of the temperature controller is 450 K, so there is initially no coolant flow. When the temperature controller setpoint is reached and exceeded, the temperature controller output signal OPTC drops from 100 to 50%, which opens the cooling water valve. As OPTC continues to decrease, the output signal from the override controller OPOR decreases. When this signal is lower than the signal from the pressure controller (OPPC), the low selector (LS) passes the lower of the two signals to the hydrogen feed

```
% Program "chafedbatch.m"
%  Nonlinear dyanmic simulation of batch reactor: aniline + H2 = CHA;  Tr=450 K;
%  Fed-batch operation; 90% full o aniline at t=0;
% temp controller  manipulates CW; pressure controller manipulates H2; jacket area
clear
% Calculate vapor pressure constants
vpa2=20;ta2=342+273;vpa1=1;ta1=184+273;
vpcha2=382/14.7;tcha2=300+273;vpcha1=167/14.7;tcha1=250+273;
bvpa=ta2*ta1*log(vpa2/vpa1)/(ta1-ta2);avpa=log(vpa2)-bvpa/ta2;
bvpcha=tcha2*tcha1*log(vpcha2/vpcha1)/(tcha1-tcha2);avpcha=log(vpcha2)-bvpcha/tcha2;
% Parameter values
ca0=8.01;k0=40000;e=46.49e6;t0=300;u=851;lambda=-190e6;
roe=801;ma=93;mcha=99;mh2=2;cp=3137;cj=4183;roej=1000;
dr=2; vrtotal=pi*(dr^2)*dr/2;areaj=2*pi*dr^2;areahx=areaj;vj=.1*areahx;areatotal=areahx;
% Initial conditions;
vr=0.90*vrtotal;tj=350;tr=350;ca=8.01;ch2=0.00;ccha=0;time=0;
vrtr=vr*tr;vrca=vr*ca;vrch2=vr*ch2;vrccha=vr*ccha;
% Controller settings
kctc=0.05;tauitc=60*10; sptc=450;kcpc=0.5;tauipc=60*10;sppc=10;
np=0;erinttc=0;erintpc=0;tplot=0;tcin=300;
%  Maximum flows of H2 and CW
fjmax=0.005;%  flow in cu m/sec
fh2max=0.035; % flow is kmol/sec
delta=0.5;trlag1=tr;trlag2=tr;tstop=7*3600;taum=30;
% Integration loop
while time<tstop
   if vr>vrtotal;break;end;
   if ca<8.01*.001;break;end;
% Calculate pressure from compositions and temperature
   xa=ca/(ca+ch2+ccha);xh2=ch2/(ca+ch2+ccha);xcha=ccha/(ca+ch2+ccha);
      psh2=7300;psa=exp(avpa+bvpa/tr);pscha=exp(avpcha+bvpcha/tr);pr=xa*psa+xh2*psh2+xcha*pscha;
 % Temperature Controller  sets  CW flow
      errortc=(sptc-trlag2); optc=0.7+kctc*errortc+erinttc;
 % Override Controller pinches fh2 if optc goes below 0.5
      opor=1;if optc<0.5;opor=optc*4;end;if opor<0;opor=0;end;if opor>1;opor=1;end
            if optc<0.5; optc=0.5;end; if optc>1;optc=1;end; fj=fjmax*(1-optc)*2;
 % Pressure controller sets H2 feed
     errorpc=(sppc-pr)/5;oppc=0.4+kcpc*errorpc+erintpc;
     if oppc>1;oppc=1;end;if oppc<0;oppc=0;end;fh2=fh2max*oppc;
 %Low selector
   if opor<oppc;fh2=fh2max*opor;end
   if vr>vrtotal;fh2=0;end
% store data for plotting
if time>=tplot;   np=np+1;timep1(np)=time/60;trp1(np)=tr;tjp1(np)=tj;fjp1(np)=fj*1000;
   cap1(np)=ca;ch2p1(np)=ch2;cchap1(np)=ccha;vrp1(np)=vr;fh2p1(np)=fh2;prp1(np)=pr;
   optcp(np)=optc;oppcp(np)=oppc;oporp(np)=opor;  tplot=tplot+10;end
areahx=vr*areatotal/vrtotal;q=u*areahx*(tr-tj);k=k0*exp(-e/tr/8314);
% Derivative evaluations
dvr=fh2*mh2/roe;dvrca= -vr*k*ca*ch2;dvrch2= fh2-3*vr*k*ca*ch2;dvrccha= +vr*k*ca*ch2;
dtj=fj*tcin/vj -fj*tj/vj+q/(cj*roej*vj);dvrtr= fh2*t0*mh2/roe-lambda*vr*k*ca*ch2/roe/cp - q/(cp*roe);
dtrlag1=(tr-trlag1)/taum;dtrlag2=(trlag1-trlag2)/taum;
%  Integration
time=time+delta;vr=vr+dvr*delta;vrca=vrca+dvrca*delta;vrch2=vrch2+dvrch2*delta;
vrccha=vrccha+dvrccha*delta;vrtr=vrtr+dvrtr*delta;
tj=tj+dtj*delta;trlag1=trlag1+dtrlag1*delta;trlag2=trlag2+dtrlag2*delta;
tr=vrtr/vr;ca=vrca/vr;ch2=vrch2/vr;ccha=vrccha/vr;
if optc>0.5;if optc<1;erinttc=erinttc+errortc*kctc*delta/tauitc;end;end
if opor>oppc;erintpc=erintpc+errorpc*kcpc*delta/tauipc;end;
end
subplot(4,2,1);plot(timep1,trp1);grid;ylabel('TR (K)');
title(' CHA Fed Batch; TR=450; areaj;  kctc=0.05; taum=30 ')
subplot(4,2,2);plot(timep1,fjp1);grid;ylabel('Fj (kg/sec)');
subplot(4,2,3);plot(timep1,prp1);grid;ylabel('PR (atm)');
subplot(4,2,4);plot(timep1,1000*fh2p1);grid;ylabel('FH2 (mol/sec)');
subplot(4,2,5);plot(timep1,tjp1);grid;ylabel('TJ (K)');
subplot(4,2,6);plot(timep1,ch2p1);grid;ylabel('CH2 (kmol/cu m)');
subplot(4,2,7);plot(timep1,cap1,timep1,cchap1,'--');grid;ylabel('A/CHA (kmol/cu m)');xlabel('Time (min)');
subplot(4,2,8);plot(timep1,oppcp,timep1,oporp,'--');grid;ylabel('OPPC/OPOR');xlabel('Time (min)');
```

Figure 4.40 Matlab program for fed-batch hydrogenation reactor.

valve. This cuts back on hydrogen to maintain reactor temperature close to its setpoint. Pressure is not controlled and drops. The plot in the lower right corner of Figure 4.41 shows the output signals of the pressure controller (OPPC that is at 100%) and the override controller (OPOR that drops to about 75%). The hydrogen feed (second plot from the top in the right side of Fig. 4.41) is reduced to prevent the temperature from going too high.

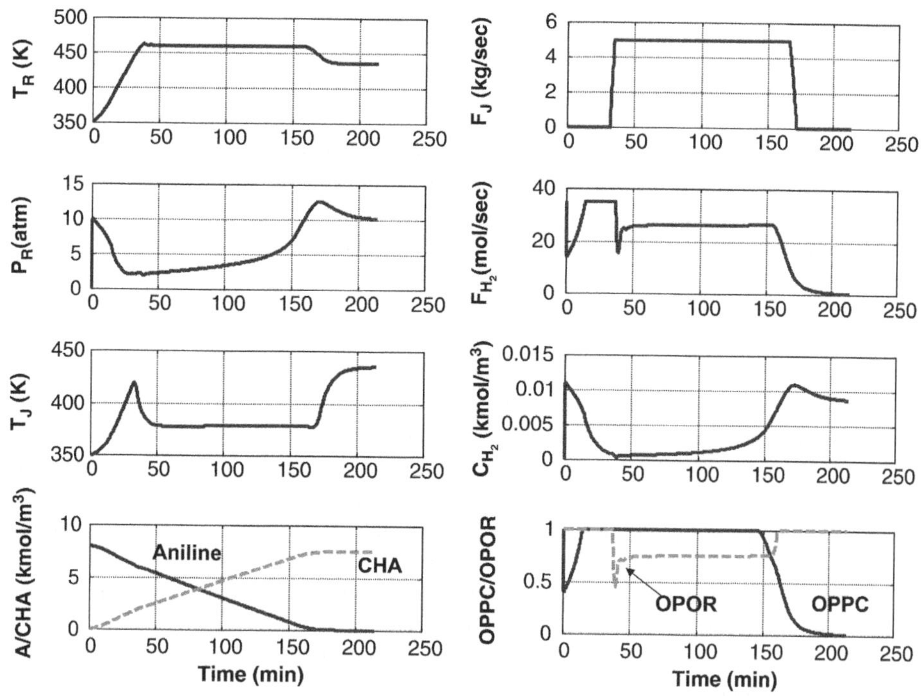

Figure 4.41 Fed-batch hydrogenation reactor trajectories.

After about 170 min, most of the aniline has been consumed (plot at the bottom of the left column in Fig. 4.41). The hydrogen concentration increases (which increases reactor pressure and reduces the hydrogen feed) and the reaction rate slows (which reduces the heat generation). The reactor temperature drops, cutting the cooling water. The batch is stopped when the conversion of aniline reaches 99.9% at 214 min.

4.7 BATCH TML REACTOR

Large quantities of tetramethyl lead (TML) were used back in the 1960s as an antiknock additive in high-octane gasoline before they were banned because of air pollution problems. TML was produced in a batch reactor that had the interesting feature of requiring the removal of a byproduct during the batch. The main reaction involves a solid sodium–lead alloy and liquid methyl chloride:

$$NaPb + 4CH_3Cl \xrightarrow{k_1} Pb(CH_3)_4 + 4NaCl \tag{4.29}$$

The solid is in the form of flakes that have significant variability in particle size. The reaction is highly exothermic. It uses a catalyst of aluminum chloride and is conducted at about 383 K and 25.8 atm, but both temperature and pressure vary during the batch cycle. Since TML is an explosive, some toluene is added with the reactor charge to serve as a moderator.

The solid alloy, liquid methyl chloride, catalyst, and toluene are charged to the reactor vessel at about 320 K. A slight excess of methyl chloride is charged to ensure complete conversion of the sodium–lead alloy. Excess methyl chloride also prevents reactor temperatures from getting too high as the high-boiling TML is produced and the low-boiling methyl chloride is consumed. The vapor pressures of methyl chloride and toluene give an initial reactor pressure of 11.2 atm. Steam is introduced to the jacket of the reactor to heat the contents. When the temperature approaches 363 K (pressure is 18 atm), the exothermic reaction begins to generate significant heat and cooling is required.

Autorefrigeration is used with chilled water flowing through an external condenser, which is mounted high enough above the reactor so that gravity flow can be used to return the liquid back to the reactor. The vapor boiled off the reactor is mostly methyl chloride because TML and toluene have much higher boiling points.

A pressure controller with a time-varying setpoint is used to control the reactor. The setpoint is ramped from 18 to 25.8 atm at a normal rate of 0.11 atm per minute. The controller output signal goes to three control valves. A split-range value setup is used. The controller output signal is initially at 100%, and the steam valve is wide open. As the pressure rises and the output signal drops to 83%, the steam valve (AO) is completely closed. As the pressure continues to rise, dropping the output signal from 83 to 50%, the chilled water valve (AC) goes from closed to completely open. Thus the chilled water valve operates over only 33% of the full scale.

The third valve is on a vent stream from the condenser. This vent is needed because there is a second undesirable reaction that produces ethane:

$$2Na + 2CH_3Cl \xrightarrow{k_2} C_2H_6 + 2NaCl \tag{4.30}$$

The ethane is much lighter than the methyl chloride, so it accumulates in the condenser and acts essentially like an inert substance that "blankets" the condenser. The effect of the inert substance can be considered to reduce either (1) the bubblepoint temperature, thus reducing the differential temperature driving force and reducing heat transfer, or (2) the effective heat transfer area. Either effect is a reduction in heat transfer. So if the ethane is not vented off during the batch, the pressure cannot be controlled even with the chilled water valve wide open.

On the other hand, there are negative effects of venting. Some methyl chloride is lost in the vent. This must be recovered and recycled, which involves capital and energy expenses. In addition, the loss of methyl chloride raises reactor temperature, which can produce reaction rates that are too high. Runaway reactions can occur that result in disk ruptures, which represent both safety and pollution problems.

So it is important to vent off some vapor to get rid of the ethane, but not too much. The vent valve (AC) is split-range, so that it is closed when the pressure controller output signal is 83% of full scale. It is wide open when the controller output is at 0%. The sizing of the steam, chilled water, and vent valves is critical to the safe and efficient operation of this batch reactor. Figure 4.42 gives a sketch of the reactor, the controller, the setpoint generator, and the three control valves.

The plant reactor is a 1.9-m^3 (500-gal) horizontal vessel with a horizontal agitator, which stirs the solid and liquid phases in the reactor. The vapor generated by boiling the liquid flows up to the top of the condenser and is condensed in a "downflow" configuration. The vent stream comes off the side of the bottom head of the condenser. The liquid is

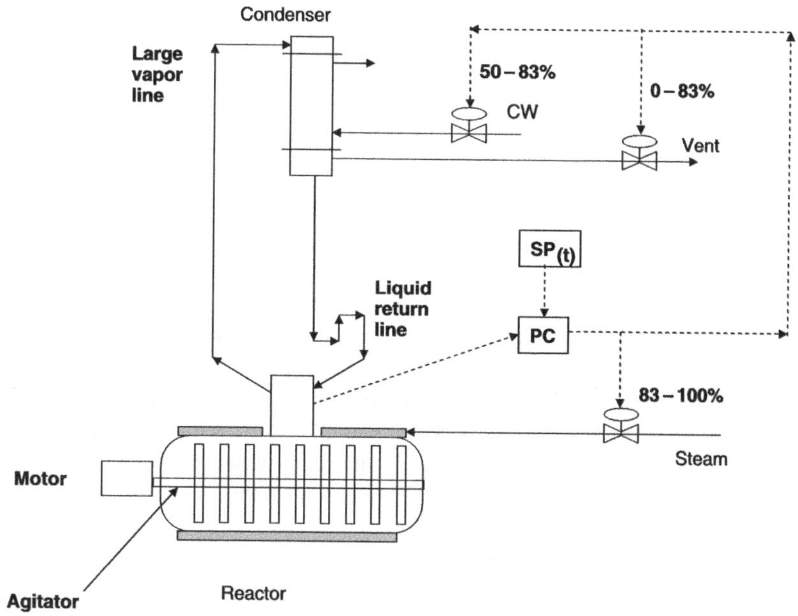

Figure 4.42 TML batch reactor.

returned through a U-leg seal to the middle of the reactor vessel. Typical batch times are about 100 min. Conversion is 98% and TML yield is about 88%.

The kinetics of the two reactions show that the desirable TML reaction has a higher activation energy than does the undesirable ethane reaction. Therefore the yield of TML is maximized by running the reaction at high temperatures. Of course, an additional and very significant advantage of high-temperature operation is shorter batch times (increased production rate).

However, high temperatures mean large reaction rates and high heat removal requirements that can exceed the heat transfer capacity of the condenser. Thus there is an optimum pressure–time trajectory that maximizes yield while not violating heat removal constraints. The normal ramp rate of the pressure controller setpoint is about 0.11 atm/min.

The variability of the sizes of the solid sodium–lead alloy particles introduces another factor. The smaller the particle size, the more surface area is exposed to reaction, which increases the effective reaction rate. Big particles react slowly; small particles react quickly. So, for the same pressure–time trajectory, significantly different heat load–time requirements can occur. If the alloy charge contains very small particles, a runaway reaction can occur if the pressure controller setpoint is ramped up too quickly.

Back in the 1960s you could sell all the TML you could produce, so there was a large incentive to increase production rates. A research program was initiated to increase TML production. The first step was to go to the laboratory and obtain good kinetic data. The normal laboratory run in a Parr bomb operated at constant temperature, and yields were typically higher than those obtained in the pilot plant reactor and in the plant reactor. Nonisothermal experiments were performed to follow the same temperature trajectory as seen in the plant, and these produced similar yields.

A dynamic mathematical model was developed to match the laboratory data. Very simple first-order kinetics and perfect mixing were assumed. The predictions of the model were found to fit experiments conducted in the 0.02-m^3 (5-gal) pilot plant reactor quite well. The final yields and conversions checked well, but more importantly, the time-dependent heat transfer rates predicted by the model and measured in the pilot plant were in close agreement.

Then the possibility of operating with higher pressure ramp rates was explored on the model to see the potential benefits in increased yield and shorter batch time. Pilot plant tests again confirmed model predictions.

The next step was to compare model predictions with plant data. Yield, conversions, and batch time results were similar, but there was a significant difference in the peak in the heat removal rate. A typical peak heat transfer rate predicted by the model (and confirmed in the pilot plant) with the 0.11 atm/min pressure setpoint ramp rate was 315 kW at about 50 min into the batch when the pressure was 21.4 atm. With the same pressure setpoint ramp rate, the peak heat-removal load in the plant was 500 kW, which occurred at about 40 min when the pressure was 20 atm. The reason for this large difference was, at the time, unclear.

Despite this unexplained discrepancy, plant tests were run with higher pressure setpoint ramp rates. Yield was increased, and batch time was shortened. However, the plant reactor experienced more "hot" reactions and more disk ruptures when the higher ramp rate was used.

So it was back to the model and the pilot plant to conduct more tests to explain the discrepancy between plant data and pilot plant data. The plant reactor was a horizontal vessel 1.8 m in length. All the liquid from the condenser was returned to the middle of the vessel. So the liquid had to flow 0.9 m to get to the end of the vessel. The pilot plant reactor was 0.406 m in length with the liquid from the condenser returned at the middle of the vessel. So the liquid had to flow only 0.2 m to get to the end of the vessel.

This difference in flow length suggested that the problem might be imperfect mixing. The model was modified to include this effect, and hotspots at the ends of the vessel and higher peak heat loads were predicted. Experiments were run in the pilot plant reactor with the liquid returned to one end of the vessel instead of the middle. This lengthening of the flow path from 0.2 to 0.4 m produced wide (50 K) temperature differences between the "wet" and "dry" zones even with the vapor temperature showing normal values. Peak heat loads were also increased. In fact, a runaway reaction (disk rupture) was produced by reducing the agitator speed.

This insight provided by the model and confirmed by the pilot plant experiments led to modifications in the large plant reactor. Liquid was returned to both ends and to the middle of the vessel. These simple plumbing modifications permitted the plant to use higher pressure setpoint ramp rates and at the same time reduced the frequency of disk ruptures. The higher TML yields and the shorter batch times increased production rates.

This study is a good example of the synergistic effect of combining fundamental mathematical modeling with small-scale and large-scale experimentation.

4.8 FED-BATCH REACTOR WITH MULTIPLE REACTIONS

The fed-batch reactors considered in Sections 4.1, 4.2, and 4.6 all involved single reactions. The control problem was to prevent reaction temperature runaways. In fed-batch

TABLE 4.1 Parameters of Fed-Batch Reactor with Multiple Reactions

k_1 at 40°C	0.099 m^3 kmol^{-1} min^{-1}	k_2 at 40°C	0.0736 m^3 kmol^{-1} min^{-1}
E_1	2×10^4 kJ/kmol	E_2	5×10^4 kJ/kmol
λ_1	-1×10^5 kJ/kmol	λ_2	-1.5×10^5 kJ/kmol
U_A	37 kJ min^{-1} C^{-1}		
$C_{A(t=0)}$	0.5 kmol/m^3	C_{B0}	5 kmol/m^3
Total V_R	7.5 m^3	$V_{R(t=0)}$	5 m^3
$T_{F,\text{in}}$	40°C	ρ	1000 kg/m^3
$T_{C,\text{in}}$	10°C	L_R/D_R	1
c_p	4.2 kJ kg^{-1} C^{-1}		

systems with multiple reactions, the issues of conversion and yield can also become important. In this section we consider such a system. The numerical example is based on a paper by Marchetti et al.[5]

There are two reactions with different reaction rates and activation energies. The first reaction produces the desired product C from the reaction of components A and B:

$$A + B \xrightarrow{k_1} C \tag{4.31}$$

An initial charge of component A (5 m^3 with concentration 0.5 kmol/m^3) is placed in a 7.5-m^3 reactor. Component B is fed into the reactor at a flowrate $F_{(t)}$ (m^3/min), which can be a function of time. The initial concentration of B in the reactor is zero, but the concentration builds up with time as pure B is fed ($C_{B0} = 5$ kmol/m^3). Table 4.1 gives the values of the kinetic and process parameters.

The second reaction produces an undesirable product D from the reaction of component B:

$$2B \xrightarrow{k_2} D \tag{4.32}$$

The activation energy of the second reaction is larger than the first, so low reactor temperatures favor the yield of the desired component C. However, low reactor temperatures reduce the conversion of A for a fixed batch time.

The second reaction rate is strongly dependent on the concentration of B in the reactor, so high concentrations of B increase the rate of formation of the undesirable component D. Therefore, selectivity is increased by feeding slowly. However, a slow feed flowrate produces a long batch time for a given conversion of A.

Both of the reactions are exothermic ($\lambda_1 = -1 \times 10^5$ kJ/kmol and $\lambda_2 = -1.5 \times 10^5$ kJ/kmol). Heat is transferred to the jacket surrounding the vessel. If an aspect ratio of 1 is assumed, the diameter of the vessel is 2.12 m and the total jacket area is 14.14 m^2. Cooling water at 10°C is fed to the jacket at a maximum flowrate of 0.2 m^3/min. A temperature controller attempts to control the reactor temperature T_R at 40°C by manipulating cooling water flowrate F_{CW}. A circulating cooling water system is assumed, so the water in

[5]A. Marchetti, M. Amrhein, B. Chachuat, and D. Bonvin, Scale-up of batch processes via decentralized control, *ADCHEM* IFAC, 2006, p. 221.

the jacket is perfectly mixed at T_J. The thickness of the jacket is 0.05 m, and the volume in the jacket is (0.05 m)(14.14 m^2).

Two 1-min temperature measurement lags are included in the temperature control loop. A 50°C temperature transmitter span is used. Controller gain is 10, and integral time is 10 min.

4.8.1 Equations

The nonlinear dynamic model of this fed-batch reactor consists of a total mass balance, component balances for three components, an energy balance for the liquid in the reactor, and an energy balance for the cooling water in the jacket:

Total mass balance (assuming constant densities; m^3/min):

$$\frac{dV_R}{dt} = F \tag{4.33}$$

Component balance for A (kmol/min):

$$\frac{d(V_R C_A)}{dt} = -V_R k_1 C_A C_B \tag{4.34}$$

Component balance for B (kmol/min):

$$\frac{d(V_R C_B)}{dt} = F C_{B0} - V_R k_1 C_A C_B - 2V_R k_2 (C_B)^2 \tag{4.35}$$

Component balance for D (kmol/min):

$$\frac{d(V_R C_D)}{dt} = V_R k_2 (C_B)^2 \tag{4.36}$$

Reactor energy balance (kJ/min):

$$\frac{d(V_R T_R)}{dt} = F T_{F,\text{in}} - \frac{\lambda_1 V_R k_1 C_A C_B}{\rho c_p} - \frac{\lambda_2 V_R k_2 (C_B)^2}{\rho c_p} - \frac{U A_{\text{hx}} (T_R - T_J)}{\rho c_p} \tag{4.37}$$

Cooling jacket energy balance (kJ/min):

$$\frac{dT_J}{dt} = \frac{F_{\text{CW}} (T_{c,\text{in}} - T_J)}{V_J} + \frac{U A_{\text{hx}} (T_R - T_J)}{V_J \rho_J c_J} \tag{4.38}$$

The heat transfer area A_{hx} varies with time because the volume of liquid in the vessel increases as feed is added. The instantaneous heat transfer area is calculated from the ratio of the instantaneous volume to the total volume:

$$A_{\text{hx}(t)} = A_{\text{total}} \frac{V_{R(t)}}{V_{R\text{total}}} \tag{4.39}$$

Figure 4.43 gives a Matlab program for the simulation of this fed-batch reactor.

```
% Program "fedbatchmarchetti.m"
%  Nonlinear dyanmic simulation of batch reactor with two reactions
%         A+B=C  and   2B = D Tr=40 C; Vr=7.5 cu m
%  units: cu m, sq m, C, kmol, kg, kJ, min
%  ca(t=0)=0.5 kmol/cu m; k in units of cu m/(kmol-min)
clear
cb0=5;tcin=10; tfin=40;roe=1000;ma=50;mb=50;mc=100;md=100;cp=4.2;cj=4.183;roej=1000;
%  Aspect ratio = 1
vrtotal=7.5;dr=(vrtotal*4/pi)^0.3333;areaj=pi*dr^2;areahx=areaj;vj=.05*areahx;
areatotal=areahx; trss=40;u=37/areaj  % u in kJ/min-C-sq m)
% Initial conditions
vr=5;ca=0.5;cb=0;cc=0;cd=0;tr=40;tj=40;time=0;erint=0;initiala=vr*ca;feedtotal=0;
f=2/1000 ;fss=f      % flow in cu m/min
slope=1/300/1000;    % slope of ramp
vrtr=vr*tr;vrca=vr*ca;vrcb=vr*cb;vrcc=vr*cc;vrcd=vr*cd;
% kinetics for two reactions
k1base=0.099;e1=2e4;k10=k1base*exp(e1/(273+40)/8314);lambda1=-1e5;
k2base=0.0736;e2=5e4;k20=k2base*exp(e2/(273+40)/8314);lambda2=-1.5e5;
% Controller settings
kc=10;taui=10; fcwmax=0.2; % cw in cu m/min
np=0;tplot=0;delta=0.01;trlag1=trss;trlag2=trss;taum=1;trsp=trss;tstop=300;
% Integration loop
while  time<tstop;
   if ca<0.01;break;end
      error=(trsp-trlag2)/50;op=0.7+kc*error+erint;if op>1; op=1;end; if op<0;op=0;end;
   %  AC valve action
   fcw=fcwmax*(1-op);          if vr>vrtotal;f=0;end
%ramp feed
   f=fss-slope*time;if f<0;f=0;end

% store data for plotting
if time>=tplot;   np=np+1;timep(np)=time;trp(np)=tr;tjp(np)=tj;fp(np)=f; fcwp(np)=fcw;
   cap(np)=ca;cbp(np)=cb;ccp(np)=cc;cdp(np)=cd; vrp(np)=vr;tplot=tplot+0.1;end
areahx=vr*areatotal/vrtotal;q=u*areahx*(tr-tj);
k1=k10*exp(-e1/(tr+273)/8314);k2=k20*exp(-e2/(tr+273)/8314);
% Derivative evaluations
dvr=f;dvrca= -vr*k1*ca*cb;
dvrcb= f*cb0-vr*k1*ca*cb-2*vr*k2*(cb^2);
dvrcc= vr*k1*ca*cb;dvrcd=vr* k2*cb^2;
dtj=(fcw*tcin)/vj -fcw*tj/vj+q/(cj*roej*vj);
dvrtr= f*tfin-lambda1*vr*k1*ca*cb/roe/cp - lambda2*vr*k2*(cb^2)/roe/cp - q/(cp*roe);
dtrlag1=(tr-trlag1)/taum;dtrlag2=(trlag1-trlag2)/taum;
%  Integration
time=time+delta;vr=vr+dvr*delta;vrca=vrca+dvrca*delta;vrcb=vrcb+dvrcb*delta;
vrcc=vrcc+dvrcc*delta;vrcd=vrcd+dvrcd*delta;
vrtr=vrtr+dvrtr*delta;tj=tj+dtj*delta;trlag1=trlag1+dtrlag1*delta;trlag2=trlag2+dtrlag2*delta;
tr=vrtr/vr;ca=vrca/vr;cb=vrcb/vr;cc=vrcc/vr;cd=vrcd/vr;
%  anti-reset windup
if op<1; if op>0; erint=erint+delta*error*kc/taui;end;end;
end
conversiona=(initiala-vr*ca)/initiala,selectivity=cc/(cc+2*cd)
subplot(4,2,1);plot(timep,trp);grid;ylabel('TR (C)');
title('A+B=C;2B=D;  F0=2; U=37 kJ/min/K/sq m; L/D=1; Kc=10; taui=10')
subplot(4,2,2);plot(timep,fcwp);grid;ylabel('Fcw (cu m/min)');
subplot(4,2,3);plot(timep,fp*1000);grid;ylabel('F (L/min)');
subplot(4,2,4);plot(timep,vrp);grid;ylabel('VR (cu m)');
subplot(4,2,5);plot(timep,tjp);grid;ylabel('TJ (C)');
subplot(4,2,6);plot(timep,cap,timep,ccp);
axis([0 300 0 0.5]);grid;ylabel('CA/CC (kmol/cu m)');
subplot(4,2,7);plot(timep,cbp);grid;ylabel('CB (kmol/cu m)');xlabel('Time (min)');
subplot(4,2,8);plot(timep,cdp);grid;ylabel('CD (kmol/cu m)');xlabel('Time (min)');
```

Figure 4.43 Fed-batch reactor Matlab program.

4.8.2 Effect of Feed Trajectory on Conversion and Selectivity

Figure 4.44 shows how variables change during a 300-min batch. The feed flowrate is ramped down from an initial value of 2 L/min to 1 L/min at the end of the batch. The temperature controller increases the flowrate of cooling water as the reactor temperature

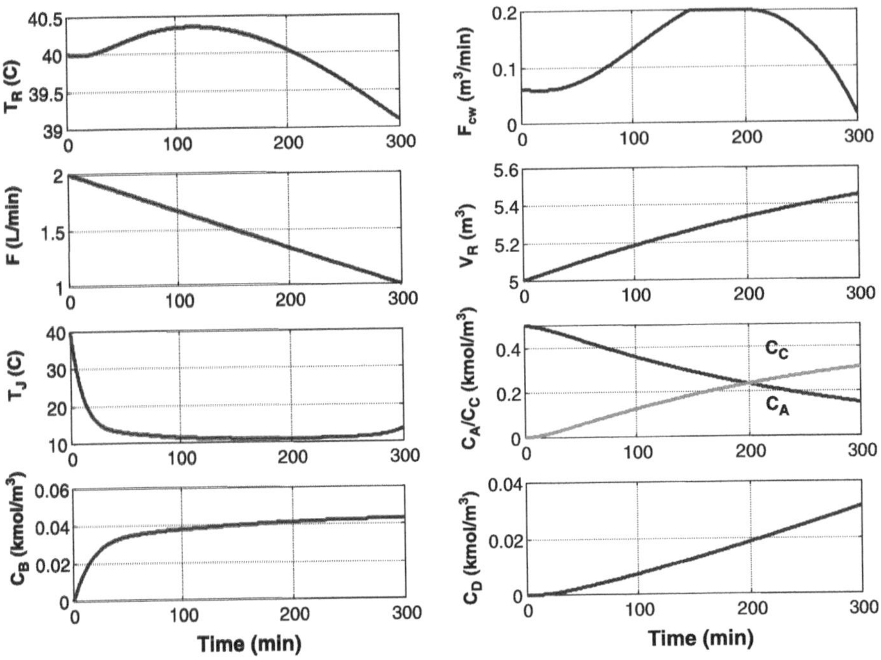

Figure 4.44 Fed-batch reactor with consecutive reactions; F ramped from 2 to 1 L/min.

rises. The cooling water valve is wide open at about 110 min. The jacket temperature drops to almost the supply inlet temperature of 10°C in about 40 min because of the high flowrate of cooling water. The concentration of B builds up, component A is consumed, and component C is produced. However, as the concentration of B gets larger, the undesirable product D begins to be produced.

Using this ramped feed flow policy permits fairly tight temperature control. The resulting conversion and selectivity using this feed strategy can be calculated from the initial amount of A charged to the reactor and the final amounts of C and D in the reactor at the end of the batch.

$$\text{Conversion of A} = \frac{V_{R(t=0)}C_{\text{A}(t=0)} - V_{R(t=300)}C_{\text{A}(t=300)}}{V_{R(t=0)}C_{\text{A}(t=0)}}$$

$$= \frac{(5\,\text{m}^3)(0.5\,\text{kmol/m}^3) - (5.45\,\text{m}^3)(0.151\,\text{kmol/m}^3)}{(5\,\text{m}^3)(0.5\,\text{kmol/m}^3)} = 0.671 \quad (4.40)$$

$$\text{Selectivity of C relative to D} = \frac{C_{\text{C}(t=300)}}{C_{\text{C}(t=300)} + 2C_{\text{D}(t=300)}}$$

$$= \frac{0.3077}{0.3077 + 2(0.0311)} = 0.831 \quad (4.41)$$

Note that *selectivity* is defined as the ratio of the moles of B that have gone into the production of the desired product C to the total moles of B that have been consumed in producing both C and D. These results show that a slow feedrate keeps the concentration

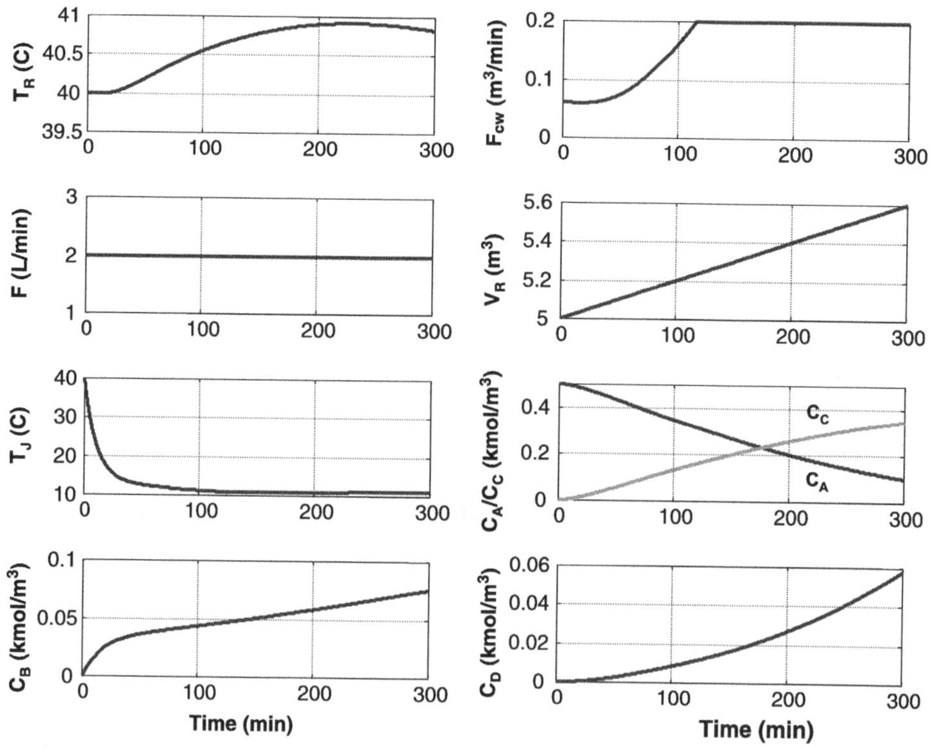

Figure 4.45 Fed-batch reactor with consecutive reactions; $F = 2\ \mathrm{L/min}$.

of B low and keeps the temperature near 40°C. These effects produce little undesirable D, but conversion is not very high.

Figure 4.45 shows what happens if a constant high feed flowrate is maintained throughout the batch (at $2\ \mathrm{L/min}$). The temperature increases steadily despite having a high cooling water flowrate and a low jacket temperature. The higher temperatures give a higher conversion (from 67.2 to 77.2%):

$$\text{Conversion of A} = \frac{V_{R(t=0)}C_{\mathrm{A}(t=0)} - V_{R(t=300)}C_{\mathrm{A}(t=300)}}{V_{R(t=0)}C_{\mathrm{A}(t=0)}}$$

$$= \frac{(5\ \mathrm{m}^3)(0.5\ \mathrm{kmol/m}^3) - (5.6\ \mathrm{m}^3)(0.1017\ \mathrm{kmol/m}^3)}{(5\ \mathrm{m}^3)(0.5\ \mathrm{kmol/m}^3)} = 0.772 \tag{4.42}$$

However, the selectivity decreases substantially (from 83.1 to 74.8%). This is partially due to somewhat higher temperatures. But more importantly, it is due to the higher concentration of B in the reactor. Compare the C_{B} trajectory of Figure 4.44 (reaching a peak of about $0.043\ \mathrm{kmol/m}^3$ at the end of the batch) with the same trajectory in Figure 4.45 (reaching a peak of about $0.075\ \mathrm{kmol/m}^3$ at the end of the batch). The corresponding concentrations of D are $0.0311\ \mathrm{kmol/m}^3$ at the end of the batch

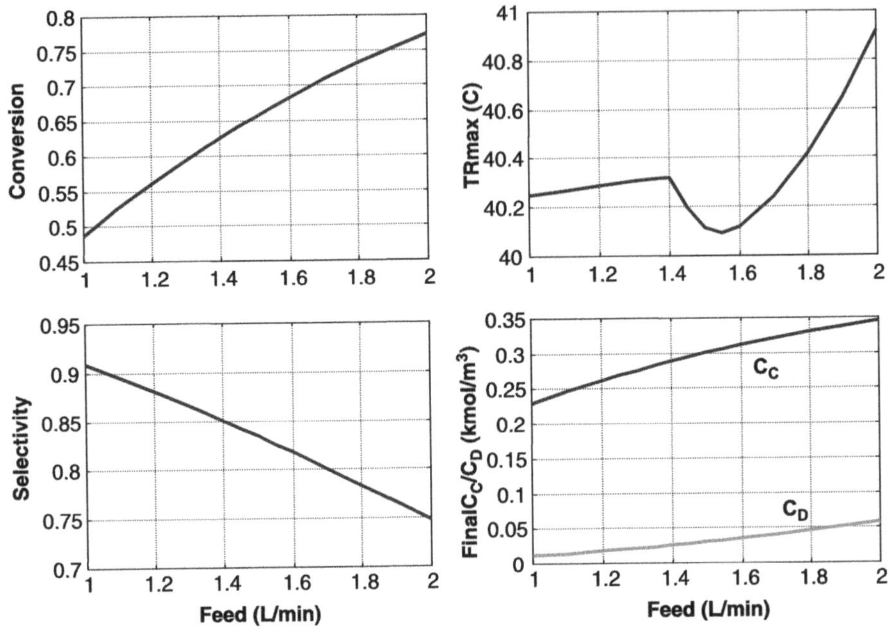

Figure 4.46 Effect of feed flowrate; $T_{\text{batch}} = 300$ min.

for the ramped feed flowrate in Figure 4.44 and 0.0579 kmol/m^3 for the higher feed flowrate in Figure 4.45:

$$\text{Selectivity of C relative to D} = \frac{C_{\text{C}(t=300)}}{C_{\text{C}(t=300)} + 2C_{\text{D}(t=300)}}$$

$$= \frac{0.3447}{0.3447 + 2(0.0597)} = 0.748 \qquad (4.43)$$

Figure 4.46 gives results for a range of feed flowrates when a constant feed flowrate policy is employed. The batch time is fixed in these results at 300 min. High feed flowrates produce high concentrations of B, high maximum temperatures, and high conversions, but result in low selectivities.

4.8.3 Batch Optimization

The optimum operation of this fed-batch reactor involves finding the trajectory of feed versus time that maximizes (or minimizes) some economic performance criterion. Of course, the batch time is also an operating optimization variable.

For example, suppose that the conversion of component A is specified to be 75% and the selectivity is specified to be 76%. These are determined from the overall economics of the process, which include the cost of raw materials, the value of products, and the cost of separation. Fixing conversion determines the batch time; that is, we run until 75% of the reactant A initially charged to the reactor has been consumed. Selectivity will depend on the feed flowrate trajectory.

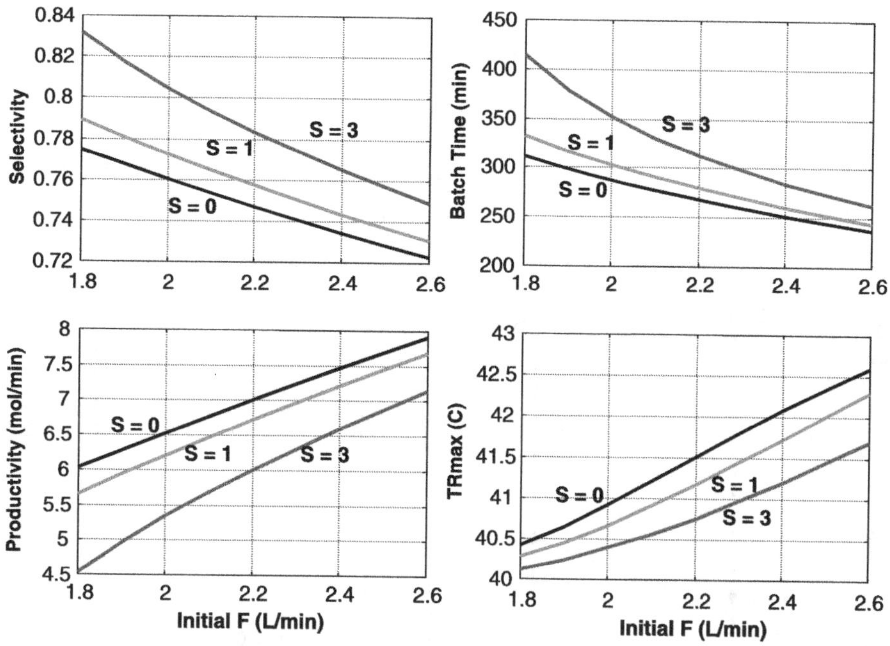

Figure 4.47 Effect of feed trajectory; conversion = 75%.

For simplicity, we restrict ourselves to having a feed flowrate that starts at some value F_0 and ramps down with a constant slope S. This practical approach to trajectory optimization is recommended by Smith and Choong[6] for batch processes. We want to find the values of F_0 and S that achieve the desired conversion and selectivity. There will be many pairs of values that will satisfy the two criteria. Each will have a different batch time and a different amount of C produced.

Our objective function is the rate of production of C, which is the amount of C produced divided by the batch time:

$$\text{Productivity} = \frac{V_{R(t_{\text{batch}})} C_{C(t_{\text{batch}})}}{t_{\text{batch}}} \tag{4.44}$$

Figure 4.47 gives results for a range of initial feed flowrates and ramp slopes. The units of S are 10^{-6} L/min^2. The conversion for all these results is fixed at 75%. Higher initial feed flowrates give shorter batch times and higher productivity but lower selectivity. Higher slopes give higher selectivity but longer batch times and smaller productivity.

Two alternative operating strategies are shown on Figure 4.48*a* that both achieve the specified 75% conversion and 76% selectivity. With an initial feed flowrate of 2.0 L/min and zero slope, the productivity is 6.52 mol C/min with a batch time of 287 min. With an initial feed flowrate of 2.48 L/min and a slope of 3×10^{-6} L/min^2, the productivity is 6.82 mol C/min with a batch time of 275 min.

Figure 4.48*b* gives results for higher slopes and initial flowrates. The productivity with a slope of 5×10^{-3} L/min^2 is 6.91 mol C/min; with a slope of 10×10^{-3} L/min^2, it is

[6]R. Smith and L. Choong, Optimizing batch operations, *Chem. Eng. Prog.*, **102**, 31–36 (Jan. 2006).

6.96 mol C/min; with a slope of 15×10^{-3} L/min^2, it is 6.72 mol C/min. So the *optimum* feed trajectory has an initial feed flowrate of 3.36 L/min, a slope of 10×10^{-6} L/min^2, and a batch time of 260 min.

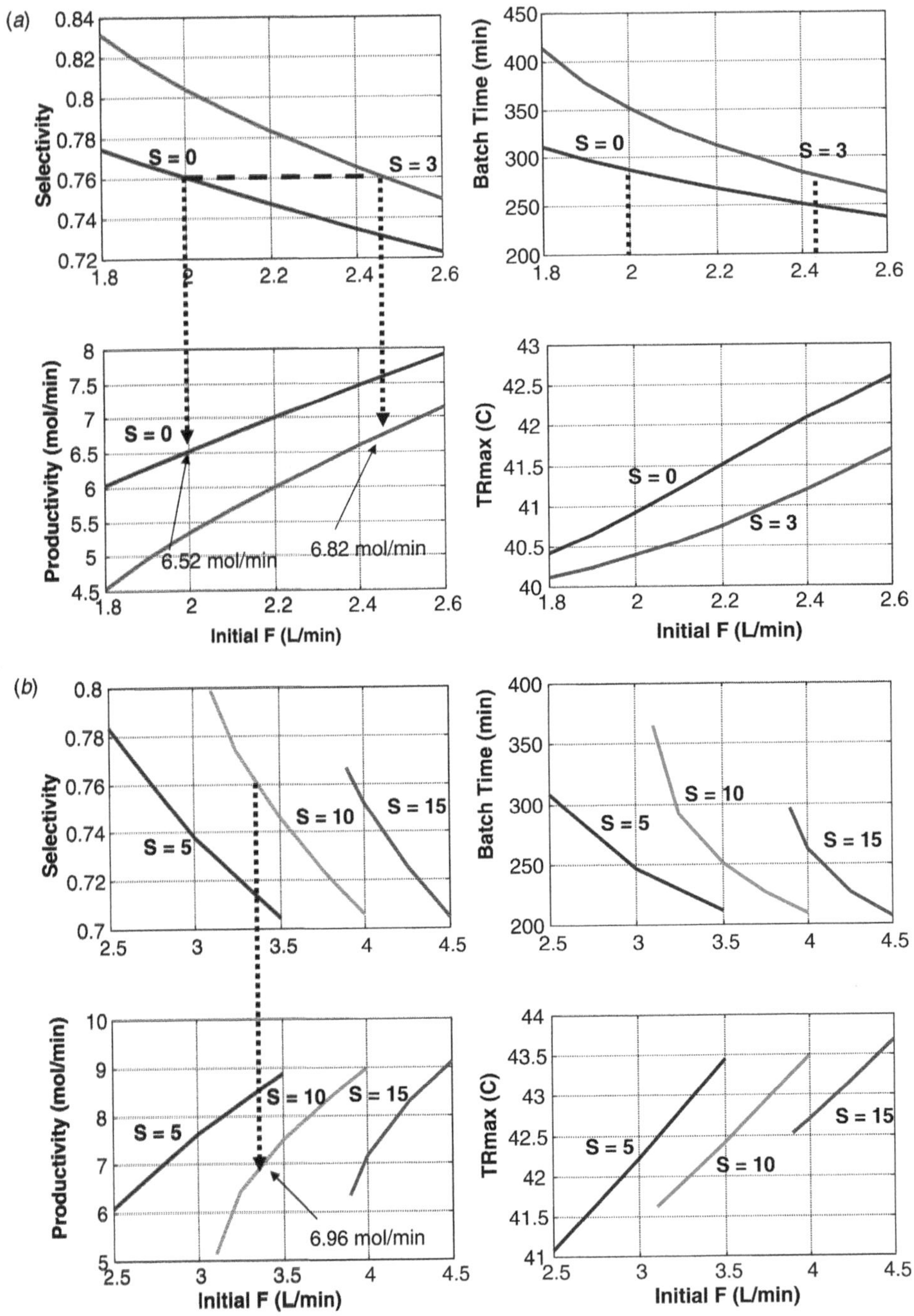

Figure 4.48 Selectivity = 76%, conversion = 75%.

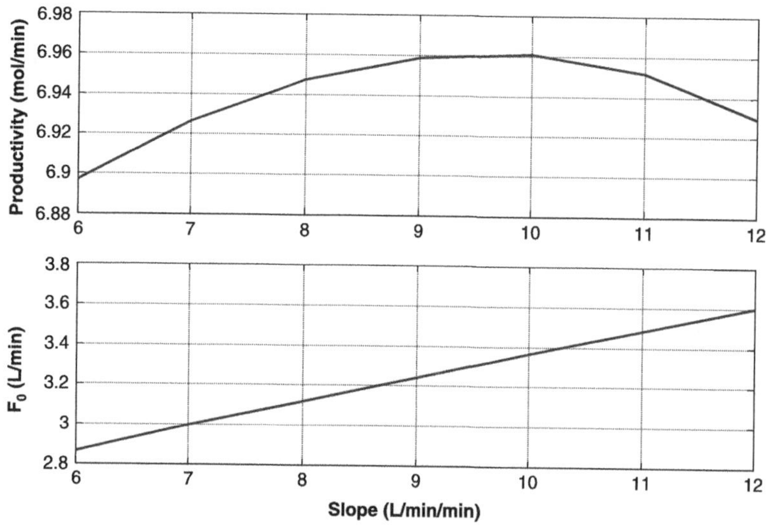

Figure 4.49 Slope and initial feed flowrate for maximum productivity; conversion = 75%, selectivity = 76%.

The top graph in Figure 4.49 shows how the slope of the ramp affects productivity for the 75% conversion and 76% selectivity base case. The bottom graph gives the corresponding initial feed flowrate.

The effects on the optimum policy of changing the conversion and selectivity specifications from the base case are shown in Figures 4.50 and 4.51. In Figure 4.50, the conversion is increased to 80%. The optimum slope decreases from 10 to 6.5×10^{-6} L/min^2, the optimum initial feed flowrate decreases from 3.36 to 3.07 L/min, and productivity decreases from 6.97 to 5.873 mol C/min.

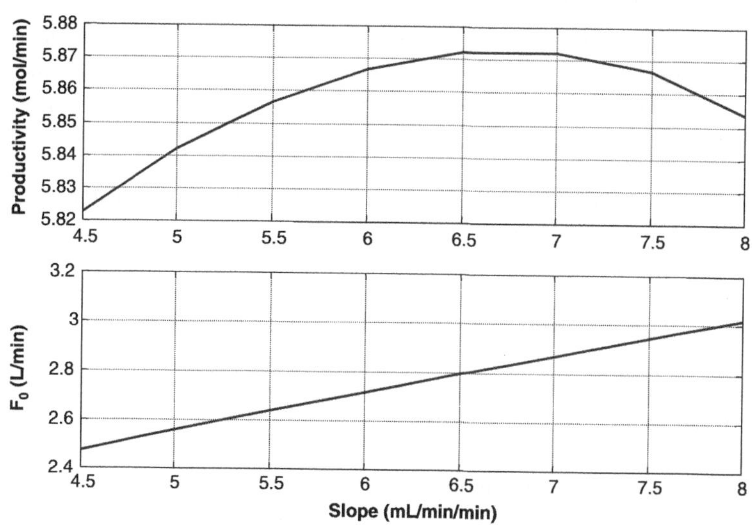

Figure 4.50 Slope and initial feed flowrate for maximum productivity; conversion = 80%, selectivity = 76%.

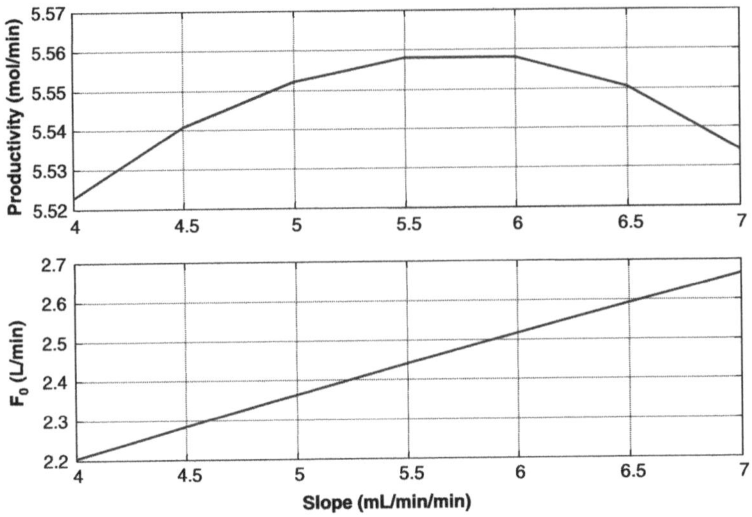

Figure 4.51 Slope and initial feed flowrate for maximum productivity; conversion = 75%, selectivity = 80%.

In Figure 4.51, the selectivity is increased to 80%. The optimum slope decreases from 10 to 5.5×10^{-6} L/min^2, the optimum initial feed flowrate decreases from 3.36 to 2.44 L/min, and productivity decreases from 6.96 to 5.557 mol C/min. These results show that demanding higher performance results in lower productivity.

The process variable trajectories are given in Figure 4.52 for the base case when this optimum feed strategy is used.

4.8.4 Effect of Parameters

There are a number of kinetic and equipment parameters used in the numerical example. The size of the reactor, the overall heat transfer coefficient, the heats of reaction, the specific reaction rates, and the activation energies are among the most important. All of these were explored.

If it is important to maintain good temperature control and heat transfer is limiting, we would expect that the feed flowrate could be increased if the heat transfer coefficient is larger, if the heats of reaction are lower, or if the heat transfer area is larger. However, faster feed flowrates increase the concentration of B in the reactor and produce more undesirable D. The result is that, for specified conversion and selectivity constraints, the optimum feed trajectory depends very little on all of parameters except the ratio of the base reaction rates k_1/k_2 at 40°C.

As expected, increasing this ratio increases productivity for the same conversion and selectivity. Figure 4.53 illustrates the effect. The base value of k_2 at 40°C is reduced by a factor of 2 from the value shown in Table 4.1. Keeping the conversion at 75% and the selectivity at 76%, the maximum productivity increases from 6.96 mol C/min for the base case to 13.5 mol C/min for the smaller k_2 case.

Figure 4.54 shows how variables change during the batch with this optimum feed flowrate trajectory for the case with smaller k_2. These results should be compared with

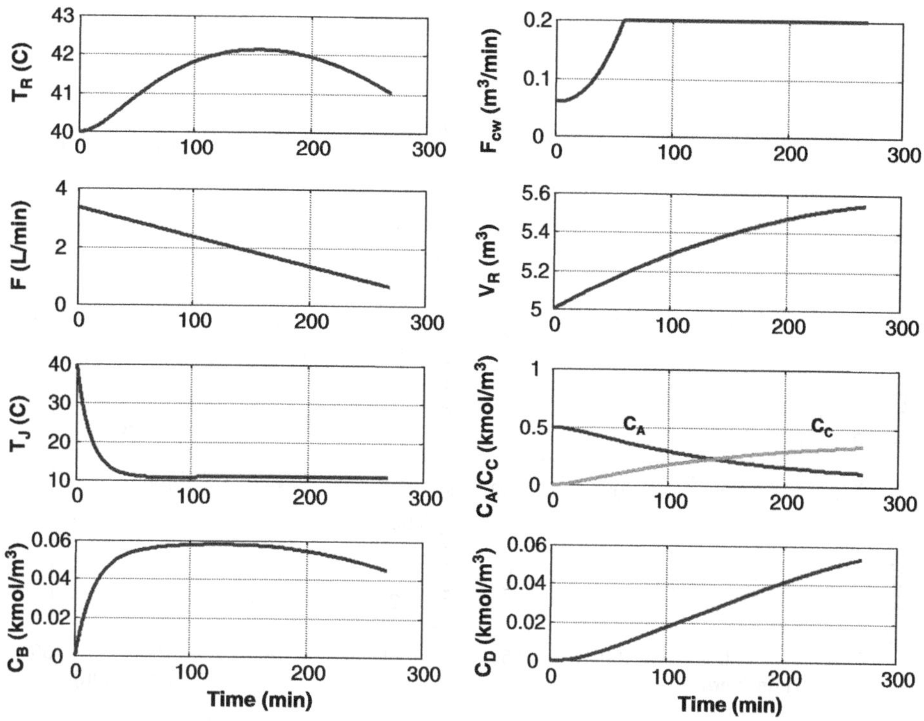

Figure 4.52 Fed-batch reactor with optimum feed strategy; base case; $F_0 = 3.36\ \mathrm{L/min}$, slope $= 10 \times 10^{-3}\ \mathrm{L/min^2}$; productivity $= 6.96$ mol C/min.

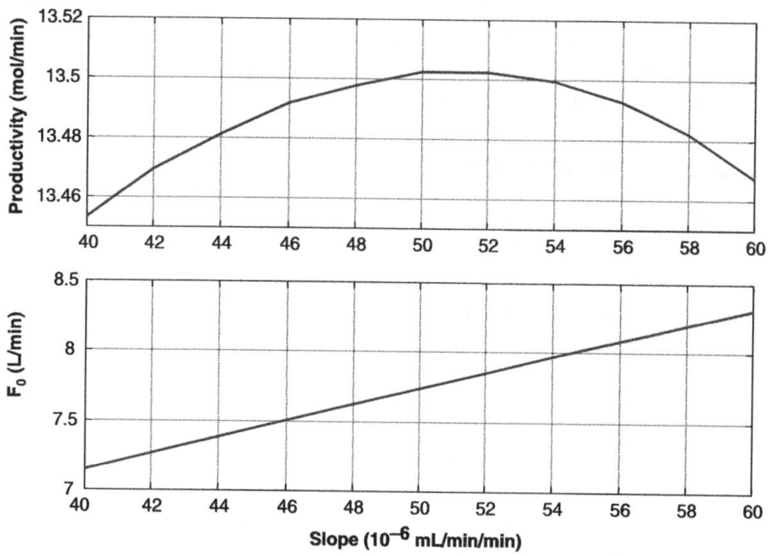

Figure 4.53 Slope and initial feed flowrate for k_2*0.5; conversion = 75%, selectivity = 76%.

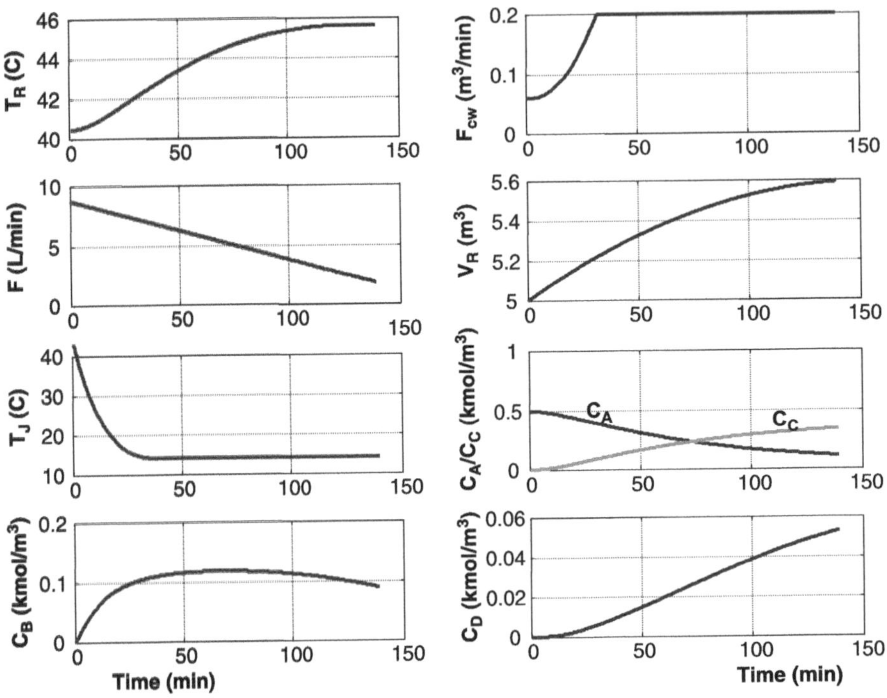

Figure 4.54 Fed-batch reactor with optimum feed strategy; 0.5^*k_2; $F_0 = 7.74$ L/min, slope $= 50 \times 10^{-6}$ L/min², productivity = 13.5 mol C/min.

those given in Figure 4.52 for the base case. The batch time is reduced to 139 minutes, which yields the higher productivity. Reactor temperatures climb to almost 46°C, but over this fairly small temperature range, the difference in activation energies does not impact selectivity significantly. The concentrations of B are higher, but the shorter batch time gives the same final concentration of D at the end of the batch.

Increasing the conversion and selectivity constraints to 85% for the same smaller value of k_2 reduces the productivity, as shown in Figure 4.55.

4.8.5 Consecutive Reaction Case

The chemistry considered up to this point in this section has two *simultaneous* reactions in which the undesirable product depends on the concentration of one of the reactants. The methods discussed can be applied to other chemistry. Suppose that the reactions are such that the desired component can react further to form an undesirable product in a *consecutive* reaction system:

$$\begin{aligned} A + B &\xrightarrow{k_1} C \\ C + B &\xrightarrow{k_2} D \end{aligned} \tag{4.45}$$

We would expect that this type of system would inherently have much more unfavorable yields and conversions because the desired product C can be consumed as its

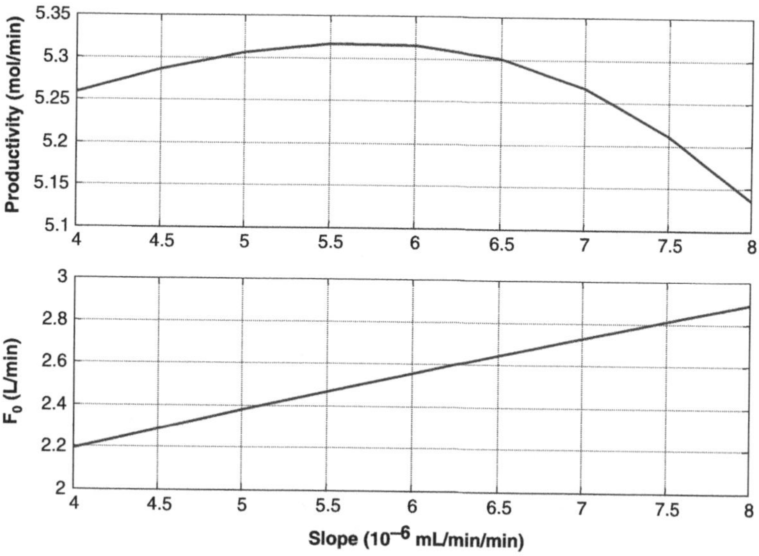

Figure 4.55 Slope and initial feed flowrate for k_2*0.5; converstion = 85%, selectivity = 85%.

concentration builds up. The optimum design of the fed-batch system would feature lower conversions and lower selectivities.

In this consecutive reaction case, the component balance equations for B, C, and D become

Component balance for B (kmol/min):

$$\frac{d(V_R C_B)}{dt} = FC_{B0} - V_R k_1 C_A C_B - V_R k_2 (C_B)(C_C) \tag{4.46}$$

Component balance for C (kmol/min):

$$\frac{d(V_R C_C)}{dt} = +V_R k_1 C_A C_B - V_R k_2 (C_B)(C_C) \tag{4.47}$$

Component balance for D (kmol/min):

$$\frac{d(V_R C_D)}{dt} = V_R k_2 (C_B)(C_C) \tag{4.48}$$

Figure 4.56 gives some results for this consecutive reaction system. The same kinetic parameters are used. The feed flowrate is constant at 2 L/min. If the batch time is 300 min, the conversion is high (73.9%), but, as expected, the selectivity is very poor (39.5%). Stopping the batch earlier, for example, at 150 min, gives a higher selectivity (67.6%) but a lower conversion (42.1%). Figure 4.57 shows this tradeoff between conversion and selectivity over a range of batch times.

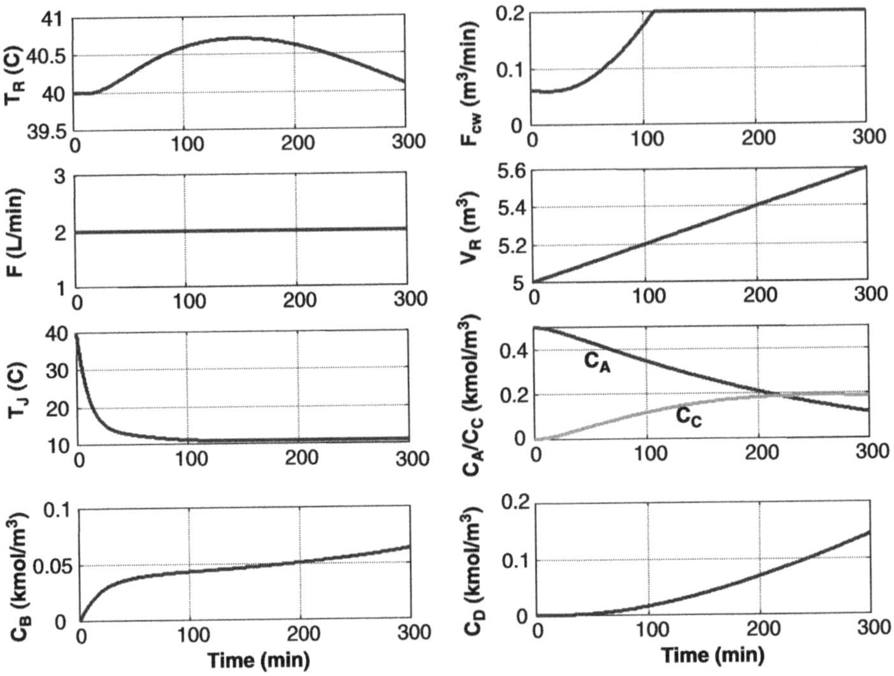

Figure 4.56 Consecutive reactions; $F_0 = 2$ L/min; conversion = 73.9%, selectivity = 39.5%.

This fed-batch example illustrates some of the fascinating aspects of design and operating a batch reactor system. The issue of dynamic controllability dominates in most systems, but the economics of the operation can also be strongly affected by the impact of control on conversion and yield.

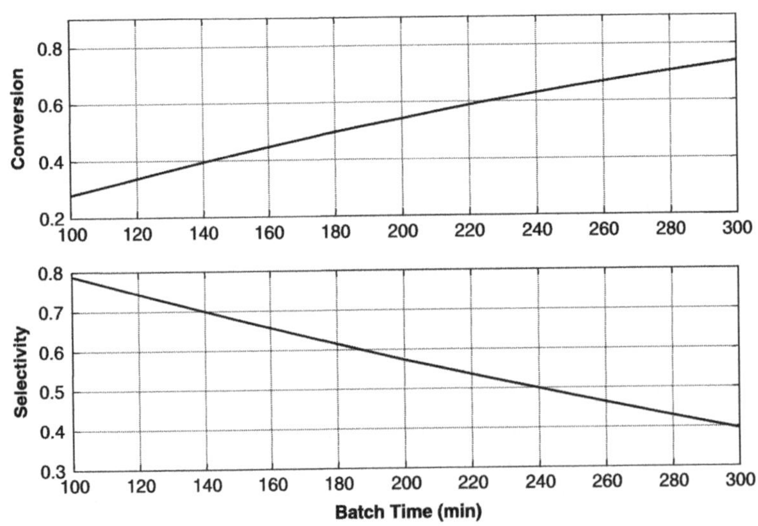

Figure 4.57 Consecutive reactions; effect of batch time on conversion and selectivity; $F_0 = 2$ L/min.

4.9 CONCLUSION

In this chapter we have studied a variety of batch reactors. Their inherent dynamic nature and their many design and operating parameters present challenges to the engineer involved in both their design and their operation.

CHAPTER 5

STEADY-STATE DESIGN OF TUBULAR REACTOR SYSTEMS

5.1 INTRODUCTION

In the next two chapters of this book we turn to the chemical reactor that is probably the most challenging: the tubular or plug flow reactor. The inherent distributed nature of the unit (variables change with axial and radial position) gives rise to complex behavior, which is often counterintuitive and difficult to explain. The increase in the number of independent variables makes the development and solution of mathematical models more complex compared to the perfectly mixed CSTR and batch reactor.

There is also a wider variety of reactor and system types for tubular reactors. Many operate adiabatically, while others are heated or cooled. Multiple tubular reactors in series with intermediate heating or cooling are quite common. The most common industrial use of tubular reactors is in systems where a solid catalyst is required. The catalyst is installed in beds or inside tubes in the shell of the reactor vessel, and the process reacting fluid (gas or liquid) flows through the fixed catalyst.

There are five fundamental differences between CSTRs and tubular reactors. The first is the variation in properties with axial position down the length of the reactor. For example, in an adiabatic reactor with an exothermic irreversible reaction, the maximum temperature occurs at the exit of the reactor under steady-state conditions. However, in a cooled tubular reactor, the peak temperature usually occurs at an intermediate axial position in the reactor. To control this peak temperature, we must be able to measure a number of temperatures along the reactor length.

The second difference is the dynamic response to disturbances or changes in manipulated variables. In a perfectly mixed CSTR, a change in an input variable has an immediate effect on variables in the reactor. In a tubular reactor it takes time for the disturbance to work its way through the reactor to the exit. Therefore there are very significant dynamic lags and deadtimes between changes made at the inlet of the reactor and

Chemical Reactor Design and Control. By William L. Luyben

temperatures and compositions further down the length of the reactor. The absence or presence of a solid catalyst phase in the reactor makes a tremendous difference in the time constants of the openloop response because of the thermal inertia of the catalyst. The mass of the solid is much larger than the mass of process fluid if the reaction is gas-phase.

The third difference is the issue of heat transfer in nonadiabatic reactors. Ideally we would like to be able to control the temperature at each axial position down the reactor. However, it is mechanically very difficult to achieve independent heat transfer at various axial positions. About all that can be done is to have the cooling/heating medium flow either cocurrent or countercurrent to the direction of the process flow. The only two variables that can be manipulated are the flowrate of the medium and its inlet temperature. The former is the normal manipulated variable. The result is that only a single temperature can be controlled, which can be the peak temperature or the exit temperature. However, because of the significant dynamics of the tubular reactor, the control of these temperatures is sometimes quite difficult and tight control cannot be achieved in the face of load disturbances.

The fourth difference is the impact of feed temperature. In a CSTR the feed temperature has little effect unless we are trying to control reactor temperature by manipulating feed as discussed in Chapter 3. In a tubular reactor, however, the feed temperature is a very important design and control variable. Consider the design of an adiabatic tubular reactor in which a certain per-pass conversion is desired. For given reactor inlet compositions, this fixes the temperature rise through the reactor. Now, if a low inlet temperature is selected, the reaction rates in the front end of the reactor will be slow, so a large reactor will be required. As the reactor inlet temperature is increased, the required reactor size decreases, but the reactor exit temperature goes up directly with the inlet temperature for a fixed conversion. In most systems, there is a maximum allowable temperature, as limited by a number of factors, such as catalyst decay, undesirable side reactions, and materials of construction. Therefore this means that there is a maximum inlet temperature, so the size of the reactor is fixed.

Remember this is for a fixed per-pass conversion and inlet composition and flowrate. If the desired conversion is decreased, the inlet temperature can be increased and the reactor size reduced. Of course, this means that more unreacted material in the reactor effluent must be recovered and recycled. Keep in mind the difference between the "per pass" conversion, which is a crucial design optimization variable, and the "overall" conversion of the entire process, which is typically quite close to 100% because the cost of raw materials imposes a severe economic penalty on a process with low overall conversion. A lower per-pass conversion translates into a higher recycle flow with its associated higher energy and capital costs. But chemical process economics usually dictate that capital investment can be justified to save energy, and both energy and capital can be expended to reduce losses of raw materials and/or products.

The discussion above points out one of the most important tradeoffs in chemical reactor design. The smaller the reactor size, the larger the recycle flowrate. This reactor/recycle tradeoff dominates the steady-state economics of the design of tubular reactor systems, as we will illustrate in several examples in this chapter. It also has a major impact on dynamic control, as we will see in the next chapter.

The fifth difference between a CSTR and a tubular reactor is pressure drop. A CSTR has essentially no pressure drop. A tubular reactor can have very substantial pressure drop, which is very important in gas-phase systems with gas recycle because of compression costs. Pressure drop can be reduced by using shorter tubes and/or larger-diameter tubes. But shorter tubes means that more parallel tubes are required, which leads to a more expensive larger-diameter shell and can worsen flow maldistribution problems. Larger-diameter tubes provide less heat transfer area per volume and can degrade dynamic control. So another engineering tradeoff exists between reactor physical dimensions and compression costs.

All of the factors discussed above contribute to the complexity of designing and controlling tubular reactors.

5.2 TYPES OF TUBULAR REACTOR SYSTEMS

This chapter presents a comparison of the steady-state economics of four alternative tubular reactor systems. The entire process will be considered, not just the reactor in isolation, because the optimum economic steady-state design can be determined only for the entire plant. The type of recycle, the phase of the reaction, and the heat transfer configuration all affect the optimum design.

5.2.1 Type of Recycle

The recovery of unreacted components in the reactor exit stream can produce a liquid recycle or a gas recycle. Liquid recycle systems typically involve distillation columns to separate products from the material to be recycled. The recycle stream is usually a liquid that can be inexpensively pumped back to the reactor. If the reaction is liquid-phase, the liquid recycle is added to the liquid fresh feedstreams. If the reaction is vapor-phase and the recycle is liquid, a vaporizer is used. The major cost with liquid recycle is energy, in both the separation step and the vaporizer, if it is required.

Gas recycles usually involve a liquid–vapor phase separation in which the lighter reactants come off the top of the flash drum as vapor and are compressed so that they can be recycled back to the reactor inlet. A gas recycle almost always implies a gas-phase reaction. The major costs with gas recycle are compressor capital investment and compressor work.

5.2.2 Phase of Reaction

The reactor can operate with either a liquid-phase reaction or a gas-phase reaction. In both types, temperature is very important. With a gas-phase reaction, the operating pressure is also a critical design variable because the kinetic reaction rates in most gas-phase reactions depend on partial pressures of reactants and products. For example, in ammonia synthesis ($N_2 + 3H_2 \Leftrightarrow 2NH_3$), the gas-phase reactor is operated at high pressure because of LeChatelier's principle, namely that reactions with a net decrease in moles should be run at high pressure. The same principle leads to the conclusion that the steam–methane reforming reaction to form synthesis gas ($CH_4 + H_2O \Leftrightarrow CO + 3\ H_2$) should be conducted at low pressure.

The pressure level also affects compression costs of feed and product streams. Steam–methane reformers are operated at fairly high pressure (20 atm) despite LeChatelier's principle because (1) methane is usually available at high pressure, (2) water can be inexpensively pumped to the required pressure before vaporizing, and (3) the product hydrogen is required at high pressure (60 atm) and there are many moles of hydrogen produced. The negative effect on the reaction rates of operating at high pressure is compensated for by running at high temperatures in the fuel-fired reforming furnace.

5.2.3 Heat Transfer Configuration

Because feed temperature is important, some type of feed preheating is usually an important part of the plant design. A feed-effluent heat exchanger (FEHE) is the standard configuration for an exothermic reaction conducted in an adiabatic reactor. The reactor effluent is at a high enough temperature to provide heat to the cold reactor feedstream, which consists of the fresh feeds plus recycles. As we will illustrate quantitatively in Chapter 7, the FEHE can present serious dynamic control problems because of the positive feedback of energy. Bypassing of some of the cold feed around the FEHE is used to control reactor inlet temperature. Figure 5.1 shows a typical single-stage adiabatic reactor with a FEHE. Many systems have both a FEHE and a trim utility heat-input unit (furnace for high-temperature reactions or steam heat exchanger). The furnace is usually part of the process because it is needed for startup to achieve the initial high temperature required for the reaction to begin. Therefore it is available to provide a second manipulated variable for improved control.

If the system consists of a series of adiabatic reactors, there are two basic configurations. The first has heat exchangers or furnaces between each of the reactors to cool or heat the reactor effluent before it enters the next reactor. The second configuration uses "cold shot" cooling. Some of the cold reactor feed is bypassed around the upstream reactor(s) and mixed with the hot effluent from the reactor to lower the inlet temperature to the downstream catalyst bed.

Some tubular reactors are not adiabatic but are internally cooled or heated. The reactor is a series of parallel tubes (in a tube-in-shell heat exchanger or in a furnace) with cooling or heating medium surrounding the tubes. The steam–methane reforming furnace is an important example of this type. The catalyst is inside the furnace tubes, which are heated by radiant and convective heat transfer from the burning fuel to provide the heat of reaction for the highly endothermic reaction. The vinylacetate reactor is another important example. The exothermic, high-temperature reaction (acetic acid + propylene + oxygen → vinyl acetate + water) is conducted in the tubes of a reactor with steam generated on the shell side.

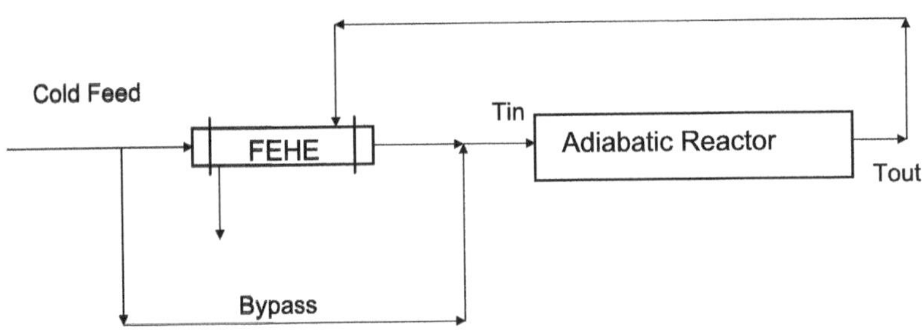

Figure 5.1 PFR with feed effluent heat exchanger (FEHE).

5.3 TUBULAR REACTORS IN ISOLATION

It is useful to initially examine the tubular reactor as an isolated unit so that some insight can be gained about the effects of various design and operating parameters on its inherent behavior. The equations describing the steady-state operation of a tubular reactor are presented and illustrated for a specific numerical example. Both adiabatic and nonadiabatic tubular reactors are considered.

The flow patterns, composition profiles, and temperature profiles in a real tubular reactor can often be quite complex. Temperature and composition gradients can exist in both the axial and radial dimensions. Flow can be laminar or turbulent. Axial diffusion and conduction can occur. All of these potential complexities are eliminated when the "plug flow" assumption is made. A plug flow tubular reactor (PFR) assumes that the process fluid moves with a uniform velocity profile over the entire cross-sectional area of the reactor and no radial gradients exist. This assumption is fairly reasonable for adiabatic reactors. But for nonadiabatic reactors, radial temperature gradients are inherent features. If tube diameters are kept small, the plug flow assumption in more correct. Nevertheless the PFR can be used for many systems, and this idealized tubular reactor will be assumed in the examples considered in this book. We also assume that there is no axial conduction or diffusion.

5.3.1 Adiabatic PFR

As a numerical example, we consider a gas-phase exothermic irreversible reaction with two reactants and one product that occurs in a PFR packed with a solid catalyst:

$$A + B \xrightarrow{k} C \tag{5.1}$$

The rate of reaction depends on the partial pressures of the reactants and is expressed as kmol per second of C generated per kilogram of catalyst.

$$\begin{aligned} \mathcal{R} &= k_{(T)} P_A P_B \\ &= y_A y_B P^2 k_0 e^{-E/RT} \end{aligned} \tag{5.2}$$

where k is the specific reaction rate (kmol s^{-1} bar^{-2} kg catalyst^{-1}), P_j is the partial pressure of component j (bar), P is the total pressure (bar), and y_j is the mole fraction of component j in the gas. All of the variables change with axial position: $T_{(z)}$, $y_{j(z)}$, and $P_{(z)}$. Table 5.1 gives values of kinetic parameters and physical properties.

The ordinary differential equations describing a steady-state adiabatic PFR can be written with axial length z as the independent variable. Alternatively the weight of catalyst w can be used as the independent variable. There are three equations: a component balance on the product C, an energy balance, and a pressure drop equation based on the Ergun equation. These equations describe how the molar flowrate of component C, temperature T, and the pressure P change down the length of the reactor. Under steady-state conditions, the temperature of the gas and the solid catalyst are equal. This may or may not be true dynamically:

$$\frac{dF_C}{dw} = \mathcal{R} \tag{5.3}$$

TABLE 5.1 Kinetic Parameters and Physical Properties

Heat of reaction	λ	kJ/kmol	−23,237
Activation energy	E	kJ/kmol	69,710
Preexponential factor	k_0	kmol s^{-1} bar^{-2} (kg catalyst)$^{-1}$	0.19038
Heat capacities	c_{pA}	kJ kmol^{-1} K^{-1}	30
	c_{pB}	kJ kmol^{-1} K^{-1}	40
	c_{pC}	kJ kmol^{-1} K^{-1}	70
Molecular weights	M_A	kg/kmol	15
	M_B	kg/kmol	20
	M_C	kg/kmol	35
Catalyst porosity	ε		0.4
Catalyst particle size	D_p	m	0.002
Catalyst bulk density	ρ_{cat}	kg/m^3	2000
Gas viscosity	μ	kg m^{-1} s^{-1}	0.18×10^{-4}
Ratio heat capacities	γ		1.312
Heat-transfer coefficients	U_{FEHE}	kW m^{-2} K^{-1}	0.142
	U_{HX}	kW m^{-2} K^{-1}	0.227

$$\frac{dT}{dw} = \frac{-\lambda \Re}{c_{pA}F_A + c_{pB}F_B + c_{pC}F_C} \tag{5.4}$$

Here, λ is the heat of reaction (kJ/kmol), F_j is the molar flowrate of component j at any axial position, and c_{pj} is the molar heat capacity of component j

$$\frac{dP}{dw} = \frac{-fL_R\rho V^2}{D_p W_{cat} 10^5} \tag{5.5}$$

where f is the friction factor given in Eq. (5.6), L_R is the length of the reactor (m), ρ is the gas density (kg/m^3), V is the superficial gas velocity (m/s), D_p is the catalyst particle diameter (m), and W_{cat} is the total weight of catalyst in the reactor (kg):

$$f = \frac{(1-\varepsilon)}{\varepsilon^3}\left[1.75 + 150\frac{(1-\varepsilon)}{\text{Re}}\right] \tag{5.6}$$

Here, ε is the catalyst porosity and Re is the particle Reynolds number (Re $= D_pV\rho/\mu$) and μ is viscosity (kg m^{-1} s^{-1}). The pressure and temperature change with axial position, so the friction factor changes somewhat down the length of the reactor.

As the gas flows down the reactor and the reaction proceeds, the molar flowrate F_C increases, F_A decreases, F_B decreases, P decreases, and T increases. If the reactor inlet contains no C, the initial conditions are $F_{C(w=0)} = 0$ and $T_{(w=0)} = T_{in}$. The equations are integrated from $w = 0$ to $w = W_{cat}$ (the total kg of catalyst in the reactor).

Using the reaction stoichiometry, the molar flowrates of A and B are calculated at any axial position from the known molar flowrate of C and the known amounts of A and B fed to the reactor:

$$\begin{aligned} F_{A(w)} &= F_{A,in} - F_{C(w)} \\ F_{B(w)} &= F_{B,in} - F_{C(w)} \end{aligned} \tag{5.7}$$

```
% Program "adiabaticss.m"
% Solves ODE for adiabatic PFR at steady state
% Given reactor inlet conditions and dr
%  calculate lr = 10*dr and wmax
%  generates profiles of yA yC, T and P
clear
% Parameters
k0=0.19038;e=69710;lambda=-23237;dp=0.003;roecat=2000;porosity=0.4;
cpa=30;cpb=40;cpc=70;ma=15;mb=20;mc=35;viscosity=0.18e-4;
dw=1;wplot=0;yar=0.5;ybr=0.5;np=0;
dr=1.71;lr=10*dr;areacs=pi*(dr^2)/4;wmax=lr*areacs*roecat;
% Initial conditions
w=0;fc=0;fa0=0.12;fb0=0.12;fr=1.2;temp=445;p=50;fain=fr*yar+fa0;fbin=fr*ybr+fb0;
% Integration loop
while w<wmax
   k=k0*exp(-e/8.314/temp);
   fa=fain-fc;fb=fbin-fc;ya=fa/(fa+fb+fc);yb=fb/(fa+fb+fc);yc=fc/(fa+fb+fc);
   rate=k*ya*yb*p^2;    roegasmolar=p/(temp*1.01325*82.057e-3);
   flowvol=(fa+fb+fc)/roegasmolar;
   maverage=ya*ma+yb*mb+yc*mc;roegas=roegasmolar*maverage;
   vel=flowvol/areacs;reynolds=dp*vel*roegas/viscosity;
   friction=(1-porosity)*(1.75+150*(1-porosity)/reynolds)/(porosity^3);
% Save for plotting
   if w>=wplot;np=np+1;wp(np)=w/1000;tempp(np)=temp;pp(np)=p;yap(np)=ya*100;
      ycp(np)=yc*100;wplot=wplot+10;end
     % Derivative evaluation
   dfcdw=rate;dtempdw=-lambda*rate/(cpa*fa+cpb*fb+cpc*fc);
   dpdw=-friction*lr*roegas*(vel^2)/(dp*wmax*1e5);
%  Integration
   fc=fc+dfcdw*dw;temp=temp+dtempdw*dw;p=p+dpdw*dw;w=w+dw;
   if p<1;break;end
end
clf
subplot(2,2,1);plot(wp,tempp);grid;ylabel('T (K)');title('Adiabatic');
subplot(2,2,2);plot(wp,ycp);grid;ylabel('yC (%)');
subplot(2,2,3);plot(wp,yap);grid;ylabel('yA (%)');xlabel('w  (1000 kg)');
subplot(2,2,4);plot(wp,pp);grid;ylabel('P (bar)');xlabel('w  (1000 kg)');
```

Figure 5.2 Matlab program for adiabatic PFR.

Figure 5.2 gives a Matlab program that numerically integrates the three ordinary differential equations (ODEs) for an adiabatic plug flow reactor. Figure 5.3 shows some typical temperature, composition, and pressure profiles down the length of the reactor (expressed as kg of catalyst). The numerical case shown in Figure 5.3 is for an adiabatic reactor that is 1.7 m in diameter and 17 m in length. The total feed to the reactor is the sum of a large gas recycle stream (1.2 kmol/s with a composition $y_{RA} = 0.5$ and $y_{RB} = 0.5$ mole fraction) and two fresh feedstreams (each 0.12 kmol/s of the two pure reactants).

The effect of reactor inlet temperature is shown in Figure 5.3. For an inlet temperature of 446 K, the reaction rate is small. Therefore there is only a small increase in temperature and little consumption of the reactants (low conversion). However, a quite small increase in inlet temperature to 448 K results in very rapid increases in temperature and conversion. With an inlet temperature of 450 K, the reactants are essentially completely consumed. The adiabatic temperature rise is about 330 K! This example illustrates one of the difficult problems associated with tubular reactors. They can be very sensitive to reactor inlet temperature. The problem is analogous to that seen in earlier chapters in CSTRs that are designed for low conversions. The reactor inlet stream contains high concentrations of both reactants, so there is plenty of "fuel" to generate a runaway reaction. If the maximum temperature limitation in the system is 550 K, this runaway could do real damage to the catalyst or result in a vessel meltdown.

The sensitivity can be reduced by using the concept of a "limiting reactant" to provide some composition self-regulation. This is the same idea we applied in CSTR systems with

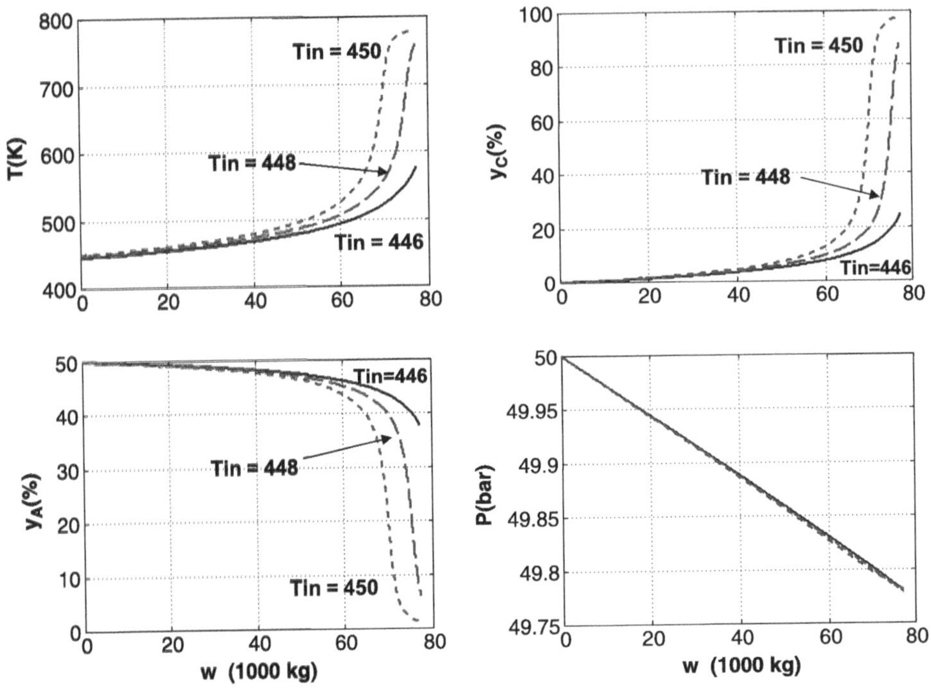

Figure 5.3 Adiabatic PFR; effect of T_{in}; $D_R = 1.7$; $y_{RA} = 0.5$.

high activation energies (high-temperature sensitivity). It is quite frequently used in tubular reactor systems. Figure 5.4 gives profiles for different compositions of the recycle stream. A larger reactor is used (2.4 m diameter, 24 m length) in which an inlet temperature of 445 K results in a runaway when the concentration $y_{RA} = 0.5$ (and $y_{RB} = 0.5$). As the concentration is lowered, the adiabatic temperature rise decreases. With $y_{RA} = 0.125$ mole fraction A (which means that the concentration of B is $y_{RB} = 0.875$), the reactor temperature rises to only about 550 K. This example illustrates that the reactor feed concentrations can be adjusted to limit the adiabatic temperature rise, even with high per-pass conversion of the limiting reactant.

In the results presented in Figure 5.4, the reactor is very large (2.5 m in diameter and 25 m in length) and contains a large amount of catalyst (240,000 kg). But it provides almost complete conversion. Suppose that the catalyst and the reactor vessel are very expensive, which makes reactor size very important. Figure 5.5 gives results for different-size reactors with a recycle composition $y_{RA} = 0.125$ mole fraction A. The diameter is kept at 2.5 m so that pressure drop is small, but the length of the reactor is reduced. The exit temperature T_{out} and the conversion are given for three different reactor inlet temperatures T_{in} as functions of reactor length. As the inlet temperature is increased, the length of the reactor required to achieve a given conversion decreases. But the reactor exit temperature also increases as the inlet temperature increases.

Typical adiabatic reactor design involves finding the "best" values of recycle concentration, inlet temperature, reactor size, recycle flowrate, and per-pass conversion. Of course, not all of these parameters are independent. When the entire process is studied

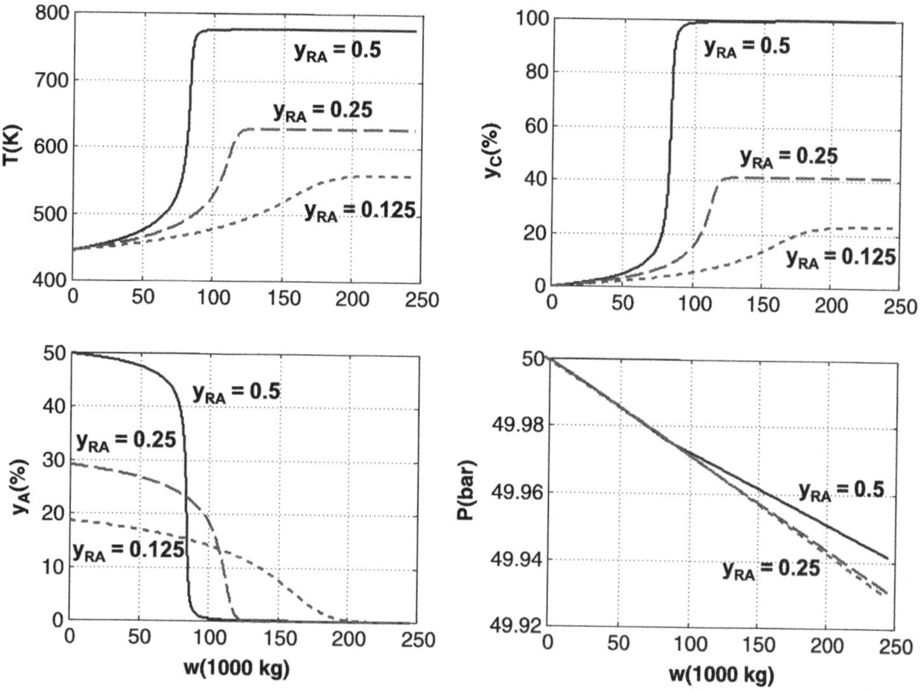

Figure 5.4 Adiabatic PFR; effect of reactant concentration; $D_R = 2.5$; $T_{in} = 445$ K.

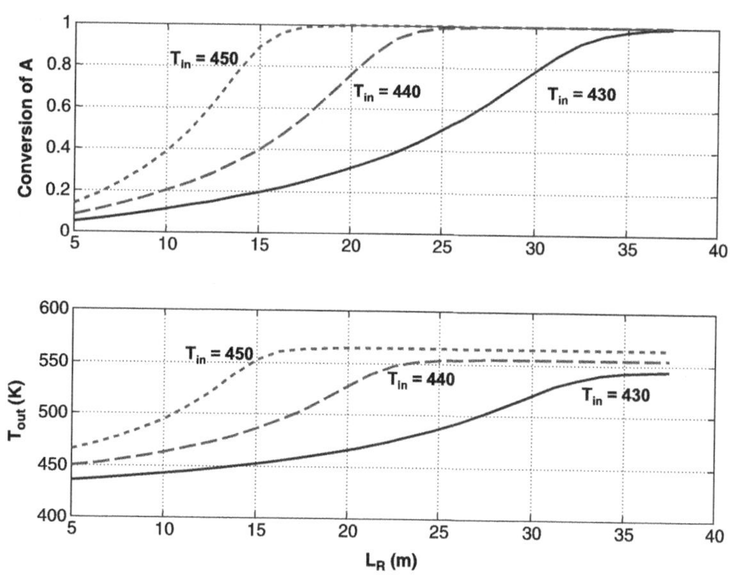

Figure 5.5 Adiabatic PFR; effect of length and T_{in}; $y_{RA} = 0.125$; $D_R = 2.5$.

later in this chapter, we will discuss the important issue of design degrees of freedom, which is the number of variables that can be specified to completely define the system.

5.3.2 Nonadiabatic PFR

In many tubular reactors cooling or heating occurs as the process fluid flows through the reactor. This produces a major difference between an adiabatic and a nonadiabatic tubular reactor. In an *adiabatic* reactor, with an exothermic irreversible reaction, the maximum steady-state temperature occurs at the end of the reactor. In a *cooled* reactor, the maximum steady-state temperature usually occurs at some axial position part way down the reactor. Thus the temperature does not change monotonically with length.

The component balance and pressure drop equation do not change, but the energy balance contains a heat transfer term

$$\frac{dT}{dw} = \frac{-\lambda \mathfrak{R} - 4U(T - T_{st})/(\rho_{cat} D_{tube})}{c_{pA}F_A + c_{pB}F_B + c_{pC}F_C} \tag{5.8}$$

where λ is the heat of reaction (kJ/kmol), U is the overall heat transfer coefficient (kJ s^{-1} m^{-2} K^{-1}), ρ_{cat} is the density of the catalyst (kg/m^3), D_{tube} is the tube diameter (m), and T_{st} is the temperature (K) of the cooling medium at any axial position. In most high-temperature tubular reactors that require cooling, boiler feedwater is fed to the shell side of the reactor, and steam is generated. In this case the temperature does not change with axial position. If a co-current or countercurrent cooling or heating fluid is used in the shell, an energy balance for this material must be included in the model.

Note that the diameter of the reactor tube D_{tube} appears explicitly in Eq. (5.8). This indicates that this parameter is important in the design of a nonadiabatic tubular reactor, which is certainly the case. Geometry would tell us that a small-diameter tube will provide more heat transfer area per unit volume. The total volume of all the parallel tubes is set by the total amount of catalyst required, which is determined by the inlet temperature, the temperature profile, and the conversion specified. These same specifications fix the total amount of heat that must be removed. A small-diameter tube gives more heat transfer area and requires smaller differential temperature driving forces, so the cooling medium temperature T_{st} can be higher (higher-pressure steam generated).

However, a small-diameter tube gives more pressure drop for a given flowrate through each tube and a given tube length. Of course, a larger number of parallel tubes that are shorter can be used to keep pressure drop at a reasonable level, but this increases the shell diameter of the reactor, which increases the cost. Mechanical problems also limit the minimum tube diameter. Typical tube diameter in cooled tubular reactors is 0.03 m. Typical tube diameter in a furnace-fired heated tubular reactor is 0.15 m.

The overall heat transfer coefficient U (kW m^{-2} K^{-1}) depends on the velocity through the tubes:

$$U = 0.01545 + \frac{0.6885 \times 10^{-6}}{D_p} \mathrm{Re} \tag{5.9}$$

Figure 5.6 gives a Matlab program for a non-adiabatic tubular reactor. The reactor inlet temperature is assumed to be the same as the steam temperature ($T_{st} = T_{in} = 477$ K in the

```
% Program "cooledss.m"
% Solves ODE's for cooled PFR at steady state
% Given reactor inlet conditions, dtube,ltube, ntube, Tst, Tin and dp
clear
% Parameters
k0=0.19038;e=69710;lambda=-23237;dp=0.003;roecat=2000;porosity=0.4;
cpa=30;cpb=40;cpc=70;ma=15;mb=20;mc=35;viscosity=0.18e-4;tst=477;
dtube=0.111;ntube=190;ltube=8.51;
dw=1;wplot=0;yar=0.5;ybr=0.5;np=0;
areacstube=pi*(dtube^2)/4;wmaxtube=ltube*areacstube*roecat;wmax=wmaxtube*ntube;
% Initial conditions
w=0;fc=0;fa0=0.12;fb0=0.12;fr=0.38;temp=477;p=50;fain=fr*yar+fa0;fbin=fr*ybr+fb0;
% Integration loop
while w<wmax
    fa=fain-fc; fb=fbin-fc; ya=fa/(fa+fb+fc); yb=fb/(fa+fb+fc); yc=fc/(fa+fb+fc);
    k=k0*exp(-e/8.314/temp);  rate=k*ya*yb*p^2;
    roegasmolar=p/(temp*1.01325*82.057e-3); flowvoltube=(fa+fb+fc)/roegasmolar/ntube;
    maverage=ya*ma+yb*mb+yc*mc;  roegas=roegasmolar*maverage;
    vel=flowvoltube/areacstube;  reynolds=dp*vel*roegas/viscosity;
    friction=(1-porosity)*(1.75+150*(1-porosity)/reynolds)/(porosity^3);
    u=0.01545+0.6885e-6*reynolds/dp;
% Save for plotting
    if w>=wplot;np=np+1;wp(np)=w/1000;tempp(np)=temp;pp(np)=p;yap(np)=ya*100;
       ycp(np)=yc*100;wplot=wplot+10;end
% Derivative evaluation
dfcdw=rate;
dpdw=-friction*ltube*roegas*(vel^2)/(dp*wmax*1e5);
dtempdw=(-lambda*rate-4*u*(temp-tst)/(roecat*dtube))/(cpa*fa+cpb*fb+cpc*fc);
%  Integration
fc=fc+dfcdw*dw;temp=temp+dtempdw*dw;p=p+dpdw*dw;w=w+dw;
if p<10;break;end
end
clf
subplot(2,2,1);plot(wp,tempp);grid;ylabel('T (K)');title('Cooled; Tst=477');
subplot(2,2,2);plot(wp,ycp);grid;ylabel('yC (%)');
subplot(2,2,3);plot(wp,yap);grid;ylabel('yA (%)');xlabel('w  (1000 kg)');
subplot(2,2,4);plot(wp,pp);grid;ylabel('P (bar)');xlabel('w  (1000 kg)');
```

Figure 5.6 Matlab program for cooled PFR.

numerical case used in the program). The flowrate of the recycle is 0.38 kmol/s. The reactor has 190 tubes, 0.111 m in diameter and 8.51 m in length. The total catalyst in the reactor is 31,290 kg. Figure 5.7 shows how variables change down the length of the reactor (in terms of catalyst weight). The process gas temperature starts at 477 K and hits a peak of 502 K about halfway down the reactor. The exit temperature is about 496 K. Product C is produced and reactant A is consumed (as is reactant B, not shown). The per-pass conversion is 41.5%. The pressure drop through the reactor is about 0.7 bar.

Figure 5.8 illustrates the sensitivity of the cooled reactor to changes in the inlet temperature. The steam temperature is held constant at 447 K. The inlet temperature is increased from 447 K by 10 and 20 K. The peak temperature increases by about the same amount. These results demonstrate that the cooled reactor is much less sensitive to changes in the inlet temperature than is the adiabatic reactor. This is one of its important advantages. Note that the exit temperature actually decreases slightly when the inlet temperature increases. As expected, more product is produced at the higher inlet temperature. Also note that the location of the peak temperature moves toward the inlet end of the reactor as the inlet temperature is increased.

Figure 5.9 shows the effects of changing the steam temperature (from 477 to 500 K) for different tube diameters. For the base case tube diameter of 0.111 m, raising the steam temperature (and the reactor process inlet temperature) has a drastic effect on the peak temperature. It exceeds the maximum temperature limit of 500 K by over 120 K. Of course, conversion is much greater. The other two curves show what happens when the

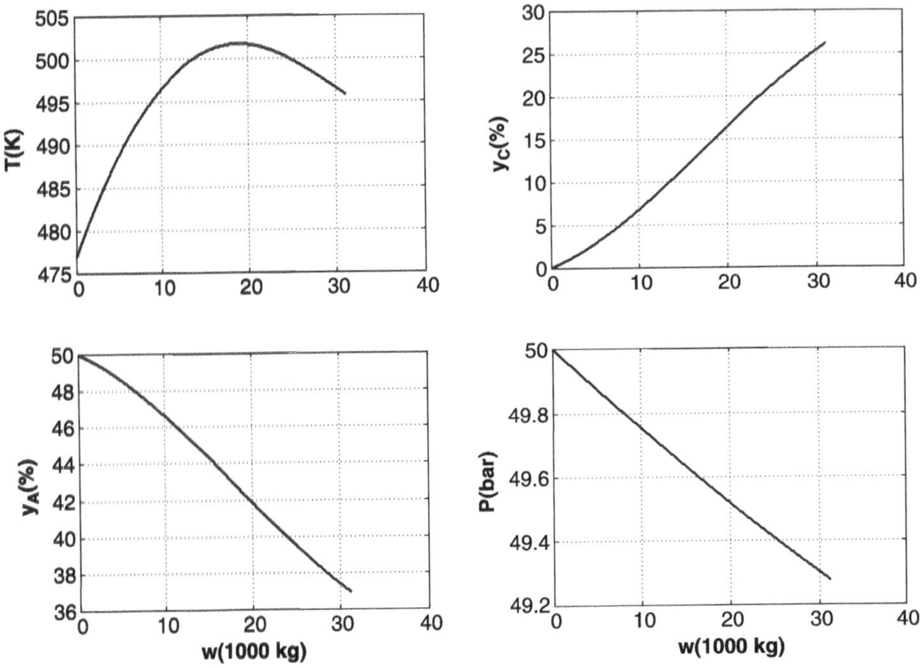

Figure 5.7 Cooled reactor profiles; $D_{\text{tube}} = 0.111$; $T_{\text{st}} = 477$ K.

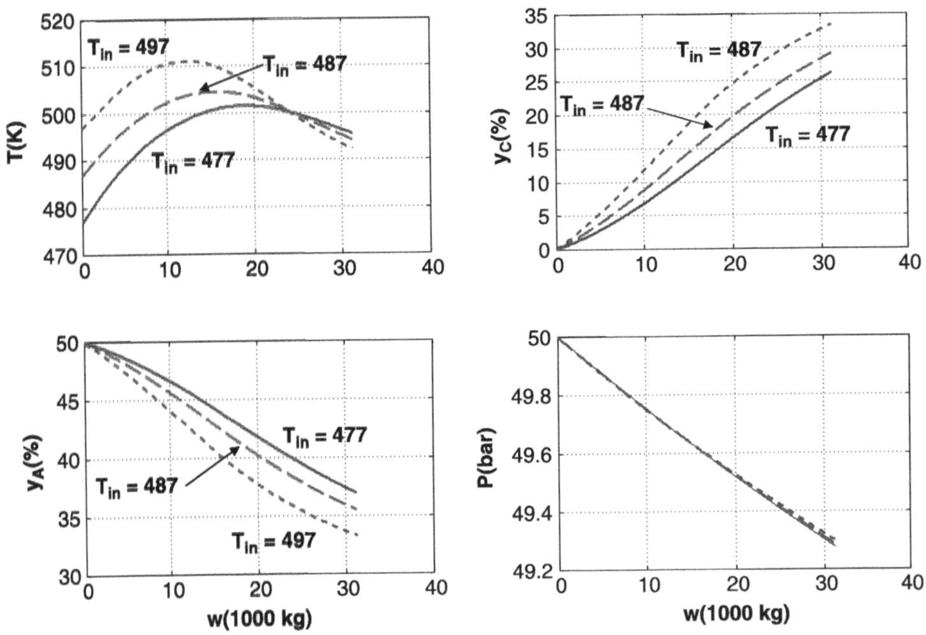

Figure 5.8 Effect of T_{in}; $D_{\text{tube}} = 0.111$; $T_{\text{st}} = 477$ K.

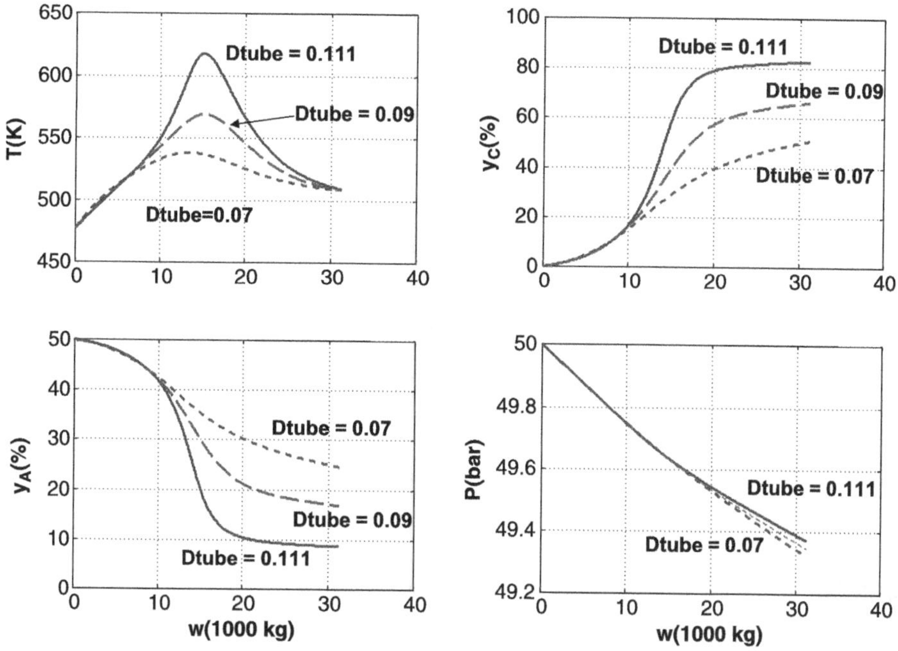

Figure 5.9 Cooled reactor profiles; $D_{tube} = 0.111/0.09/0.07$; $T_{st} = 500$ K.

tubes diameter is decreased. The total weight of catalyst and the tube length are kept constant, so the number of tubes increases (190 for $D_{tube} = 0.111$, 280 for $D_{tube} = 0.09$, and 477 for $D_{tube} = 0.07$). The smaller-diameter tubes provide more heat transfer area for the same tube volume, so the peak temperatures are lower (as is the conversion). Note that the pressure drop is slightly higher. This is due to the lower conversion. The reaction consumes 2 mol of reactants and produces 1 mol of product, so the molar flowrate and velocity is higher for lower conversion.

Figure 5.10 gives a Matlab program that adjusts the steam temperature to maintain the peak temperature at 500 K. A guess of $T_{st} = T_{in}$ is made, the equations are integrated down the reactor, and the peak temperature is found by using the Matlab *max* function ("*tempmax = max(tempp)*"). Then an interval-halving convergence method is used to adjust the steam temperature to drive the peak temperature to the maximum. Results are shown in Figure 5.11 for the base case tube diameter of 0.111 m. The steam temperature required to give a peak temperature of 500 K is $T_{st} = 476$ K.

Results for other tube diameters are given in Figure 5.12. In these cases the length of the tubes and the total amount of catalyst are fixed, so there are more tubes as diameter is reduced.

The smaller the tube diameter, the higher the steam temperature (and the reactor inlet temperature). The resulting steam temperatures for tube diameters of 0.111, 0.06, and 0.03 m are 476, 485, and 492 K, respectively. This is the result of the larger heat transfer areas. The total heat transfer areas for the three tube diameters are 564, 1043, and 2086 m^2, respectively.

The numbers of tubes for the three tube diameters are 190, 650, and 2601, respectively. There is an order-of-magnitude increase in the number of tubes as tube diameter is

```
% Finds steam temp that gives 500 peak temp
clear
% Parameters
k0=0.19038;e=69710;lambda=-23237;dp=0.003;roecat=2000;porosity=0.4;
cpa=30;cpb=40;cpc=70;ma=15;mb=20;mc=35;viscosity=0.18e-4;
dtube=0.111;ntube=190;ltube=8.51;
areacstube=pi*(dtube^2)/4;wmaxtube=ltube*areacstube*roecat;wmax=wmaxtube*ntube
errorpeak=10;tst=477;dtst=2;flagm=-1;flagp=-1;
% Loop to vary steam temp
while errorpeak>0.01
     dw=10;wplot=0;yar=0.5;ybr=0.5;np=0;
% Initial conditions
w=0;fc=0;fa0=0.12;fb0=0.12;fr=0.38;temp=tst;p=50;fain=fr*yar+fa0;fbin=fr*ybr+fb0;
% Integration loop
while w<wmax
   fa=fain-fc; fb=fbin-fc; ya=fa/(fa+fb+fc); yb=fb/(fa+fb+fc); yc=fc/(fa+fb+fc);
   k=k0*exp(-e/8.314/temp);  rate=k*ya*yb*p^2;
   roegasmolar=p/(temp*1.01325*82.057e-3); flowvoltube=(fa+fb+fc)/roegasmolar/ntube;
   maverage=ya*ma+yb*mb+yc*mc;  roegas=roegasmolar*maverage;
   vel=flowvoltube/areacstube;  reynolds=dp*vel*roegas/viscosity;
   friction=(1-porosity)*(1.75+150*(1-porosity)/reynolds)/(porosity^3);
   u=0.01545+0.6885e-6*reynolds/dp;
% Save for plotting
if w>=wplot;np=np+1;
  wp(np)=w/1000;tempp(np)=temp;pp(np)=p;yap(np)=ya*100; ycp(np)=yc*100;wplot=wplot+10;end
% Derivative evaluation
dfcdw=rate;  dpdw=-friction*ltube*roegas*(vel^2)/(dp*wmax*1e5);
dtempdw=(-lambda*rate-4*u*(temp-tst)/(roecat*dtube))/(cpa*fa+cpb*fb+cpc*fc);
%  Integration
fc=fc+dfcdw*dw;temp=temp+dtempdw*dw;p=p+dpdw*dw;w=w+dw;end  % end of integration
   temppeak=max(tempp);errorpeak=abs(temppeak-500);
   if temppeak>500;if flagm>0;dtst=dtst/2;end;tst=tst-dtst;flagp=1;end;
   if temppeak<500;if flagp>0;dtst=dtst/2;end;tst=tst+dtst;flagm=1;end;
  end  % end of loop to vary tst ("while")
clf
subplot(2,2,1);plot(wp,tempp);grid;ylabel('T (K)');
title('Cooled; Tst=500; Dtube=.111');
subplot(2,2,2);plot(wp,ycp);grid;ylabel('yC (%)');
subplot(2,2,3);plot(wp,yap);grid;ylabel('yA (%)');xlabel('w  (1000 kg)');
subplot(2,2,4);plot(wp,pp);grid;ylabel('P (bar)');xlabel('w  (1000 kg)');
```

Figure 5.10 Matlab program to find steam temperature for 500 K peak temperature.

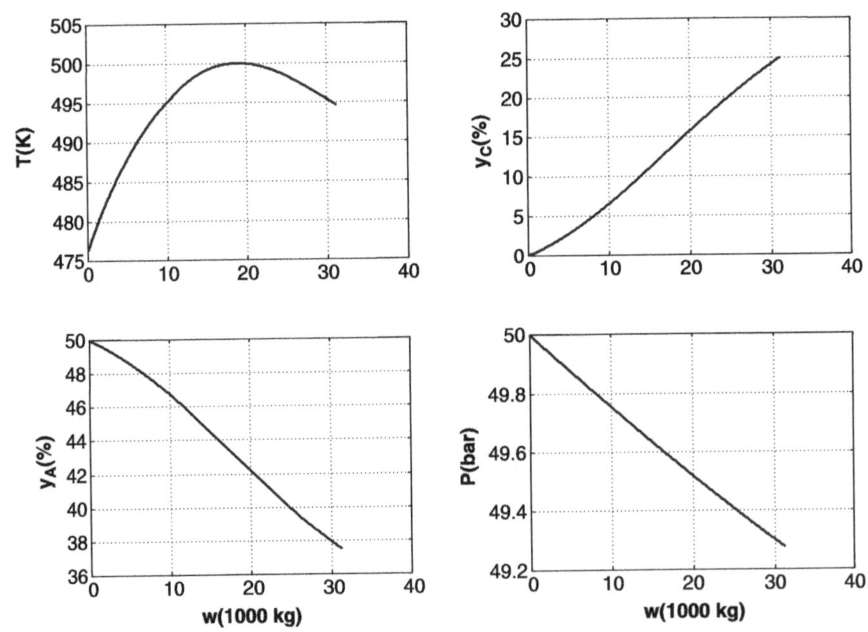

Figure 5.11 Steam temperature = 476 K to give a 500 K peak temperature; $D_{\text{tube}} = 0.111$.

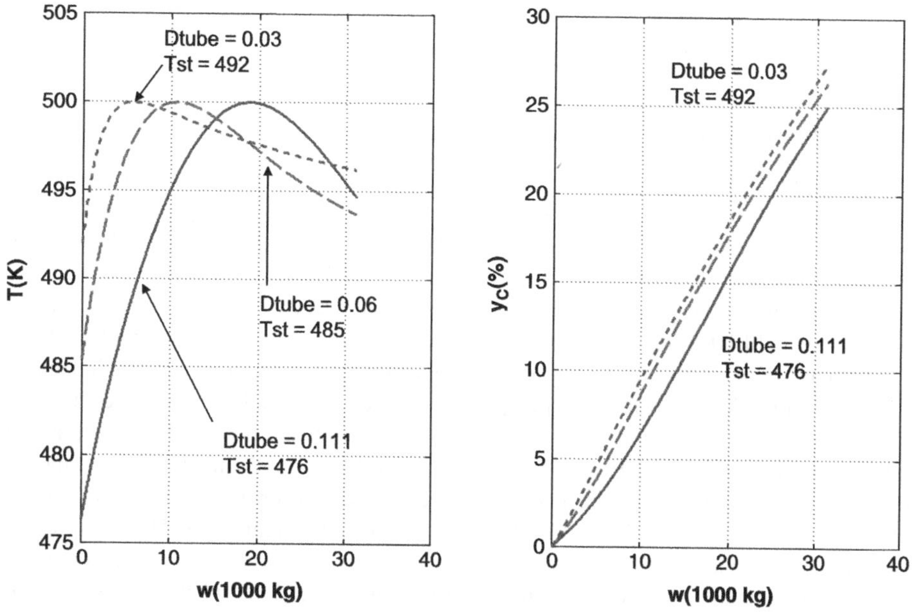

Figure 5.12 Peak temperature = 500 K; $D_{tube} = 0.111/0.06/0.03$.

changed from 0.111 to 0.03 m. The corresponding diameters of the reactor vessels required for the number of tubes and diameters are 7, 10, and 14 m, respectively. These are huge reactors! This calculated vessel diameter assumes that the spacing between tubes is half a tube diameter.

Remember, the tube length is set at 8.51 m so that the pressure drop is small. Increasing tube length would reduce the number of tubes and vessel size at the expense of an increase in compression cost due to the higher pressure drop.

The profiles given in Figure 5.12 show that the larger tube diameters and lower steam and inlet temperatures result in the smaller reaction rates in the front part of the bed, so the conversion is smaller.

Now that some insight has been gained in looking at adiabatic and cooled tubular reactors in isolation, we are ready to explore an entire process that contains reactors, compressors, separators, heat exchangers, makeup feedstreams, a recycle stream, and a product stream. This material is taken from the paper by Jaisathaporn and the author.[1]

5.4 SINGLE ADIABATIC TUBULAR REACTOR SYSTEMS WITH GAS RECYCLE

We start by studying the steady-state design and economics of a process with a single adiabatic reactor. The design considers the entire plantwide process: reactor, heat exchangers, gas recycle compressor, preheat furnace, condenser, and separator. The economic objective function is total annual cost, which includes annual capital cost (reactor, catalyst, compressor, and heat exchangers) and energy cost (compressor work and furnace fuel).

[1]P. Jaisathaporn and W. L. Luyben, *Ind. Eng. Chem. Res.*, **42**, 3304. (2003).

Optimum operating conditions and equipment sizes are found. The effect of catalyst cost on the steady-state design is considered.

5.4.1 Process Conditions and Assumptions

The chemistry is the exothermic, irreversible, gas-phase reaction A + B → C, which is conducted in the presence of a solid catalyst in a tubular reactor. The production rate of C is fixed at 0.12 kmol/s for all designs. This means that the two gas-phase fresh feedstreams are both 0.12 kmol/s. Their temperature is 313 K.

Figure 5.13 gives the flowsheet of the single adiabatic reactor process. A feed-effluent heat exchanger (FEHE) is used to preheat the reactor feedstream by recovering heat from the hot reactor effluent stream. A furnace provides additional heat when it is needed and is necessary during plant startup. A bypass valve between the FEHE and furnace gives an additional manipulated variable for reactor inlet temperature control. The reactor exit stream is cooled in the FEHE and a condenser. Product C and unreacted reactants A and B are separated in a flash drum in which perfect separation is assumed (no A or B in the liquid product stream L_C and no C in the gas recycle stream F_R). The compressor operates with a discharge pressure P_R of 50 bar and an efficiency of 75%. The compressor discharge temperature is calculated during the design calculations when the compressor suction pressure is known.

A maximum reactor temperature of 500 K is used in this study. This maximum temperature occurs at the exit of the adiabatic reactor under steady-state conditions. Plug flow is assumed with no radial gradients in concentrations or temperatures and no axial diffusion or conduction.

Gas pressure drops in each heat exchanger and in the furnace are assumed to be 0.5 bar at steady-state design conditions. The pressure drops over both the tube side and the shell side of the FEHE are each 0.5 bar. The design pressure drop over the control valve between the FEHE and the furnace is also 0.5 bar. Reactor pressure drop is calculated using the Ergun equation and therefore varies with design parameters (reactor diameter, reactor length, and flowrate through the reactor). The reactor vessel is designed with a length-to-diameter ratio of 10.

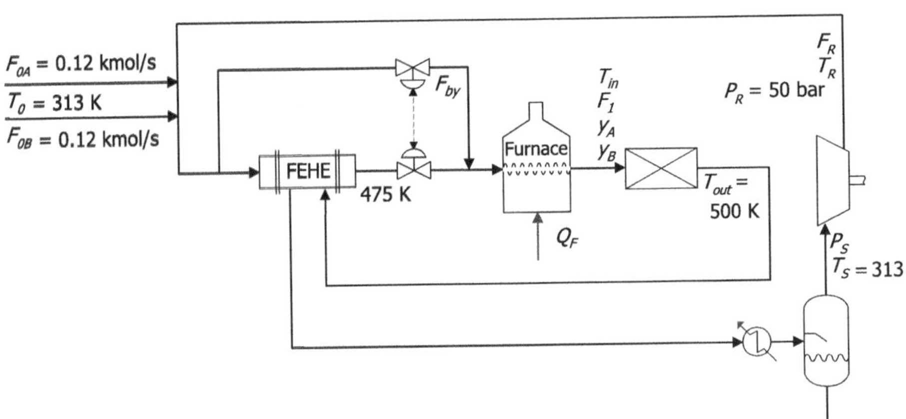

Figure 5.13 Single adiabatic reactor process.

The energy requirement of the furnace is zero for this adiabatic reactor case since the reactor exit stream at maximum temperature can provide enough heat. The design of the FEHE and the amount of bypassing are determined by the reactor inlet temperature T_{in}. A temperature difference of 25 K is assumed for the hot end of the FEHE. The hot reactor effluent enters the hot side of the FEHE at the high-temperature limit of 500 K, so the cold-side exit stream is 475 K. If the specified reactor inlet temperature is less than 475 K, bypassing is used. A fairly low overall heat transfer coefficient of 0.142 kJ s^{-1} m^{-2} K^{-1} is used in the FEHE because of the gas–gas heat transfer.

One important physical property assumption is made about heat capacities. The mass heat capacities of all components are assumed to be the same (2 kJ kg^{-1} K^{-1}). This means that the product of the mass flowrate and the mass heat capacity is constant for any stream and equal to the sum of the product of the component molar flowrates times the corresponding molar heat capacities. Thus, despite the fact that molar flowrates of individual components vary down the length of the reactor, the term $F\sum_{j=A}^{C} y_j c_{pj}$ is constant, where F is the total molar flowrate, y_j is the mole fraction of component j, and c_{pj} is the molar heat capacity of component j. This relationship is used in the design procedures discussed below to calculate the inlet flowrate from an energy balance around the reactor.

5.4.2 Design and Optimization Procedure

This process has two design optimization variables. We select the reactor inlet temperature T_{in} and the ratio of the reactant concentrations in the recycle stream y_{RA}/y_{RB}. Therefore the optimization problem involves a two-dimensional search for the values of these two variables that minimize the total annual cost of the process. The steps in the design procedure are detailed below:

1. Pick a reactor inlet temperature T_{in} and a y_{RA}/y_{RB} ratio (the design variables to be optimized). This fixes the reactor inlet composition and temperature.
2. Calculate the molar flowrate F_{in} of the gas entering the reactor using an energy balance around the reactor given as follows:

$$F_{in} = \frac{-0.12\lambda}{(T_{out} - T_{in})\sum_{j=A}^{C}(y_j c_{pj})} \tag{5.10}$$

   The conversion is known, so the heat generated is known: $(-\lambda)(0.12$ kmol s). Thus, given the inlet temperature (T_{in}) and the outlet temperatures ($T_{out} = 500$ K), the inlet flowrate to the reactor can be calculated, using the fact that the term $F\sum_{j=A}^{C} y_j c_{pj}$ applies to both the inlet flow and the outlet flow and the inlet compositions are known. Once the inlet flowrate is known, the outlet molar flowrate F_{out} is given by

$$F_{out} = F_{in} - 0.12 \tag{5.11}$$

   Then component balances are used to calculate the reactor exit compositions.
3. Knowing all the conditions at the reactor inlet, integrate down the length of the reactor, using the ordinary differential equations given in Eqs. (5.3)–(5.5), until

the temperature is 500 K. This gives the amount of catalyst required, the composition profiles in the reactor, and the reactor pressure drop.

4. Now all flowrates, including the recycle flowrate F_R, and pressures are known, so the compressor work can be calculated. All flowrates and temperatures throughout the entire process are now known, so heat exchanger areas can be calculated from energy balances.
5. Calculate the size and cost of the reactor, heat exchanges, furnace, condenser, and compressor.
6. Evaluate the TAC of the process. Table 5.2 gives the economic data and equations used in evaluating equipment capital costs and energy costs.
7. Vary T_{in} and y_{RA}/y_{RB} until the minimum TAC is obtained. The *fminsearch* function in Matlab is used to search for the best values of the two design optimization variables.

The optimum designs for two different values of the cost of catalyst are determined ($5/kg and $100/kg). Catalyst is assumed to represent a capital cost. With a payback period of 3 years, the catalyst is replaced every 3 years.

TABLE 5.2 Economic Data

Compressor power

$$W_{comp} = \frac{RT_S F_R \gamma}{0.75(\gamma - 1)}\left[\left(\frac{P_R}{P_S}\right)^{(\gamma-1)/\gamma} - 1\right]$$

$$T_R = T_S\left(\frac{P_R}{P_S}\right)^{(\gamma-1)/\gamma}$$

Capital costs

Single-bed reactor cost $= 0.035D_R^{1.066}\,L_R^{0.082}$ ($\$10^6$)
Multibed reactor cost $= 0.035D_R^{1.066}\,L_R^{0.082}\,(1 + 0.25N_R)$ ($\$10^6$)
Cooled reactor cost $= 2(0.0073)A_R^{0.65}$ ($\$10^6$)
Catalyst cost = \$5 or \$100 kg^{-1}
$\text{Log}_{10}(\text{furnace cost}) = -2.49 + 0.762\,\log_{10}(Q_F)$ ($\$10^6$)
Heat exchanger/condenser cost $= 0.0073A_R^{0.65}$ ($\$10^6$)
Compressor cost $= 0.00433W_{comp}^{0.82}$ ($\$10^6$)

Energy costs

Furnace energy cost $= 0.000150Q_F$ ($\$10^6\ \text{year}^{-1}$)
Compressor work $= 0.000613W_{comp}$ ($\$10^6\ \text{year}^{-1}$)
Steam credit $= 0.348F_{st}$ CF ($\$10^6\ \text{year}^{-1}$)
where $\text{CF} = -0.00024P_{st}^2 + 0.0219P_{st} + 0.634$
$P_{st} = 0.0023T_{st}^2 - 1.84T_{st} + 373$

Total annual cost (TAC)

Capital cost = reactor + catalyst + furnace + heat exchangers + compressor
Operating cost = furnace energy cost + compressor work − steam credit

$$\text{Total annual cost} = \frac{\text{capital cost}}{3} + \text{operating cost}$$

5.4.3 Results for Single Adiabatic Reactor System

Figure 5.14 shows how several important variables change as the value of the reactor inlet temperature T_{in} is varied. At each point the optimum ratio y_{RA}/y_{RB} in the reactor feedstream is found. As the inlet temperature increases, the recycle flowrate must increase to maintain the same reactor exit temperature of 500 K. This increases reactor pressure drop and decreases per-pass conversion. The amount of catalyst required to achieve 100% overall conversion decreases initially because the higher temperature gives higher reaction rates. However, as the inlet temperature continues to increase and the recycle flowrate becomes large, the reactor size must be increased to keep reactor pressure drop from becoming too large. These competing effects give a minimum in the catalyst requirement–inlet temperature curve.

The more expensive the catalyst, the higher the optimum recycle flowrate (and reactor pressure drop). We are trading off recycle costs with reactor costs. No bypass flow is needed when the inlet temperature is 475 K or higher. The optimum y_{RA}/y_{RB} ratio decreases as recycle flowrate increases because the costs associated with heat transfer and compression are lower with more B in the gas because of its higher molar heat capacity.

Figure 5.15 illustrates the effect of changing the reactor inlet temperature on the economics. All the costs are in terms of annual costs ($ per year) using a payback period of 3 years for capital investment. The only energy cost is compressor work. There is a minimum in the total annual cost (TAC) curve. For the expensive catalyst, the optimum

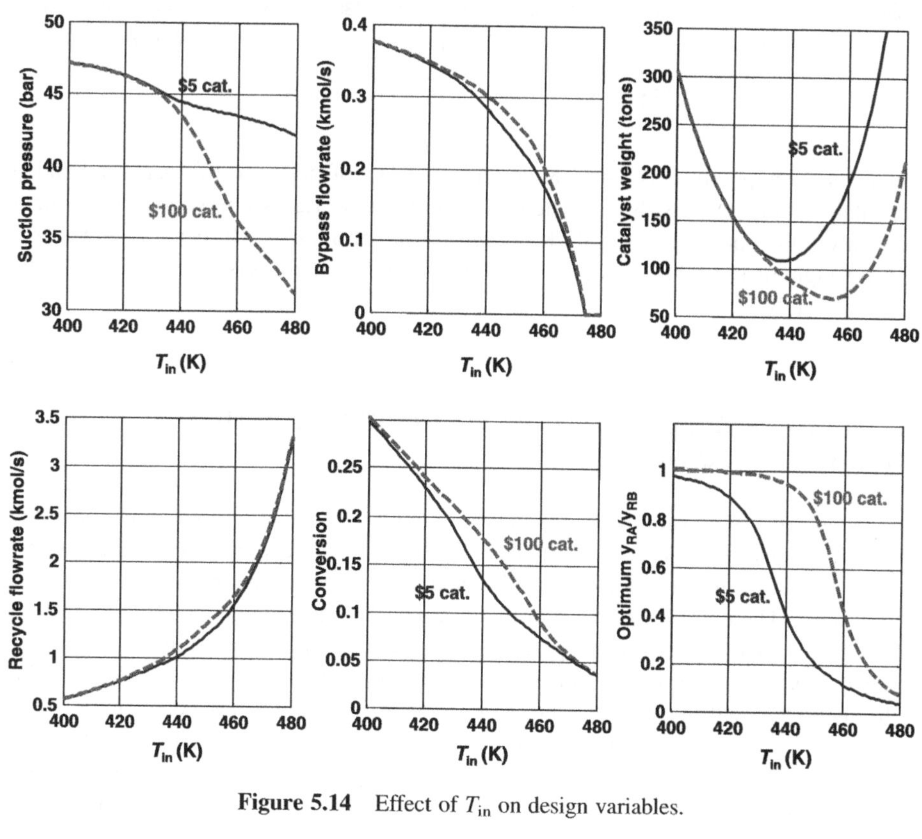

Figure 5.14 Effect of T_{in} on design variables.

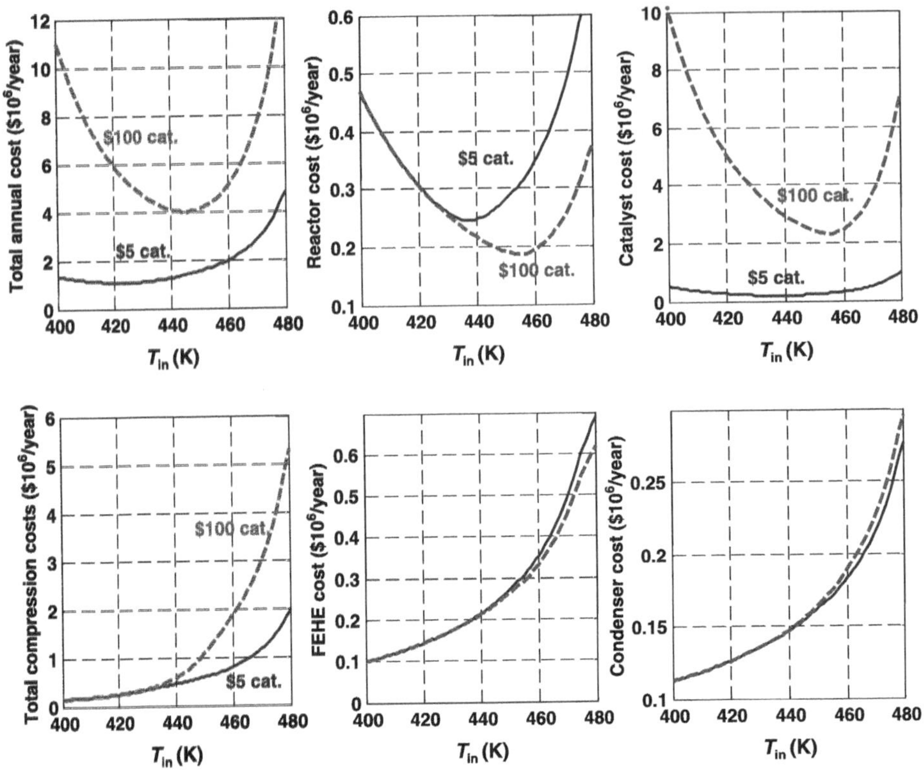

Figure 5.15 Effect of T_{in} on economics.

reactor inlet temperature is 445.1 K, giving a TAC of 3.99×10^6 per year and a total catalyst weight of 78,780 kg. The recycle flowrate is 1.20 kmol/s. For the inexpensive catalyst, the optimum is 422.47 K, giving a TAC of 1.07×10^6 per year and a total catalyst weight of 142,500 kg. The recycle flowrate of 0.78 kmol/s is much less than in the expensive catalyst case.

5.5 MULTIPLE ADIABATIC TUBULAR REACTORS WITH INTERSTAGE COOLING

The second process considered is one in which multiple adiabatic reactors are used that operate in series and use heat exchangers between each reactor to cool the process gas to the desired inlet temperature to the next downstream reactor. Figure 5.16 gives the flowsheet. There are N_R reactors in the process. The inlet temperatures T_n to each of the adiabatic reactors are design optimization variables. The reactor outlet streams are cooled in steam-generating interstage heat exchangers. Boiler feedwater (BFW) is fed on the shell side, and the steam generated is considered a credit in the economics of the process. The value of the steam is a function of the pressure at which it is generated, which depends on the temperature on the process side (hot side) of the heat exchanger. The steam temperature is assumed to be 25 K lower than the process exit temperature

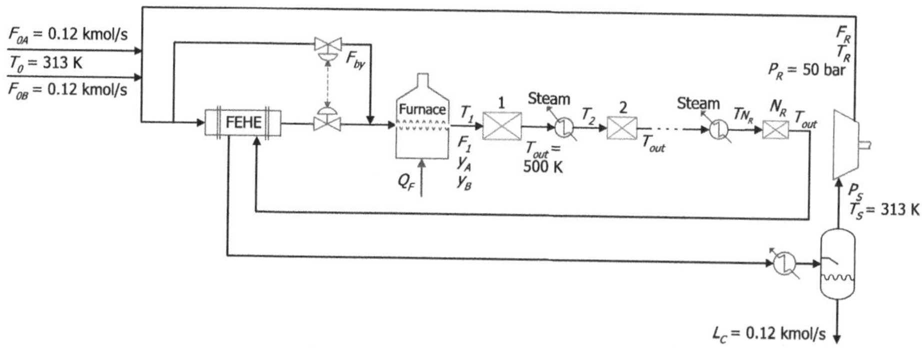

Figure 5.16 Adiabatic reactors with interstage cooling.

(the inlet temperature of the downstream reactor). Note that the reactors can have different inlet temperatures, different diameters, and different lengths. An overall heat transfer coefficient of 0.227 kJ s^{-1} m^{-2} K^{-1} is used in the steam-generating intercoolers.

5.5.1 Design and Optimization Procedure

This process has $N_R + 1$ design optimization variables. We select the reactor inlet temperatures T_n and the ratio of the reactant concentrations in the recycle stream y_{RA}/y_{RB}. The steps in the design procedure are detailed below:

1. Specify the number of reactors in the system N_R.
2. Pick reactor inlet temperatures $T_1, T_2, \ldots, T_{NR}$ and the y_{RA}/y_{RB} ratio (all $N_R + 1$ variables to be optimized).
3. Calculate the inlet flowrate F_1 to the first reactor using an energy balance Eq. (5.12) that is similar to that used for the single adiabatic reactor [Eq. (5.10)] but modified to account for partial conversion in each reactor.

$$F_1 = \frac{-0.12\lambda}{\left(\sum_{k=1}^{N_R} \Delta T_k\right)\left(\sum_{j=A}^{C} (y_j c_{pj})_1\right)} \tag{5.12}$$

The ΔT_k term for the kth reactor is its outlet temperature (500 K) minus the inlet temperature T_k of that reactor. The fraction of the total conversion of the entire reaction section (0.12 kmol/s) that occurs in the nth reactor is proportional to the ratio of the temperature difference in that reactor to the sum of all the temperature differences over the N_R reactors. Then the flowrates in and out of each reactor can be calculated using Eq. (5.13). Component balances then yield all composition throughout the process.

$$F_{\text{out},n} = F_1 - \frac{0.12 \sum_{k=1}^{n} \Delta T_k}{\sum_{k=1}^{N_R} \Delta T_k} \tag{5.13}$$

4. Integrate down the length of each reactor until the temperature is 500 K and obtain the amount of catalyst required for each reactor.

5. Calculate the sizes and costs of all equipment.
6. Evaluate the TAC of the process.
7. Vary $T_1, T_2, \ldots, T_{NR}$ and y_{RA}/y_{RB} until the minimum TAC is obtained.

5.5.2 Results for Multiple Adiabatic Reactors with Interstage Cooling

A contour plot given in Figure 5.17 shows how TAC varies in a two-stage adiabatic reactor system with interstage cooling. The reactant ratio y_{RA}/y_{RB} is fixed at unity in this figure, so there are two design optimization variables, the inlet temperatures of the two reactors T_1 and T_2.

Complete results with various numbers of reactors in the system are given in the reference paper. For the expensive catalyst case, the optimum design has three reactors and a TAC of $\$3.41 \times 10^6$ per year. This is lower than the TAC of a single adiabatic reactor system ($\$3.99 \times 10^6$ per year). The inlet temperatures to the three reactors are 448.1, 456.8, and 466.5 K, respectively. The optimum y_{RA}/y_{RB} ratio is 0.96. The total amount of catalyst in all three reactors is 76,980 kg. The reactor sizes are slightly different: 27,700/26,000/23,300 kg in reactors 1, 2 and 3, respectively. The recycle flowrate of 0.38 kmol/s is a factor of 3 less than in the single reactor case. This occurs because the interstage cooling and smaller conversions in each stage permit smaller flowrates through the reactors without exceeding the maximum temperature limitation.

For the inexpensive catalyst case, the optimum design has four reactors and a TAC of $\$0.74 \times 10^6$ per year. This is lower than the TAC of a single adiabatic reactor system ($\$1.07 \times 10^6$ per year). The inlet temperatures to the four reactors are 438.9, 445.7, 453.3, and 461.1 K, respectively. The optimum y_{RA}/y_{RB} ratio is 0.98. The total amount of catalyst in the four reactors is 105,520 kg. The reactor sizes are slightly different: 26,000/26,100/26,200/27,200 kg in reactors 1, 2, 3, and 4, respectively. The recycle flowrate of only 0.16 kmol/s, which 20% of the recycle flowrate in the single reactor case.

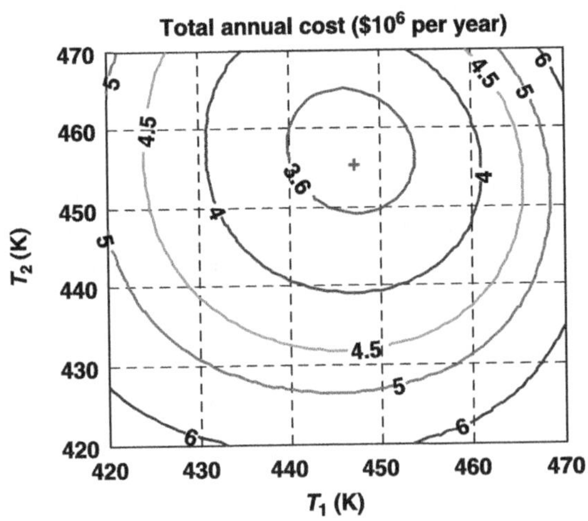

Figure 5.17 Two adiabatic rectors with interstage cooling.

5.6 MULTIPLE ADIABATIC TUBULAR REACTORS WITH COLD-SHOT COOLING

The third process considered is one in which multiple beds of catalyst are installed in a reactor vessel. Each bed operates adiabatically. Some of the cold feed is added at the inlet of each catalyst bed and mixed with the hot gas coming from the upstream bed to reduce the temperature T_n of the inlet gas to the downstream bed n to the desired value. The flowsheet is shown in Figure 5.18.

The catalyst beds are mounted in a single reactor vessel because it is more economical than using multiple vessels. The spacing between beds is set at 1 m. The length-to-diameter aspect ratio of the vessel is 10. Because a multibed reactor must have internal piping, flow distributors, and bed supports, a multi-bed reactor vessel is more expensive than a simple vessel. We assume that each additional bed increases reactor capital cost by about 25%, as shown in Table 5.2.

Note that there is no steam generated in the cold-shot system as there is in the interstaged cooled system, so there is no economic credit for steam generation.

5.6.1 Design–Optimization Procedure

This process has N_R+1 design optimization variables. We select the reactor inlet temperatures T_n and the ratio of the reactant concentrations in the recycle stream y_{RA}/y_{RB}. The steps in the design procedure are detailed below:

1. Specify N_R.
2. Pick $T_1, T_2, \ldots, T_{NR}$ and the y_{RA}/y_{RB} ratio (all $N_R + 1$ variables to be optimized).
3. Guess the temperature of the cold-shot stream T_{CS}. This is the temperature of the stream after the recycle gas (at an initially unknown temperature and flowrate coming from the compressor) is combined with the two fresh feedstreams at 313 K. An iteration loop is used throughout this step until the guessed value and the calculated value are sufficiently close.
4. Solve a set of $2N_R$ linear equations [Eqs. (5.14)–(5.17)] for the unknown variables that are the rates of generation of C in each reactor $R_{C,n}$ ($n = 1, \ldots, N_R$), the feed flowrate of the first reactor F_1, and flowrates of cold-shot streams entering the other

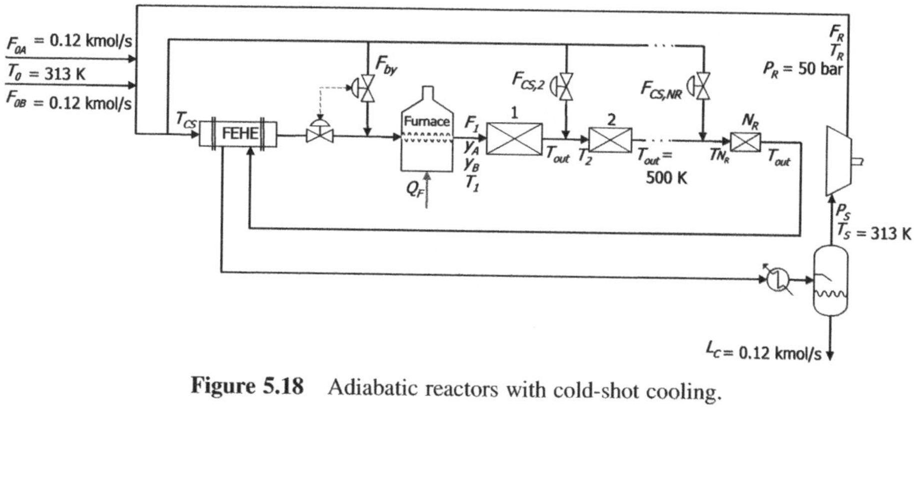

Figure 5.18 Adiabatic reactors with cold-shot cooling.

reactors $F_{\mathrm{CS},m}$ $(m = 2, \ldots, N_R)$ (see Fig. 5.18). Equation (5.14) expresses the total generation of C in all reactors:

$$\sum_{n=1}^{N_R} R_{C,n} = 0.12\,\text{kmol/s} \tag{5.14}$$

Equation (5.15) gives the energy balance around each reactor:

$$\sum_{j=1}^{n} F_j = \frac{-\lambda R_{C,n}}{(T_{\mathrm{out}} - T_n)\sum_{j=A}^{C}(y_j c_{pj})_1} \quad (n = 1, \ldots, N_R) \tag{5.15}$$

Equation (5.16) gives the energy balance at the reactor inlet where the cold stream $F_{\mathrm{CS},m}$ added before the mth reactor is mixed with the hot-stream F_j leaving the previous reactor at $T_{\mathrm{out}} = 500$ K:

$$\sum_{j=1}^{m-1} F_j(T_{\mathrm{out}} - T_m) = F_{\mathrm{CS},m}(T_m - T_{\mathrm{CS}}) \quad (m = 2, \ldots, N_R) \tag{5.16}$$

The outlet flowrate of each reactor is calculated using Eq.(5.17). Component balances are used to calculate all composition throughout the process:

$$F_{out.n} = \sum_{j=1}^{n} F_j - \sum_{j=1}^{n} R_{C,j} \tag{5.17}$$

The resulting equations are linear in the unknown variables, so an analytical solution is easily obtained. For example, for the case with three adiabatic beds ($N_R = 3$) in the reactor vessel, there are six equations and six unknowns ($R_{\mathrm{C}1}$, $R_{\mathrm{C}2}$, $R_{\mathrm{C}3}$, F_1, $F_{\mathrm{CS},2}$ and $F_{\mathrm{CS},3}$):

$$R_{\mathrm{C}1} + R_{\mathrm{C}2} + R_{\mathrm{C}3} = 0.12 \tag{5.18}$$

$$F_1 = \frac{-\lambda R_{\mathrm{C}1}}{(T_{\mathrm{out}} - T_1)(y_{R\mathrm{A}} c_{p\mathrm{A}} + y_{R\mathrm{B}} c_{p\mathrm{B}} + y_{R\mathrm{C}} c_{p\mathrm{C}})_1} \tag{5.19}$$

$$F_1 + F_{\mathrm{CS},2} = \frac{-\lambda R_{\mathrm{C}2}}{(T_{\mathrm{out}} - T_2)(y_{R\mathrm{A}} c_{p\mathrm{A}} + y_{R\mathrm{B}} c_{p\mathrm{B}} + y_{R\mathrm{C}} c_{p\mathrm{C}})_1} \tag{5.20}$$

$$F_1 + F_{\mathrm{CS},2} + F_{\mathrm{CS},3} = \frac{-\lambda R_{\mathrm{C}3}}{(T_{\mathrm{out}} - T_3)(y_{R\mathrm{A}} c_{p\mathrm{A}} + y_{R\mathrm{B}} c_{p\mathrm{B}} + y_{R\mathrm{C}} c_{p\mathrm{C}})_1} \tag{5.21}$$

$$F_1(T_{\mathrm{out}} - T_2) = F_{\mathrm{CS},2}(T_2 - T_{\mathrm{CS}}) \tag{5.22}$$

$$F_1(T_{\mathrm{out}} - T_3) + F_{\mathrm{CS},2}(T_{\mathrm{out}} - T_3) = F_{\mathrm{CS},3}(T_3 - T_{\mathrm{CS}}) \tag{5.23}$$

5. Integrate down the length of each reactor until the temperature is 500 K and obtain the amount of catalyst required for each reactor.
6. Calculate the compressor discharge (recycle) temperature T_R, and then calculate T_{CS}.
7. Update T_{CS} and return to step 4. When this iterative loop has converged, continue.

8. Calculate the sizes and costs of all equipment.
9. Evaluate the TAC of the process.
10. Vary $T_1, T_2, \ldots, T_{NR}$ and y_A/y_B until the minimum TAC is obtained.

5.6.2 Results for Adiabatic Reactors with Cold-Shot Cooling

For the expensive catalyst, the optimum economic steady-state design has a vessel with seven beds. The TAC is $2.03 × 10^6$ per year, which is about half that of the single adiabatic reactor system. The recycle flowrate is 0.66 kmol/s, which is more than that of the interstage-cooled system because of the use for cold feed to provide cooling. The total catalyst in all the beds is 33,800 kg. The optimum bed inlet temperatures range from $T_1 = 475$ K to $T_7 = 486.9$ K. The optimum y_{RA}/y_{RB} ratio is 0.994.

For the inexpensive catalyst, the optimum economic steady-state design has a vessel with four beds. The TAC is $0.78 × 10^6$ per year. The recycle flowrate is 0.45 kmol/s. The total catalyst in all the beds is 57,300 kg. The optimum bed inlet temperatures range from $T_1 = 453.1$ K to $T_4 = 466.6$ K. The optimum y_{RA}/y_{RB} ratio is 0.96.

The optimum seven-bed reactor for the expensive catalyst requires six cold-shot streams, which control six temperatures. All of these originate from the single gas recycle stream. As we will see in Chapter 6, the temperature loops are quite interactive and lead to control difficulties. As a result, the dynamic controllability of a suboptimum design with only three beds will be studied in Chapter 6. The three-bed design has a higher total annual cost ($2.37 × 10^6$ per year vs. $2.03 × 10^6$ per year for the seven-bed design). But its dynamic performance is much better. This is an example of the commonly encountered conflict between steady-state economics and dynamic controllability. The recycle flowrate is higher (0.79 vs. 0.66 kmol/s), and the total catalyst is larger (44,200 vs. 33,800 kg). Reactor diameter increases from 1.48 to 1.53 m.

5.7 COOLED REACTOR SYSTEM

The final tubular reactor system considered is one in which a single cooled reactor is used. Figure 5.19 shows the flowsheet. The cooled reactor, which is assumed to be simply a shell-and-tube heat exchanger, has catalyst packed inside the parallel tubes. Steam is generated on the shell side, serving as a coolant. The liquid level in the shell is controlled by bringing in BFW to keep the tubes covered. The steam-side temperature is constant at all axial locations in the reactor because the BFW is vaporized at a constant temperature. The temperature of the steam is assumed to be equal to the reactor inlet temperature, and the temperature of the BFW is assumed to be equal to the steam temperature.

This process has four design degrees of freedom. The four design optimization variables used are the diameter of the tubes D_{tube}, the length of the tubes L_{tube}, the flowrate of the process gas fed into one tube F_{tube}, and the y_{RA}/y_{RB} ratio.

The diameter of the reactor tubes is a design optimization variable, but a maximum limit of 0.12 m is used to avoid mechanical and heat transfer difficulties. The cross-sectional area of the reactor vessel is assumed to be twice the total cross-sectional area of all the reactor tubes, that is, the volume in the shell outside the tubes is equal to the total tube volume.

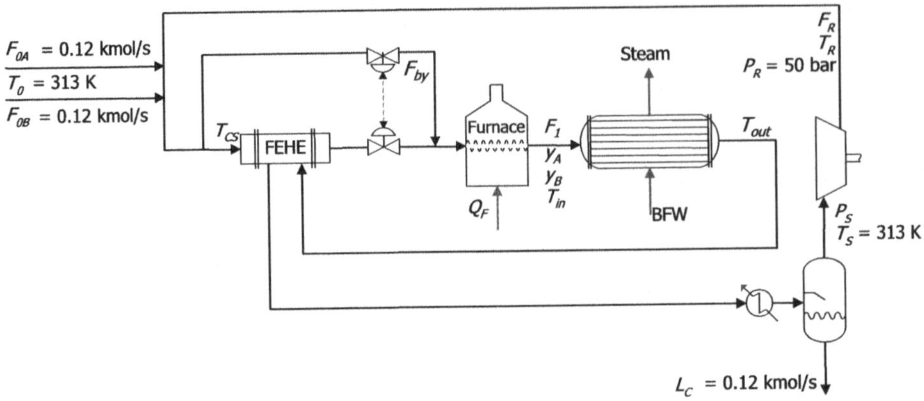

Figure 5.19 Cooled reactor system.

5.7.1 Design Procedure for Cooled Reactor System

1. Pick the tube diameter D_{tube}, the length L_{tube}, the inlet flowrate per tube F_{tube}, and the y_A/y_B ratio (four variables to be optimized).
2. Calculate the amount of catalyst per tube W_{tube} from the volume of the tube and the bulk density of the catalyst (2000 kg/m^3).
3. Guess $T_{\text{in}} = T_{\text{st}}$ (inlet temperature = steam temperature).
4. Integrate down the length of the reactor tube to obtain the temperature profile and the generation of C per tube using Eq. (5.8), which is the energy balance with heat transfer.
5. If the peak temperature is not equal 500 K, adjust $T_{\text{in}} = T_{\text{st}}$ and then return to step 4.
6. Calculate the number of tubes N_{tube} and the total amount of catalyst needed for the total generation of C = 0.12 kmol/s.
7. Calculate the sizes and costs of all equipment.
8. Evaluate the TAC of the process.
9. Vary D_{tube}, L_{tube}, F_{tube}, and y_A/y_B until the minimum TAC is obtained.

5.7.2 Results for Cooled Reactor System

For both the expensive and inexpensive catalysts, the cooled reactor system has the lowest total annual cost among all the alternative systems. This occurs for two reasons:

1. This system requires less catalyst (smaller reactor) because it operates at a higher average temperature, which results from having a higher inlet temperature. However, the capital cost of the cooled reactor is higher per unit of volume than for the simple adiabatic reactor vessels.
2. This system has a lower recycle flowrate because cooling down the length of the reactor reduces the problem of exceeding the maximum temperature limitation. An adiabatic reactor requires higher flowrates so as not to exceed this high-temperature constraint. The lower recycle flowrate results in a smaller compressor, smaller heat exchangers, and lower compression energy cost. The operating cost advantage is offset somewhat by the need to provide energy in the furnace.

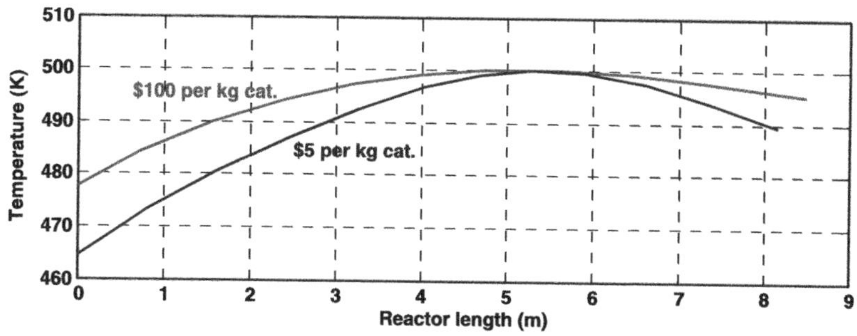

Figure 5.20 Cooled reactor temperature profiles.

For the expensive catalyst, the optimum steady-state design has a TAC of $\$1.57 \times 10^6$ per year. The reactor has 190 tubes, 0.111 m in diameter and 8.51 m in length, containing a total weight of catalyst of 31,380 kg. The inlet (and steam temperature) is 477.3 K. The recycle flowrate is 0.38 kmol/s. The y_{RA}/y_{RB} ratio is 1.017.

For the inexpensive catalyst, the optimum steady-state design has a TAC of $\$0.46 \times 10^6$ per year. The reactor has 226 tubes, 0.12 m in diameter and 8.13 m in length, containing a total weight of catalyst of 41,500 kg. The inlet (and steam temperature) is 464.2 K. The recycle flowrate is 0.22 kmol/s. The y_{RA}/y_{RB} ratio is 1.029.

Figure 5.20 shows the temperature profiles in the cooled tubular reactor for the optimum designs with the two catalysts. The optimum recycle flowrate is larger with the expensive catalyst, as expected, which yields an optimum inlet temperature that is higher.

It is important to note that the heat transfer coefficient U in the expensive catalyst case is almost twice that in the cheap catalyst case. This is due to the larger recycle flowrate and higher velocity in the tubes. The heat transfer areas A_R of the two designs are only 20% different because the inlet temperature of the expensive catalyst design is higher, which provides less temperature differential. The UA_R product has an important effect on the dynamic controllability of this system, which will be studied in Chapter 6.

5.8 TUBULAR REACTOR SIMULATION USING ASPEN PLUS

Tubular reactors can be simulated using Aspen Plus. Several configurations are available: constant-temperature reactor, adiabatic reactor, reactor with constant coolant temperature, reactor with countercurrent flow of coolant, and reactor with co-current flow of coolant. The isothermal reactor cannot be exported into Aspen Dynamics because it is not possible to dynamically control the temperature at all axial positions. Therefore only the last four types will be discussed.

The numerical example is the chlorination of propylene. There are two vapor-phase, irreversible, exothermic reactions:

$$C_3H_6 + Cl_2 \xrightarrow{k_1} CH_2 = CH - CH_2Cl + HCl$$

$$C_3H_6 + Cl_2 \xrightarrow{k_2} CH_2Cl - CHCl - CH_3 \qquad (5.24)$$

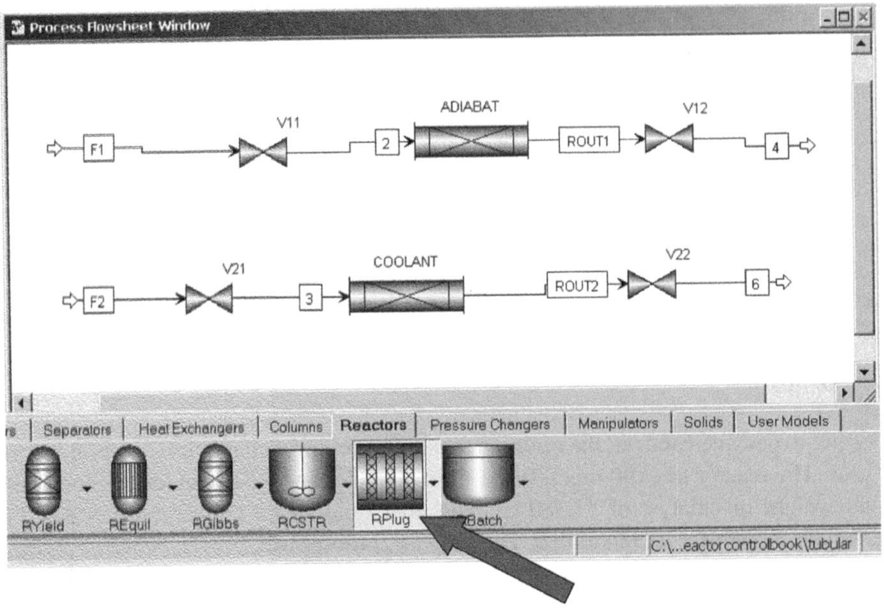

Figure 5.21 Installing RPlug tubular reactors.

The first produces allyl chloride and hydrogen chloride with a reaction rate $\Re_1$ (kmol s^{-1} m^{-3}). The second produces 1,2-dichloropropane with a reaction rate $\Re_2$ (kmol s^{-1} m^{-3}).

$$\Re_1 = k_1 P_{C_3} P_{Cl_2} = \left(8.992 \times 10^{-8} e^{-15111/RT}\right) P_{C_3} P_{Cl_2} \tag{5.25}$$

$$\Re_2 = k_2 P_{C_3} P_{Cl_2} = \left(5.107 \times 10^{-12} e^{-3811/RT}\right) P_{C_3} P_{Cl_2} \tag{5.26}$$

where partial pressures P_j are in pascals and activation energies are in cal/mol.

The tubular reactor in Aspen Plus is called "RPLUG" and is installed on the flowsheet as shown in Figure 5.21. Two different tubular reactors with their feed and product streams are shown. The five possible types of reactors are listed on the *Specifications* page tab when *Setup* under the reactor block is clicked.

5.8.1 Adiabatic Tubular Reactor

An adiabatic reactor is selected, as shown in the upper window in Figure 5.22. Clicking the *Configuration* page tab, opens the lower window in Figure 5.22, on which the size of the reactor is set. Figure 5.23 shows how reactions are selected and pressure is set at 3 atm. If a catalyst is present, the forms shown on Figure 5.24 are filled out. The catalyst does not affect the steady-state design if the volume that it occupies is taken into account. However, the catalyst has a very significant effect on the dynamics of the reactor, as we will demonstrate in Chapter 6. For the numerical example, the reactor is specified to have a diameter of 1 m and a length of 10 m. The feed to the reactor is 0.025 kmol/s at 400 K with 10 mol% chlorine and 90 mol% propylene.

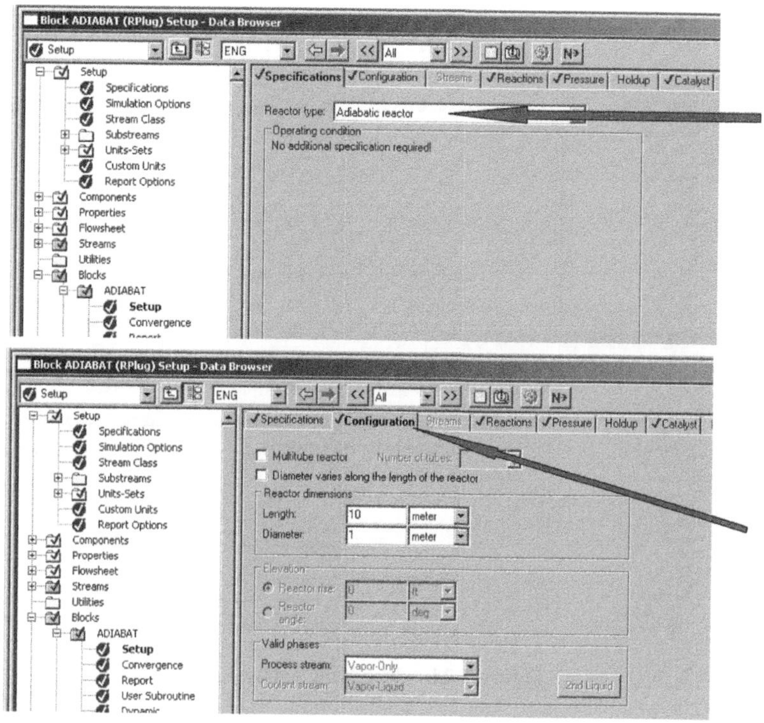

Figure 5.22 Specifying reactor type.

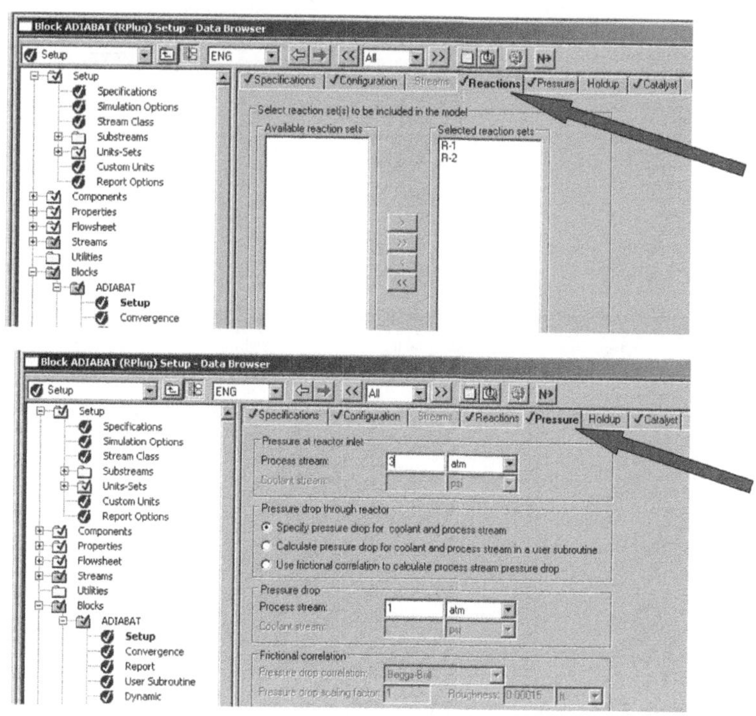

Figure 5.23 Specifying reactions and pressure.

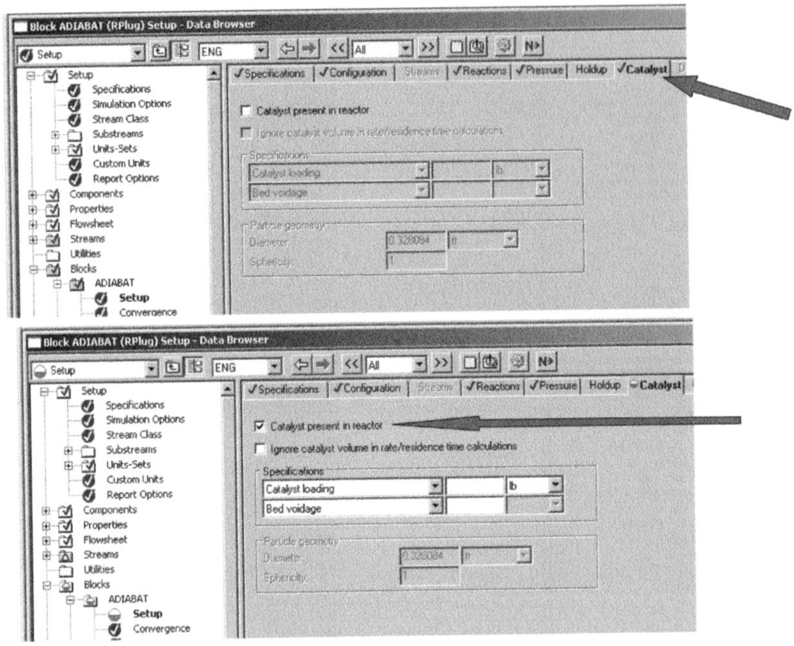

Figure 5.24 Specifying catalyst.

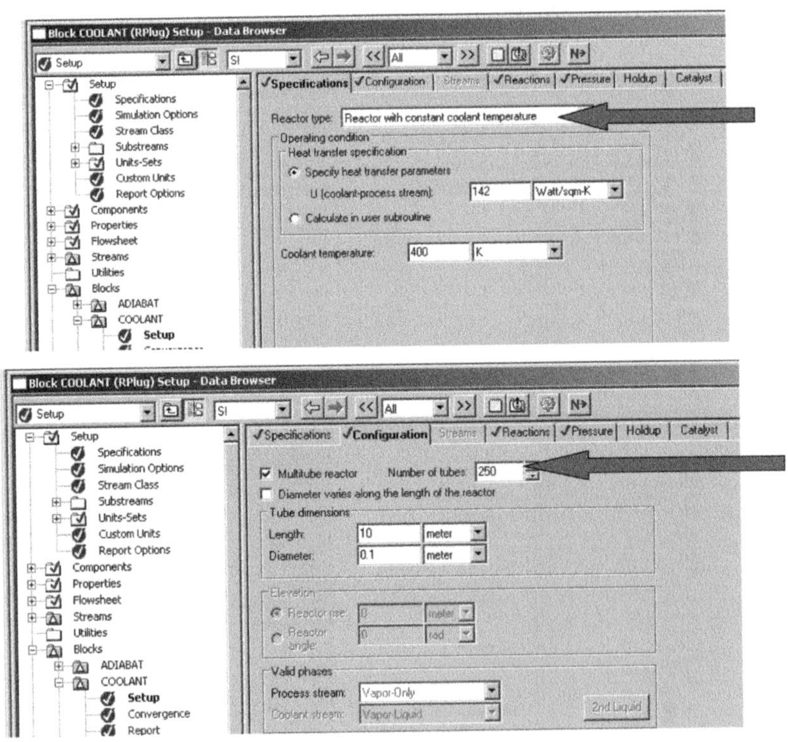

Figure 5.25 Selecting a reactor with constant coolant temperature.

5.8.2 Cooled Tubular Reactor with Constant-Temperature Coolant

This type is selected in the top window shown in Figure 5.25 on the *Specifications* page tab. The overall heat transfer coefficient ($U = 142$ W K^{-1} m^{-2}) and the coolant temperature (400 K) are specified. Clicking the *Configuration* page tab opens the window shown at the bottom of Figure 5.25. The number of tubes (250), their length (10 m), and their diameter (0.1 m) are specified.

Aspen Plus solves the ordinary differential equations down the length of the reactor. Plots of the profiles can be obtained by using the *Plot Wizard*. The number of points plotted can be adjusted from the default value of 10 by clicking on *Report* under the reactor block and setting the number of points as shown in Figure 5.26. The number of points selected does not affect the steady-state results in Aspen Plus. However, it has an important effect on the dynamic model used when the case is exported to Aspen Dynamics. A "lumped" model is used in Aspen Dynamics to approximate the distributed system, which is represented rigorously by partial differential equations. The number of lumps used in Aspen Dynamics is the number specified in the *Report* page tab. The larger this number, the more accurate the lumped model approximates the real system. Unless there are very sharp temperature or composition profiles in the reactor, a 20- or 30-lump model works well for most systems. We will return to this issue in Chapter 6.

Results for the two reactors are given in Figure 5.27. In the adiabatic reactor, temperature increases down the reactor as chlorine is consumed and products are formed. In the constant-coolant temperature reactor, temperature reaches a maximum of 411 K at about 0.6 m from the inlet. Note that very little allyl chloride is formed in the cooled reactor because the temperatures are low. In the adiabatic reactor, the high temperatures increase the allyl chloride reaction rates because of the higher activation energy.

5.8.3 Cooled Reactor with Co-current or Countercurrent Coolant Flow

Figure 5.28 shows the flowsheet and the *Specifications* page tab for a cooled tubular reactor with co-current flow of coolant. The flowrate and temperature of the coolant

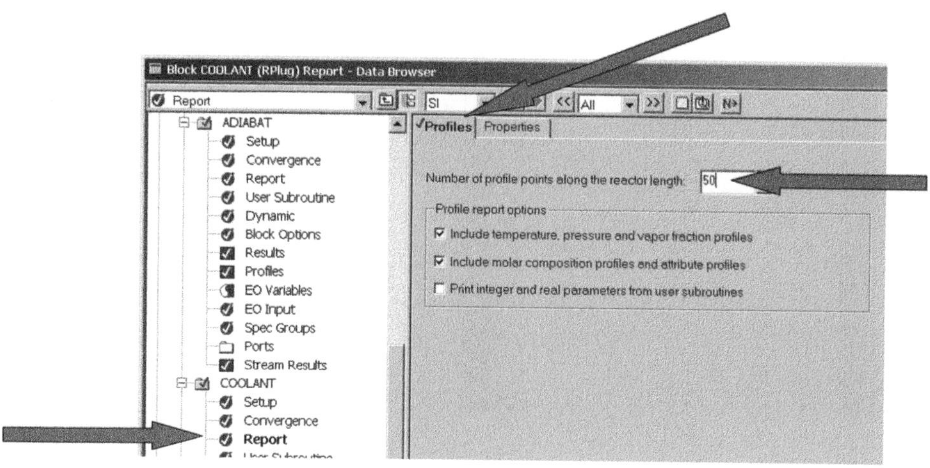

Figure 5.26 Increasing number of points in plot.

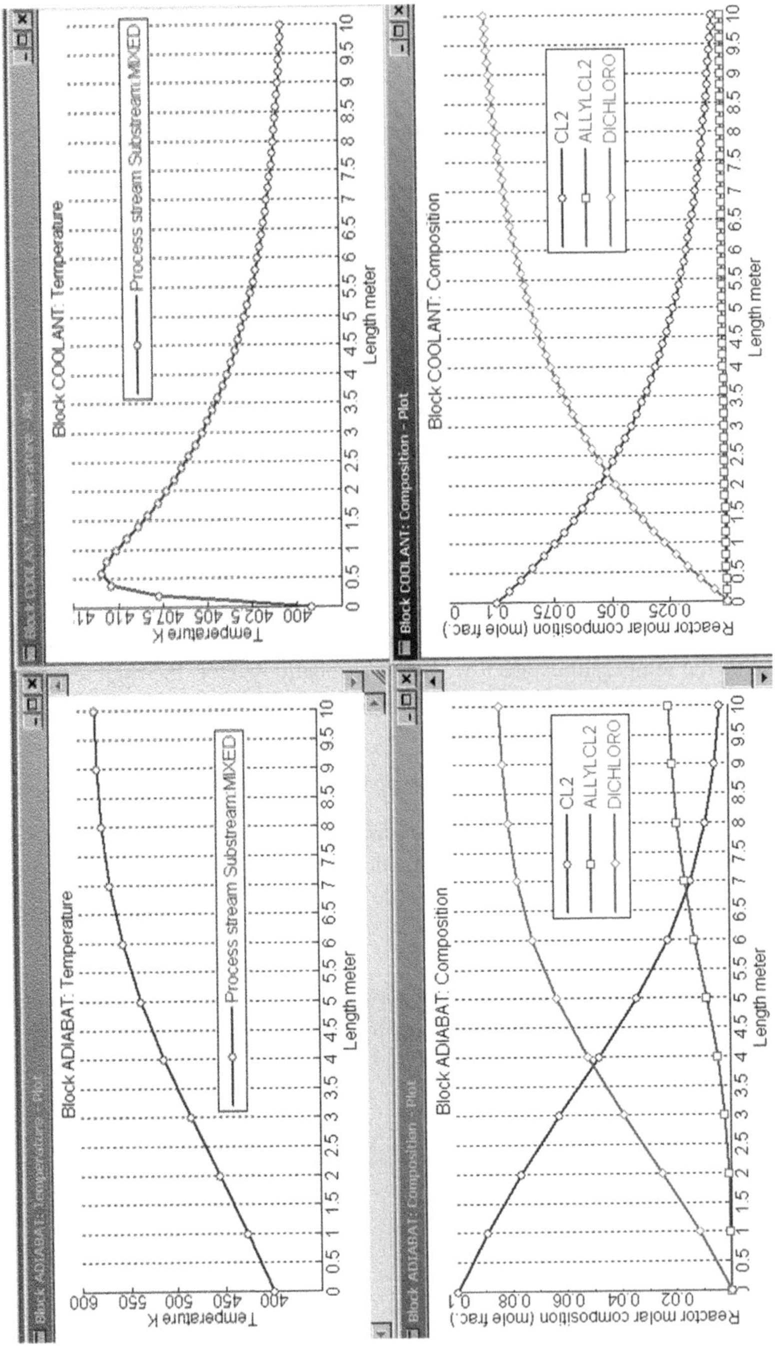

Figure 5.27 Temperature and composition profiles.

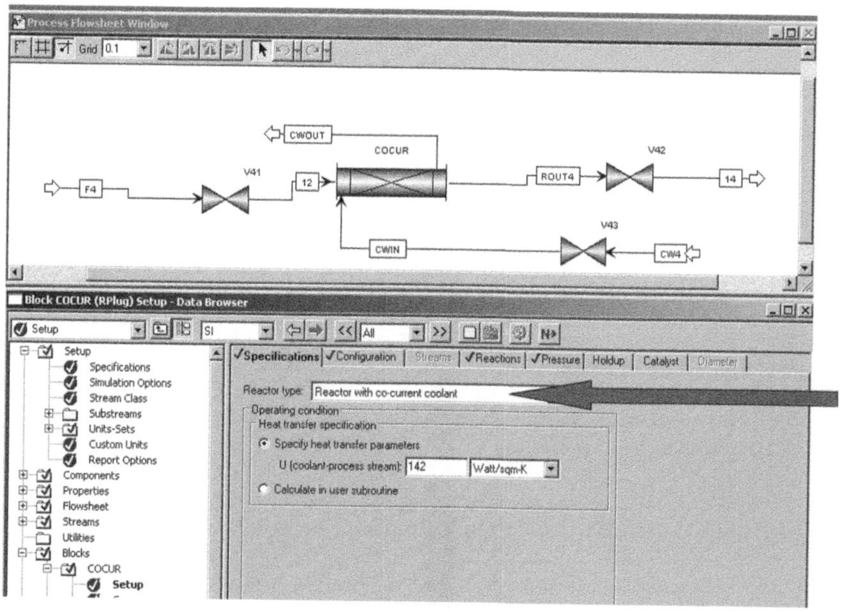

Figure 5.28 Co-current cooling.

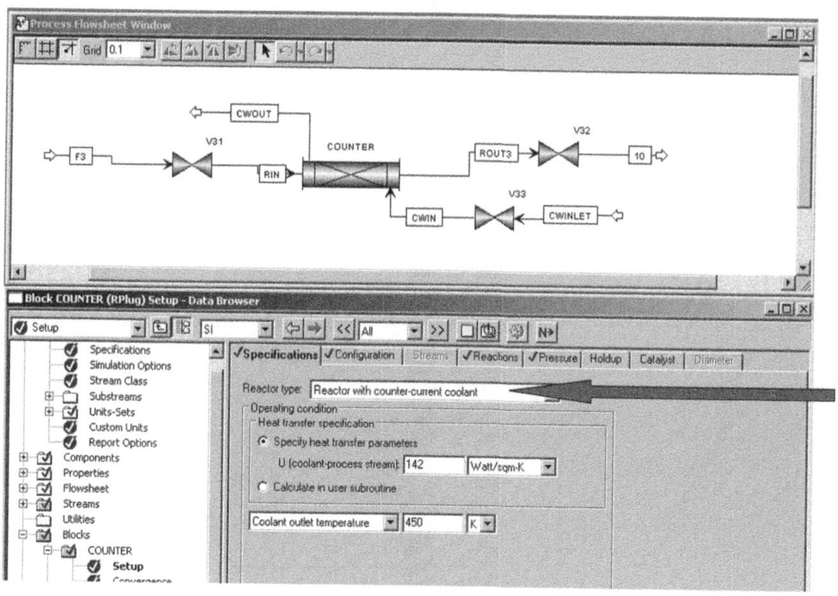

Figure 5.29 Countercurrent cooling.

entering the reactor at the same end as the process are specified to be 0.05 kmol/s and 350 K. Figure 5.29 gives the flowsheet and the *Specifications* page tab for counter-current flow. The flowrate and temperature of the coolant *leaving* the reactor at the same end as the process are specified to be 0.05 kmol/s and 450 K. Figure 5.30 gives temperature and composition profiles for the two cases.

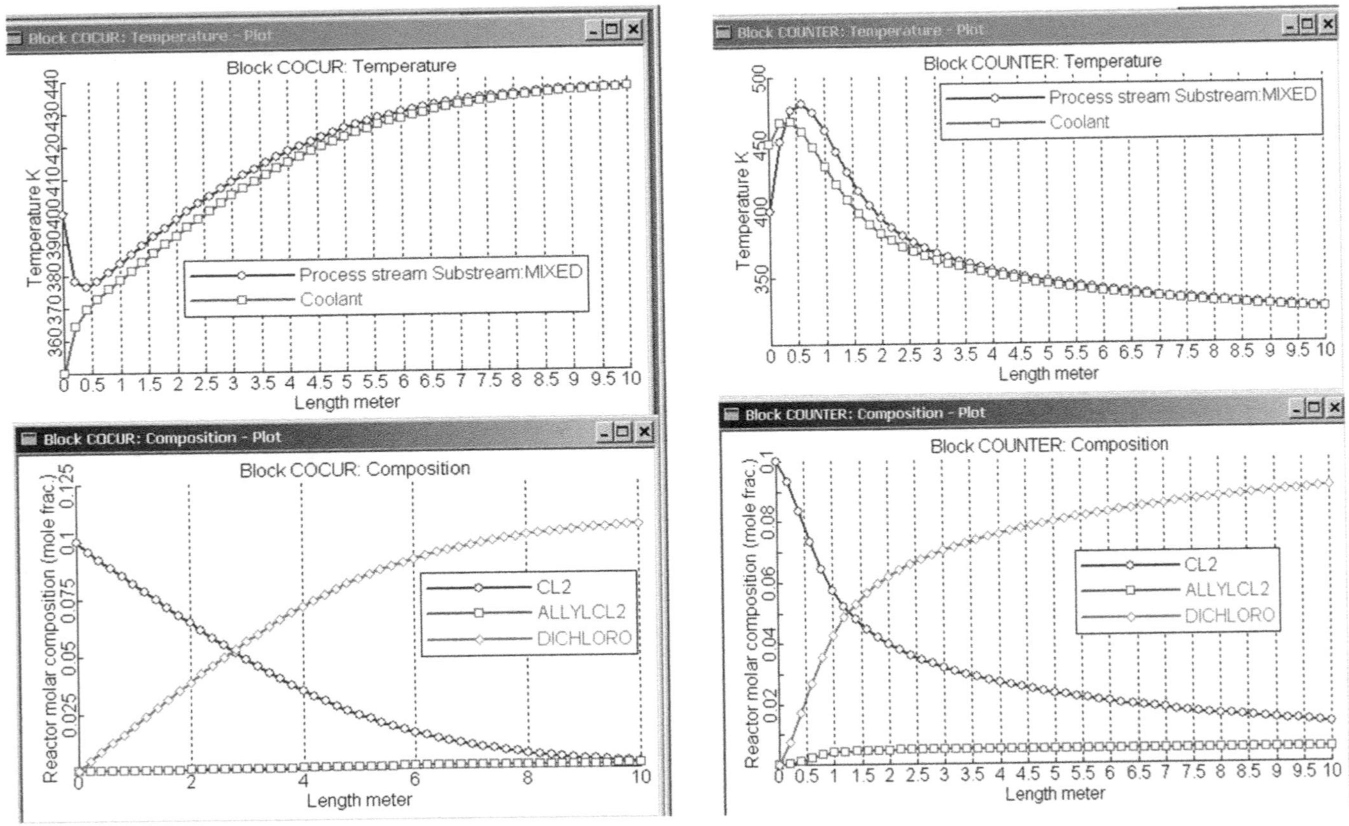

Figure 5.30 Temperature and composition profiles.

5.9 CONCLUSION

In this chapter we have discussed the several types of tubular reactors and some important aspects of their steady-state design. In all tubular reactors, the inlet temperature plays a significant role in the design of the reactor system. Higher inlet temperatures result in smaller reactors for the same per-pass conversion, but also result in higher exit temperatures.

The design of tubular reactor systems is dominated by the classical tradeoff between reactor size and recycle flowrate. Gas phase systems are particularly affected because of the high cost of compression.

In adiabatic reactors, the importance of the adiabatic temperature rise has been emphasized. It can be reduced by decreasing the concentration of the limiting reactant in the feed. Adiabatic reactors with high feed concentrations can be very sensitive to feed temperature changes.

The complex dynamics of tubular reactors are considered in Chapter 6.

CHAPTER 6

CONTROL OF TUBULAR REACTOR SYSTEMS

6.1 INTRODUCTION

The distributed nature of the tubular plug flow reactor means that variables change with both axial position and time. Therefore the mathematical models consist of several simultaneous nonlinear partial differential equations in time t and axial position z. There are several numerical integration methods for solving these equations. The "method of lines" is used in this chapter.[1]

The four types of tubular reactor systems designed in Chapter 5 are investigated for dynamic controllability in this chapter. The four flowsheets are given in Figures 6.1–6.4 with stream conditions and equipment sizes shown. These are the optimum economic flowsheets for the expensive catalyst cases. A three-bed cold-shot system is shown, but a seven bed system is the optimum steady-state design. As we will show, the seven bed system is uncontrollable.

6.2 DYNAMIC MODEL

The reactor is modeled by three partial differential equations: component balances on A and B [Eqs. (6.1) and (6.2)] and an energy balance [Eq. (6.3) for an adiabatic reactor or Eq. (6.4) for a cooled reactor]. The overall heat transfer coefficient U in the cooled reactor in Eq. (6.4) is calculated by Eq. (6.5) and is a function of Reynolds number Re, Eq. (6.6). Equation (6.7) is used for pressure drop in the reactor using the friction factor f given in Eq. (6.8). The dynamics of the momentum balance in the reactor are neglected because they are much faster than the composition and temperature dynamics. A constant

[1]See W. E. Schiesser, *The Numerical Method of Lines*, Academic Press, 1991.

Chemical Reactor Design and Control. By William L. Luyben

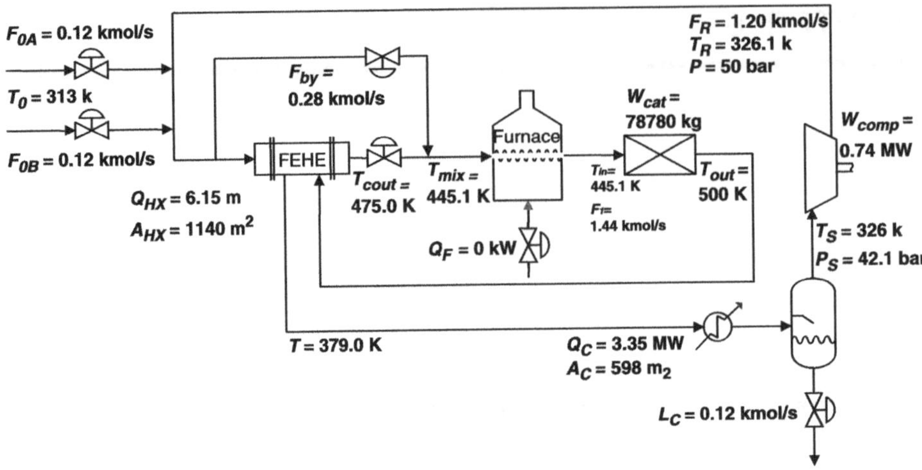

Figure 6.1 Single-stage adiabatic reactor system.

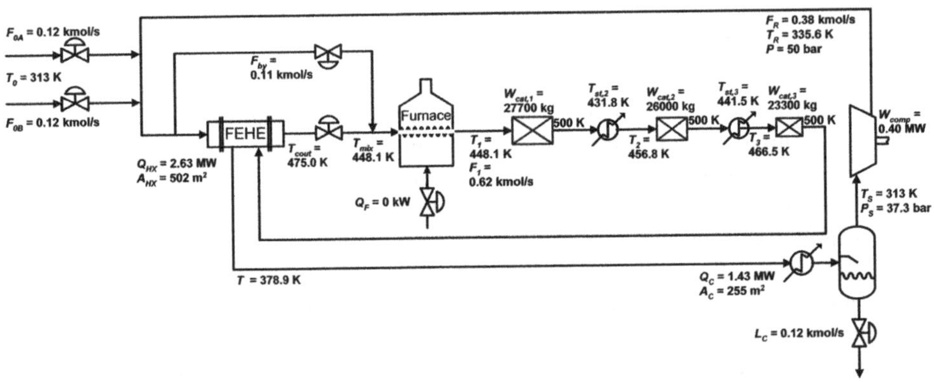

Figure 6.2 Three-stage adiabatic reactors with interstage cooling system.

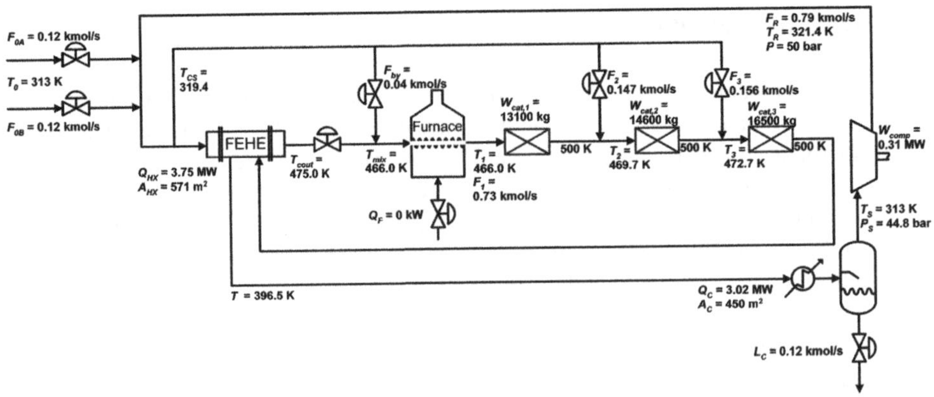

Figure 6.3 Three-stage adiabatic reactors with cold-shot cooling system.

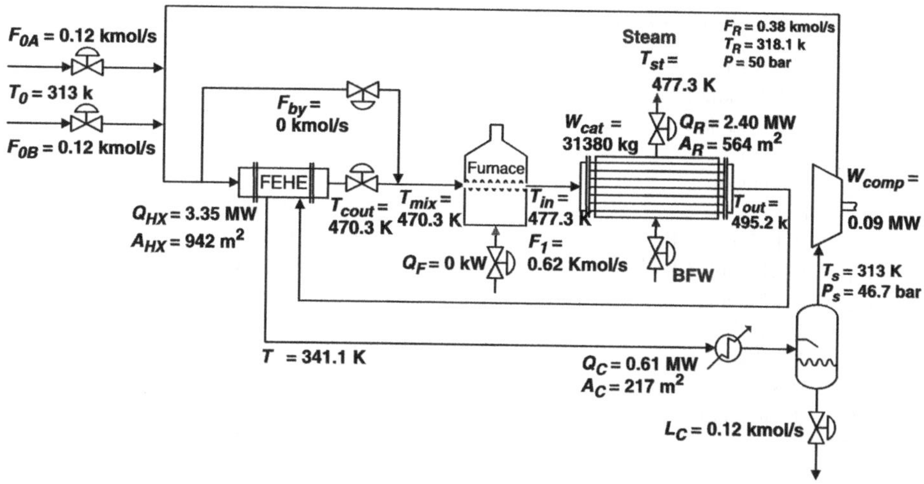

Figure 6.4 Cooled reactor system.

mass flow through the reactor is assumed. The reaction rate r'_C is based on the volume of the reactor and is calculated by Eq. (6.9)

$$\varepsilon \frac{\partial y_A}{\partial t} = -v \frac{\partial y_A}{\partial z} - \frac{(1 - y_A) r'_C}{C} \tag{6.1}$$

$$\varepsilon \frac{\partial y_B}{\partial t} = -v \frac{\partial y_B}{\partial z} - \frac{(1 - y_B) r'_C}{C} \tag{6.2}$$

$$\rho_{cat} c_{cat} \frac{\partial T}{\partial t} = -c \rho v \frac{\partial T}{\partial z} - \lambda r'_C \tag{6.3}$$

$$\rho_{cat} c_{cat} \frac{\partial T}{\partial t} = -c \rho v \frac{\partial T}{\partial z} - \lambda r'_C - \frac{4U}{D_{tube}} (T - T_{st}) \tag{6.4}$$

$$U = 0.01545 + \frac{0.6885 \times 10^{-6}}{D_P} \mathrm{Re} \tag{6.5}$$

$$\mathrm{Re} = \frac{D_P v \rho}{\mu} \tag{6.6}$$

$$\frac{dP}{dz} = \frac{-f \rho v^2}{D_P 10^5} \tag{6.7}$$

$$f = \frac{(1 - \varepsilon)}{\varepsilon^3} \left(1.75 + 150 \frac{(1 - \varepsilon)}{\mathrm{Re}} \right) \tag{6.8}$$

$$r'_C = \alpha \rho_{cat} e^{(-E/RT)} y_A y_B P^2 \tag{6.9}$$

where ε = voidage of bed
v = gas superficial velocity (m/sec)
C = gas concentration (kmol/m^3)
c = heat capacity of gas (kJ kg^{-1} K^{-1})
D_{tube} = diameter of reactor tube (m)
D_P = catalyst-pellet diameter (m)

TABLE 6.1 Process Parameters

$\alpha = 0.19038$ kmol s^{-1} bar^{-2} kg · cat^{-1}	$M_C = 35$ kg/kmol
$E = 69710$ kJ/kmol	$F_{0A} = 0.12$ kmol/s
$\lambda = -23237$ kJ/kmol	$F_{0B} = 0.12$ kmol/s
$D_P = 0.003$ m	$T_0 = 313$ K
$\rho_{cat} = 2000$ kg/m^3	$L_C = 0.12$ kmol/s
$\varepsilon = 0.4$	$P_R = 50$ bar
$c_{pA} = 30$ kJ kmol^{-1} K^{-1}	$T_S = 313$ K
$c_{pB} = 40$ kJ kmol^{-1} K^{-1}	$\Delta T_{HX} = 25$ K
$c_{pC} = 70$ kJ kmol^{-1} K^{-1}	$\Delta P_{HX} = 0.5$ bar
$c_g = 2$ kJ kg^{-1} K^{-1}	$\gamma = 1.312$
$c_{cat} = 0.5$ kJ kg^{-1} K^{-1}	$U_{FEHE} = 0.142$ kJ s^{-1} m^{-2} k^{-1}
$M_A = 15$ kg/kmol	$U_{HX} = 0.227$ kJ s^{-1} m^{-2} k^{-1}
$M_B = 20$ kg/kmol	

Table 6.1 lists the parameters used.

The system pressure P is computed from a total material balance on all the units in the gas loop [Eq. (6.10)] with an average temperature $T_{av} = 500$ K and average molecular weight $M_{av} = 20$ kg/kmol. The gas volume V_{gas} is fixed and includes the gas volumes of the FEHE, furnace, reactor, condenser, and flash drum. The diameter of tubes in all heat exchangers is 0.0254 m. Equal volumes on shell and tube sides are assumed. The volume of the furnace is computed from the furnace heat duty and is 1.35 m^3 per MW. The diameter of the flash drum is calculated using an F factor of 0.79 (in SI units) and a height-to-diameter ratio of 2. Table 6.2 lists equipment gas volume for the four optimum alternative tubular reactor systems.

$$\frac{dP}{dt} = \frac{RT_{av}(M_A F_{0A} + M_B F_{0B} - M_C L_C)}{V_{gas} M_{av}} \tag{6.10}$$

The feed-effluent heat exchanger is assumed to be single-pass, countercurrent shell-and-tube design. Three partial differential equations are used for the temperatures of gas on tube side, gas on shell side, and the tube metal: Eqs. (6.11), (6.12), and (6.13), respectively. The overall heat transfer coefficients on both tube and shell sides, U_t and U_s, are constant and are equal to 0.284 kJ s^{-1} m^{-2} K^{-1}. Equal heat transfer area per volume is assumed for the shell and tube sides ($A_t/V_t = A_s/V_s$) and is 157 m^2/m^3, based

TABLE 6.2 Equipment Gas Volume

Equipment	1-Adiabatic	3-Adiabatic Interstage HX	3-Adiabatic Cold-Shot	Cooled
FEHE	15.38	6.75	7.77	11.93
Furnace	1.37	0.59	0.69	0.88
Interstage HXs	—	2.29	—	—
Reactors	15.5	15.10	8.69	6.27
Condenser	3.8	1.59	2.81	1.37
Flash drum	24.09	3.61	11.64	2.87
Total (m^3)	60.14	29.93	31.60	23.32

on 0.0254 m tube diameter. The density of tube material ρ_m is 7700 kg/m^3. The heat capacity of the tube c_m is 0.5 kJ kg^{-1} K^{-1}. The thickness of the tube th_m is 3.048×10^{-3} m:

$$\frac{\partial T_t}{\partial t} = -v_t \frac{\partial T_t}{\partial z} + \frac{U_t A_t (T_m - T_t)}{\rho_t c V_t} \tag{6.11}$$

$$\frac{\partial T_s}{\partial t} = v_s \frac{\partial T_s}{\partial z} - \frac{U_s A_s (T_s - T_m)}{\rho_s c V_s} \tag{6.12}$$

$$\frac{\partial T_m}{\partial t} = \frac{U_s (T_s - T_m) - U_t (T_m - T_t)}{\rho_m c_m th_m} \tag{6.13}$$

Equation (6.14) is used for the temperature of gas in the tubes of the intermediate heat exchanger used in the multistage adiabatic reactor system. Heat is transferred from the hot gas to saturated water on the shell side generating steam at temperature T_{st}. The overall heat transfer coefficient U_{HX} is constant and is equal to 0.227 kJ s^{-1} m^{-2} K^{-1}. The heat transfer area per volume A_{HX}/V_{HX} is 157 m^2/m^3:

$$\frac{\partial T}{\partial t} = -v \frac{\partial T}{\partial z} + \frac{U_{HX} A_{HX} (T_{st} - T)}{\rho c_g V_{HX}} \tag{6.14}$$

The system with ordinary and partial differential equations (ODEs and PDEs) is programmed in Matlab. The method of lines is used to solve the PDEs. The partial derivatives in z are computed by using a first-order finite difference approximation. The number of grid points in z used for the reactor in the single-stage configurations is 51. The number of grid points in z used for each reactor in multistage configurations is 31. The number of grid points in z used for each heat exchanger is 21. Matlab's *ode23* is used for numerical integration.

6.3 CONTROL STRUCTURES

Figures 6.5–6.8 show the control structures used for the four alternative tubular reactor systems. In Figures 6.6 and 6.7 (the multistage adiabatic systems with interstage

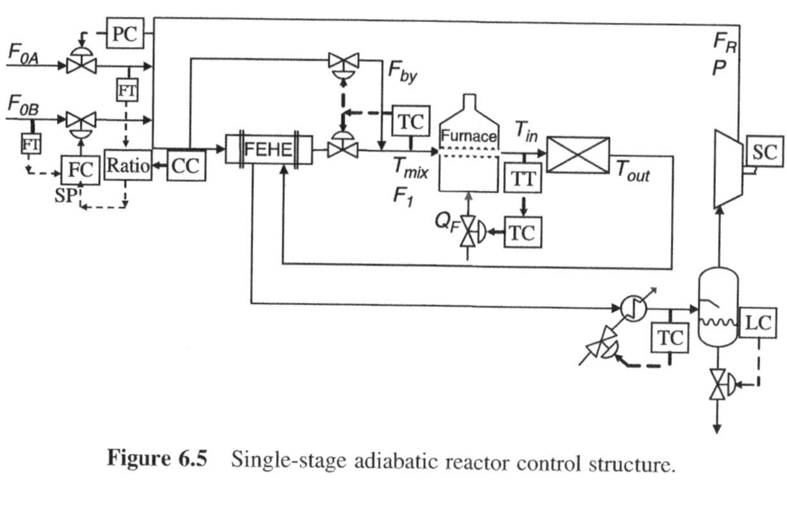

Figure 6.5 Single-stage adiabatic reactor control structure.

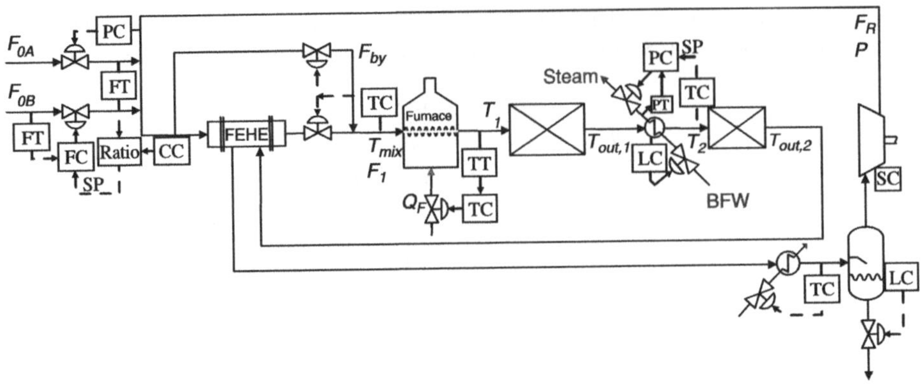

Figure 6.6 Adiabatic reactors with interstage cooling control structure.

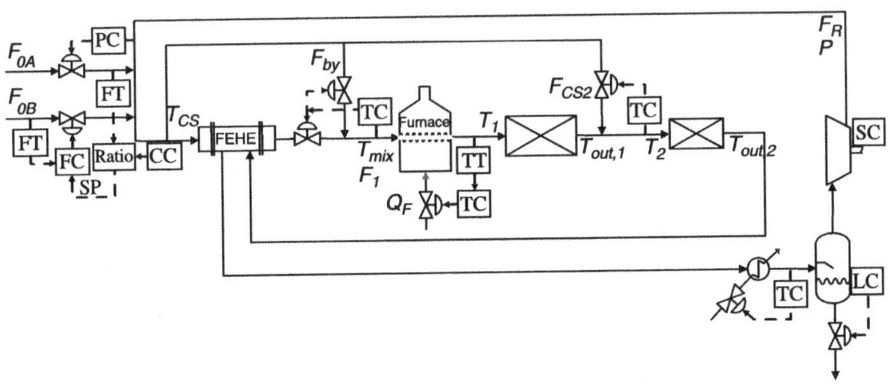

Figure 6.7 Adiabatic reactors with cold-shot cooling control structure.

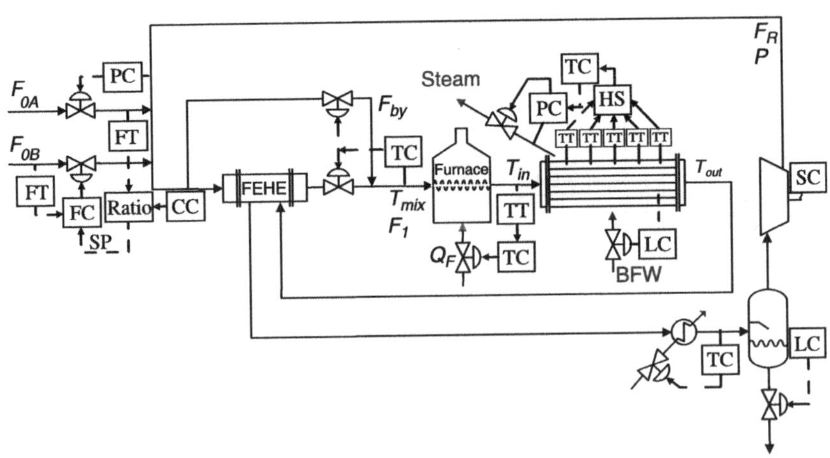

Figure 6.8 Cooled reactor control structure.

cooling and cold-shot cooling), only two reactor sections are shown so as not to clutter the diagram too much. The remaining reactor sections have the same control structures.

All of these systems have some common control loops. The system pressure is controlled by manipulating the fresh feed of A (F_{0A}). The concentration controller with ratio control is used to control reactor inlet gas composition by manipulating the fresh feed of B (F_{0B}). Bypassing (F_{by}) around the FEHE is used to control gas mixture temperature T_{mix}. Reactor inlet temperature (T_{in} or T_1) is controlled by manipulating the furnace heat input Q_F. The setpoints of these two temperature controllers are the same, and the controller output signals are split-ranged so that bypassing and furnace heat input cannot occur simultaneously.

The liquid level in the flash drum is controlled by manipulating liquid product L_C. A temperature controller manipulates heat removal in the condenser to hold drum temperature constant. Compressor speed is controlled, which is assumed to fix the recycle gas flowrate F_R.

The multistage adiabatic reactor system with interstage cooling has a controller on the inlet temperature of the second reactor as shown in Figure 6.6. The temperature is controlled by manipulating the steam pressure in the steam-generating heat exchanger. The level in the steam generator is controlled by manipulating boiler feedwater (BFW) makeup flow.

Figure 6.7 shows the control structure of the multistage adiabatic reactor system with cold-shot cooling for the two-reactor case. The inlet temperature of the second reactor T_2 is controlled by manipulating the cold-shot flowrate F_{CS2}. Note that changing the cold-shot flow F_{CS2} produces an opposite change in the gas flow fed to the first reactor F_1 because the total recycle flowrate F_R is constant. For the multibed systems, the inlet temperature T_n of each bed is controlled by manipulating cold-shot flow at the inlet of that bed F_{CSn}. Changing any of these flows affects the flow to the first reactor. The effect of this interaction on controllability is important and will be discussed in the next section. It represents a basic problem with cold-shot cooling when the total gas recycle flowrate is fixed, which is the situation in most practical systems because of compressor limitations.

Figure 6.8 shows the control structure of the cooled reactor system. Five temperature transmitters at different axial locations are used in the reactor. One of the transmitters is located where the steady-state peak temperature occurs. Two other transmitters are placed upstream and two downstream of this peak location. The distance between each transmitter is 10% of the length of the reactor. A high selector receives the five temperature transmitter signals and sends the highest signal to a peak temperature controller. The controller controls the peak temperature by manipulating the steam pressure (or equivalently, the steam temperature T_{st}). The liquid level in the shell side of the reactor is controlled by manipulating liquid BFW flow.

Perfect control is assumed for the composition/ratio controller, level controllers, gas condenser temperature controller, and speed controller.

6.4 CONTROLLER TUNING AND DISTURBANCES

Ziegler–Nichols (ZN) and Tyreus–Luyben (TL) PI tunings are evaluated. Ultimate gain and frequency are obtained by performing relay–feedback tests. Temperature control loops have three 20-s lags. The pressure control loop has two 30-s lags. There is a

TABLE 6.3 Controller Tuning for Single-Stage Adiabatic Reactor System

	ZN Tuning		TL Tuning	
Control Loop	K_C	τ_I (min)	K_C	τ_I (min)
P,F_{0A}	0.83	2.35	0.57	6.19
T_{mix},F_{by}	−2.76	0.99	−1.90	2.62
T_1,Q_F	11.27	2.56	7.75	6.77

TABLE 6.4 Controller Tuning for Three-Stage Adiabatic Reactor System with Interstage Cooling

	ZN Tuning		TL Tuning	
Control Loop	K_C	τ_I (min)	K_C	τ_I (min)
P,F_{0A}	0.37	2.40	0.25	6.32
T_{mix},F_{by}	−2.82	0.99	−1.94	2.62
T_1,Q_F	11.24	2.56	7.73	6.77
$T_2,T_{st,2}$	5.19	1.06	3.57	2.79
$T_3,T_{st,3}$	5.80	1.05	3.99	2.78

120-s lag in the furnace heat input. The pressure transmitter span is 10 bars. Temperature transmitter spans are 100 K. The furnace is capable of increasing the inlet gas temperature by 30 K. Tables 6.3–6.7 give the controller tuning parameters used in the various systems.

The openloop behaviors of the alternative systems are investigated by using ±20% step changes in recycle flowrate F_R. In the openloop tests, the pressure controller, the temperature controllers, and the composition controller are all on manual.

In the closedloop tests, all controllers are on automatic. The disturbances in the adiabatic reactor systems are setpoint changes in the inlet temperature and changes in recycle flowrate, F_R. The cooled reactor system has the same disturbances with an additional disturbance

TABLE 6.5 Controller Tuning for Seven-Stage Adiabatic Reactor System with Cold-Shot Cooling

	ZN Tuning		TL Tuning	
Control Loop	K_C	τ_I (min)	K_C	τ_I (min)
P,F_{0A}	0.44	2.09	0.25	6.32
T_{mix},F_{by}	−2.65	0.99	−1.94	2.62
T_1,Q_F	11.28	2.56	7.75	6.77
T_2,F_{CS2}	−13.02	1.01	−8.95	2.66
T_3,F_{CS3}	−12.47	1.05	−8.57	2.77
T_4,F_{CS4}	−13.20	1.06	−9.08	2.79
T_5,F_{CS5}	−13.77	1.06	−9.47	2.79
T_6,F_{CS6}	−14.36	1.06	−9.87	2.79
T_7,F_{CS7}	−15.00	1.06	−10.31	2.80

TABLE 6.6 Controller Tuning for Three-Stage Adiabatic Reactor System with Cold-Shot Cooling

	ZN Tuning		TL Tuning	
Control Loop	K_C	τ_I (min)	K_C	τ_I (min)
P,F_{0A}	0.53	2.18	0.36	5.75
T_{mix},F_{by}	−2.70	0.99	−1.85	2.61
T_1,Q_F	11.26	2.56	7.74	6.77
T_2,F_{CS2}	−7.21	1.01	−4.95	2.67
T_3,F_{CS3}	−7.96	1.01	−5.47	2.67

TABLE 6.7 Controller Tuning for Cooled Reactor System

	ZN Tuning		TL Tuning	
Control Loop	K_C	τ_I (min)	K_C	τ_I (min)
P,F_{0A}	0.40	2.22	0.27	5.86
T_{mix},F_{by}	−5.24	1.00	−3.60	2.63
T_1,Q_F	11.22	2.57	7.71	6.78
T_{peak},T_{st}	2.67	2.76	1.84	7.28

in the peak temperature setpoint change. All closedloop disturbances are assumed to enter the system as 5-min ramps from the initial steady-state value to a new value. The temperature setpoint change or recycle flowrate change is made such that the production rate is increased or decreased by 25%. Since each reactor has a different gain ($\Delta T_{out}/\Delta T_{in}$), setpoint changes in the *inlet* temperatures of multiple adiabatic reactors are made such that the reactor *exit* temperature of each reactor reaches the same temperature when the new steady-state conditions are attained.

6.5 RESULTS FOR SINGLE-STAGE ADIABATIC REACTOR SYSTEM

Openloop Response The openloop responses of a single adiabatic tubular reactor system to $\pm 20\%$ step changes in recycle flowrate F_R are shown in Figure 6.9. The solid lines represent increases in recycle flow and the dashed lines, decreases. The results show that the system produces limit cycle behavior, alternating between high temperatures and low temperatures. This type of dynamic response is called "openloop-unstable" behavior in this chapter.

When the recycle flowrate is increased, the temperatures throughout the reactor decrease (see T_{out} in Fig. 6.9) because higher flowrate through the reactor provides more thermal sink for the heat being generated by reaction, so the adiabatic temperature rise is smaller. The reaction rate decreases, and the pressure starts to increase. This gradual increase in pressure eventually starts to increase the reaction rate (at about 40 min). The higher reaction rate increases reactor temperatures, and the reactor exit temperature goes above 500 K at about 50 min. The pressure and then decreases because of the high reaction rate. The system finally exhibits limit cycle behavior with a period of about

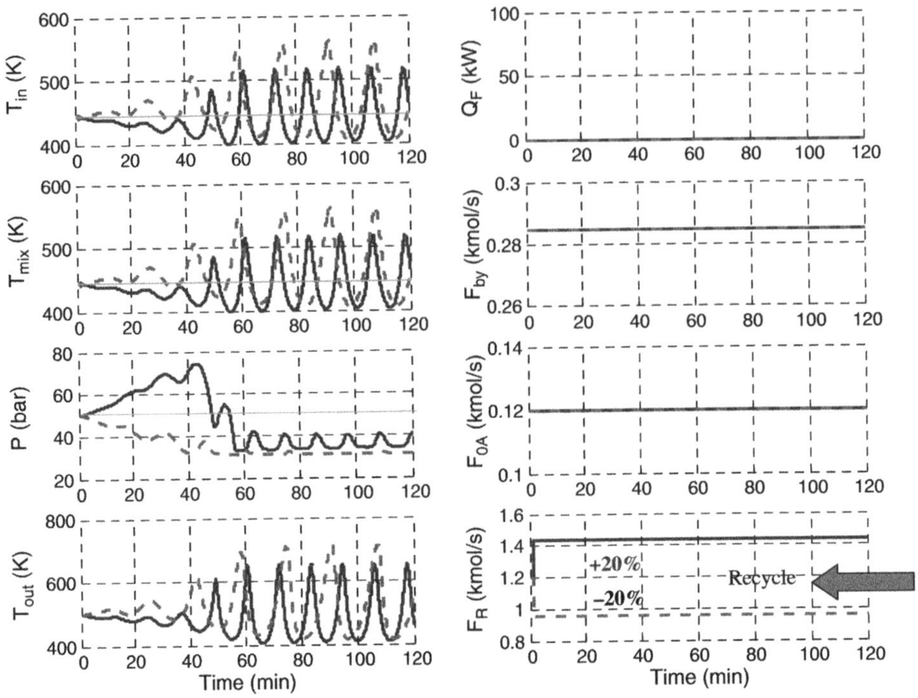

Figure 6.9 Openloop; single-stage adiabatic system.

12 min and with very large changes in temperatures and pressure. A decrease in recycle flowrate causes an increase in the reactor temperature and results in a similar oscillating (limit cycle) behavior. The fundamental reason for the openloop instability is the positive feedback between the adiabatic reactor (which has a large reactor gain $\Delta T_{out}/\Delta T_{in}$) and the FEHE. The limit-cycle behavior results from the nonlinearity of the system.

Closedloop Response Figure 6.10 shows the response of the closedloop system to a change in the setpoint of the reactor inlet T_{in} temperature controller. The setpoint is ramped up by 2.8 K over 5 min. The TL tuning constants (dashed lines) give oscillatory responses, but ZN tuning (solid lines) gives tight control and no oscillation. Oscillations occur with loose tuning because the high reactor gain and the positive feedback of heat require a minimum value of the controller gain to stabilize the system. As shown earlier, the system with zero controller gain (openloop) is unstable and oscillates.

The 2.8 K increase in inlet temperature increases the production rate by 25% due to the increase in the reactor temperatures. Figure 6.11 shows the response to a ramped decrease in reactor inlet temperature of 3.7 K over 5 min. The system with ZN tuning has tighter control and is less oscillatory than the system with TL tuning. The production rate is decreased by 25% because of the reactor temperature decrease.

Figure 6.12 shows the response to a 22% ramped *increase* in the recycle flow F_R. The production rate is *decreased* by 25% because of the reactor temperature decrease. Figure 6.13 shows the response to a 13% ramped decrease in recycle. The production rate is increased by 25% as a result of the reactor temperature increase. The system oscillates with TL tuning, but ZN tuning gives tight control.

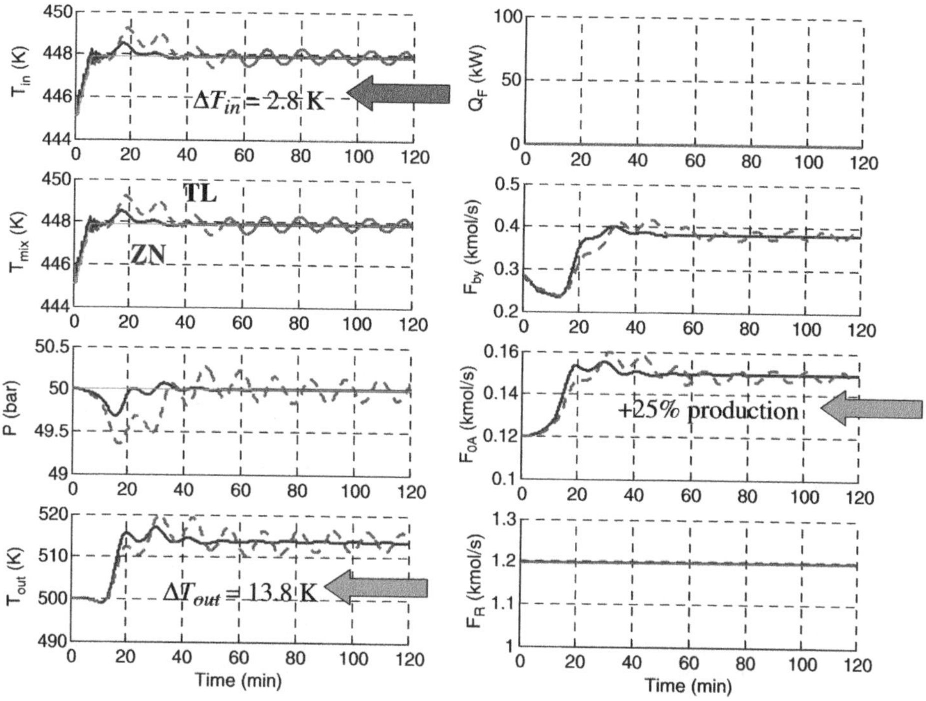

Figure 6.10 Closedloop; single-stage adiabatic system; +2.8 K T_{in}^{set}.

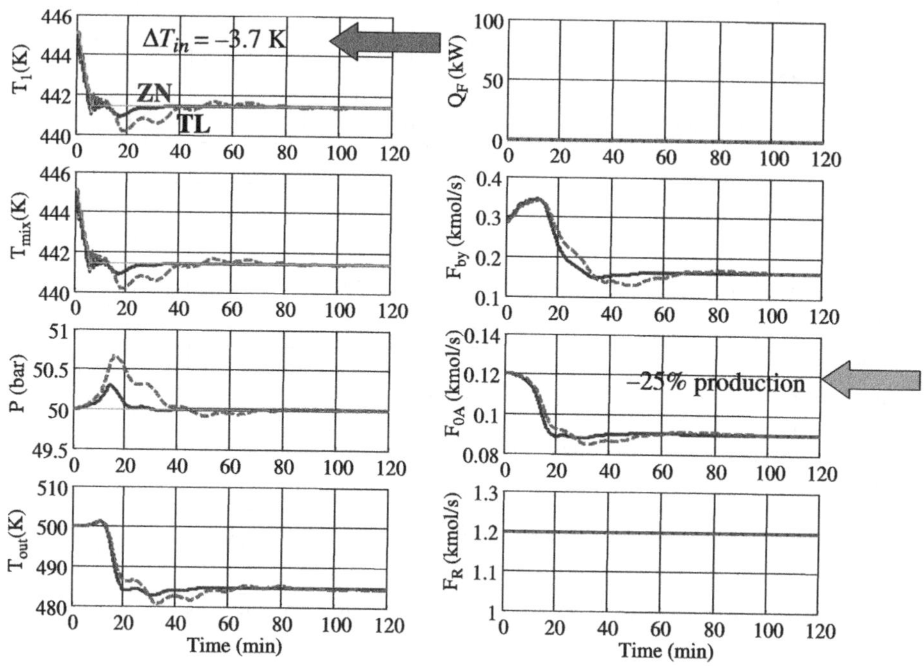

Figure 6.11 Closedloop; single-stage adiabatic system; −3.7 K T_{in}^{set}.

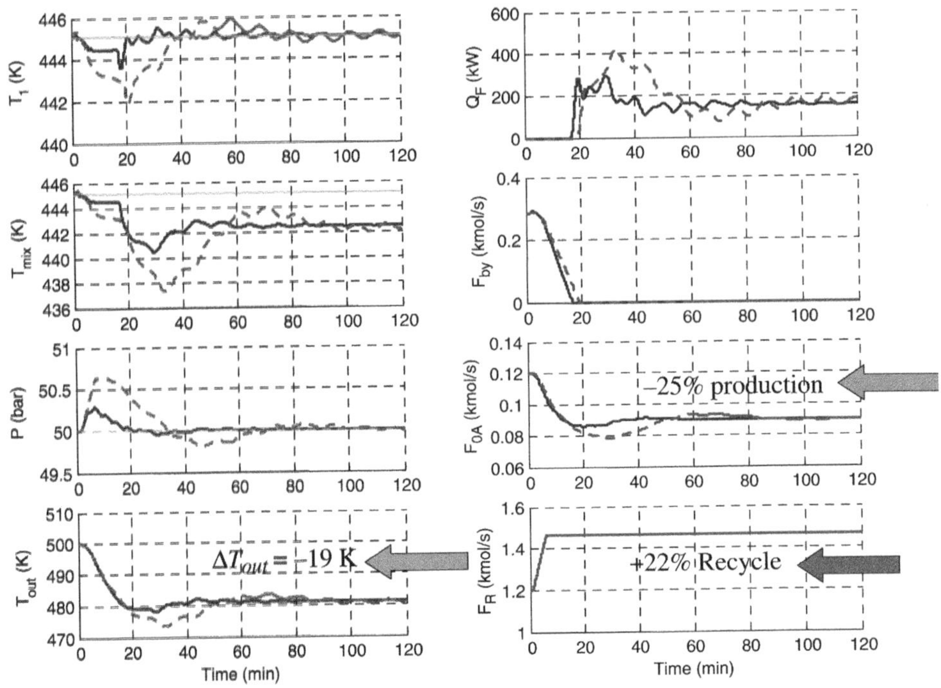

Figure 6.12 Closedloop; single-stage adiabatic system; +22% F_R.

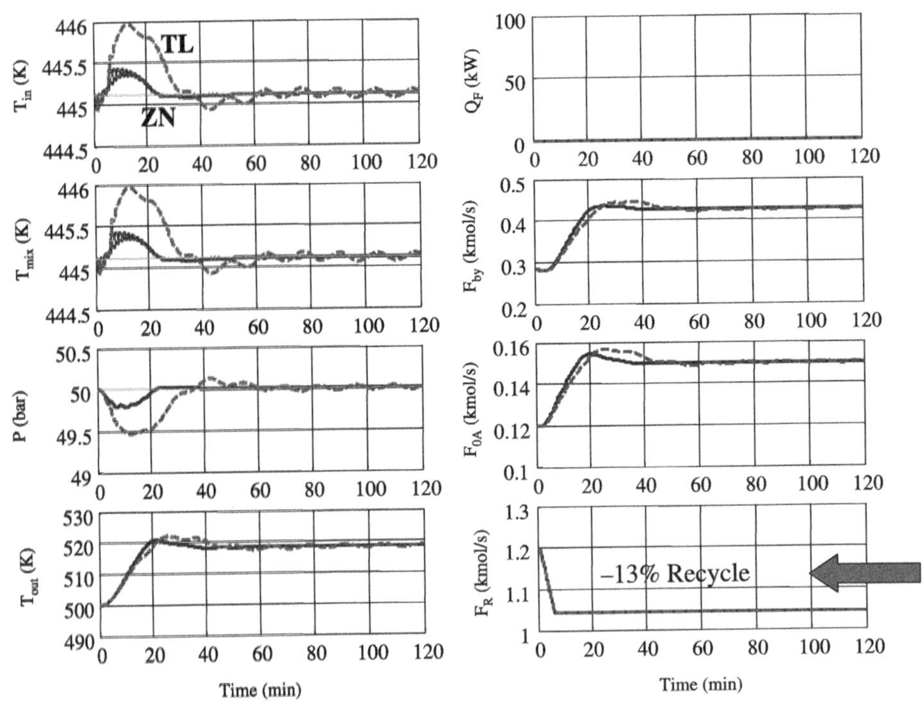

Figure 6.13 Closedloop; single-stage adiabatic system; −13% F_R.

These results show that aggressive tuning (ZN) is required to stabilize this openloop unstable system. Production rate can be effectively changed by changing either reactor inlet temperature or recycle flow.

6.6 MULTISTAGE ADIABATIC REACTOR SYSTEM WITH INTERSTAGE COOLING

Openloop Response The openloop response of a three-stage adiabatic reactor system with interstage cooling to a 20% increase in recycle flow F_R is shown in Figure 6.14. The reactor inlet and exit temperatures of each reactor are shown. This system with heat feedback and large reactor gains is openloop unstable.

These responses are similar to the response of a single adiabatic reactor system except now the upstream reactors affect the downstream reactors. For example, the increase in the exit temperature of the first reactor at 30 min produces an increase in the exit temperature of the second reactor, which then produces an even larger increase in the exit temperature of the third reactor. Thus these disturbances are amplified as they move down the reactor train. Similar results occur for a decrease in recycle flow.

Closedloop Response Figure 6.15 shows the response to increases in the three reactor inlet temperature controller setpoints. These changes are made so that the exit temperatures of all three reactors come to the same value at the new steady state (512.4 K). The required setpoint inlet temperature changes are +4.1 K for T_1, +6.3 K for T_2 and +9.0 K for T_3, which all enter as 5-min ramps starting at time equal 1 min. These changes in inlet temperature increase production rate by 25%. In this openloop-unstable system, ZN tuning gives tighter control than TL tuning.

When the reactor inlet temperatures are decreased (4.9 K for T_1, 7.1 K for T_2, and 9.6 K for T_3), the production rate is decreased by 25%, and the new steady-state exit temperature of all reactors is 485.6 K

Figure 6.16 shows results for a 20% increase in recycle flow F_R. The production rate is changed by only ~10%, which is less than observed for the single adiabatic reactor. Thus the recycle flowrate is not an effective manipulating variable for changing the production rate in this multistage adiabatic reactor system with intermediate cooling.

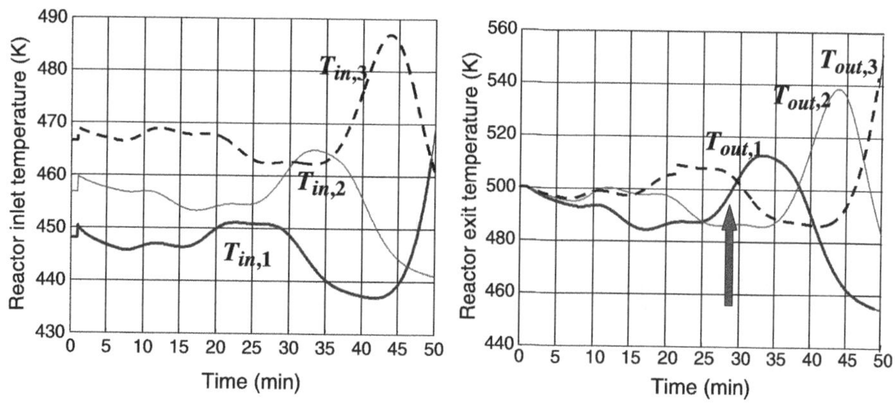

Figure 6.14 Openloop; three-stage adiabatic reactors with interstage cooling; +20% F_R.

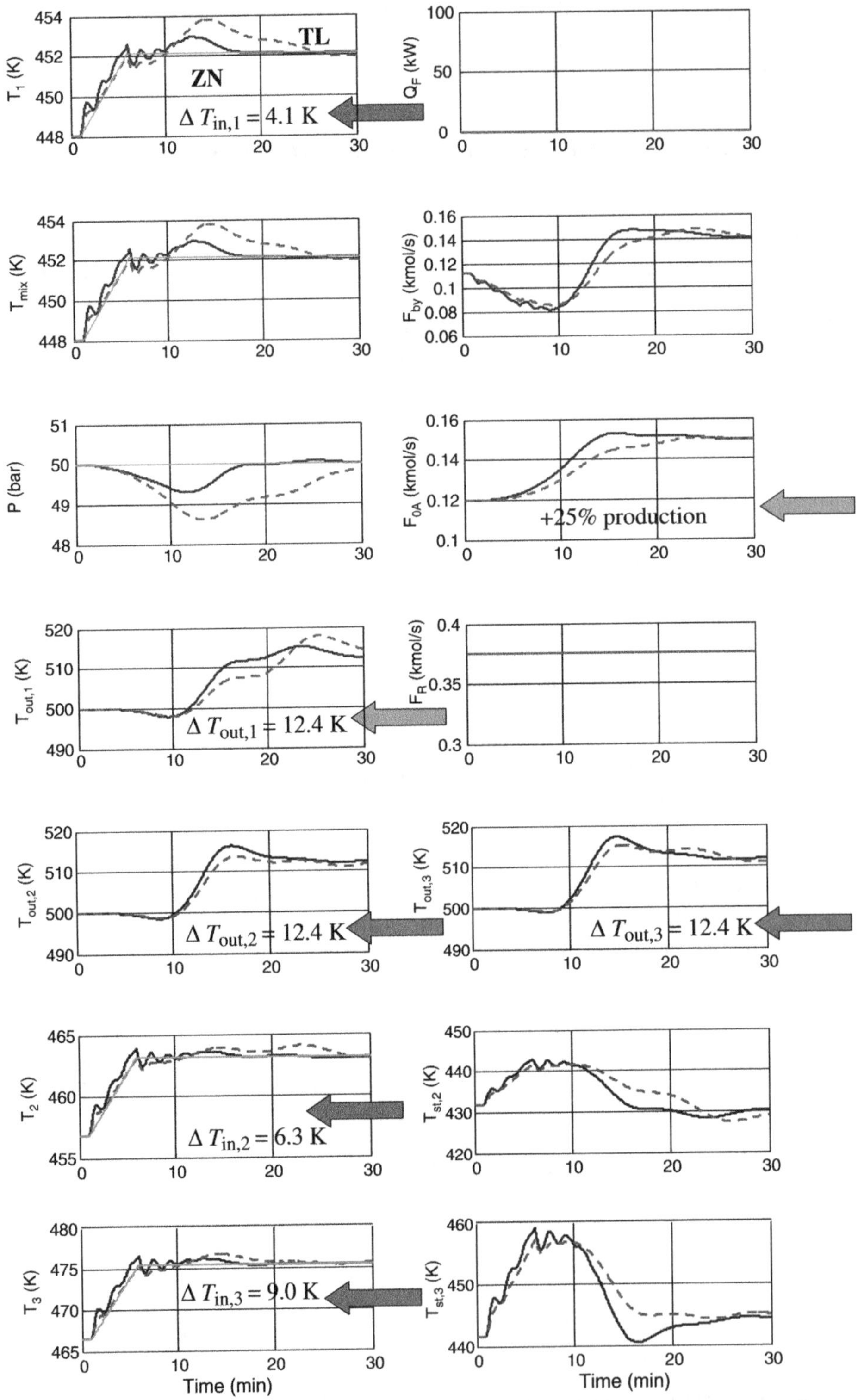

Figure 6.15 Closedloop; three-stage adiabatic reactors with interstage cooling; +4.1 T_1^{set}.

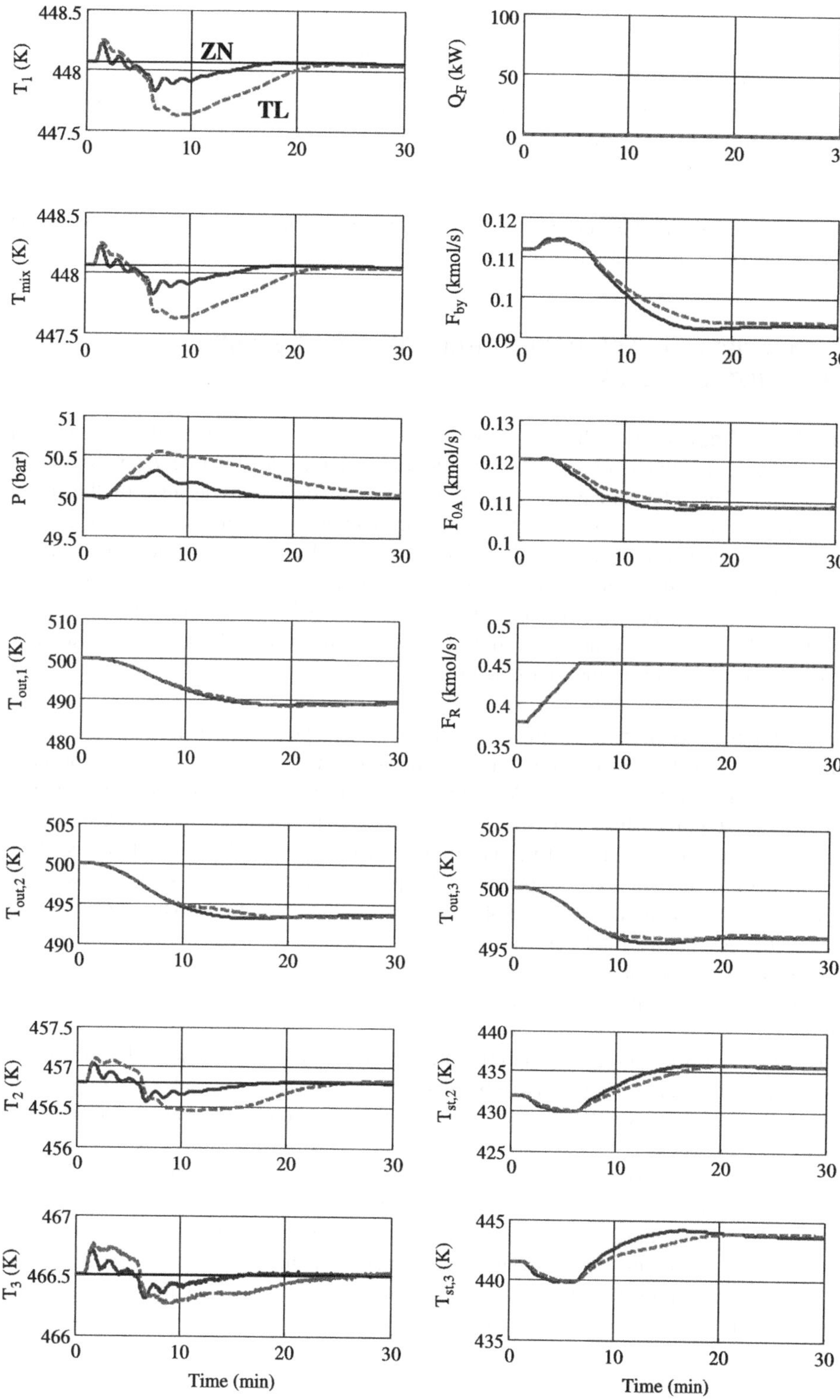

Figure 6.16 Closedloop; three-stage adiabatic reactors with interstage cooling; +20% F_R.

The new steady-state exit temperatures of each reactor are not the same. The first reactor exit temperature is more sensitive to changes in recycle flowrate. For a 20% decrease in F_R, the first reactor exit temperature increases to 521 K while the second and third reactor exit temperatures increase to only 503 and 501 K. If it is desired to maintain the same exit temperatures in all reactors, the inlet temperatures would have to be adjusted.

These results illustrate an inherent disadvantage of this system. The first reactor is the most sensitive to disturbances. The changes in its exit temperature will be greater than those of the other downstream reactors.

6.7 MULTISTAGE ADIABATIC REACTOR SYSTEM WITH COLD-SHOT COOLING

Openloop Response The openloop response of the seven-bed adiabatic reactor system with cold-shot cooling to a 20% increase in recycle flowrate F_R is shown in Figure 6.17. The inlet temperature of the first reactor decreases because of the larger flow through the preheat system. However, the reactor inlet temperatures of the other reactors increase because the cold-shot flows are fixed. Then the first reactor inlet temperature starts to increase at 5 min because of the increase in the exit temperature of the seventh reactor. The pressure initially increases gradually because of the lower temperature in the first reactor. At about 12 min, the pressure drops rapidly because of the large increase in temperature in the first reactor.

Closedloop Response Figure 6.18 shows the response to increases in the setpoints of all reactor inlet temperatures. The setpoint change is different for each reactor, as previously discussed. We could not find any choice of tuning parameters that would stabilize this seven-bed cold-shot system. Neither the aggressive ZN tuning nor the more conservative TL tuning produced a stable closedloop system. The changes in reactor exit temperatures keep increasing and cold-shot valves eventually saturate. There are six cold-shot valves in the system. Moving these valves causes changes in the inlet flowrate of the first reactor F_1 because the total recycle flow is fixed. With many beds and many cold-shot flows, the interaction between the flows and the temperatures leads to amplification of disturbances and limit cycle behavior.

There are several alternative suboptimal designs that could be considered to overcome this problem. Reducing the number of beds is explored below. A second alternative would be to design the system with a recycle flowrate that is much larger than the steady-state optimum. This would reduce the sensitivity of the first reactor (higher inlet temperature and lower reactor gain). A third alternative is to design the system for liquid cold shot. It could also be claimed that a more advanced control structure may stabilize this system. The propagation of thermal waves through the reactor beds is an inherent dynamic property of the system that is independent of the type of controller, and this behavior makes control very difficult. The situation is similar to that discussed by Shinnar et al.[2] in cooled reactor systems.

[2]R. Shinnar, F. J. Doyle, H. M. Budman, and M. Morari. Design considerations for tubular reactors with highly exothermic reactions, *AICHE J.*, **38**, 1729 (1992).

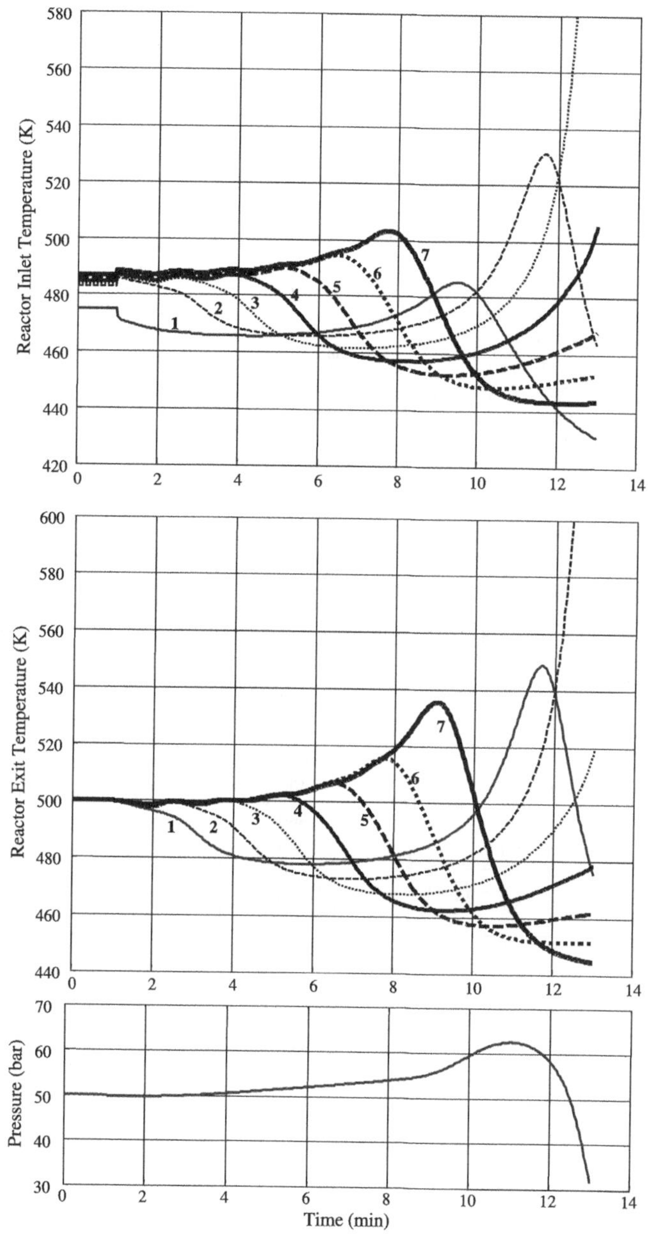

Figure 6.17 Openloop; seven-stage adiabatic reactors with cold-shot cooling; +20% F_R.

If the number of reactors is reduced, the interaction is less and the system can be made closedloop-stable. This is illustrated by considering a system with three adiabatic cold-shot reactors. The optimum steady-state design of the three-stage system is studied (steady-state conditions are given in Fig. 6.3).

The 3-stage system is still openloop unstable as shown in Figure 6.19. The same amplification and limit cycle behavior as found in the seven-stage system is observed.

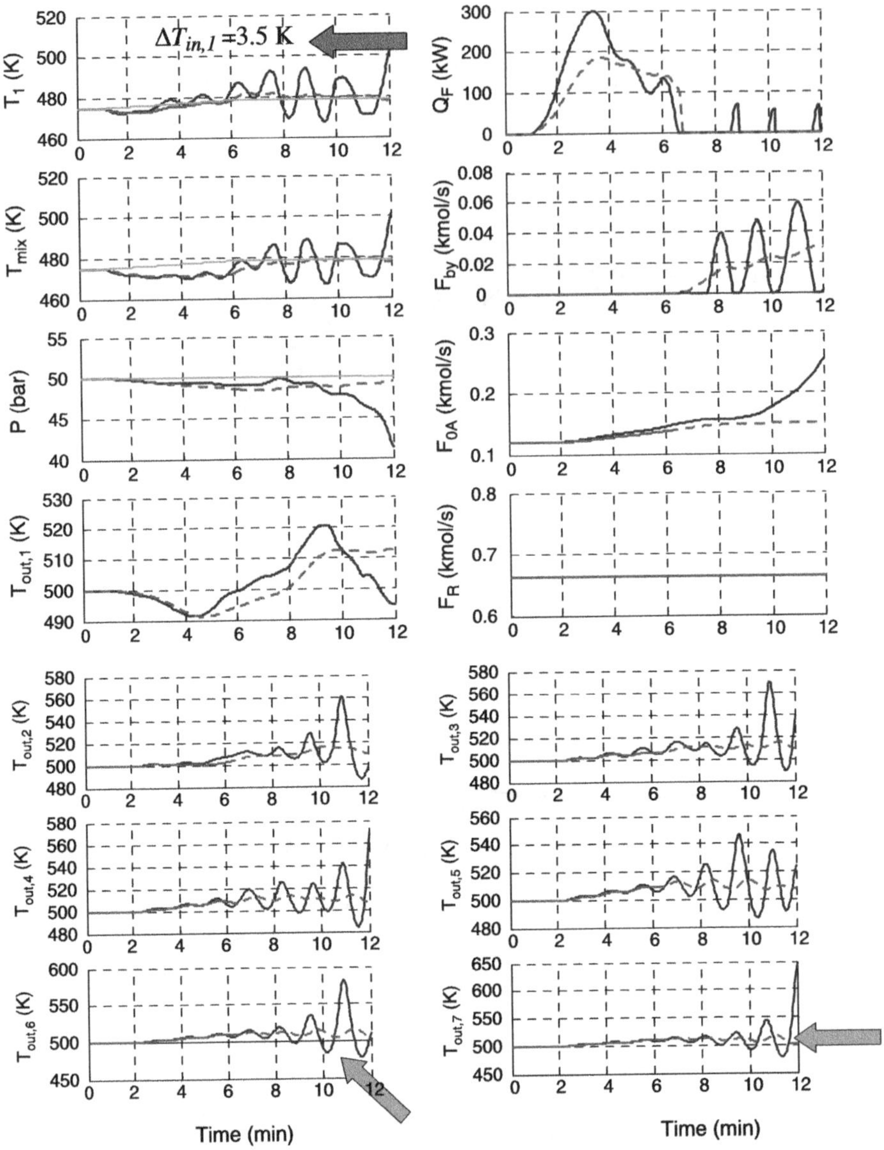

Figure 6.18 Closedloop; seven-stage adiabatic reactors with cold-shot cooling.

However, when the pressure, composition, and temperature controllers are put on automatic, the three-stage process can be made closedloop-stable. Figure 6.20 shows the response to ramped increases in the setpoints of the inlet reactor temperature controllers: 3.1 K for T_1, 4.6 K for T_2 and 5.8 K for T_3. The production rate is increased by 25%, and the new steady-state exit temperature is 510.9 K. When the setpoints of the inlet reactor temperature controllers are reduced (4.4 K for T_1, 5.9 K for T_2, and 7.1 K for T_3), the production rate is decreased by 25% and the new steady-state exit temperature is 487.3 K.

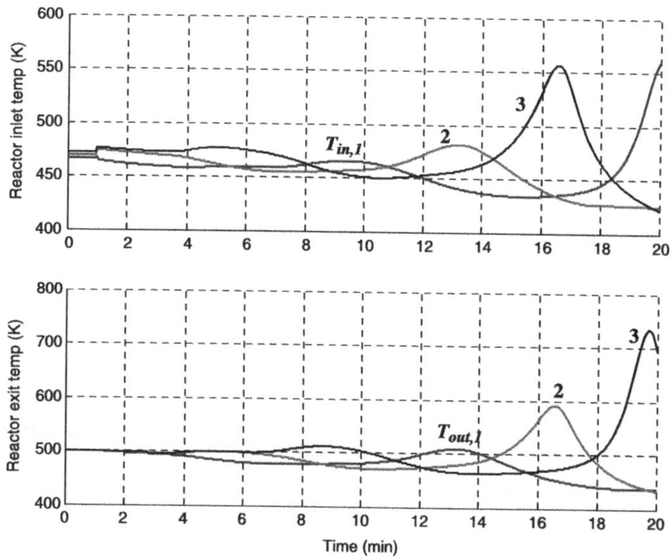

Figure 6.19 Openloop; three-stage adiabatic reactors with cold-shot cooling; +20% F_R.

Using ZN tuning gives oscillatory responses in the inlet temperatures, but using TL tuning gives effective control. Remember that ZN tuning was required for the single adiabatic reactor system, and it gave better response for the multistage reactor system with intermediate cooling. Those systems are openloop unstable and require tight control to stabilize. Large changes in the steam pressure in the intermediate coolers can be tolerated because they do not affect the rest of the process.

The cold-shot system is also openloop unstable, but it cannot tolerate large changes in the cold-shot flows because of the effect on the flow to the first reactor (with a fixed recycle flow). Therefore ZN tuning is too aggressive and gives oscillatory results for this system.

Figure 6.21 shows the results for −15% and −18% changes in recycle flow F_R, using TL tuning. When F_R is decreased 15%, the first reactor exit temperature increases significantly to 525 K. The second and third reactor exit temperatures increase to 504 and 502.5 K, respectively. The system can handle a 15% decrease in recycle flow, but when a larger decrease (18%) is made, the system goes unstable. The first reactor exit temperature increases when F_R is decreased. The cold-shot valve controlling the inlet temperature of the second reactor opens, which decreases the inlet flowrate to the first reactor even more. After about 40 min, the cold-shot valve (controlling F_{CS2} shown in Fig. 6.21) is wide open, and the system loses control of the inlet temperature. These results illustrate one of the inherent problems with cold-shot systems. If the total gas recycle is fixed, as it would be in most practical situations because of compressor limitations, changing cold-shot flows produces interaction.

The systems studied in this chapter uses gas recycle. If the recycle were liquid, the cold shot could be a liquid stream. This would mean that less cold-shot flow would be required since the latent heat of vaporization would be utilized for cooling. The gas flow to the first

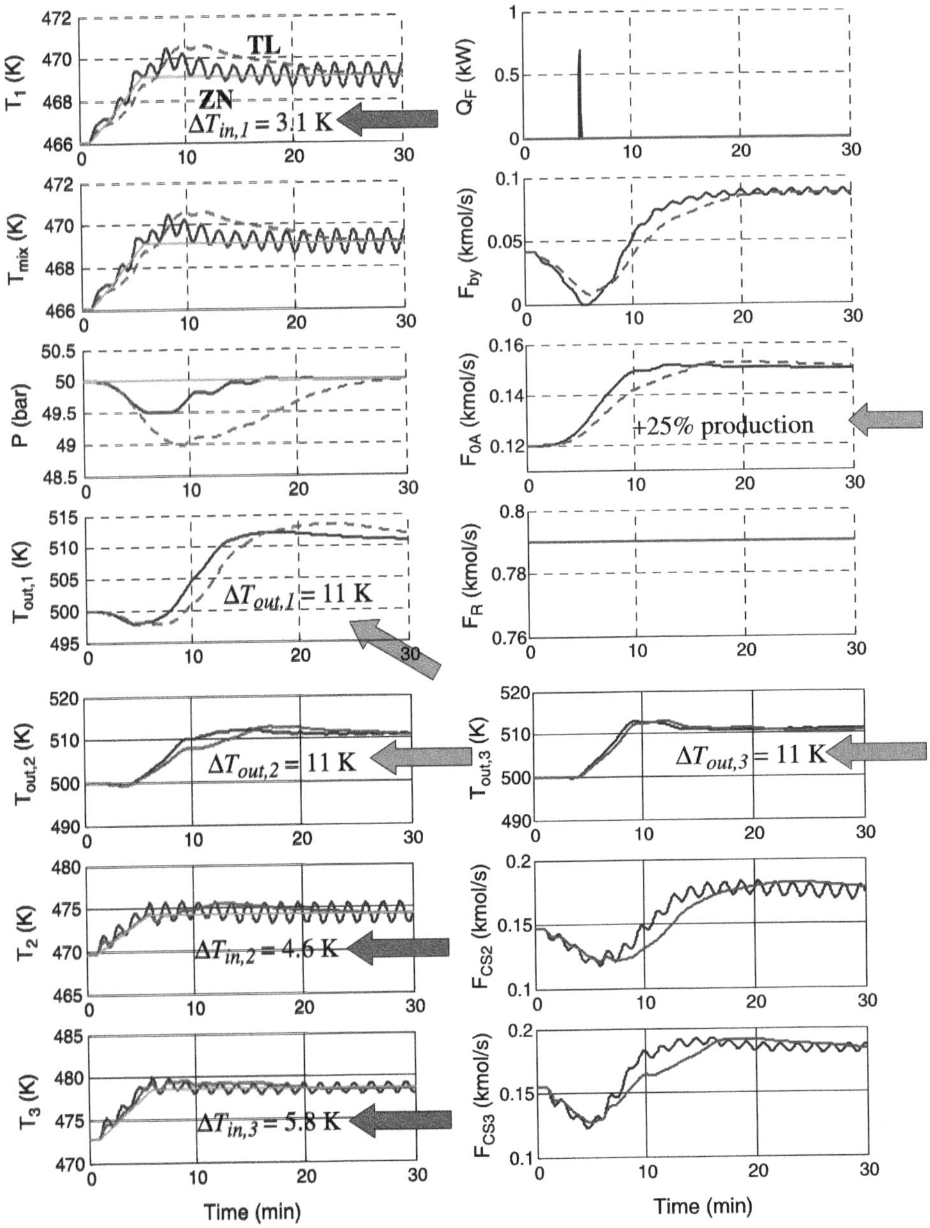

Figure 6.20 Closedloop; three-stage adiabatic reactors with cold-shot cooling; T_{in}^{set}.

reactor, probably coming from an upstream vaporizer, would not be affected by changes in the liquid cold-shot flows to the downstream reactors, which would come from a liquid surge tank somewhere in the system.

A system with liquid recycle would naturally occur when the vapor–liquid equilibrium is such that a simple flash drum cannot be used and a distillation column (or

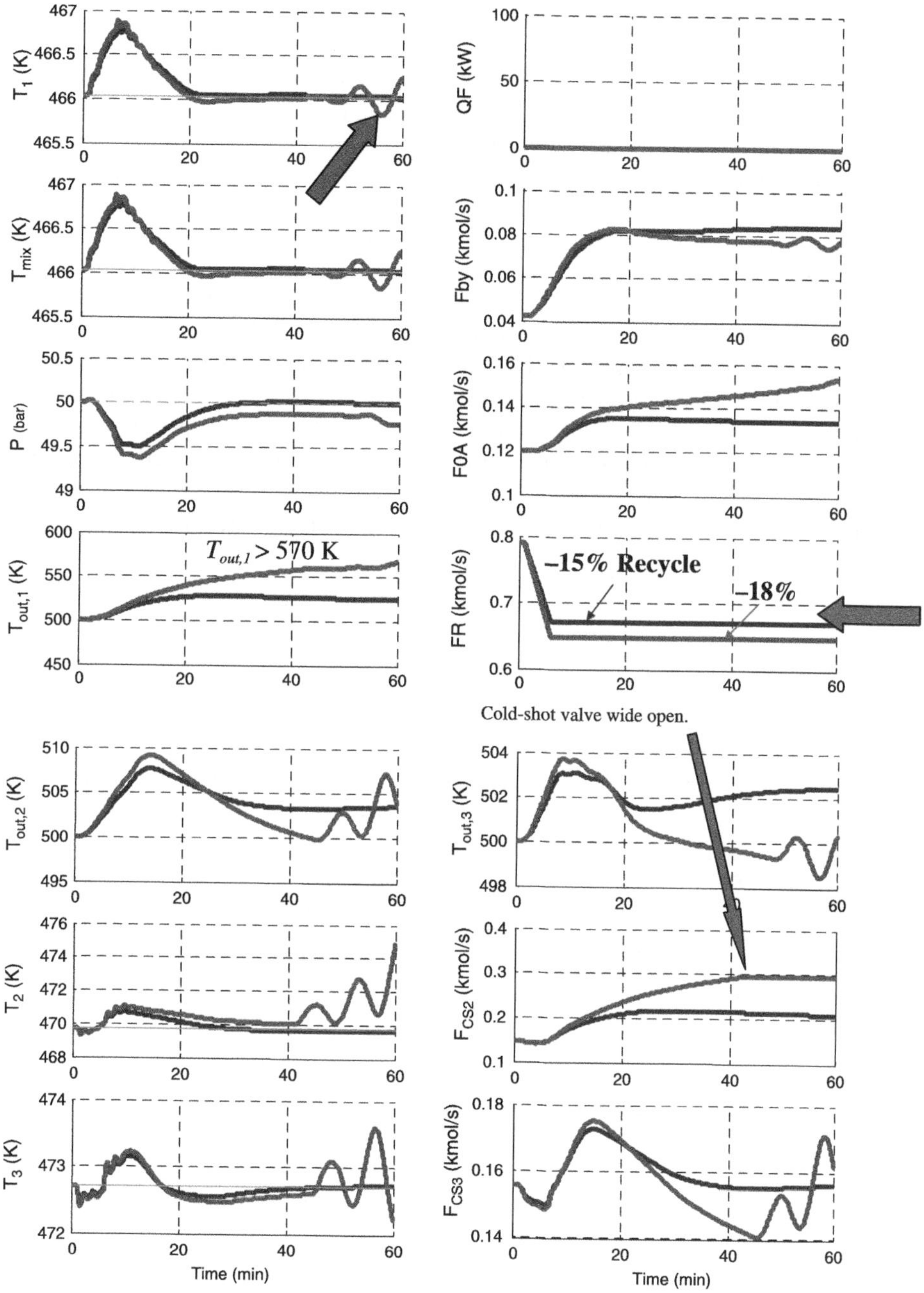

Figure 6.21 Closedloop; three-stage adiabatic reactors with cold-shot cooling; F_R changes.

some other separation unit) is required. This system would probably be more expensive from a steady-state economic standpoint because of the energy consumption in both the distillation column and in the vaporizer. However, it may have superior dynamics.

6.8 COOLED REACTOR SYSTEM

Openloop Response The openloop responses of the cooled reactor system for $\pm 20\%$ changes in recycle flowrate F_R are shown in Figure 6.22. Unlike all the adiabatic reactor systems, the cooled system is openloop-stable. The internal heat transfer provides some self-regulation so that disturbances do not grow. The reactor gain ($\Delta T_{out}/\Delta T_{in}$) is only ~ 1, compared to the adiabatic reactors with gains of 3 or 4. Cooling the reactor makes it less sensitive to changes in the inlet temperature and recycle flowrate.

Increasing recycle flow reduces the inlet, peak, and exit temperatures of the reactor. Pressure builds until the higher partial pressures of the reactants compensate for the lower specific reaction rate because of the lower temperatures. The higher velocities in the reactor tubes also increase the heat transfer coefficient, which means that the heat transfer rate does not decrease directly with the decrease in reactor temperatures. Remember, steam pressure (and temperature) is held constant in the openloop run. The net result of the various effects is that, with the fresh feed flowrates fixed, the reactor comes to a new steady-state condition, which has lower reactor temperatures but higher pressure. The net reaction rate and the heat transfer in the reactor remain the same. The

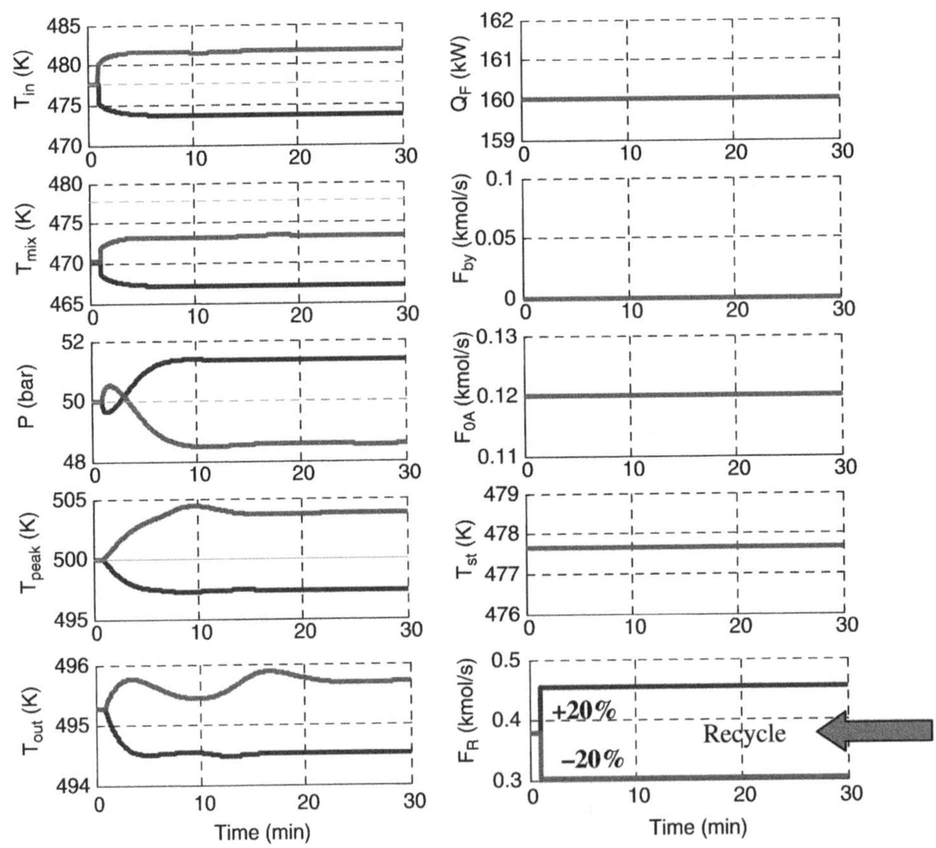

Figure 6.22 Openloop; cooled reactor; F_R changes.

decrease in temperature differential is compensated for by the increase in heat transfer coefficient.

Decreasing recycle flow has the opposite effects. Temperatures increase, pressure decreases, and the heat transfer coefficients decrease. Thus the cooled reactor system has some inherent self-regulatory properties that make it openloop-stable, at least for the set of kinetic and design parameters used in this example. Later in this chapter, a "hot" reaction system will be discussed that has a higher activation energy and larger specific reaction rate. As we will see, this new system is openloop-unstable.

Closedloop Response Figure 6.23 shows the response of the cooled reactor system to ramped increases and decreases of 10 K in the setpoint of the inlet temperature controller T_{in}. Raising the inlet temperature produces a decrease in reactor exit temperature of $\sim$2.5 K. The production rate increases by only 3%, which indicates that inlet temperature is a poor manipulated variable for production rate changes in this system.

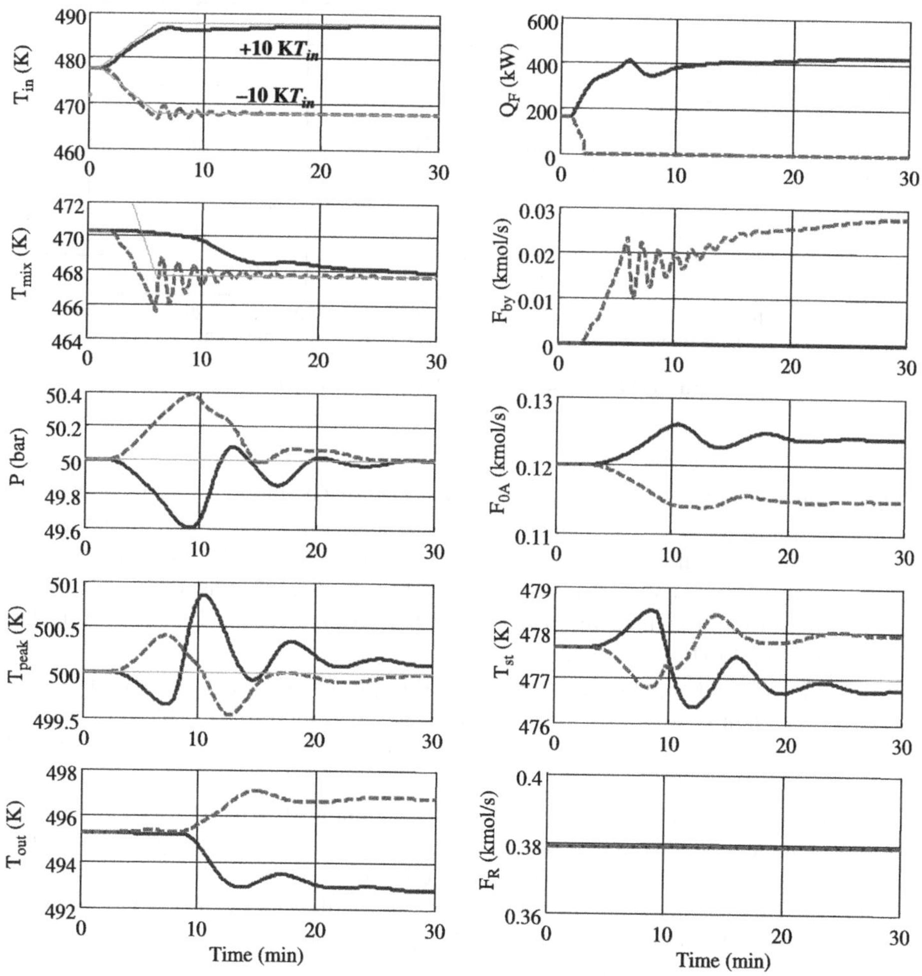

Figure 6.23 Closedloop; cooled reactor; $\pm T_{in}$.

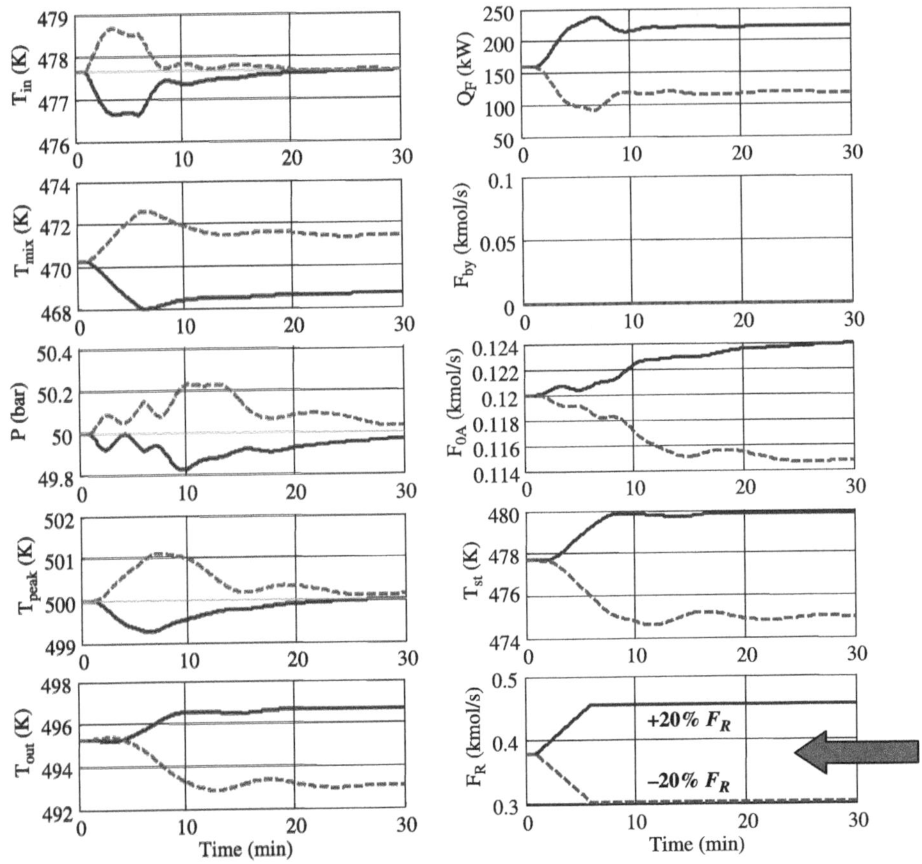

Figure 6.24 Closedloop; cooled reactor; F_R changes.

The results shown use TL tuning since the system with ZN tuning was found to be very oscillatory. Since the system is openloop-stable, aggressive controller tuning is not required. Aggressive tuning increases the interaction between the inlet temperature controller and the peak temperature controller. Changing steam temperature affects the temperature profile throughout the reactor, and the positive feedback of heat from the reactor exit to the FEHE can amplify oscillations.

When reactor inlet temperature is decreased 10 K, production rate is decreased by only 4% and the exit temperature increases by 1.6 K. Note that the furnace heat input goes to zero at about 4 min, and the inlet temperature is maintained by using bypass flow around the FEHE.

Figure 6.24 shows that the closedloop system can handle 20% increases and decreases in recycle flowrate F_R. Production rate changes by only $\sim$4%, indicating that changing recycle flow is not effective for changing production rate.

Figure 6.25 shows the responses to ramped changes of $+10.4$ K and -12.1 K in the setpoint of the peak temperature controller T_{peak}. Production rate is changed by 25%, which indicates that the peak temperature provides an effective production rate handle in the cooled reactor system.

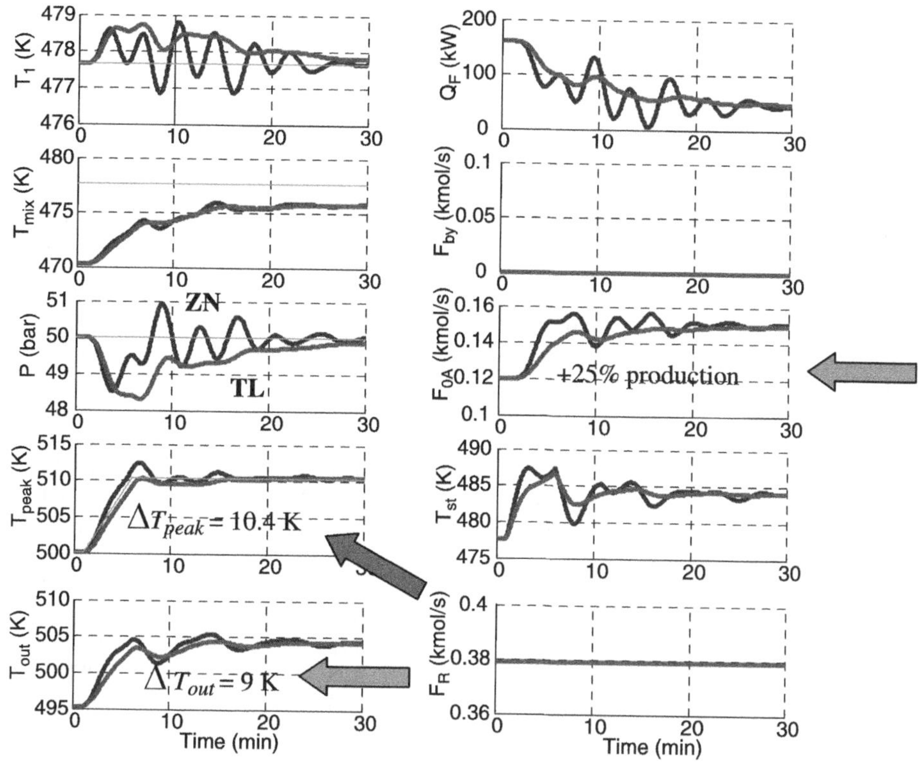

Figure 6.25 Closedloop; cooled reactor; T_{peak} changes.

6.9 COOLED REACTOR WITH HOT REACTION

The impact of reaction kinetics on the steady-state design and dynamic controllability of the cooled reactor system is studied in this section. All the results presented in previous sections used a moderate activation energy (69,710 kJ/kmol) and a moderate specific reaction rate. Now the activation energy is doubled ($E = 139{,}420$ kJ/kmol). In addition, the reaction rate at 500 K is increased by a factor of 4 (the preexponential factor $\alpha = 1.46 \times 10^7$ kmol s^{-1} bar^{-2} kg · cat^{-1}). These changes in reaction kinetics make the system highly nonlinear and very sensitive to changes in temperature.

6.9.1 Steady-State Design

Table 6.8 shows the optimum economic steady-state design for the "hot" reactor system when the catalyst cost is \$100/kg. The important steady-state design parameters for this hot reaction system are a total catalyst weight of 11,880 kg, a recycle flow of 0.27 kmol/s, a tube diameter of 0.0592 m, and a heat transfer area of 401 m^2. The design optimization variables used are the same as discussed in Chapter 5. The TAC of the optimum design is \$770,000 per year.

A comparison of this design with the moderate reaction design (Table 6.8) shows that the larger specific reaction rate produces a smaller reactor with less recycle, less heat transfer area, and lower TAC.

TABLE 6.8 Steady-State Design Results for the Cooled Reactor System

Case	Moderate Optimum	Hot Optimum	Hot Suboptimum
TAC ($\$10^6$ year^{-1})	1.57	0.77	0.84
D_{tube} (m)	0.1113	0.0592	0.0254
L_{tube} (m)	8.51	6.03	3.53
F_{tube} (kmol/s)	0.0033	0.0014	0.0002
y_A/y_B	1.017	1.024	1.025
W_{cat} (10^3 kg)	31.38	11.88	9.20
N_{tube}	190	357	2571
F_R (kmol/s)	0.38	0.27	0.33
F_{by} (kmol/s)	0	0	0
ΔP (bar)	0.75	1.23	0.50
P_s (bar)	46.75	46.27	47.00
$T_C = T_{in}$ (K)	477.33	474.73	483.42
$T_C = T_{out}$ (K)	495.25	494.83	491.09
U (kJ s^{-1} m^{-2} K^{-1})	0.246	0.36	0.31
V (m/s)	0.276	0.42	0.36
D_R (m)	2.17	1.58	1.82
A_R (m^2)	564	401	725
A_{HX} (m^2)	942	772	838
A_C (m^2)	217	178	197
W_{comp} (kW)	89.46	73.26	70.61
Q_R (10^3 kW)	2.40	2.43	2.64
Q_{HX} (10^3 kW)	3.35	2.74	2.98
Q_C (10^3 kW)	0.61	0.50	0.55
Q_F (10^3 kW)	0.15	0.09	0.34
Capital Cost ($\$10^6$)			
Reactor	0.90	0.72	1.06
Catalyst	3.13	1.19	0.92
Compressor	0.17	0.15	0.14
Furnace	0.15	0.10	0.28
FEHE	0.63	0.55	0.58
Condenser	0.72	0.64	0.68
Operating cost ($\$10^6$/year)			
Compressor	0.05	0.04	0.04
Furnace fuel	0.02	0.01	0.05
Steam	−0.41	−0.4	−0.47

The optimum design of the hot reaction system is found to be uncontrollable, as discussed in the following section. A suboptimum design using a smaller tube diameter (0.0254 m) is also shown in Table 6.8. The TAC of this system is about 10% higher than the optimum. However, the modified design provides much larger heat transfer area (725 m^2 vs. 401 m^2), which improves dynamic controllability. Note that the modified design has many more tubes (2571) than does the optimum design (357).

It is interesting to note that this suboptimal design uses less catalyst than the optimum design. This occurs because the average temperature in the reactor is higher since the inlet temperature is higher (483.4 K vs. 474.7 K). Figure 6.26 shows the temperature

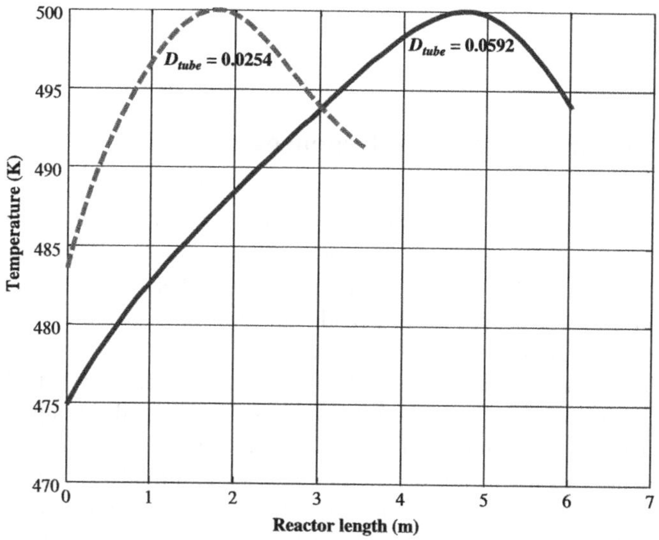

Figure 6.26 Two designs for cooled reactor with hot reaction.

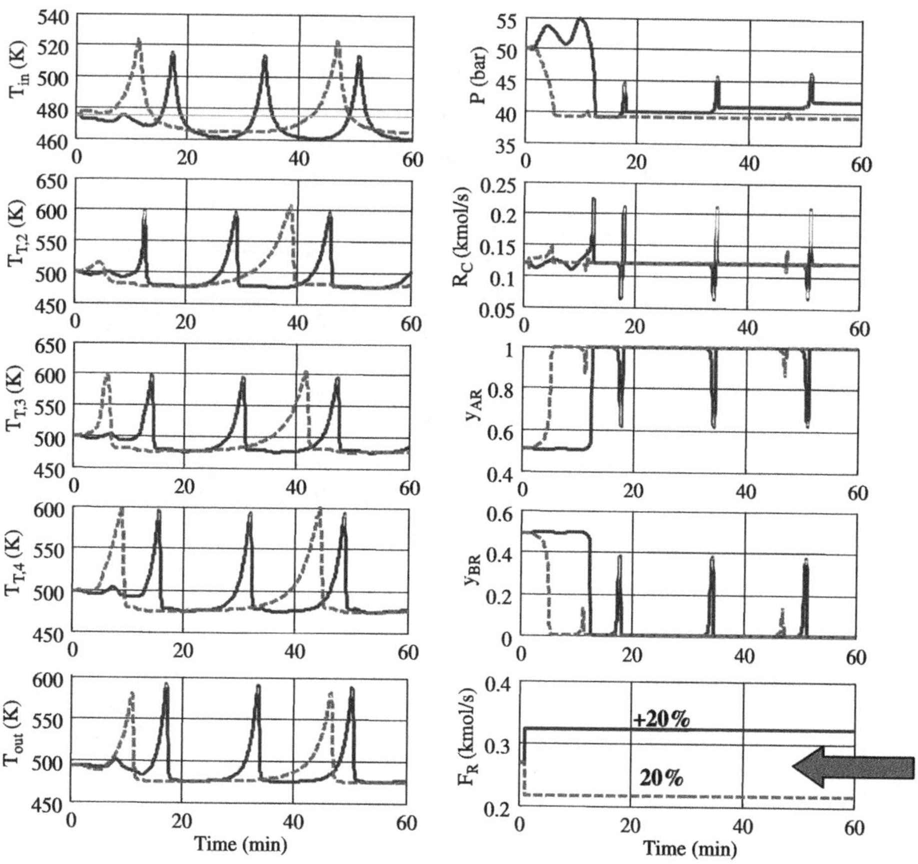

Figure 6.27 Openloop optimum cooled reactor with hot reaction; $\pm 20\%$ F_R.

profiles in the cooled tubular reactor for the optimum ($D_{tube} = 0.0592$ m) and suboptimum ($D_{tube} = 0.0254$ m) designs.

6.9.2 Openloop and Closedloop Responses

As shown in Figure 6.22, the cooled reactor with normal kinetics is openloop-stable. The cooled reactor with hot kinetics is openloop unstable, as shown in Figure 6.27. Results for two disturbances are shown: 20% decrease (dashed lines) and 20% increase (solid lines) in recycle flowrate. The system exhibits limit cycle behavior with periods that are ~20 min for an increase in recycle flow and ~40 min for a decrease in recycle flow. Temperatures go as high as 600 K in both cases.

Let us consider the increase in recycle flow. The higher flow initially decreases temperatures in the reactor. Pressure begins to build. Then the temperatures in the reactor start to increase because of the reaction rate increase due to higher pressure. At about 7 min the exit temperature increases to a value slightly above its steady-state level. The higher exit temperature increases the reactor inlet temperature through the FEHE, and this starts a temperature wave that moves down the reactor. Temperature $T_{T,2}$, which is located at about 40% of the way down the reactor, spikes first. Then

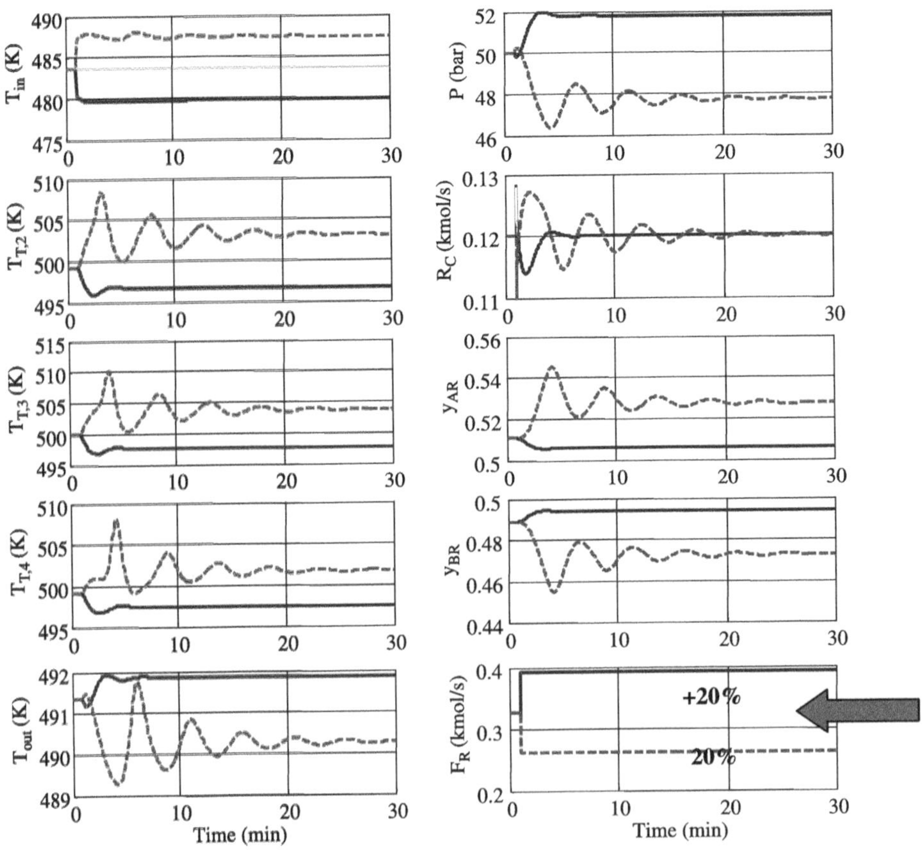

Figure 6.28 Openloop suboptimum cooled reactor with hot reaction; $\pm 20\%$ F_R.

temperature $T_{T,3}$, which is located at about 50% of the way down the reactor, spikes. The wave moves down the reactor and affects the exit temperature at about 17 min. Then the cycle repeats itself.

Decreases in recycle flow produce similar effects but in the reverse direction. The decrease in flow raises reactor temperatures, and a temperature wave starts to move down the reactor. Since the flowrate is lower, the temperature spike moves more slowly than when the recycle flowrate is increased. This explains the longer period of the cycles. These results demonstrate that the optimal design with the hot reaction is openloop unstable.

The openloop response of the suboptimal cooled reactor system (with smaller tubes) is shown in Figure 6.28. The much larger heat-transfer area makes this system openloop stable for a 20% increase or a 20% decrease in recycle flow. Note that the decrease produces an oscillatory response that eventually dies out. Remember that the system is openloop, so the oscillations are not due to the action of any controller.

However, if the decrease in recycle flow is 23%, the system becomes openloop unstable as shown in Figure 6.29. The system cannot handle large decreases in recycle flow because of its effect on the heat transfer coefficient U. At normal recycle flowrates, U is

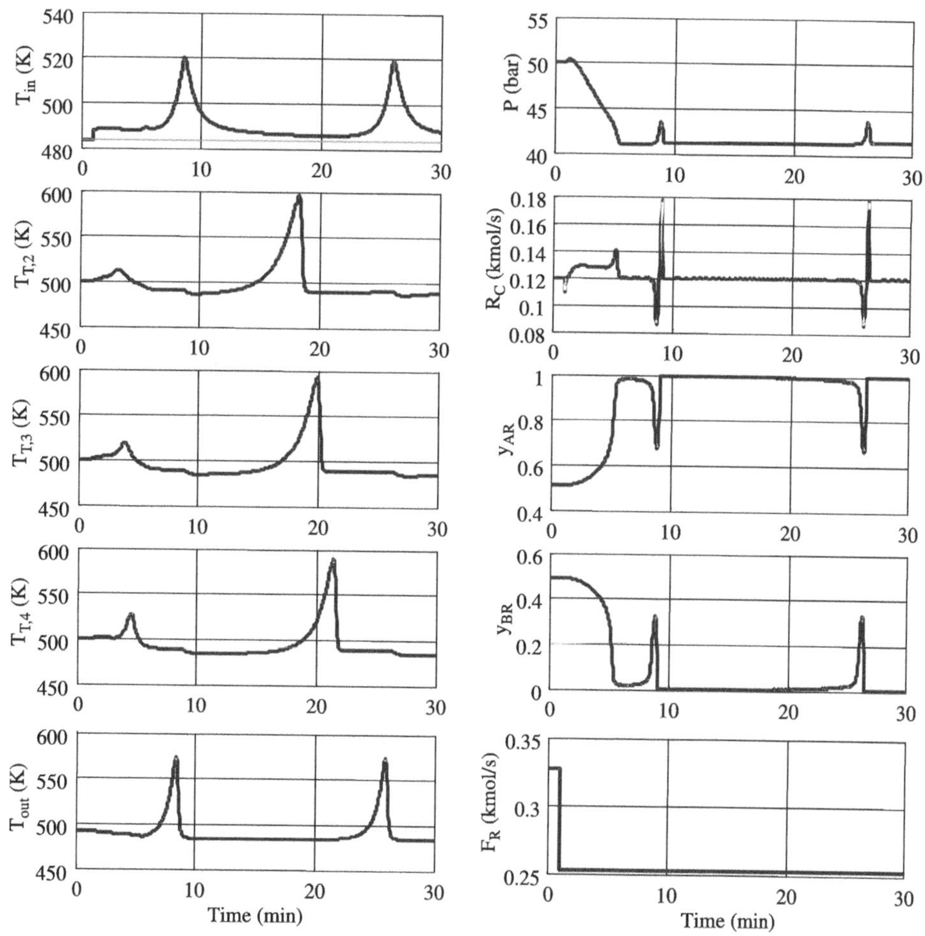

Figure 6.29 Openloop suboptimum cooled reactor with hot reaction; -23% F_R.

TABLE 6.9 Controller Tuning (Optimum Design with a Hot Reaction)

	ZN Tuning		TL Tuning	
Control Loop	K_C	τ_I (min)	K_C	τ_I (min)
P,F_{0A}	0.25	2.14	0.17	5.64
T_{mix},F_B	−5.30	1.00	−3.64	2.64
T_1,Q_F	6.96	2.81	4.79	7.42
T_{peak},T_{st}	0.42	2.63	0.29	6.93

TABLE 6.10 Controller Tuning (Suboptimum Design with a Hot Reaction)

	ZN Tuning		TL Tuning	
Control Loop	K_C	τ_I (min)	K_C	τ_I (min)
P,F_{0A}	0.29	2.05	0.20	5.42
T_{mix},F_B	−5.39	1.00	−3.71	2.64
T_1,Q_F	11.56	2.56	7.95	6.76
T_{peak},T_{st}	0.36	2.75	0.25	7.25

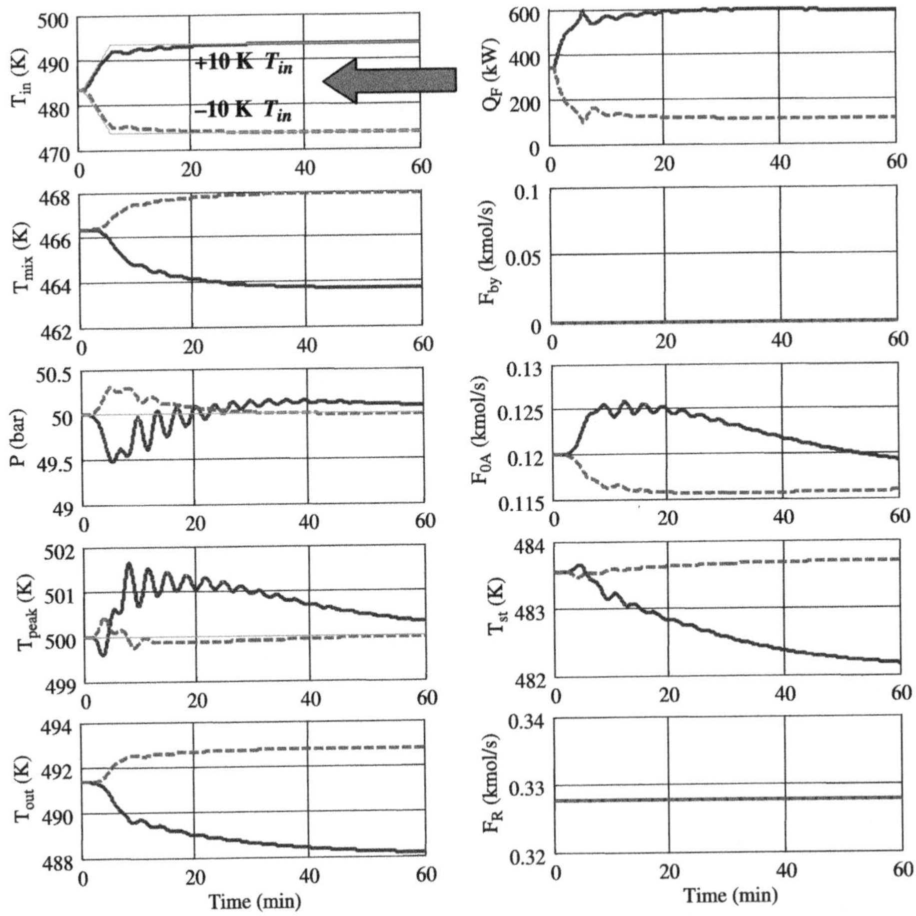

Figure 6.30 Closedloop; changes in T_{in}.

0.31 kJ s^{-1} m^{-2} K^{-1}. When the flowrate is decreased by 23%, U is about 15% lower. Control problems are expected with this system in this situation.

Table 6.9 gives the tuning parameters for the optimum design. There is difficulty performing relay–feedback tests on this optimum design because of its high sensitivity as seen in the openloop behavior. The tuning parameters are obtained using the very early period of relay–feedback tests. The system runs away if the relay–feedback tests are prolonged. The closedloop system could not be stabilized for a +5 K T_1 ramped setpoint change.

Table 6.10 gives the tuning parameters for the suboptimum design. The ZN peak temperature controller gain is 0.36 compared to 2.67 found for the cooled reactor with moderate kinetics studied in previous sections. The low controller gain of the peak temperature controller is required because of the high sensitivity of the reactor with the hot reaction. The lower gain of the pressure controller is due to the smaller gas volume of the hot reactor system.

Figure 6.30 shows the responses to a +10 K and −10 K T_1 ramped setpoint changes using TL tuning. The system with ZN tuning is unstable. The production rate is increased by only 3%.

Figure 6.31 shows the results for a 20% increase or decrease in recycle flowrate. The production rate is changed by no more than 4%. Figure 6.32 shows the response to an

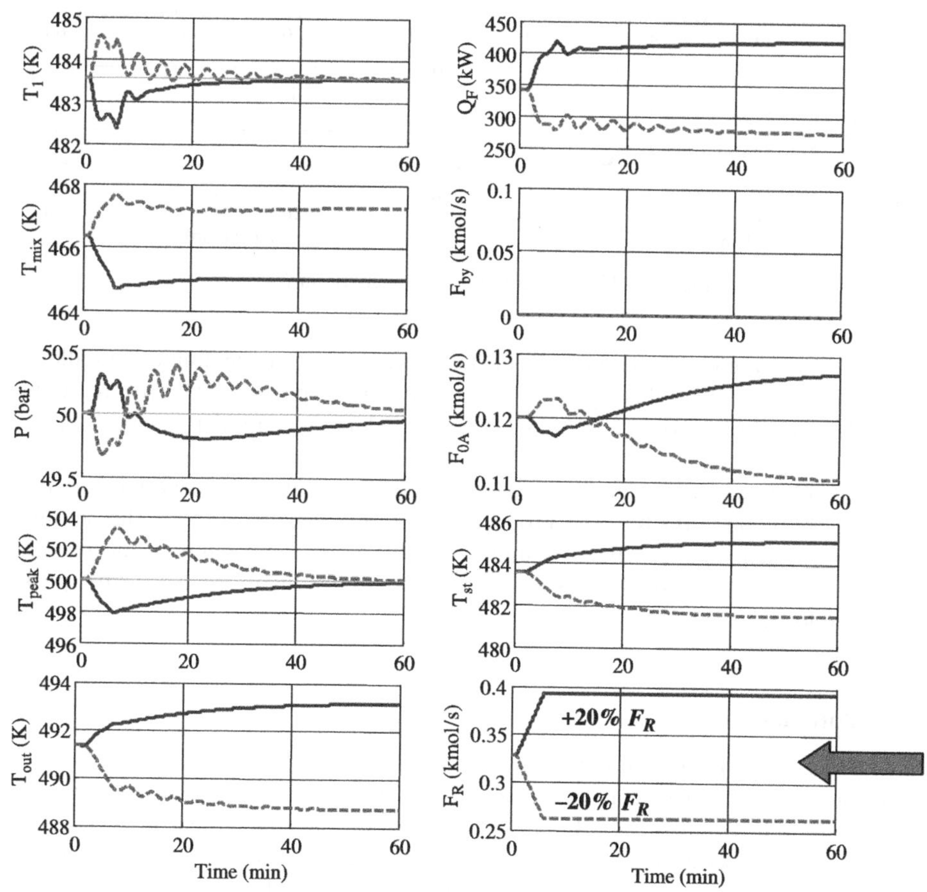

Figure 6.31 Closedloop; ±20% changes in F_R.

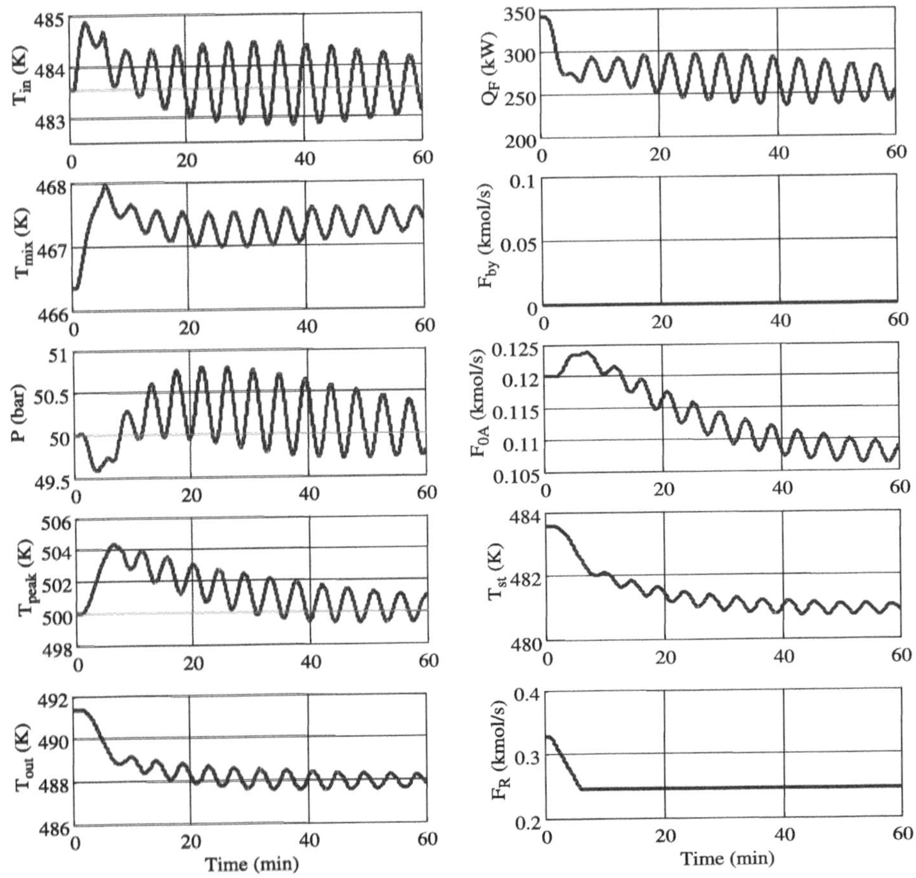

Figure 6.32 Closedloop; 25% decrease in F_R.

even greater decrease in recycle flow (25%). The system is now at the limit of closedloop stability. Remember that this system is openloop-unstable for a 23% decrease in recycle. The effect of flowrate on U is so large that adding a controller cannot stabilize the system.

Figure 6.33 shows the responses to changes in the setpoint of the peak temperature controller (+6.0 and −6.8 K). The production rate is changed by 25%. The settling time of the system is more than 40 min when the peak temperature is decreased. The long settling time is the result of low controller gain used for the peak temperature loop.

6.9.3 Conclusion

The dynamics of four different tubular reactor systems have been explored. The cooled reactor has the best dynamic response because of the internal heat transfer helps to provide some self-regulation. As discussed in Chapter 5, the cooled reactor is also the best from the standpoint of steady-state economics. This combination of both superior steady-state and dynamic performance is quite unusual in chemical processes. The typical situation is that there is a conflict between economics and dynamics.

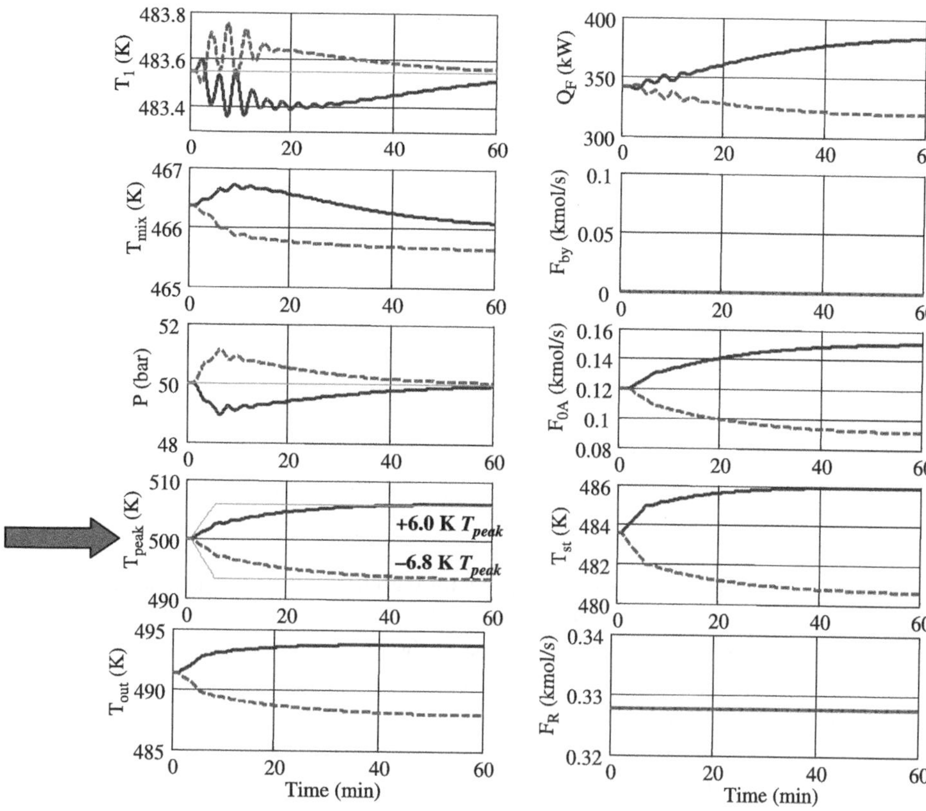

Figure 6.33 Closedloop; changes in T_{peak}.

It might be wise to point out that there can be other problems with a cooled reactor that will influence the selection of reactor type. An adiabatic reactor with a bed of catalyst is certainly mechanically straightforward to construct and maintain. Catalyst is easily loaded or discharged. A cooled reactor with multiple parallel tubes is mechanically more complex. Loading and emptying the tubes of catalyst can be difficult.

The issue of flow maldistribution (unequal flows through the parallel flow paths) can be important in some reactors.

6.10 ASPEN DYNAMICS SIMULATION

The steady-state designs of four types of tubular reactors were illustrated in Chapter 5. Now we want to explore the dynamics and control of each of these reactors.

6.10.1 Adiabatic Reactor with and without Catalyst

The numerical example used in Chapter 5 has two reactions. The feed is 0.025 kmol/s of a mixture of propylene (90 mol%) and chlorine (10 mol%) at 400 K. The adiabatic reactor is 1 m in diameter and 10 m in length, giving a total volume for reaction of $\pi(1)^2(10)/4 = 7.854\ \text{m}^3$ if the reactor is empty (containing no catalyst). The temperature of the reactor effluent leaving the reactor is 583 K.

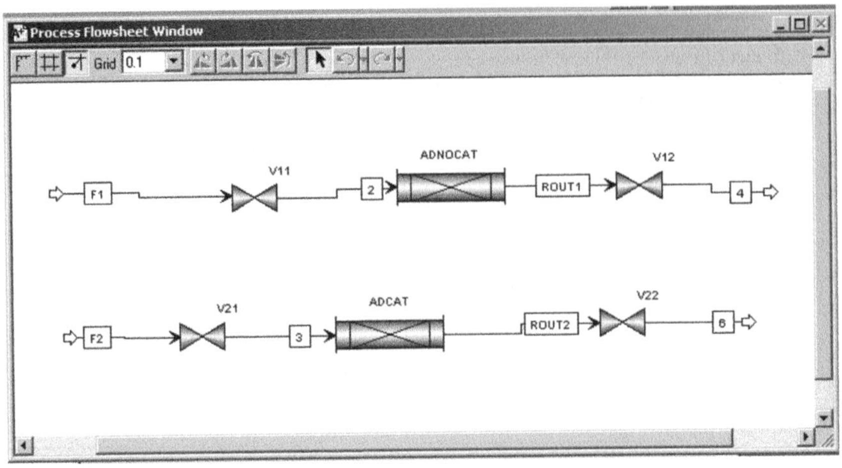

Figure 6.34 Flowsheet with two adiabatic reactors.

Two cases are considered: an adiabatic reactor with no catalyst and an adiabatic reactor with catalyst. The void volume of the catalyst is 0.5, so the total volume of the reactor with catalyst must be twice as large as the volume of the reactor without catalyst. The reactor length is increased to 20 m in this case. The density of the solid catalyst is 2000 kg/m^3, so the total amount of catalyst in the reactor is

$$\frac{\pi D^2}{4} L(\varepsilon)\rho_{cat} = \frac{\pi(1)^2}{4}(20)(0.5)(2000) = 15{,}710\,\text{kg} \tag{6.15}$$

Figure 6.34 shows the Aspen Plus flowsheet with these two adiabatic reactors installed. The empty reactor is 10 m in length. The catalyst-filled reactor is 20 m in length. The reactor effluents for the two cases are identical. Control valves are installed on the gas feedline and the gas reactor effluent line. Figure 6.35 shows the *Catalyst* page tab window under *Setup* for the reactor with catalyst. The catalyst properties are specified.

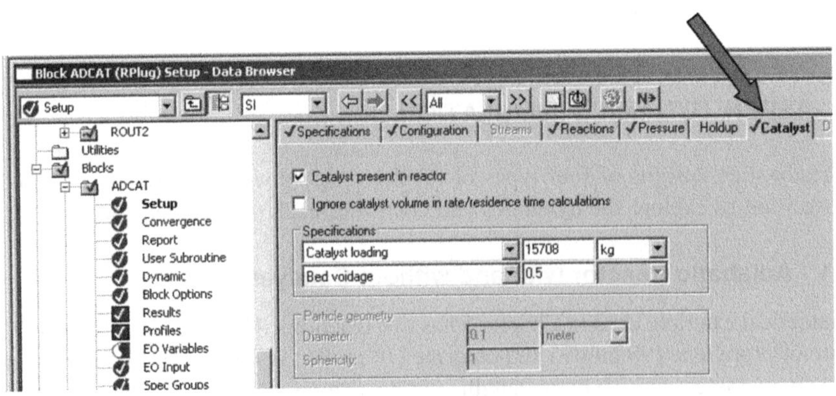

Figure 6.35 Catalyst properties.

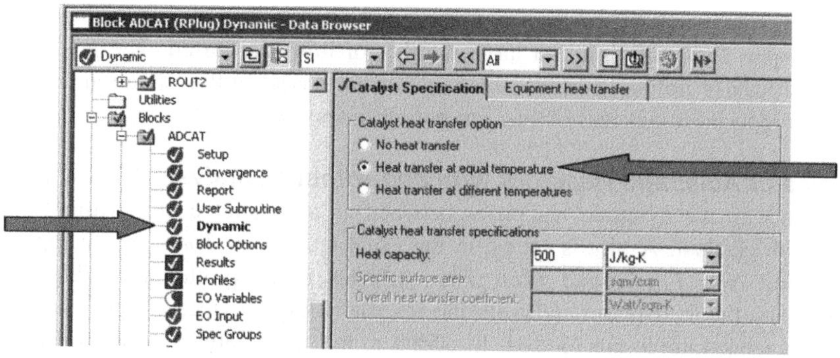

Figure 6.36 Specifying heat transfer between gas and catalyst.

Figure 6.36 shows the window that opens when *Dynamics* is selected under the reactor block. The amount of catalyst, the catalyst heat capacity, and the heat transfer coefficient between the process gas and the catalyst are specified.

The program in Aspen Plus is run and pressure-checked. It is then exported to Aspen Dynamics as a pressure-driven dynamic simulation as was done in Chapter 3 with CSTRs. The Aspen Dynamics file is opened, giving the window shown in Figure 6.37. The default control scheme has a pressure controller manipulating the valve in the reactor exit line. The simulation is run until all variables stop changing.

Effect of Number of Lumps If the number of plotting points in Aspen Plus is set at 10 (the default), the resulting exit temperature from the reactors under steady-state conditions in Aspen Dynamics is 578 K. Remember that it should be 583 K from the rigorous integration of the ordinary differential equations describing the steady-state tubular reactor that are used in Aspen Plus. Changing the number of points to 20 produces an exit temperature of 580 K. Changing the number of points to 50 produces an exit temperature of 582 K, which is very close to the correct value. Therefore a 50-lump model should be used.

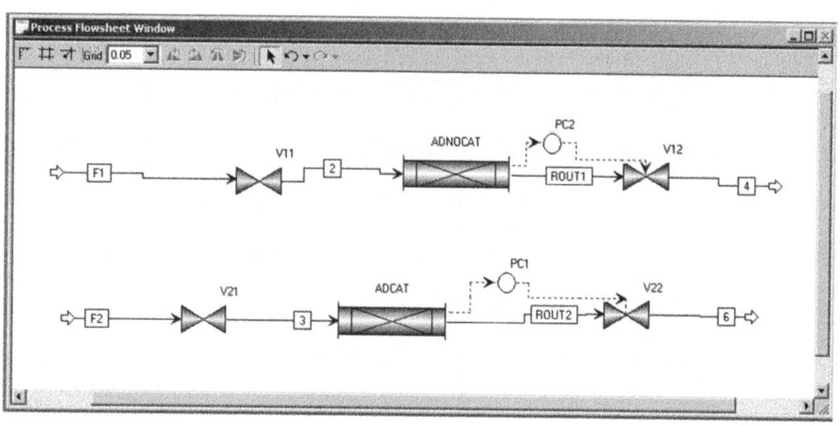

Figure 6.37 Aspen dynamics flowsheet with default controllers.

However, some numerical difficulties are sometimes encountered in running the simulation in Aspen Dynamics when a large number of lumps are used. The 50-lump case runs very slowly or not at all in the adiabatic reactor cases. In this situation the number of lumps is reduced to get reasonable computing times.

Dynamics of Adiabatic Reactors with and without Catalyst We want to show the difference in the dynamic responses between an adiabatic reactor without catalyst and one with catalyst. A plot is set up in the usual way using *Tools* on the top toolbar and selecting *New Forms* and *Plot.* The variables to be plotted are "dragged and dropped" from the feedstream and from the reactor block. The feed temperature and the reactor exit temperatures are plotted. In addition, the temperatures at several locations down the length of the reactor with catalyst are plotted: the temperatures at 5, 10, and 15 m in the 20 m reactor containing the catalyst. The models of the two adiabatic reactors use 20 lumps.

The feed temperature is changed from 400 to 420 K by selecting the feedstream and clicking *Forms* and *Manipulate*, as shown in Figure 6.38. Dynamic response results are shown in Figure 6.39. The feed temperature is increased 20 K at time = 0.1 h. The exit temperature of the reactor without catalyst $(T_{out})^{nocat}$ increases from 582 to 595 K very quickly since the gas residence time in the reactor is only 20 s.

However, the exit temperature of the reactor with catalyst $(T_{out})^{cat}$ takes about 2 hours to attain the same steady state. And in fact, it initially actually *decreases*! The "inverse response" or "wrongway effect" is caused by the thermal capacitance of the catalyst.

The dynamic responses of the internal temperature help to explain this unusual phenomenon. When the inlet temperature increases, it takes some time for the temperature at 4 m (20% down the length of the 20-m reactor) to begin to increase. When this occurs, more of the chlorine is consumed in this section of the reactor. This reduces the chlorine concentration in the gas flowing through the rest of the catalyst bed, which reduces the

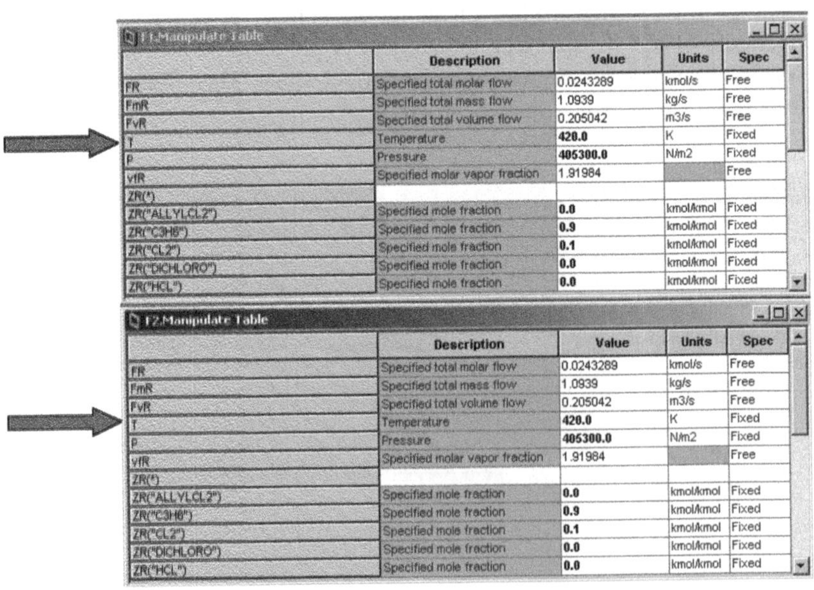

F1-Manipulate Table

	Description	Value	Units	Spec
FR	Specified total molar flow	0.0243289	kmol/s	Free
FmR	Specified total mass flow	1.0939	kg/s	Free
FvR	Specified total volume flow	0.205042	m3/s	Free
T	Temperature	420.0	K	Fixed
P	Pressure	405300.0	N/m2	Fixed
vfR	Specified molar vapor fraction	1.91984		Free
ZR(*)				
ZR("ALLYLCL2")	Specified mole fraction	0.0	kmol/kmol	Fixed
ZR("C3H6")	Specified mole fraction	0.9	kmol/kmol	Fixed
ZR("CL2")	Specified mole fraction	0.1	kmol/kmol	Fixed
ZR("DICHLORO")	Specified mole fraction	0.0	kmol/kmol	Fixed
ZR("HCL")	Specified mole fraction	0.0	kmol/kmol	Fixed

F2-Manipulate Table

	Description	Value	Units	Spec
FR	Specified total molar flow	0.0243289	kmol/s	Free
FmR	Specified total mass flow	1.0939	kg/s	Free
FvR	Specified total volume flow	0.205042	m3/s	Free
T	Temperature	420.0	K	Fixed
P	Pressure	405300.0	N/m2	Fixed
vfR	Specified molar vapor fraction	1.91984		Free
ZR(*)				
ZR("ALLYLCL2")	Specified mole fraction	0.0	kmol/kmol	Fixed
ZR("C3H6")	Specified mole fraction	0.9	kmol/kmol	Fixed
ZR("CL2")	Specified mole fraction	0.1	kmol/kmol	Fixed
ZR("DICHLORO")	Specified mole fraction	0.0	kmol/kmol	Fixed
ZR("HCL")	Specified mole fraction	0.0	kmol/kmol	Fixed

Figure 6.38 Manipulating the feed temperature.

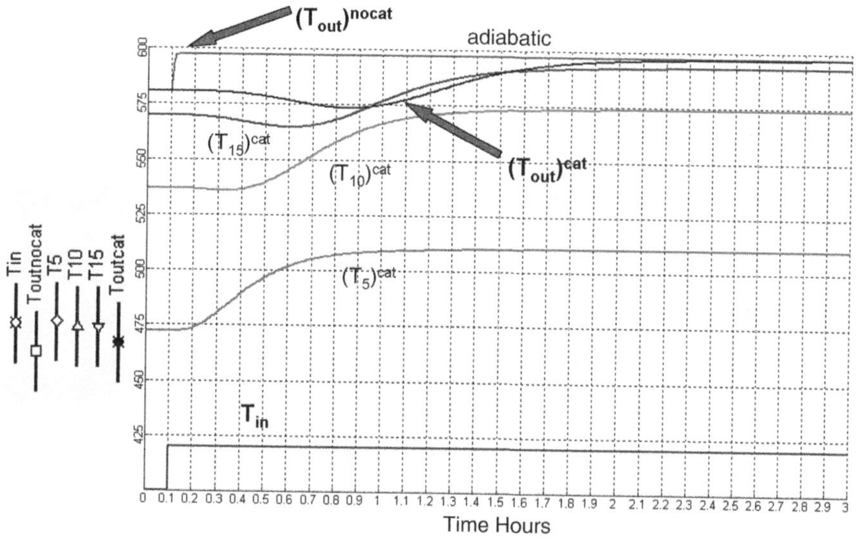

Figure 6.39 Response to feed temperature +20 K.

reaction rate. Therefore the temperatures further down the bed begin to temporarily *decrease* because less heat is being generated in these sections by the exothermic reactions. Remember that the gas is flowing through the bed very rapidly, but the temperature of the solid catalyst is changing slowly because of its large mass.

These results illustrate the drastic effect of catalyst on the dynamics of tubular reactors. It is clear that a control structure that attempts to control reactor outlet temperature by adjusting reactor inlet temperature will exhibit poor performance because of the inverse response.

6.10.2 Cooled Tubular Reactor with Coolant Temperature Manipulated

One of the three options considered in Chapter 5 was a cooled tubular reactor with a coolant temperature that is the same down the length of the bed. With this type of system, the temperature of the coolant can be used as the manipulated variable to control some variable in the reactor. We will illustrate the control of the peak temperature by using several temperature measurements at different locations and selecting the highest to feed to a temperature controller as the process variable PV signal. The output signal OP of this controller will be the coolant temperature.

The reactor tubes are filled with catalyst, so the number of tubes is increased from the 250 used in Chapter 5 for the empty reactor case to 500 tubes. The tubes are 0.1 m in diameter and 10 m in length. The coolant temperature is 400 K at steady state. A 30-lump model is used and runs with no difficulty in Aspen Dynamics.

The design value of the overall heat transfer coefficient U between the process gas and the liquid coolant used in Chapter 5 was 142 J s^{-1} K^{-1} m^{-2}. The resulting temperature profile is shown at the top of Figure 6.40. There is a sharp peak that occurs very near the inlet end of the reactor. This peak temperature is only about 5 K higher than the coolant because of the large value of U. In order to be able to demonstrate the use of a peak temperature controller, the value of U is drastically decreased to 14.2 J s^{-1} K^{-1} m^{-2}, which changes the temperature profile to that shown at the bottom of Figure 6.40. Now the

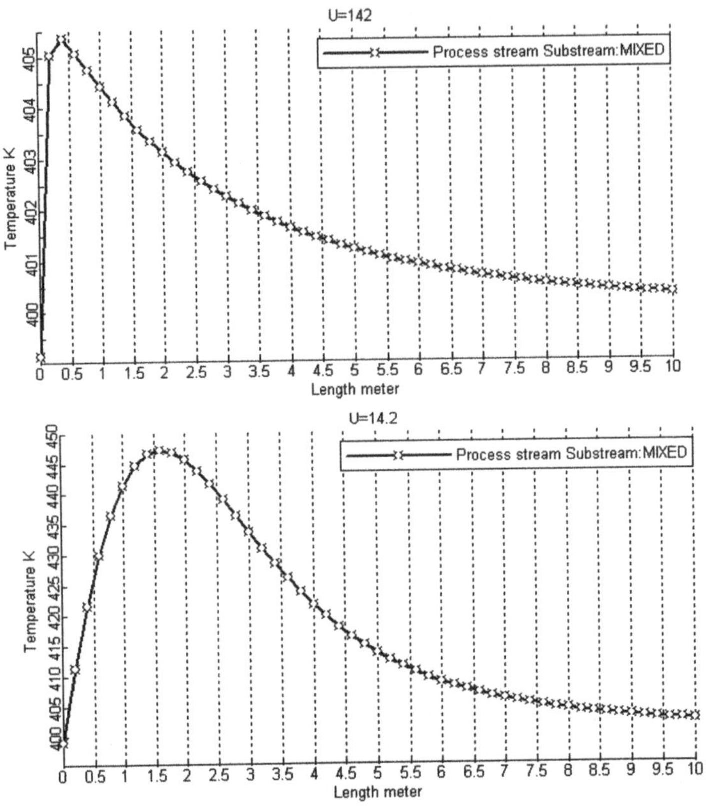

Figure 6.40 Temperature profiles for large and small *U*.

peak is about 45 K above the coolant temperature and is located at about 2 m down the 10-m reactor. A controller is installed on the flowsheet and labeled "TC_{peak}." The coolant temperature in the reactor block (called "COOLANT") is selected as its output signal OP of this controller (see Fig. 6.41).

The next issue is how to select the internal temperatures in the reactor. These are not available as normal output signals for the reactor block when you drag a control signal to the flowsheet. The temperatures can be seen by clicking the reactor icon, selecting *Forms* and *All Variables*. With a 30-lump model, there are 30 temperatures. The temperature at 2 m down the 10 m reactor is T(6).

Aspen Dynamics has the capability of using "flowsheet equations" for specifying a desired relationship. We illustrate this by setting up an equation that defines the PV signal to a temperature controller as T(6). The reactor block is *COOLANT*, so the T(6) temperature is *BLOCKS("COOLANT").T(6)*.

The upper left window shown in Figure 6.41 is the *Exploring* window. Clicking *Flowsheet* in the list produces the *Contents of Flowsheet* window, in which there is an icon with two parallel blue bars, which is labeled "Flowsheet." Double-clicking this icon opens the *Text Editor* shown at the top of Figure 6.42. The equation shown is entered to specify that the PV signal of the temperature controller TC_{peak} is the temperature T(6) from Lump 6 of the 30-lump reactor model (2 m from the inlet):

$$BLOCKS(\text{"}TCpeak\text{"}).pv = BLOCKS(\text{"}COOLANT\text{"}).T\,(6) + 273;$$

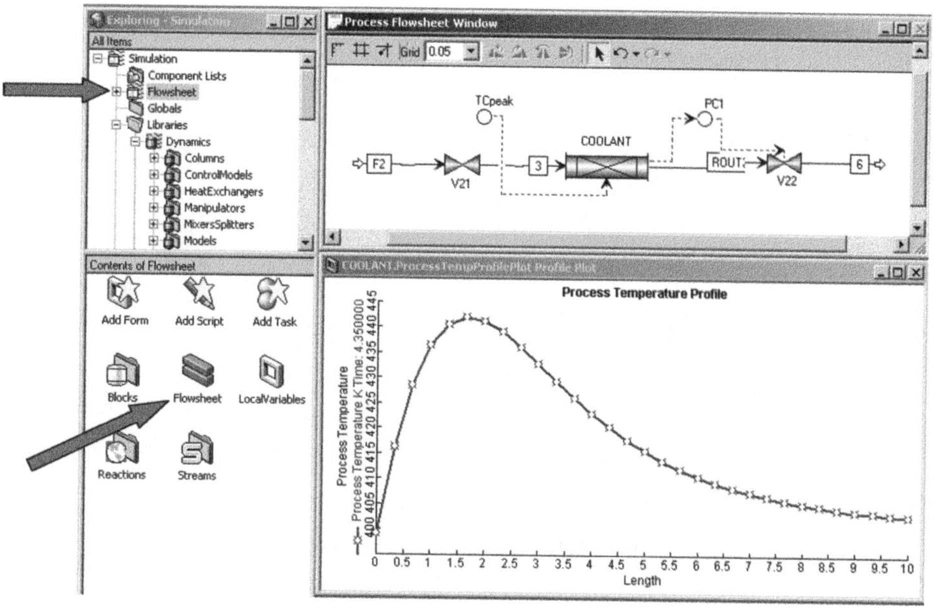

Figure 6.41 Using flowsheet equations.

The "273" converts the temperature that is picked up from the reactor (in Centigrade) to the correct Kelvin scale used in the controller. All variables in the flowsheet equations use metric units, so temperature is in degrees Centigrade. Right-click the *Text Editor* window and select *Compile*. If everything is okay, the message shown in the *Simulation Messages* window at the bottom of Figure 6.42 indicates no errors.

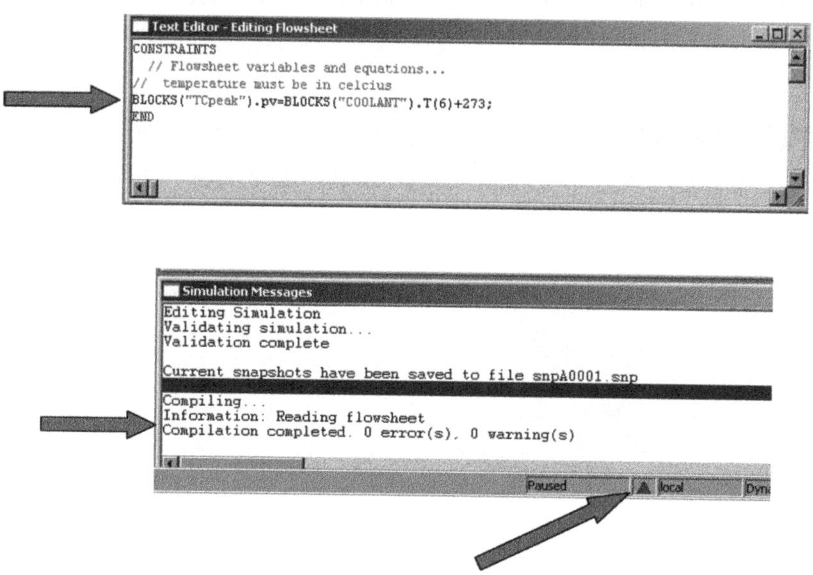

Figure 6.42 Flowsheet equations.

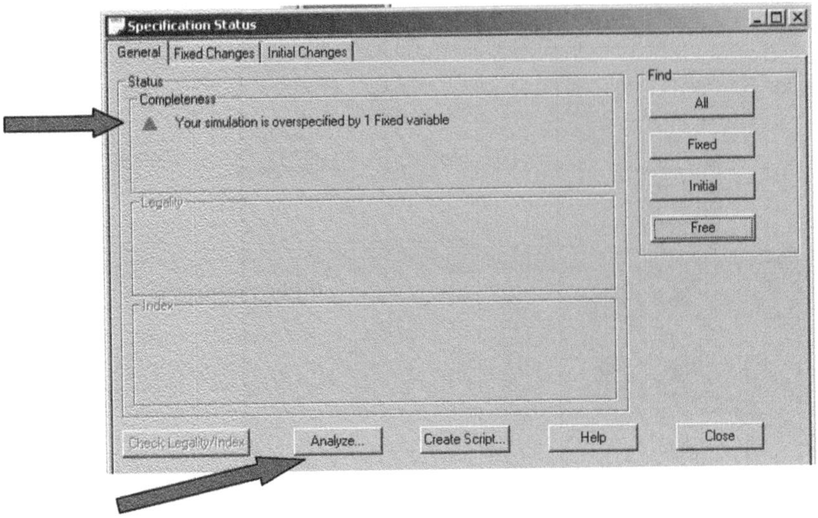

Figure 6.43 Simulation overspecified.

However, the little red triangle at the very bottom of the window appears instead of the normal green triangle. Double-clicking the triangle opens the window shown in Figure 6.43, which says that the system is overspecified by one variable, that is, that one variable must be made *Free* instead of *Fixed*. Clicking on the *Analyze* button opens the window shown in Figure 6.44. Aspen Dynamics suggests that the composition of the feed be made free. Of course, this is not what we want to do. The problem is that the PV input signal to the controller is not connected on the flowsheet (see Fig. 6.41), so the simulation assumes that this variable is *Fixed*. To change it to *Free*, click the controller icon and select *Forms* and *All Variables*. As shown in Figure 6.45, change the PV to *Free*.

An initialization run is made, and the controller is initialized and ranges provided for the input and output. The inlet temperature range for the reactor temperature is

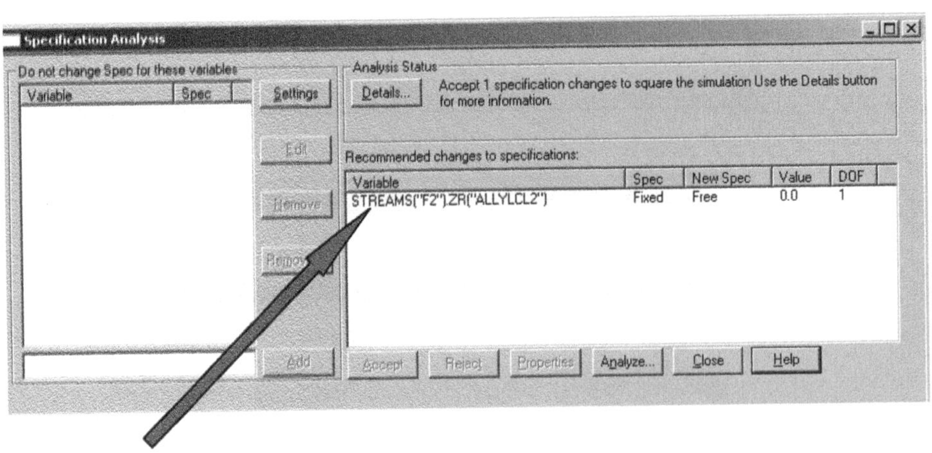

Figure 6.44 Suggested change.

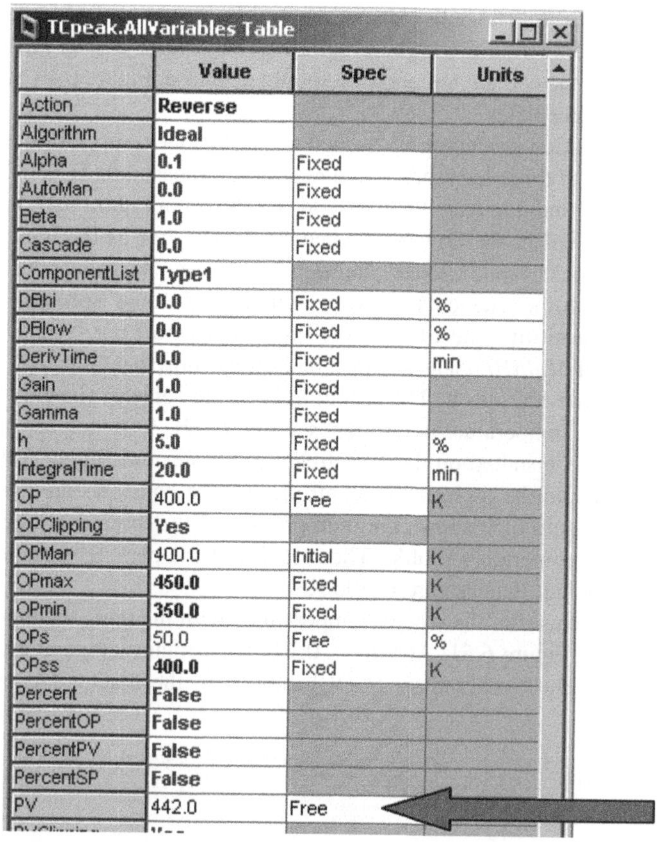
TCpeak.AllVariables Table

	Value	Spec	Units
Action	Reverse		
Algorithm	Ideal		
Alpha	0.1	Fixed	
AutoMan	0.0	Fixed	
Beta	1.0	Fixed	
Cascade	0.0	Fixed	
ComponentList	Type1		
DBhi	0.0	Fixed	%
DBlow	0.0	Fixed	%
DerivTime	0.0	Fixed	min
Gain	1.0	Fixed	
Gamma	1.0	Fixed	
h	5.0	Fixed	%
IntegralTime	20.0	Fixed	min
OP	400.0	Free	K
OPClipping	Yes		
OPMan	400.0	Initial	K
OPmax	450.0	Fixed	K
OPmin	350.0	Fixed	K
OPs	50.0	Free	%
OPss	400.0	Fixed	K
Percent	False		
PercentOP	False		
PercentPV	False		
PercentSP	False		
PV	442.0	Free	

Figure 6.45 Making *PV* a free variable.

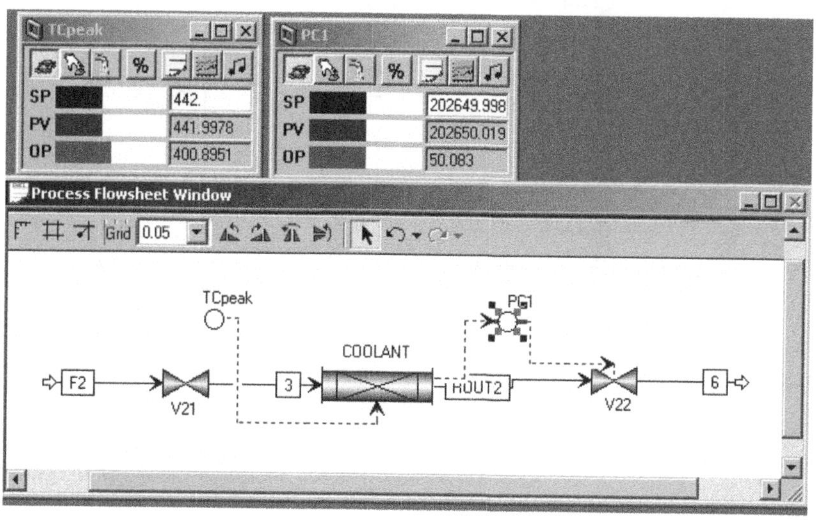

Figure 6.46 Flowsheet and controller faceplates.

specified to be 400–500 K. The outlet temperature range for the coolant temperature is set at 350–450 K. The temperature control is set up to be reverse-acting since if the peak temperature increase, the coolant temperature should decrease. Figure 6.46 gives the flowsheet and controller faceplates.

The flowsheet-equation procedure illustrated above for one temperature is repeated so that the temperatures at three locations in the reactor are measured: T(3), T(6), and T(9), corresponding to 1, 2, and 3 m from the inlet end. As shown in Figure 6.47, each temperature variable is specified in the flowsheet equation *Text Editor* as the input to a corresponding deadtime block: *dead3*, *dead6*, and *dead9*. Figure 6.48 shows the *AllVariables* window for the deadtime elements with a 1-minute deadtime specified. The inputs signals to the three deadtimes must be changed from *Fixed* to *Free*.

The control model *MultiHLoSelect* is installed on the flowsheet, and the signals from the three deadtimes are fed into it. The output of the high selector (labeled "HS") is the PV signal to the peak temperature controller (see Fig. 6.47).

A flow controller is installed on the feedstream to the reactor. Figure 6.49 shows that the action is reverse and the typical flow controller tuning constants are used ($K_C = 0.5$ and $\tau_I = 0.3$ min). Figure 6.50 shows the final flowsheet, the three controller faceplates, and the steady-state temperature profile. The setpoint of the peak temperature controller is 441 K, and the coolant temperature is 400 K.

The openloop responses of the reactor temperatures with the temperature controller on manual are shown in Figure 6.51. At time = 0.1 h, the coolant temperature is increased from 400 to 420 K. The temperatures at all locations in the reactor increase by about

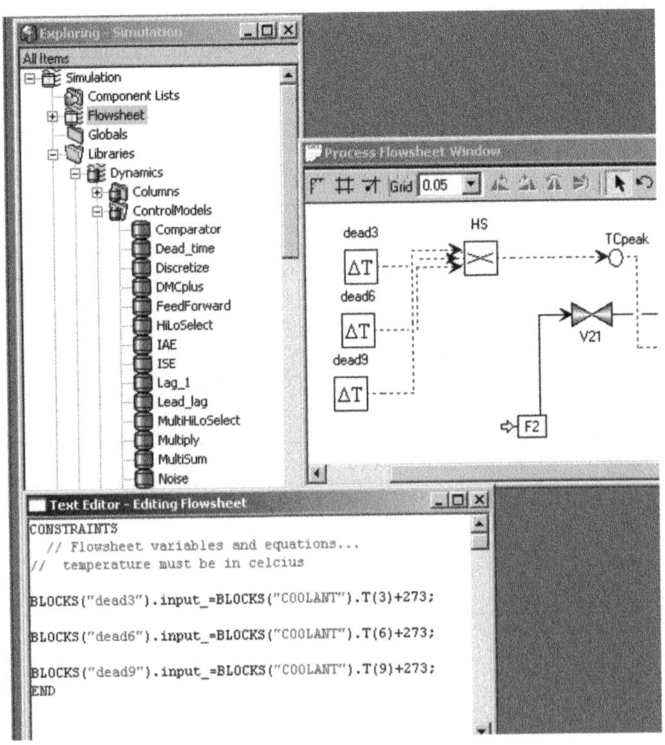

Figure 6.47 Selecting highest of three temperatures.

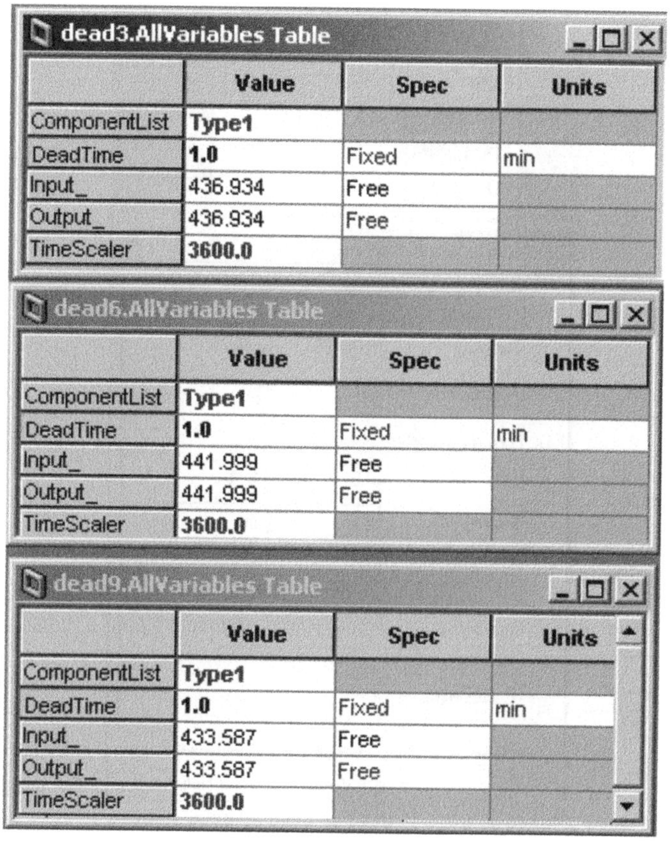

dead3.AllVariables Table

	Value	Spec	Units
ComponentList	**Type1**		
DeadTime	**1.0**	Fixed	min
Input_	436.934	Free	
Output_	436.934	Free	
TimeScaler	**3600.0**		

dead6.AllVariables Table

	Value	Spec	Units
ComponentList	**Type1**		
DeadTime	**1.0**	Fixed	min
Input_	441.999	Free	
Output_	441.999	Free	
TimeScaler	**3600.0**		

dead9.AllVariables Table

	Value	Spec	Units
ComponentList	**Type1**		
DeadTime	**1.0**	Fixed	min
Input_	433.587	Free	
Output_	433.587	Free	
TimeScaler	**3600.0**		

Figure 6.48 Specifying the deadtimes.

20 K. The location of the peak temperature remains at T(6), but the second highest temperature is now T(9). The temperature profile has shifted away from the inlet of the reactor for this increase in coolant temperature.

The catalyst-filled reactor takes almost 1 h to come to a new steady-state profile. At time = 1 h, the coolant temperature is dropped to 380 K. All temperatures decrease. The temperatures at T(3) and T(6) are almost the same, so the peak in the profile has shifted somewhat towards the inlet end of the reactor with the low coolant temperature.

To see how well the peak temperature controller works, we first perform a relay–feedback test by using the *Tune* button on the controller faceplate. The resulting Tyreus–Luyben tuning constants of the reverse acting controller are $K_C = 7.65$ and $\tau_I = 9.24$ min. Figure 6.52 shows the *Tuning* and the *Ranges* page tab windows.

Figure 6.53 gives the response of temperatures in the reactor for disturbances in feed flowrate. At time = 0.1 h, the setpoint of the feedflow controller is increased from 0.025 to 0.03 kmol/s. At 1.5 h it is dropped to 0.02 kmol/s. The peak temperature remains at 2 m, T(6), during the entire transient. It increases when more feed is added, so the controller decreases the coolant temperature to bring the peak temperature back to the setpoint. A larger temperature differential driving force is required, so the coolant temperature is lower. The reverse occurs when the feed in decreased; coolant temperature is higher.

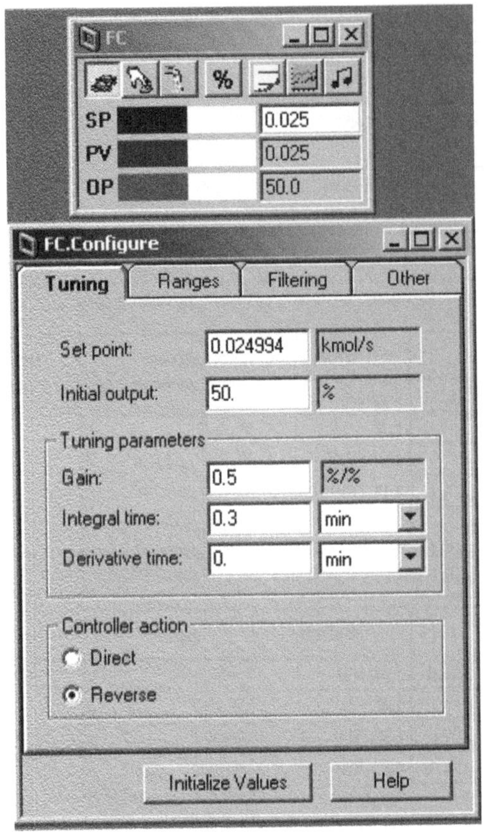

Figure 6.49 Flow controller.

Figure 6.54 shows the responses for changes in feed composition. At time = 0.1 h, the chlorine concentration in the feed is increased from 10 to 15 mol%, with a corresponding decrease in the propylene concentration. At time = 2 h, the feed composition is changed to 5 mol% chlorine. The peak temperature controller handles these very large disturbances in a stable manner, with coolant temperature changing over a wide range (380–425 K). Note that T(6) remains the peak temperature during these transients, but the peak shifts at the low feed concentrations such that T(9) is higher than T(3), specifically, the profile moves further away from the inlet of the reactor.

Figure 6.55 gives the responses for changes in feed temperature. At time = 0.1 h, the feed temperature is increased from 400 to 420 K. The temperatures near the inlet of the reactor see the disturbance first. The temperature T(3) at 1 m into the bed begins to increase after about 0.1 h, and it becomes larger than T(6) at ~0.3 h. The high selector picks up this temperature (after coming through the deadtime) and sends it to the peak temperature controller, which begins to decrease the coolant temperature. At time = 1.5 h, the feed temperature is decreased to 380 K. The T(6) temperature again becomes the highest. Note that T(9) becomes the second highest temperature instead of T(3) at the low feed temperature, that is, the profile shifts down the reactor.

These results show that the peak temperature controller works well for a variety of disturbances in this cooled tubular reactor system.

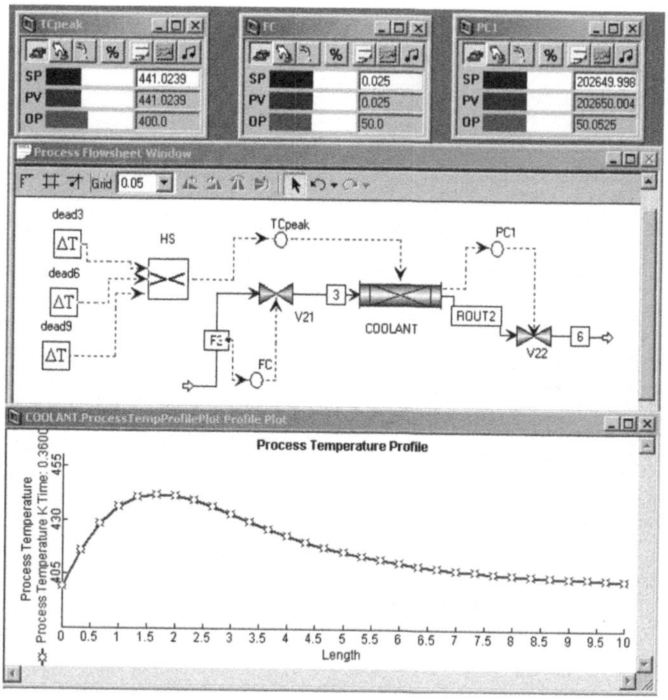

Figure 6.50 Flowsheet and controller faceplates.

6.10.3 Cooled Tubular Reactor with Co-current Flow of Coolant

The next tubular reactor system explored is one with co-current flow of coolant. The same cooled reactor used in the previous section is used. It has 500 tubes packed with 39,260 kg of catalyst. Each tube is 0.1 m in diameter and 10 m in length.

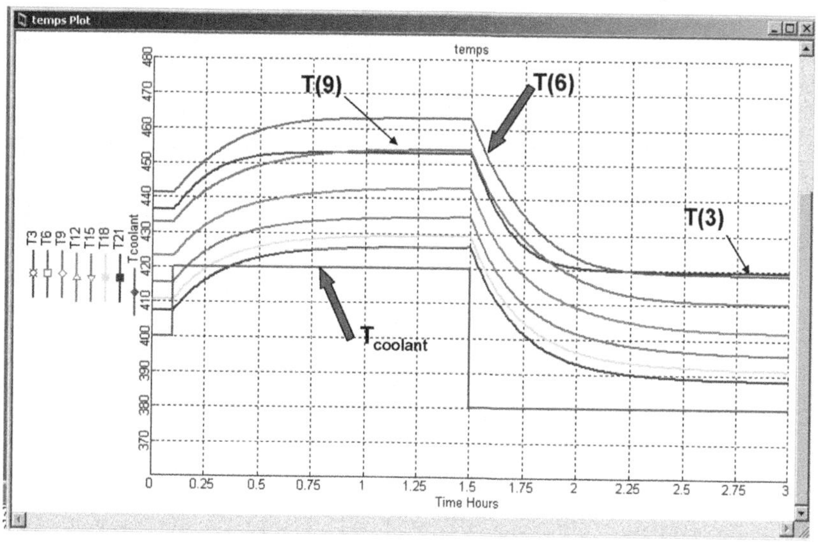

Figure 6.51 Openloop responses to changes in coolant temperature.

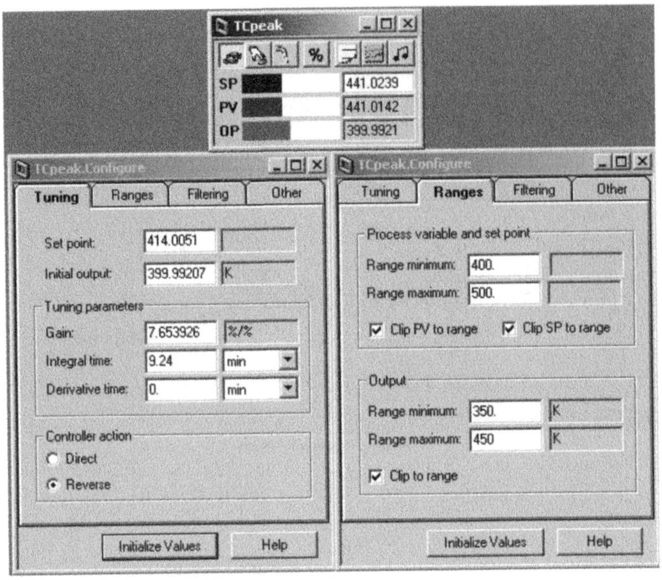

Figure 6.52 Tuning and ranges.

Two different values of the overall heat transfer coefficient U are investigated. When a U of 14.7 kJ s^{-1} m^{-2} K^{-1} is used, the temperature and composition profiles shown in Figure 6.56 result. The flowrate of the coolant is 0.05 kmol/s and its inlet temperature is 350 K. The resulting temperature of the coolant leaving the shell of the reactor is 432 K. The process stream leaves at 434 K. The differential temperature driving force at the end of the reactor is very small, even with this small value of U.

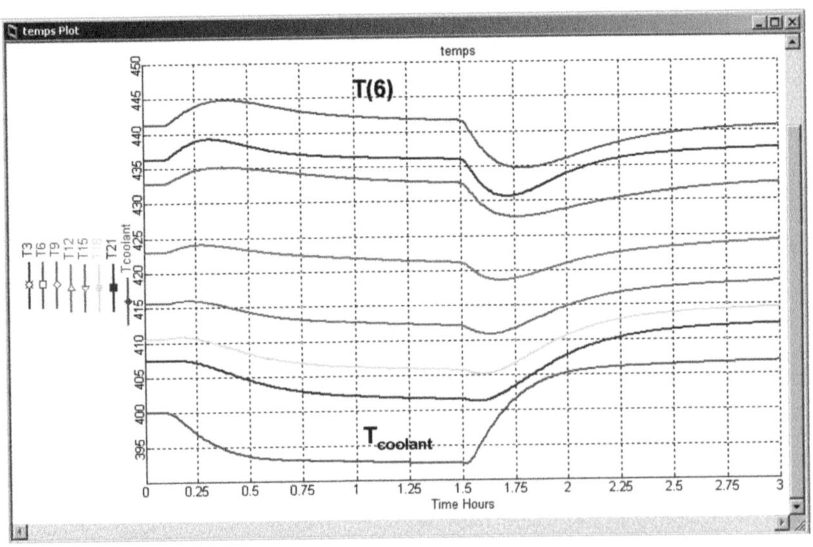

Figure 6.53 Closedloop responses to changes in feedflow.

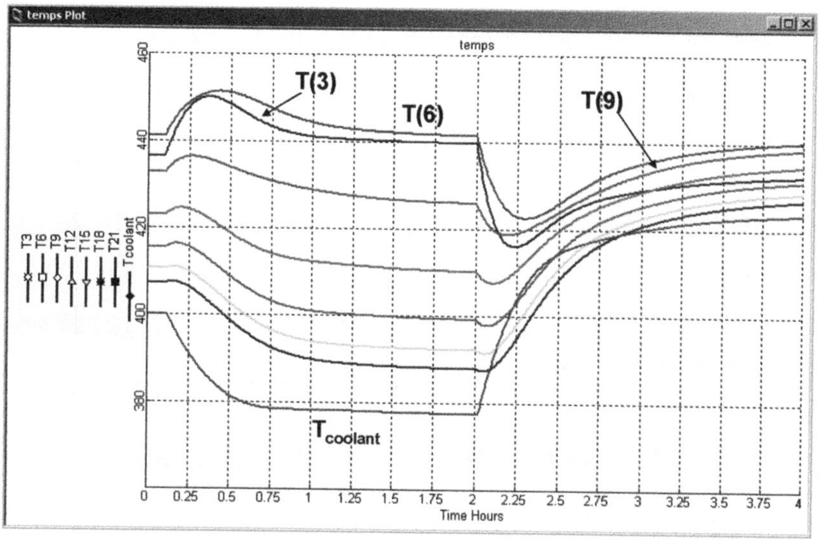

Figure 6.54 Closedloop responses to changes in feed composition.

The control structure shown in Figure 6.57 is installed on the flowsheet. The feed is flow-controlled. The outlet temperature is controlled by manipulating the coolant flowrate. Note that the OP signal is sent to *both* of the control valves on the coolant stream, opening and closing them simultaneously. The setup works in the simulations, but it is not what would be used in a real physical system. A pressure-driven simulation in Aspen Plus requires that valves be placed on both the inlet and outlet coolant streams. In a real system, the cooling water would be drawn from a supply header, which operates a fixed pressure. A single control valve would be used, either on the inlet or on the outlet, to manipulate the flowrate of coolant.

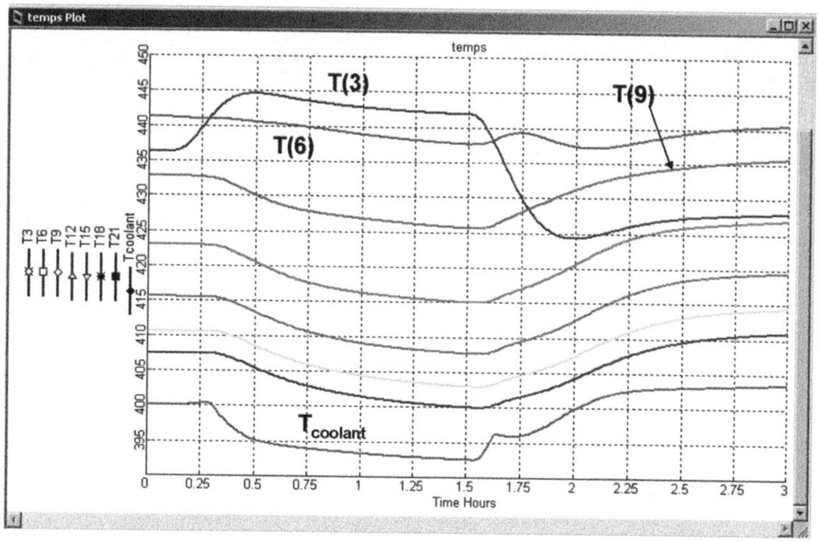

Figure 6.55 Closedloop responses to changes in feed temperature.

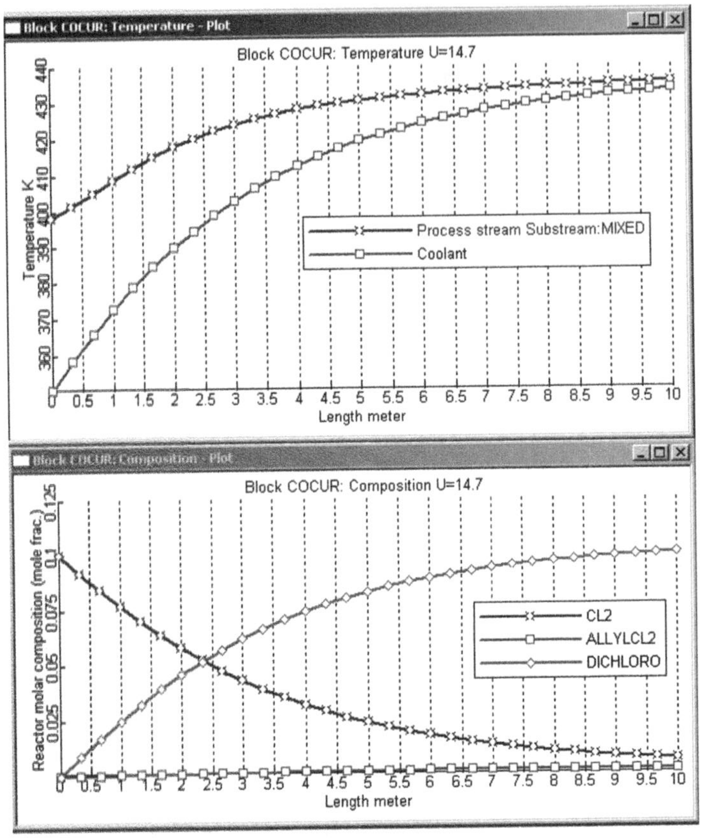

Figure 6.56 Profiles; co-current tubular reactor; $U = 14.7$.

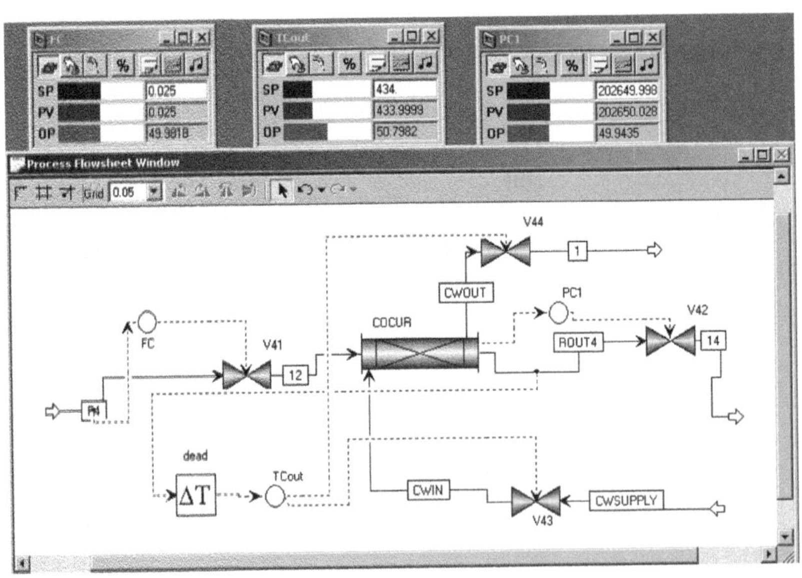

Figure 6.57 Flowsheet; co-current tubular reactor; $U = 14.7$.

The outlet temperature controller "TC_{out}" is set up to be direct-acting; if the process temperature increases, the flowrate of coolant increases. The range of the temperature transmitter is set at 400–500 K. The range of the coolant flowrate is set at 0–0.1 kmol/s. With a deadtime of 1 min, the relay–feedback test gives Tyreus–Luyben settings of $K_C = 56$ and $\tau_I = 9.24$ min. This very high controller gain is reduced to a more realistic value of 25. As the results given below demonstrate, very tight temperature control is achieved.

Figure 6.58 gives the closedloop response to changes in feed flowrate at 0.1 and 1.5 h. Figure 6.59 gives results for changes in the temperature of the feed. Note the very small scale for the T_{out} temperature. The changes in the outlet temperature are very small for all these disturbances.

When a much larger U of 147 kJ s^{-1} m^{-2} K^{-1} is used, the temperature and composition profiles shown in Figure 6.60 result. The flowrate of the coolant is 0.05 kmol/s and its temperature is 350 K. The resulting temperature of the coolant leaving the shell of the reactor is 432 K, and the process stream leaves at almost the same temperature because of the large heat transfer coefficient.

When control of the outlet temperature is attempted, it does not work. Figures 6.61 and 6.62 demonstrate that with either a high controller gain ($K_C = 10$) or with a low controller gain ($K_C = 1$), the system is unstable. This is due to the large dynamic lag this system and the pinch between the process and coolant temperatures at the exit end of the reactor.

The control structure is changed to control the temperature at 2 m from the inlet of the reactor T(6). Figure 6.62 shows the flowsheet equation *Text Editor* with the input to the deadtime block set equal to the temperature in the reactor block "COCUR" at lump 6. The input to the deadtime block is changed to *Free*.

Figure 6.63 shows this new control structure. The setpoint of the temperature controller "TC2" is 392 K, and it is direct-acting. With a deadtime of 1 min and a temperature transmitter range of 350–450 K, the relay–feedback test gives Tyreus–Luyben settings of $K_C = 14.6$ and $\tau_I = 10.6$ min. Figure 6.64 shows the response of the closedloop system

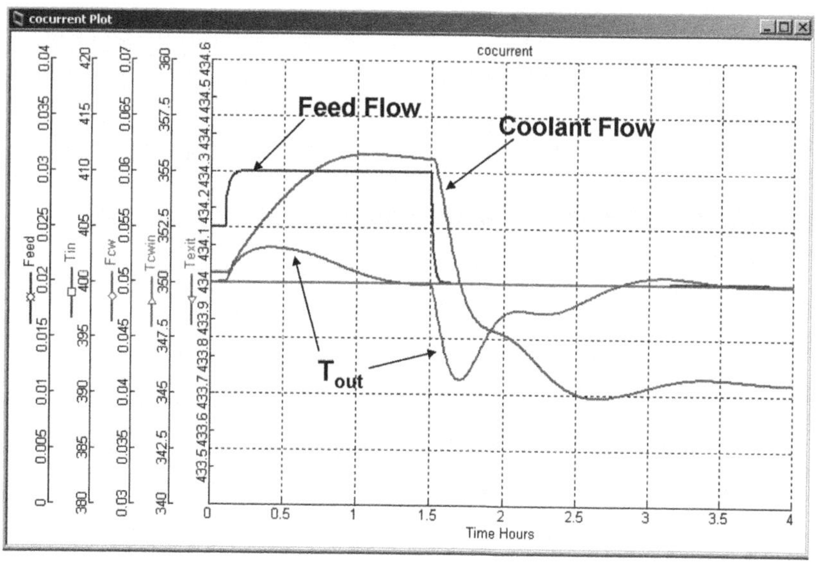

Figure 6.58 Feedrate changes; co-current tubular reactor; $U = 14.7$.

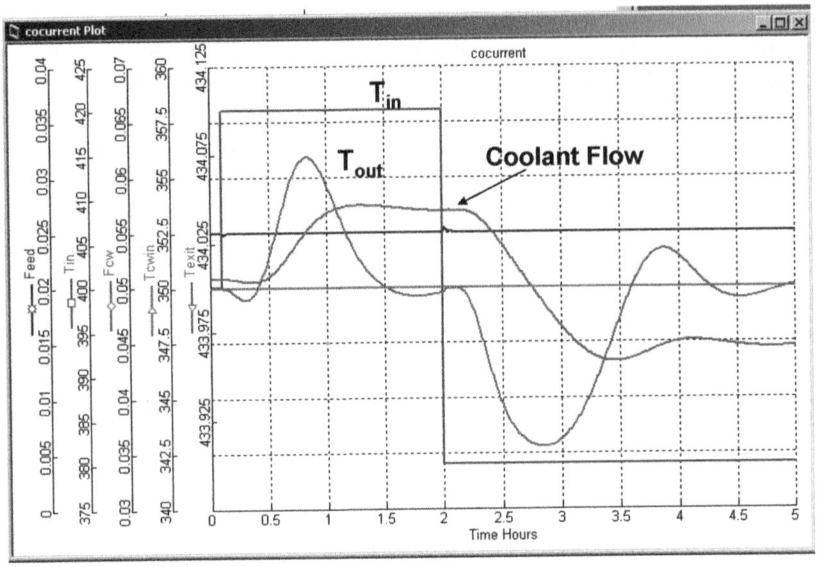

Figure 6.59 T_{in} changes; co-current tubular reactor; $U = 14.7$.

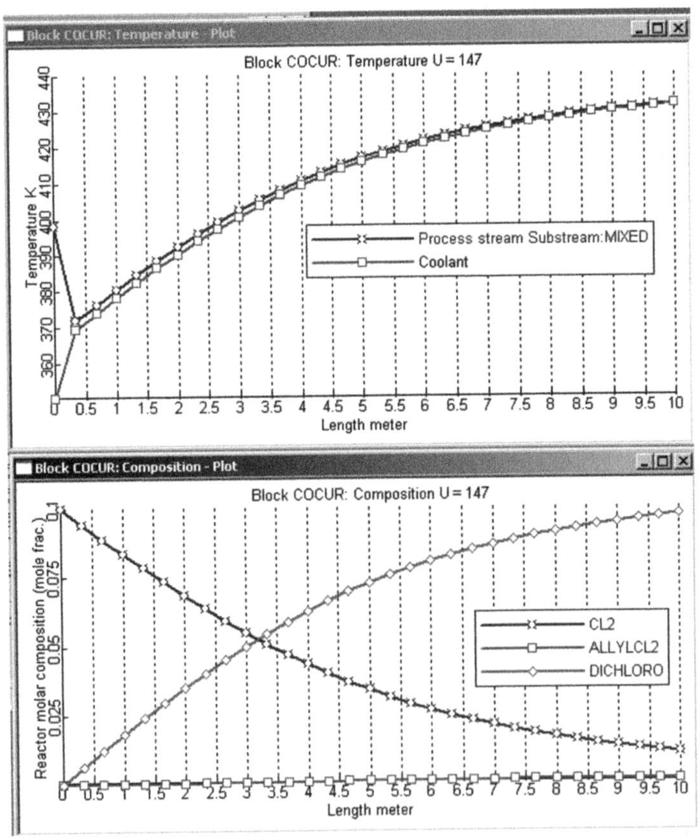

Figure 6.60 Profiles; co-current tubular reactor; $U = 14.7$.

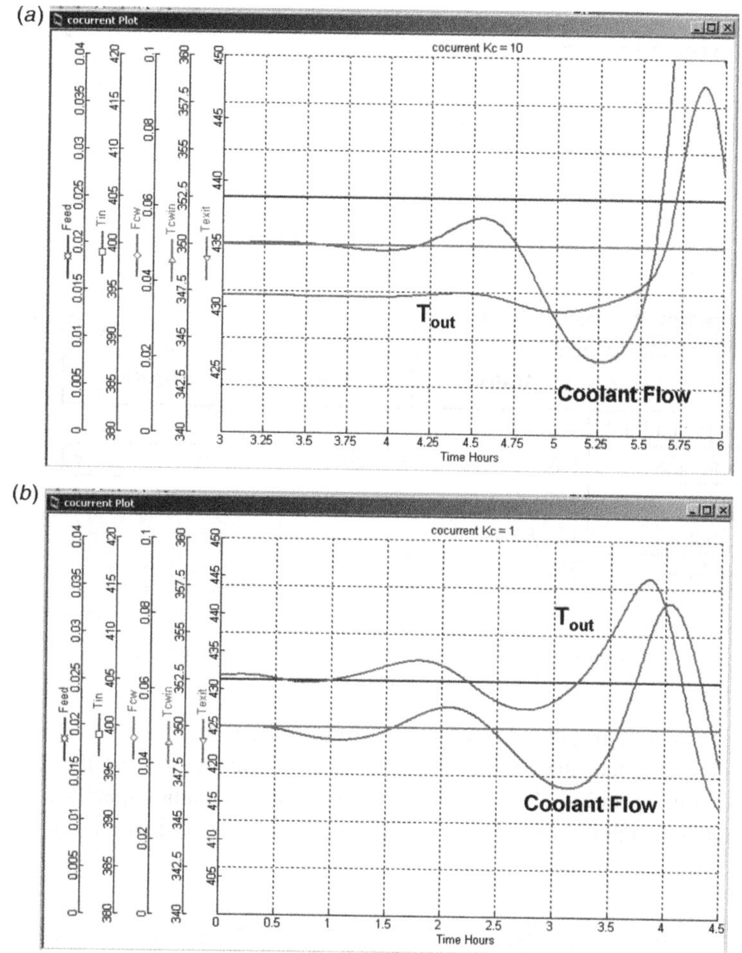

Figure 6.61 T_{out} changes; co-current tubular reactor; $U = 14.7$. (*a*) $K_C = 10$; (*b*) $K_C = 1$.

to changes in feed flowrate. The T(6) temperature is maintained within about 1 K of the setpoint for these very large disturbances. Note that the outlet reactor temperature moves much more slowly than does the T(6) temperature.

Figure 6.65 gives results for feed composition disturbances. At 0.1 h, the chlorine concentration in the feed is increased from 10 to 12.5 mol%, and at 1.5 h it is decreased to 7.5 mol%. The maximum deviation in T(6) temperature is about 2 K for the very large drop in feed composition. The temperature loop is somewhat oscillatory for the low feed composition and low coolant flow conditions, so some controller retuning would be advisable.

Figure 6.66 shows responses for changes in feed temperature T_{in} from 400 to 420 K at 0.1 h and to 380 K at 1.5 h. Temperature is effectively controlled.

6.10.4 Cooled Tubular Reactor with Countercurrent Flow of Coolant

The final tubular reactor system studied is one with countercurrent flow of coolant. This option, shown in Figure 6.67, requires that the *exit* temperature of the coolant be

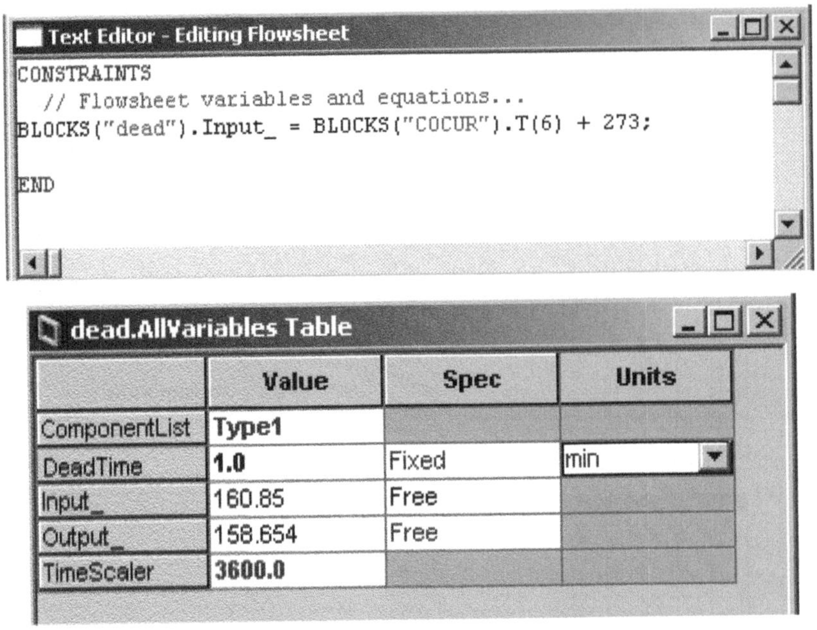

Figure 6.62 Flowsheet equations for T(6) control.

specified, which is at the process inlet end of the reactor. It is set at 450 K. However, Aspen Plus requires that the conditions of the coolant stream *entering* the reactor at the other end also be specified. This is confusing because you cannot specify both the inlet and the outlet temperatures. Figure 6.68 shows the specifications made for the

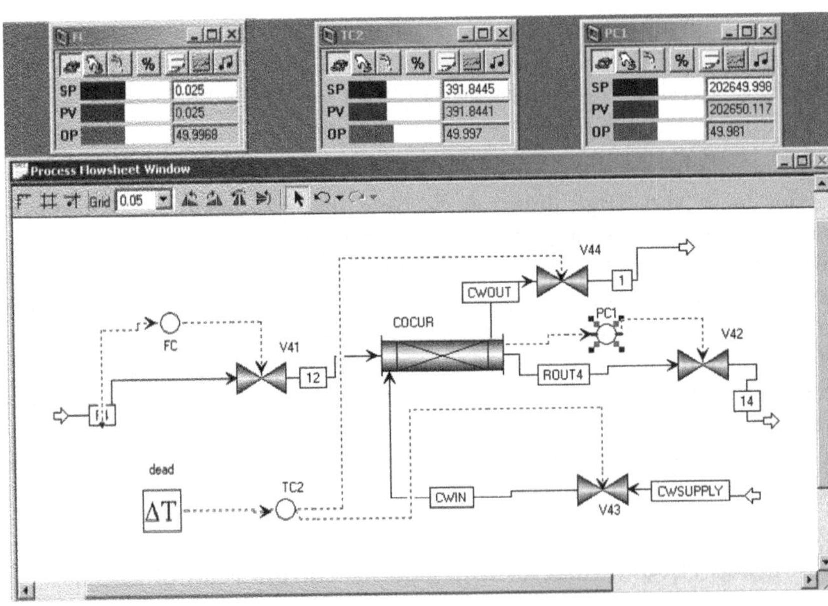

Figure 6.63 Flowsheet and controller faceplates; T(6) control.

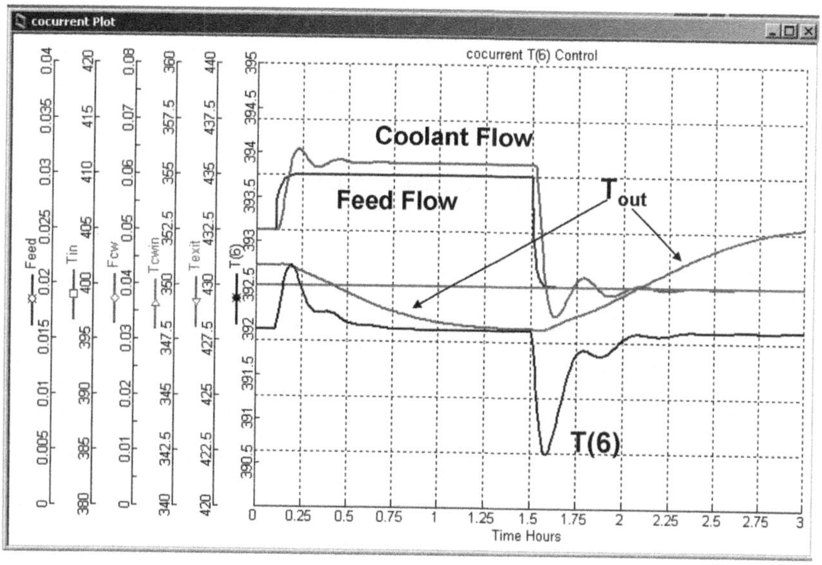

Figure 6.64 Feedrate changes; T(6) control.

"CWINLET" stream. The flowrate is set at 0.05 kmol/s, and the inlet temperature is specified to be 330 K. When the program is run, the warning message shown in Figure 6.69 indicates that the coolant inlet temperature should be 336.88 K. Despite this warning, the pressure checking and exporting to Aspen Dynamics is successful.

The heat transfer coefficient is 142 kJ s^{-1} K^{-1} m^{-2}, and the resulting temperature profiles of the process and the coolant are given in Figure 6.70. The two temperatures are pinched together very severely at the end of the reactor, where the process leaves and

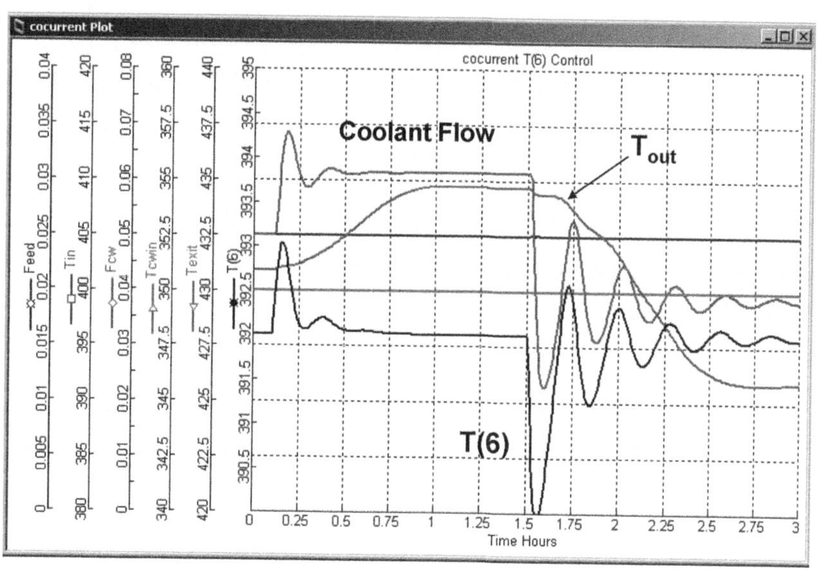

Figure 6.65 Feed composition changes; T(6) control.

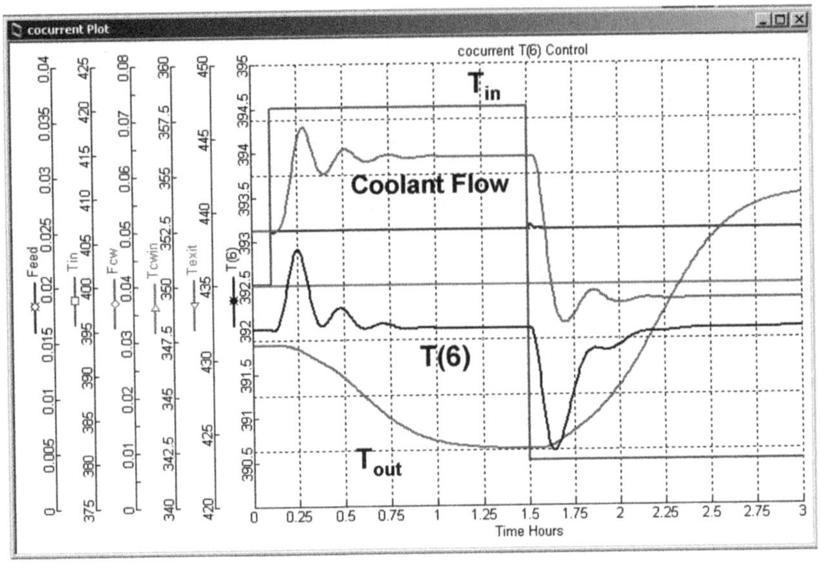

Figure 6.66 Feed temperature changes; T(6) control.

the cooling water enters. This makes the control of the exit process temperature ineffective. Therefore a temperature T(6) near the front of the reactor is selected for control. Figure 6.71 shows the flowsheet equation required for specifying the input to the deadtime element as T(6) and making it *Free*. Figure 6.72 gives the final flowsheet with the control structure and faceplates. The setpoint of the temperature controller is 391.8 K.

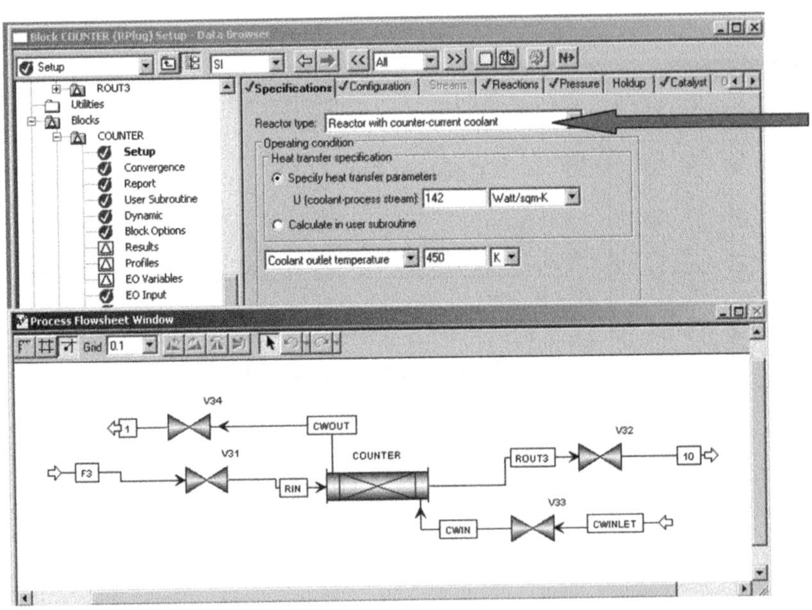

Figure 6.67 Cooled reactor with countercurrent coolant flow.

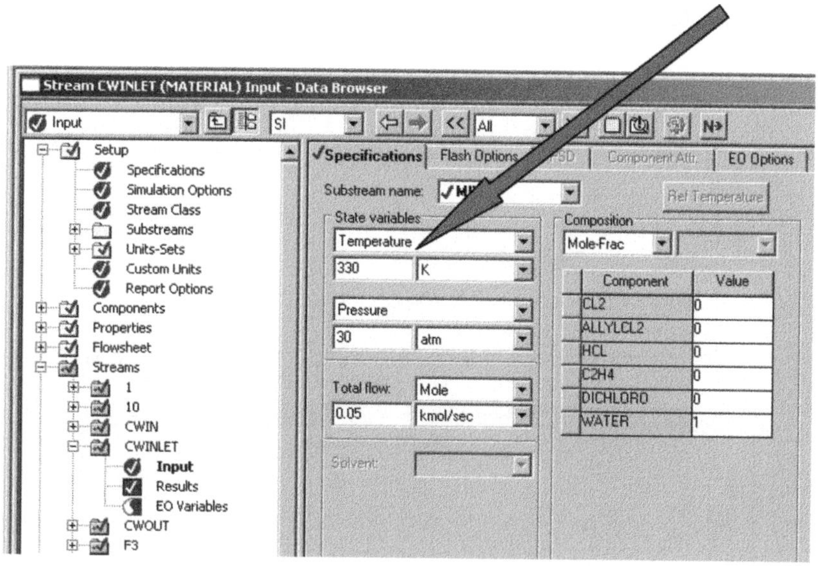

Figure 6.68 Coolant inlet temperature specified.

The system is subjected to several disturbances. The program faces numerical problems if these disturbances are too large. Figure 6.73 illustrates the erratic temperature profile that results when these numerical problems occur. The countercurrent model appears to be more prone to numerical problems than any of the other tubular reactor models. As a result the magnitudes of some of the disturbances are smaller than in the other cases.

Figure 6.74 gives result for changes in feed flowrate. At 0.1 h the setpoint of the feed-flow controller is increased from 0. 025 to 0.03 kmol/s. At 1 h it is dropped back to

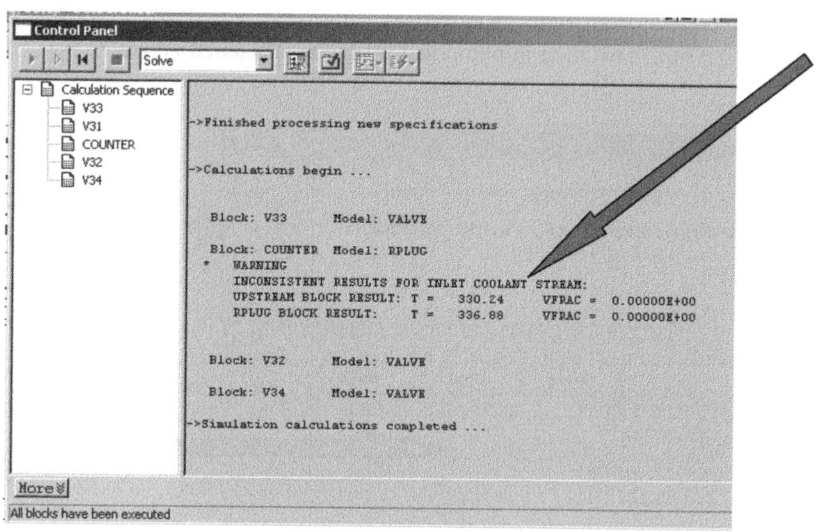

Figure 6.69 Warning message.

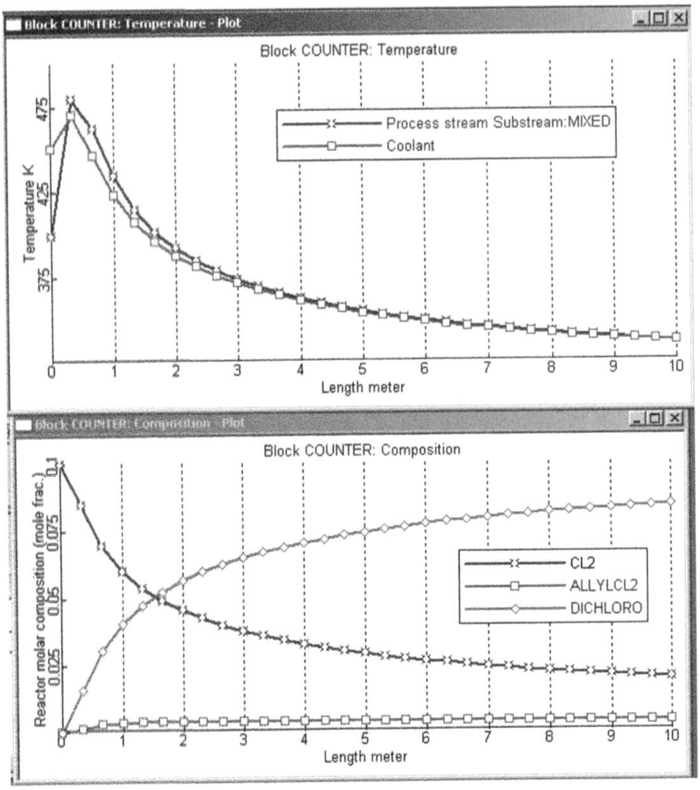

Figure 6.70 Profiles.

0.02 kmol/s. At 2 h it is dropped further to 0.015 kmol/s. The control of T(6) is quite good, with peak deviation of slightly over 1 K. Similar effective control for disturbances in feed composition is demonstrated in Figure 6.75. At 0.1 h the chlorine composition of the feed is increased from 10 to 15 mol%. At 1 h it is decreased to 7.5 mol% chlorine. A larger step to 5 mol% resulted in numerical problems discussed above.

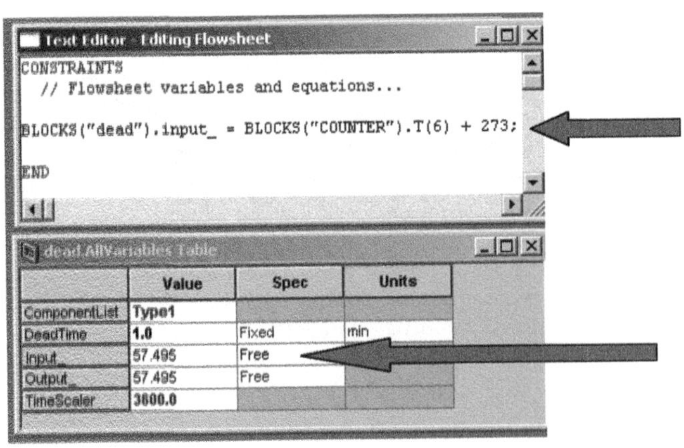

Figure 6.71 Flowsheet equation and freeing input to deadtime.

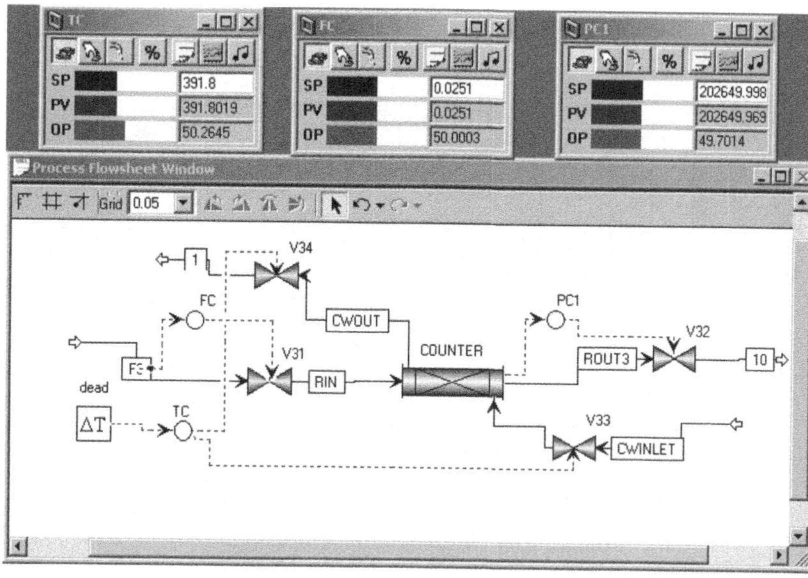

Figure 6.72 Flowsheet and faceplates.

6.10.5 Conclusions for Aspen Simulation of Different Types of Tubular Reactors

Simulation examples of four types of tubular reactors have been presented in the sections above. The adiabatic and constant-coolant temperature models are easier to set up and seem to run with fewer problems. In the adiabatic reactor the only variable that can be controlled is the inlet temperature. In the cooled reactors a temperature can be controlled by manipulating either the coolant temperature or the coolant flowrate, depending on the model.

In the constant-coolant temperature model, a temperature at a number of locations down the length of the reactor can be controlled. For the co-current and countercurrent

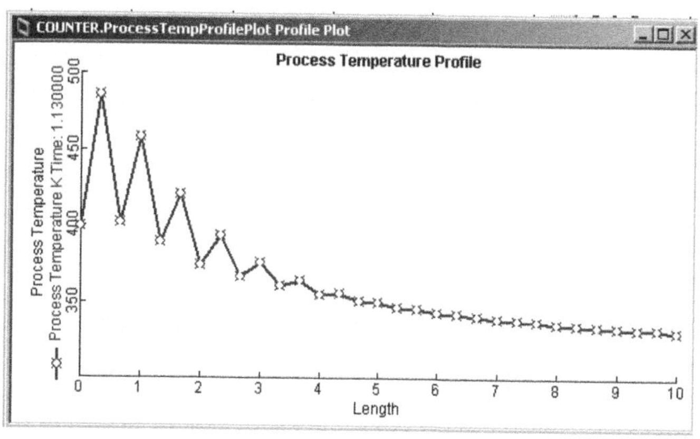

Figure 6.73 Numerical problems with large disturbance.

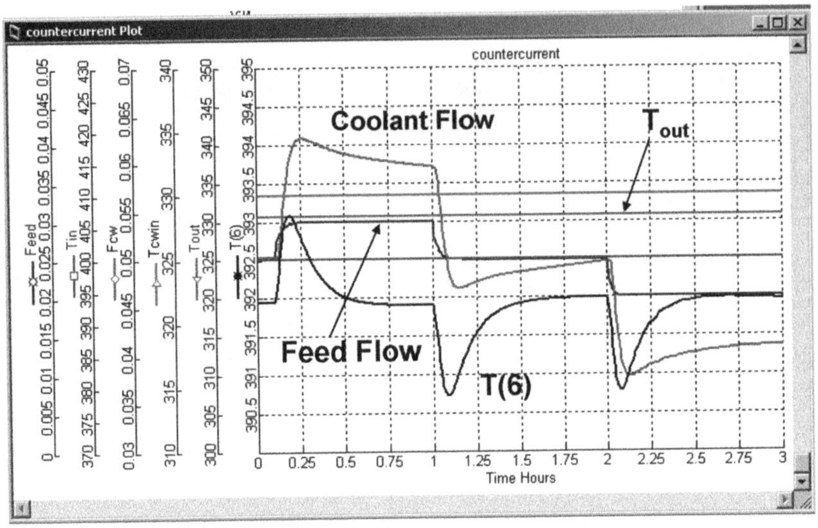

Figure 6.74 Feed flowrate disturbances.

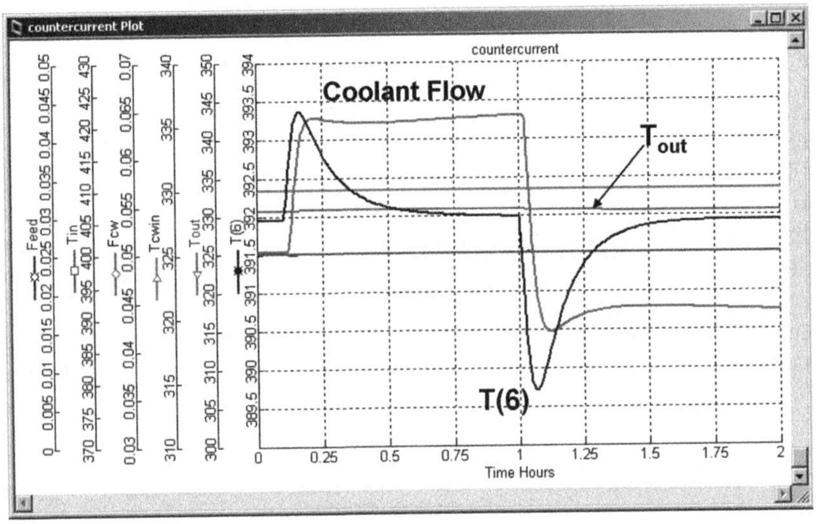

Figure 6.75 Feed composition; 10 to 15 to 7.5% C_{12}.

coolant models, the control of a temperature near the end of the reactor is ineffective because of the pinch between the process and coolant temperatures.

6.11 PLANTWIDE CONTROL OF METHANOL PROCESS

To conclude our examples of Aspen Dynamics simulation of tubular reactor systems, we study a very important industrial process for the production of methanol from synthesis

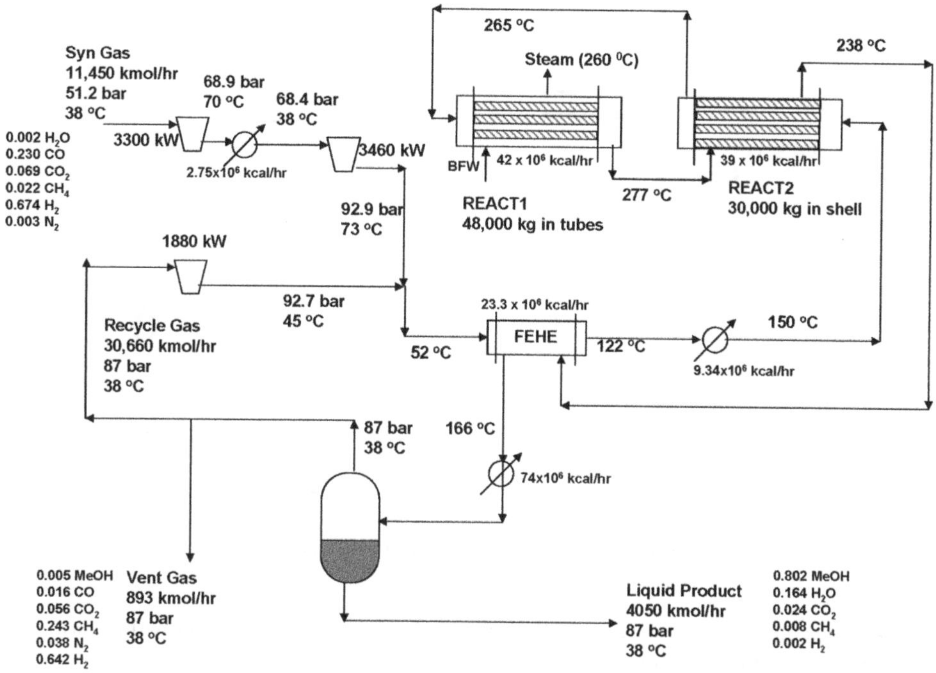

Figure 6.76 Flowsheet of methanol process.

gas. Methanol is one of the prime candidates for providing an alternative to petroleum-based liquid transportation fuels. It can be made from any hydrocarbon source, including biomass, that can be converted into synthesis gas, which is a mixture of hydrogen, carbon monoxide, and carbon dioxide. The most widely used process for making synthesis gas is steam–methane re-forming. Partial oxidation of other combustible fuels can also generate synthesis gas.

The methanol process has two cooled tubular reactors operating in series. The first is cooled by generating steam. The second is cooled by preheating the feed. There is a large gas recycle stream. Figure 6.76 gives the flowsheet with operating conditions. Figure 6.77 gives the Aspen Plus flowsheet with all pieces of equipment installed. There is recycle of both energy (in the FEHE and in the second reactor) and material (a gas recycle stream).

6.11.1 Chemistry and Kinetics

The stoichiometry of the methanol reactions involves the reaction of both carbon dioxide and carbon monoxide with hydrogen:

$$\begin{aligned} CO + 2H_2 &\Leftrightarrow CH_3OH \\ CO_2 + 3H_2 &\Leftrightarrow CH_3OH + H_2O \end{aligned} \qquad (6.16)$$

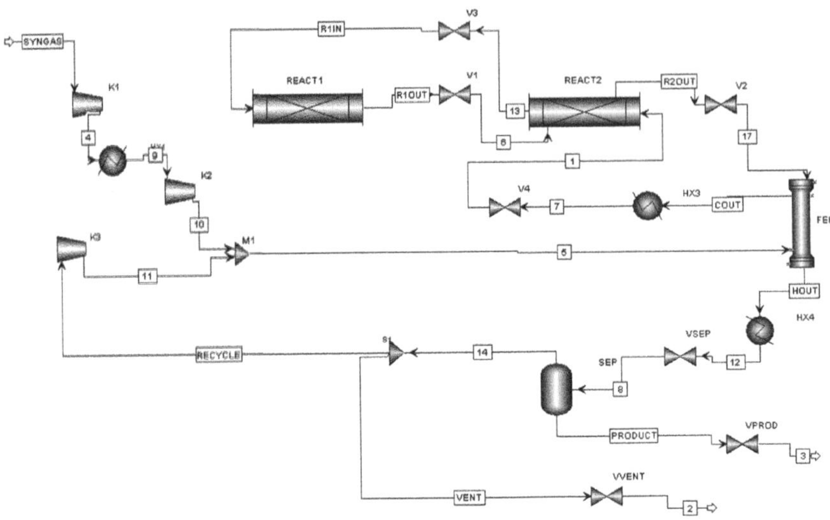

Figure 6.77 Aspen Plus flowsheet of methanol process.

The kinetics are given by vanden Bussche and Froment[3] in the following form by using the water shift reaction:

$$
\begin{aligned}
CO_2 + 3H_2 &\Leftrightarrow CH_3OH + H_2O \\
CO + H_2 &\Leftrightarrow CO + H_2O
\end{aligned}
\tag{6.17}
$$

The reactions are exothermic and use a solid catalyst. The kinetics are described by LHHW-type equations (Langmuir–Hinshelwood–Hougen–Watson). Conversion of the reaction rate equations and their units given in the original paper into the form required by Aspen Plus is a daunting task. The original data use pressures in bar and reaction rates in kmol min^{-1} kg^{-1} catalyst. These must be transformed to use pascals. This transformation was provided by Emmanuel Lejeune from Aspen Support, and this invaluable assistance is gratefully acknowledged.

As discussed in Chapter 1, the LHHW kinetic structure has the form

$$\mathfrak{R} = (\text{kinetic term})\frac{(\text{driving-force term})}{(\text{adsorption term})} \tag{6.18}$$

The reaction rate for the first reaction for the production of methanol from carbon dioxide is given as follows:

$$\mathfrak{R}_1 = (k_4 p_{CO_2} p_{H_2}) \frac{\left[1 - \dfrac{1}{K_{E1}}\left(\dfrac{p_{CH_3OH} p_{H_2O}}{p_{CO_2} p_{H_2}^3}\right)\right]}{\left[1 + k_3\left(\dfrac{p_{H_2O}}{p_{H_2}}\right) + k_1\sqrt{p_{H_2}} + k_2 p_{H_2O}\right]^3} \tag{6.19}$$

[3]K. M. Vanden Bussche and G. F. Froment, A steady-state kinetic model for methanol synthesis and the water gas shift reaction on a commercial $Cu/ZnO/Al_2O_3$ catalyst, *J. Catalysis*, **161**, 1–10 (1996).

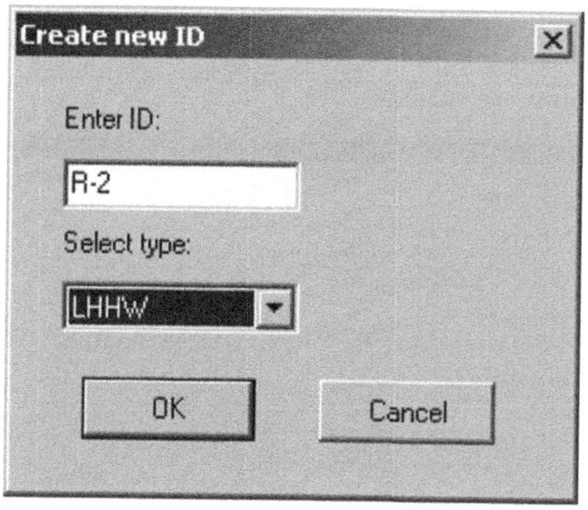

Figure 6.78 Setting up LHHW kinetics.

The reaction rate for the water shift reaction is given as follows:

$$\Re_2 = (k_5 p_{CO_2}) \frac{\left[1 - \frac{1}{K_{E2}}\left(\frac{p_{CO}p_{H_2O}}{p_{CO_2}p_{H_2}}\right)\right]}{\left[1 + k_3\left(\frac{p_{H_2O}}{p_{H_2}}\right) + k_1\sqrt{p_{H_2}} + k_2 p_{H_2O}\right]} \tag{6.20}$$

Figure 6.78 shows how the LHHW kinetics are selected. Figure 6.79 gives the *Input* window for the reactions "R-1" with the *Stoichiometry* page tab selected. Figure 6.80

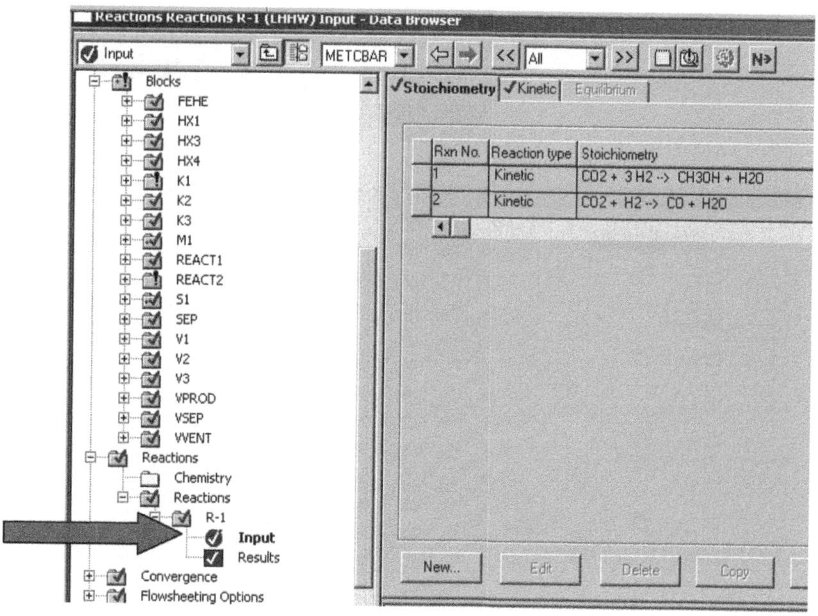

Figure 6.79 Methanol reactions.

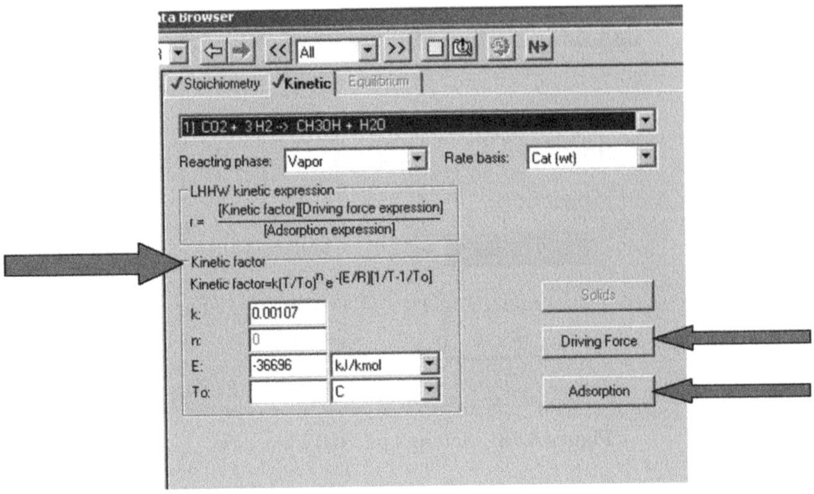

Figure 6.80 Methanol reaction; LHHW kinetic factor.

gives parameter values for the kinetic factor for the first (methanol) reaction. Clicking the *Driving Force* button opens the window shown on the left in Figure 6.81, on which the parameter values are entered. Clicking the *Adsorption Expression* button opens the window shown on the right in Figure 6.81, on which the parameter values are entered. The three terms for the second [water–gas shift (WGS)] reaction are given in Figures 6.82 and 6.83.

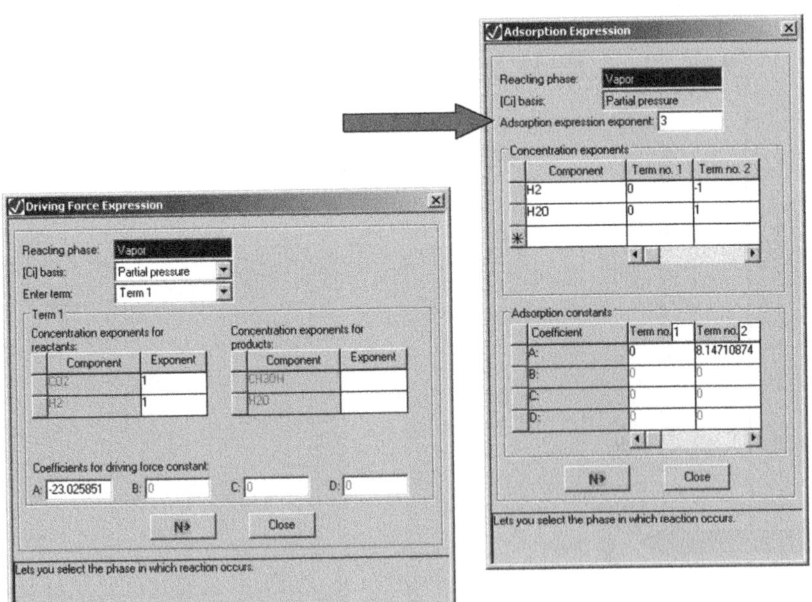

Figure 6.81 Driving force and adsorption terms for methanol reaction.

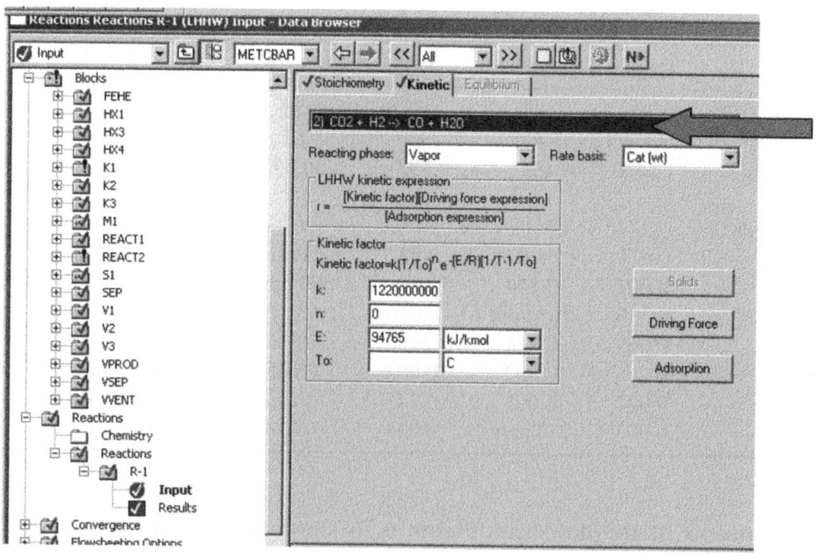

Figure 6.82 WGS reaction; LHHW kinetic factor.

6.11.2 Process Description

The flowsheet is based on a Stanford Research Institute report for the production of methanol (SRI 43C, March 2000). The flowrate of the fresh feed of synthesis gas is 11,450 kmol/h at a pressure of 51.2 bar and a temperature of 38°C. The composition is 67.4 mol% hydrogen, 23 mol% carbon monoxide, 6.9 mol% carbon dioxide, 2.2 mol% methane, and small amounts of nitrogen and water. Table 6.11 gives information on the

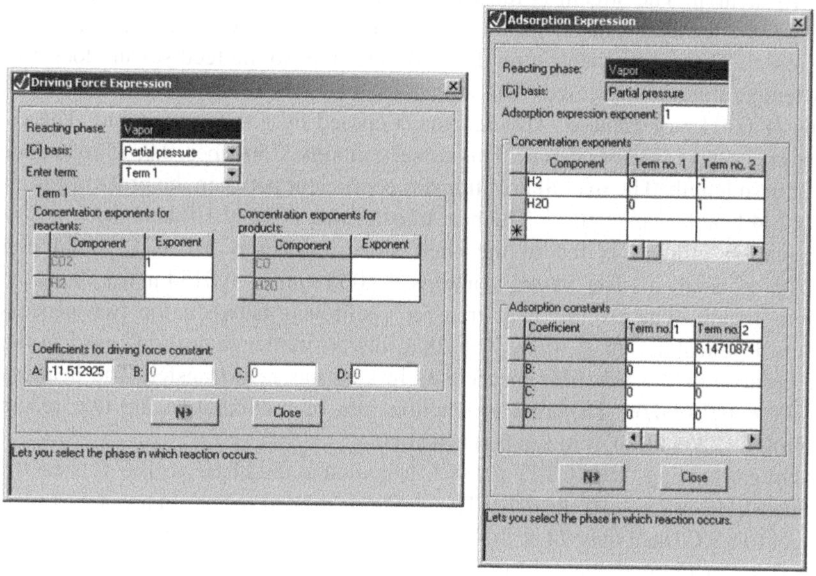

Figure 6.83 Driving force and adsorption terms for WGS reaction.

TABLE 6.11 Stream Conditions in Methanol Process

	Syngas	Product	Vent	Recycle	$R1_{in}$	$R1_{out}$	$R2_{out}$
Temperature (°C)	38	38	38	38	265	277	238
Pressure (bar)	51.2	87	87	87	91.3	89.7	87.6
Flow (kmol/h)	11,450	4051	893	30,660	42,100	37,880	35,600
Composition							
mole fraction							
H_2	0.675	0.002	0.642	0.642	0.651	0.599	0.569
H_2O	0.002	0.164	590 ppm	590 ppm	0.001	0.014	0.019
CO	0.230	100 ppm	0.016	0.016	0.074	0.040	0.014
CO_2	0.069	0.024	0.056	0.056	0.059	0.052	0.052
CH_3OH	0	0.802	0.005	0.005	0.004	0.060	0.096
CH_4	0.022	0.008	0.243	0.243	0.183	0.203	0.216
N_2	0.003	218 ppm	0.038	0.038	0.029	0.032	0.034

important streams in the flowsheet. Figure 6.76 shows the flowsheet with operating conditions and design parameters.

The synthesis gas is compressed in a two-stage compressor with interstage cooling to 92.7 bar and combined with a recycle stream that is 30,660 kmol/h with a composition that is quite different from that of the synthesis gas. The inert components (nitrogen and methane) are allowed to build up to fairly high levels (24.3 mol% methane, 3.8 mol% nitrogen) so that the losses of reactants in the vent stream are kept small. The vent stream is only 893 kmol/h.

The combined stream is preheated to 122°C in a FEHE. A heater (HX3) is installed after the FEHE so that inlet temperature of the "coolant" stream in "REACT2" can be adjusted to satisfy the energy balance when the exit temperature of the coolant stream is specified in this countercurrent tubular reactor. This temperature is 150°C, and the heat load in HX3 is 9.34×10^6 kcal/h. The stream is further preheated to 265°C in the tube side of reactor "REACT2" by the heat transfer from the reactions that are occurring in the hot shell side of this vessel. There is no catalyst on the cold tube side, so the feed stream does not react but its temperature is increased. The stream is then fed to reactor "REACT1," which contains 48,000 kg of catalyst. This reactor is cooled by generating steam. The coolant temperature is 265°C (51 bar steam). This vessel contains 3750 tubes, 0.0375 m in diameter, and 12.2 m in length. The overall heat transfer coefficient between the process gas and the steam is 244 kcal h^{-1} m^{-2} $°C^{-1}$. The heat transfer rate is 42×10^6 kcal/h.

The reactor effluent is fed to the shell side of reactor "REACT2" that contains 30,000 kg of catalyst. The vessel contains 10,000 tubes, 0.0184 m in diameter and 12.2 m in length. The overall heat transfer coefficient between the two process gas streams is 144 kcal h^{-1} m^{-2} $°C^{-1}$. Note that this is smaller than the U used in the first reactor because of the low film coefficients in gas–gas heat transfer. The heat transfer rate is 39×10^6 kcal/h. The total production rate of methanol in the two reactors is 3250 kmol/h (2260 kmol/h in the first reactor).

The hot gas leaving "REACT2" at 238°C is cooled in the FEHE to 166°C (heat transfer rate in the FEHE is 23×10^6 kcal/h). The stream is cooled further in a water-cooled heat exchanger to 38°C (heat duty 74×10^6 kcal/h). The liquid and gas phases are separated in a drum (6 m in diameter and 12 m in length), which operates at 87 bar. The liquid is removed as product and sent to a separation section for purification. This stream has a composition 80.2 mol% methanol, 16.4 mol% water, 2.4 mol% carbon dioxide, 0.8 mol% methane, and trace amounts of hydrogen and nitrogen.

The gas from the separator is split into the vent stream (893 kmol/h) and the recycle stream (30,660 kmol/h). A compressor raises the pressure of the recycle gas from 87 to 92.7 bar. Assuming efficiencies of 80%, the recycle compressor work is 1880 kW and the synthesis gas feed compressors consume a total of 6760 kW.

6.11.3 Steady-State Aspen Plus Simulation

Setting up and converging the flowsheet is fairly challenging because of the recycle of both heat and material. Both the FEHE and reactor "REACT2" feature heat transfer between process streams. There is a large recycle stream that is about twice as large as the fresh feedstream. The final converged flowsheet is shown in Figure 6.76.

The flowsheet was developed by sequentially guessing various unknown stream conditions in and out of heat exchangers and reactors. The flowrates, compositions, and temperatures of the recycle stream and the hot stream into the FEHE were guessed. The FEHE was sized by specifying an exit temperature of the cold stream of 122°C, as shown in Figure 6.84. An overall heat transfer coefficient of 144 kcal h^{-1} m^{-2} °C^{-1} is specified (Fig. 6.85) in this gas–gas exchanger.

Then a temporary heater was installed to heat the inlet stream to the first reactor to 265°C. The reactor is specified to have a constant coolant temperature (Fig. 6.86) of 260°C and a heat transfer coefficient of 244 kcal h^{-1} m^{-2} °C^{-1}. Figure 6.87 gives the *Configuration* page tab with the physical dimensions of the tubes. Figure 6.88 shows the catalyst loading.

Now the exit stream from the first reactor is known, so it is fed into the second reactor (Fig. 6.89), which is set up with countercurrent coolant. This coolant is the exit stream from the heater "HX3" (see the flowsheet in Fig. 6.77). With countercurrent flow, the *exit* temperature of the cold stream must be specified, and the program calculates the inlet temperature of the cold stream (150°C). The temperature of the stream from the "HX3" is set equal to this temperature. Figures 6.90 and 6.91 give the physical dimension and catalyst in the second reactor.

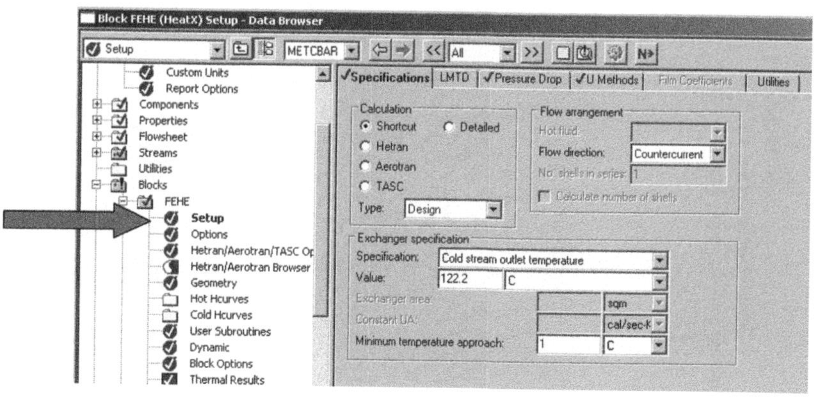

Figure 6.84 Setting up the FEHE.

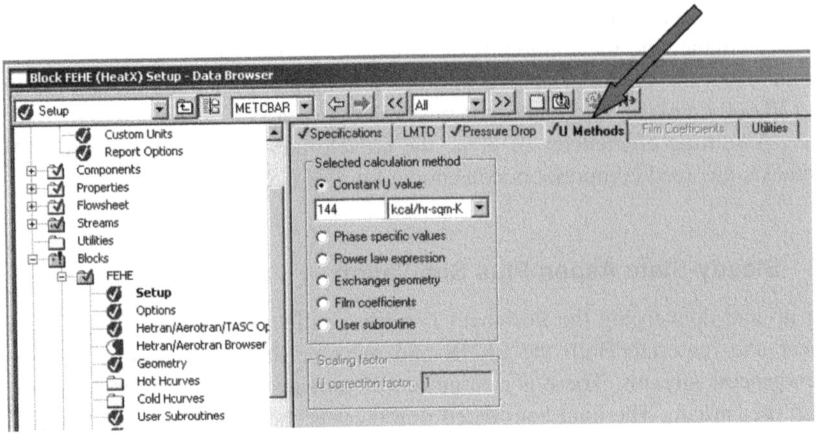

Figure 6.85 Setting the *U*.

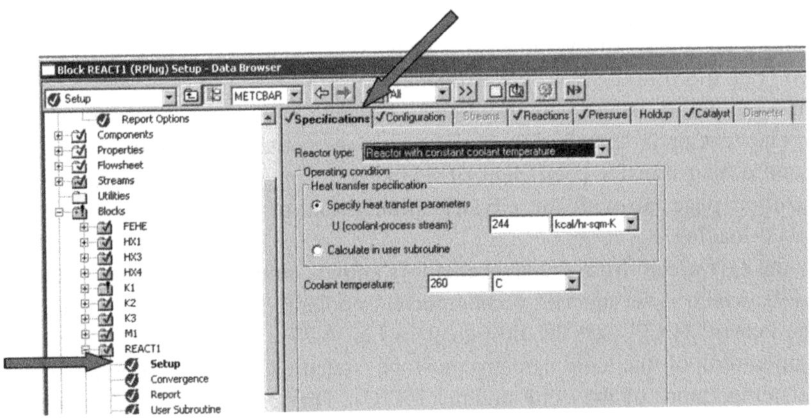

Figure 6.86 First reactor setup specifications.

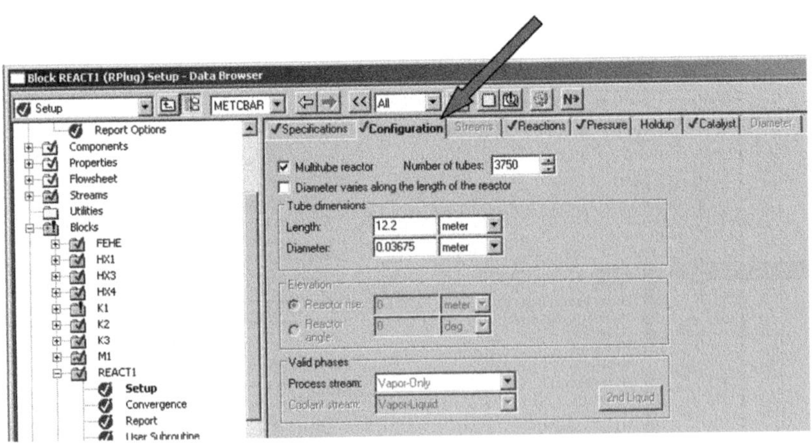

Figure 6.87 First reactor configuration.

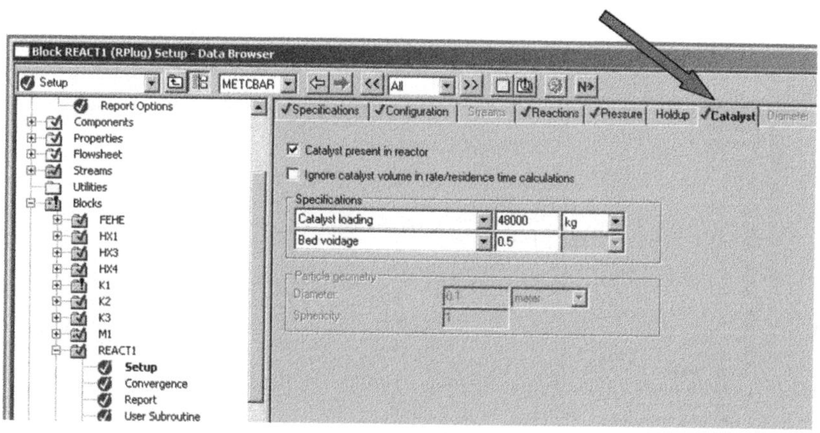

Figure 6.88 First reactor catalyst.

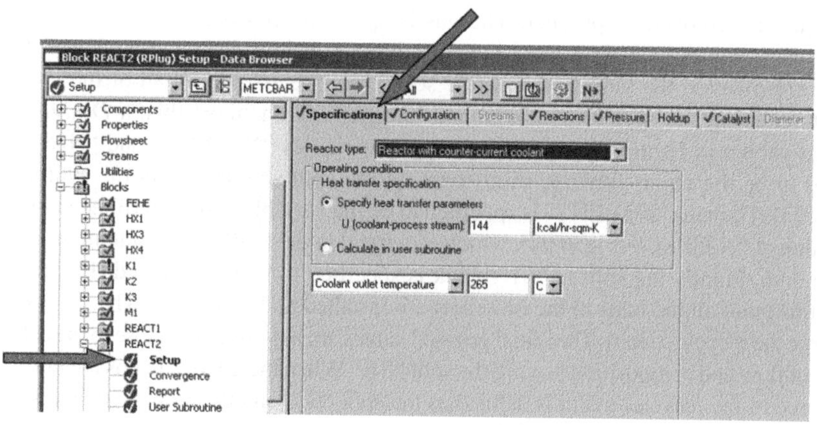

Figure 6.89 Second reactor setup specifications.

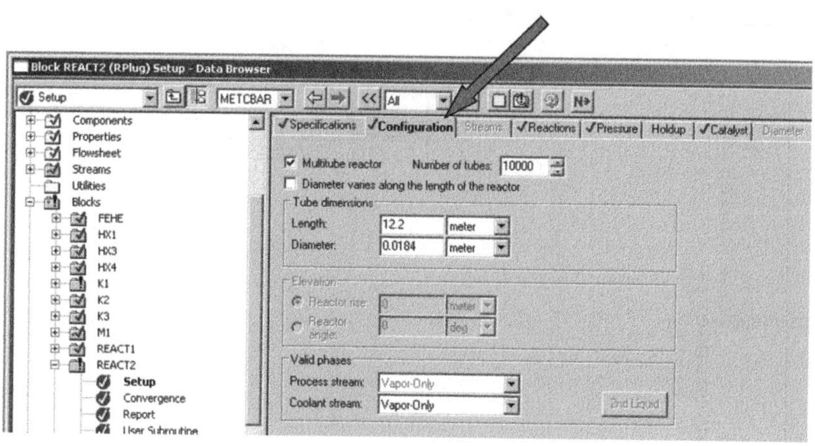

Figure 6.90 Second reactor configuration.

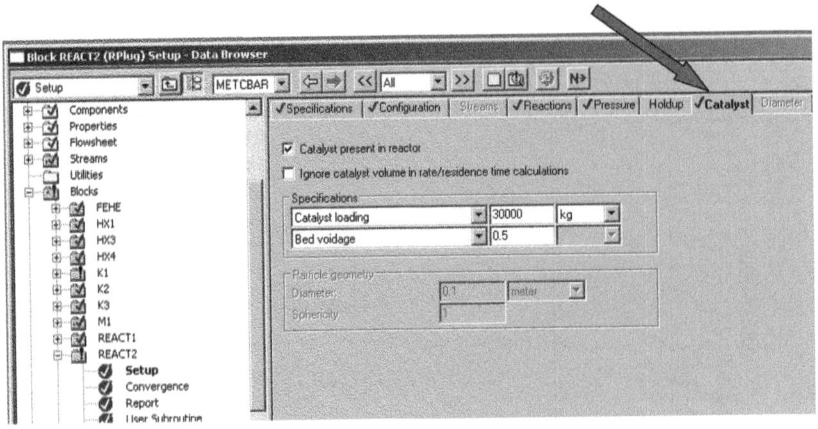

Figure 6.91 Second reactor catalyst.

The gas is cooled to 38°C in a cooler–condenser and flashed in a separator vessel. The liquid is removed as product. The gas is split between the recycle stream and the vent stream. Figure 6.92 shows that the vent flowrate is initially set to be 5% of the total flowrate. However, the vent flow must be set to remove essentially all of the inert components entering the system (methane and nitrogen). To achieve this, a *Design Spec* is used, as shown in Figures 6.93–6.96. First, the variables "N2VENT" and "N2SYN" are defined to be the molar flowrate of nitrogen in the vent stream and in the synthesis gas (Fig. 6.93). Then the "N2VENT" is specified to be the variable to be controlled at the specified value of "N2SPEC" (Fig. 6.94), which is defined in the FORTRAN statement shown in Figure 6.96. Finally the split in "S1" block is specified to be manipulated (Fig. 6.95).

At this point all the units in the flowsheet are installed and converged. The last issue is to converge the recycle stream. The initial guessed values are adjusted to be close to the calculated values of flow and composition leaving the split "S1." When these two streams are fairly close, the source of the recycle stream is defined as the split "S1" and the recycle stream is defined as a *Tear* stream. The flowsheet did not converge when the default convergence method

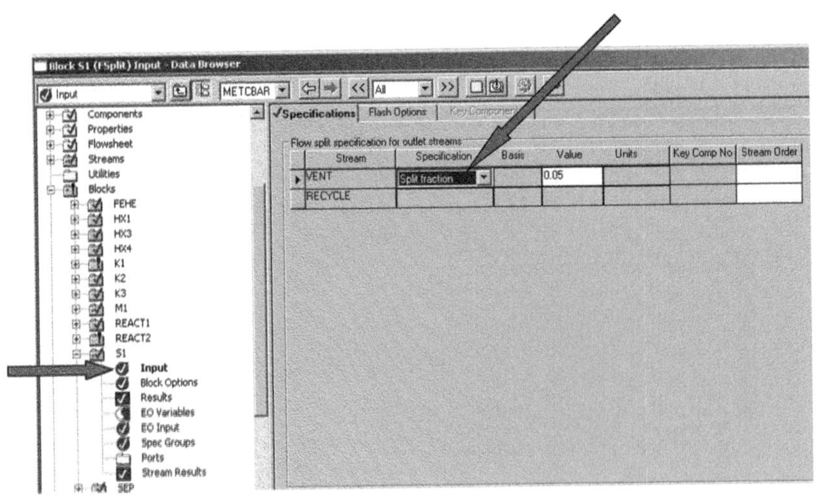

Figure 6.92 Split between recycle and vent.

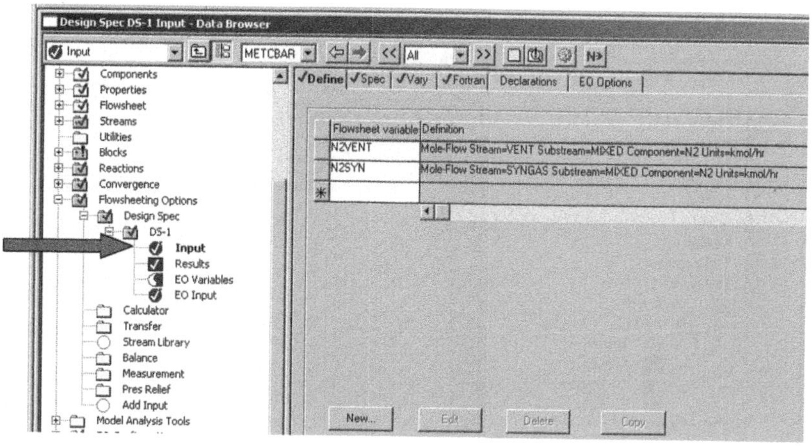

Figure 6.93 Setting up a design spec; defining variables.

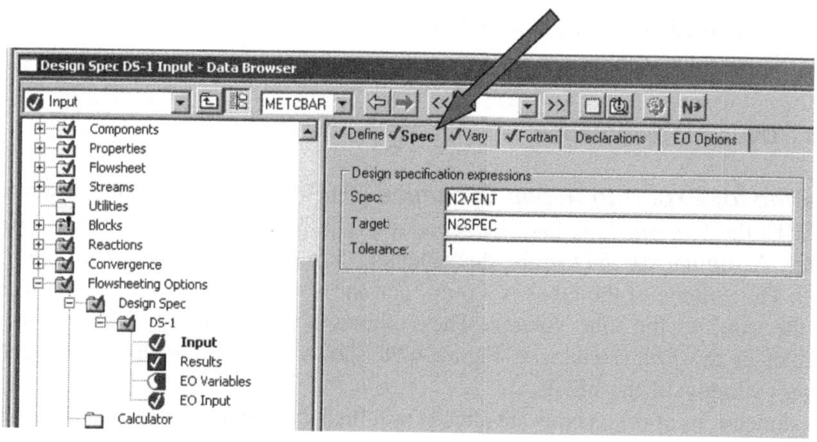

Figure 6.94 Nitrogen in vent controlled at specification.

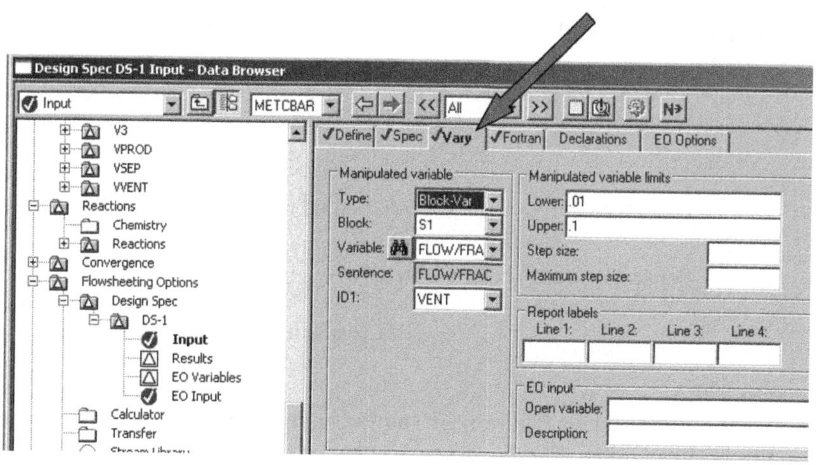

Figure 6.95 Split is varied to change vent flowrate.

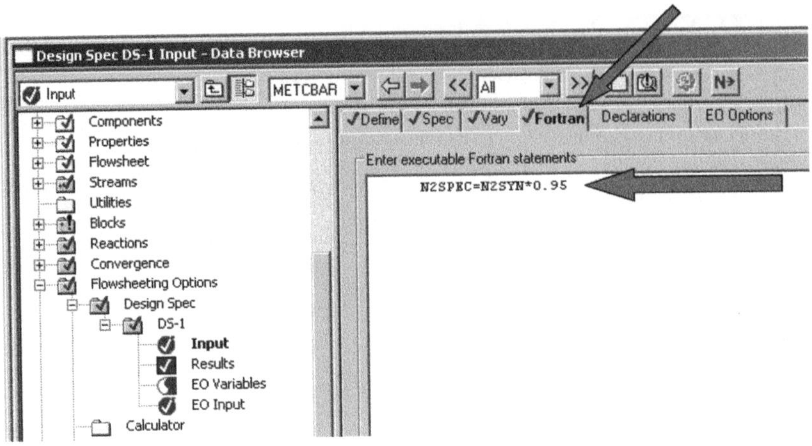

Figure 6.96 Specifying 95% removal of nitrogen.

(Wegstein) was used. Switching to the Broyden method successfully converged to flowsheet. Figure 6.97 gives the temperature and composition profiles in the two reactors.

6.11.4 Dynamic Simulation

Preparing for Export to Aspen Dynamics The dynamic units in the flowsheet are the FEHE, the two reactors, and the separator. The heat-transfer area of the FEHE is 1350 m^2. Assuming 0.0254 m tube diameter and 5 m tube length, the number of tubes is 3390. The volume of the tubes is 8.6 m^3, and the volume on the shell side is assumed to be the same as the tube volume. The volumes are split equally between the inlet and outlet of the heat exchanger. Figure 6.98 shows these parameters inserted in the *Dynamics* window of the FEHE.

The dimensions of the reactors are already specified, but the dynamics of the heat transfer between the catalyst and the process gas streams must be specified. Figure 6.99 shows this for the first reactor. The second reactor is handled in the same way.

The separator is sized by using an F factor of 0.4 (in metric units) to determine the gas velocity. The volumetric flowrate is 8200 m^3/h, the gas density is 33 kg/m^3, and an aspect ratio of 2 is used:

$$
\begin{aligned}
F\ \text{factor} &= V\sqrt{\rho_v} \\
0.4 &= V\sqrt{33} \\
\therefore V &= 0.07\ \text{m/s} \\
\text{Separator cross-sectional area} &= \frac{8200/3600}{0.07} = 32\ \text{m}^2 \\
\therefore D &= 6.4\ \text{m}
\end{aligned}
\tag{6.21}
$$

See Figure 6.100.

Note that the flowsheet in Figure 6.77 shows a number of valves that are installed between the various units. These valves are not needed for a steady-state design. However, they must be used for a pressure-driven dynamic simulation to provide some

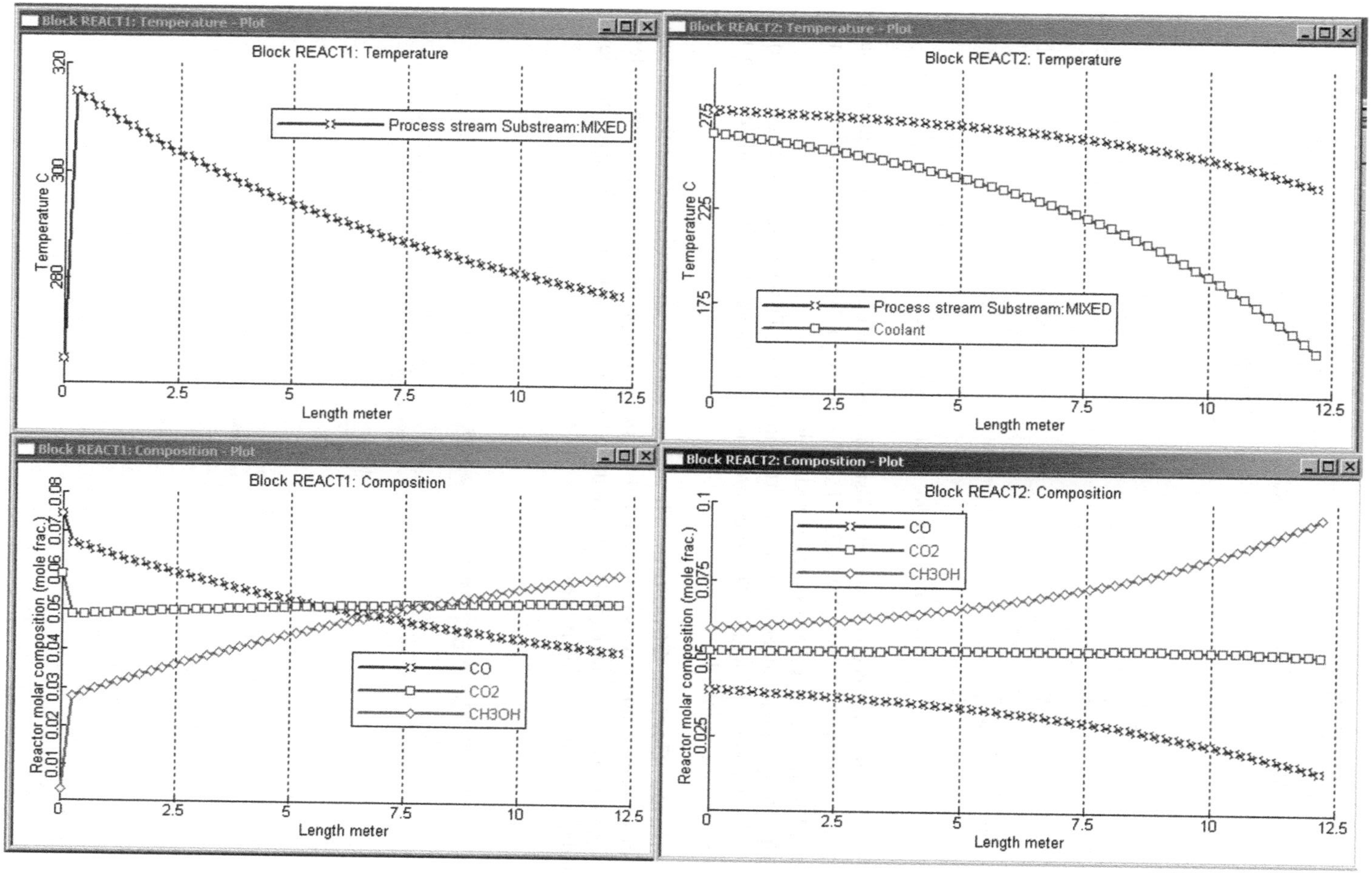

Figure 6.97 Reactor temperature and composition profiles.

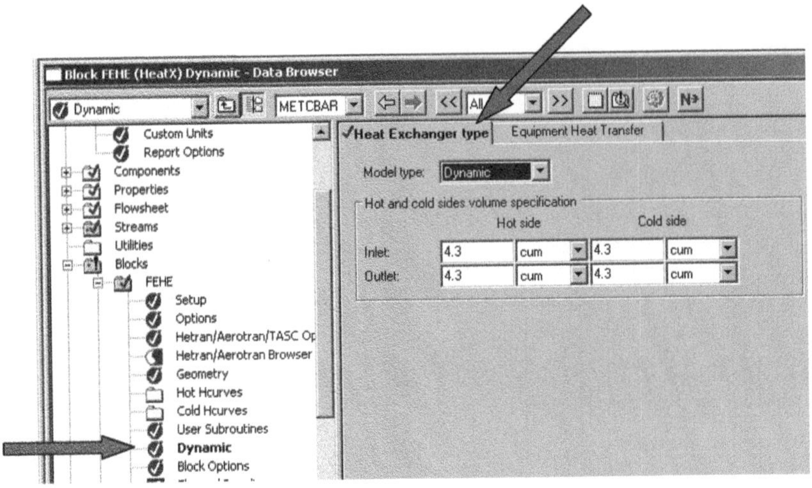

Figure 6.98 FEHE dynamics.

resistance to flow between the vessels. A small pressure drop (0.1 bar) is assigned to each of these "resistance" valves (V4, V3, V1, V2, and Vsep). When the simulation is run in Aspen Dynamics, these valves are opened wide.

The file is pressure checked and exported into Aspen Dynamics. Figure 6.101 shows the default control structure. The pressures in each reactor and the liquid level in the separator are controlled.

Plantwide Control Structure There are several alternative control structures for this process, and there is no claim that the one developed is the "best" from whatever perspective you consider. What is claimed is that it provides effective base-level regulatory control of this process.

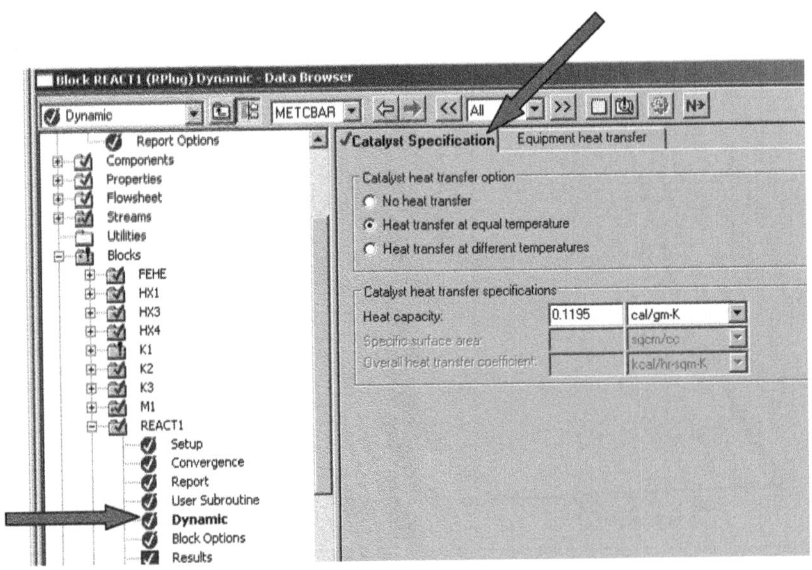

Figure 6.99 First reactor dynamics.

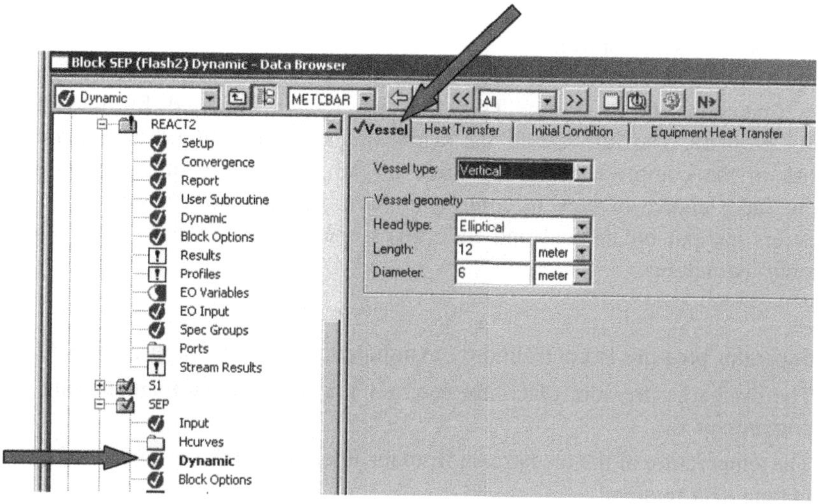

Figure 6.100 Separator dynamics.

The first question to be answered is how to control pressure in the system [the vent flow is a small stream (273 m^3/h); the total gas volume in the system is 296 m^3]:

$$\text{FEHE (both hot and cold sides)} = 17\ m^3$$

$$\text{First reactor (half of tube volume)} = 24\ m^3$$

$$\text{Second reactor (full tube volume and half of shell volume)} = 49\ m^3$$

$$\text{Separator (half of total volume)} = 206\ m^3$$

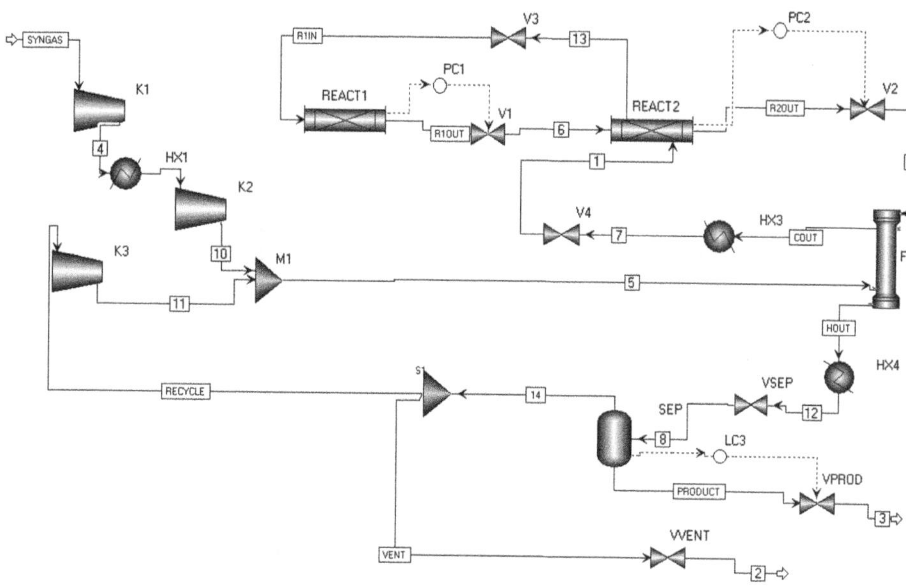

Figure 6.101 Default control structure in Aspen Dynamics.

Therefore the time constant using the vent flow to control pressure is about 1 h, which indicates poor pressure control.

On the other hand, the fresh feed flowrate of synthesis gas is large (5920 m^3/h), which gives a time constant of 3 min. Therefore the feed is selected for the control of separator pressure. The flowrate of the synthesis gas feed is adjusted by manipulating the work to the compressor K2. The rest of the control structure follows logically from this basic choice of how to control pressure. The default pressure controllers on the reactors are put on manual and the valves opened wide. Figure 6.102 shows the final control structure:

1. Separator pressure is controlled by manipulating work to compressor K2.
2. The discharge pressure of compressor K1 is controlled by manipulating work to compressor K1.
3. The temperature of the feedstream from the intercooler HX1 is controlled by manipulating heat removal.
4. The exit temperature from the heat exchanger HX3 is controlled by manipulating the heat input.
5. The exit temperature from the first reactor is controlled by manipulating the temperature of the coolant (steam temperature).
6. Separator temperature is controlled by manipulating the heat removal in heat exchanger HX4.
7. Separator liquid level is controlled by manipulating liquid product withdrawal.
8. The composition of methane in the recycle stream is controlled by manipulating the vent flowrate.
9. The power to the recycle compressor K3 is held constant, which gives an essentially constant recycle flowrate.

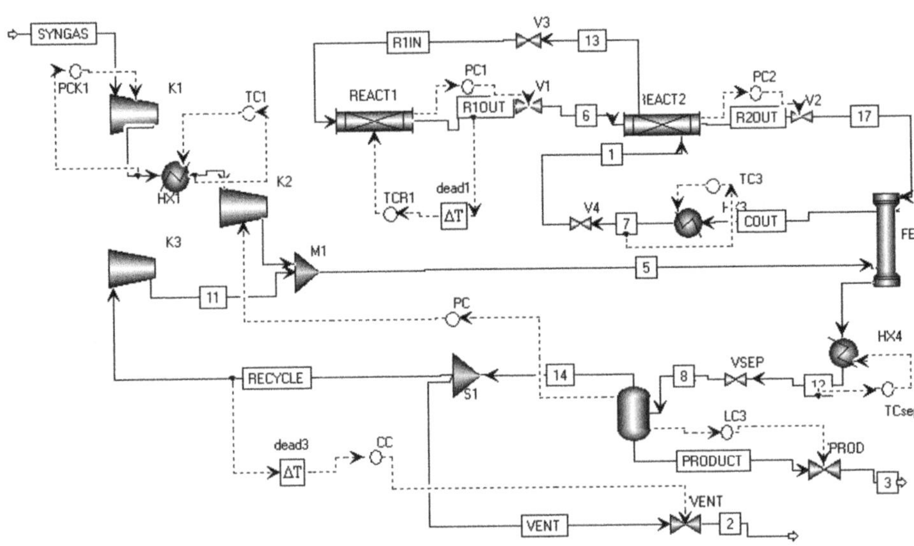

Figure 6.102 Control structure.

The default integration algorithm is Aspen Dynamics is Implicit Euler. It was found that the dynamic simulation ran quite slowly with this algorithm. For this methanol process, which is a gas-phase system, the dynamics of some of the units are quite fast while the overall dynamics of the pressure in the gas loop is somewhat slow. This mixture of dynamic time constants is handled better by using the Gear algorithm. You switch algorithms by clicking *Run* on the top toolbar and selecting *Solver Options*. The window shown in Figure 6.103 opens, on which the Gear algorithm can be selected when the *Integrator* page tab is selected.

Steady State from Dynamic Model The simulation is run out to a steady-state condition with the adjustable temperatures set to the same values as used in the steady-state Aspen Plus. These temperatures are the exit temperature from the first reactor (277°C) and the exit temperature from the heater "HX3" (150°C). The power to the recycle compressor was kept the same.

Quite remarkably, the steady state is very different. The synthesis gas feedrate is reduced by a factor of 2 (11,450 to 4990 kmol/h). Similar reductions in product and

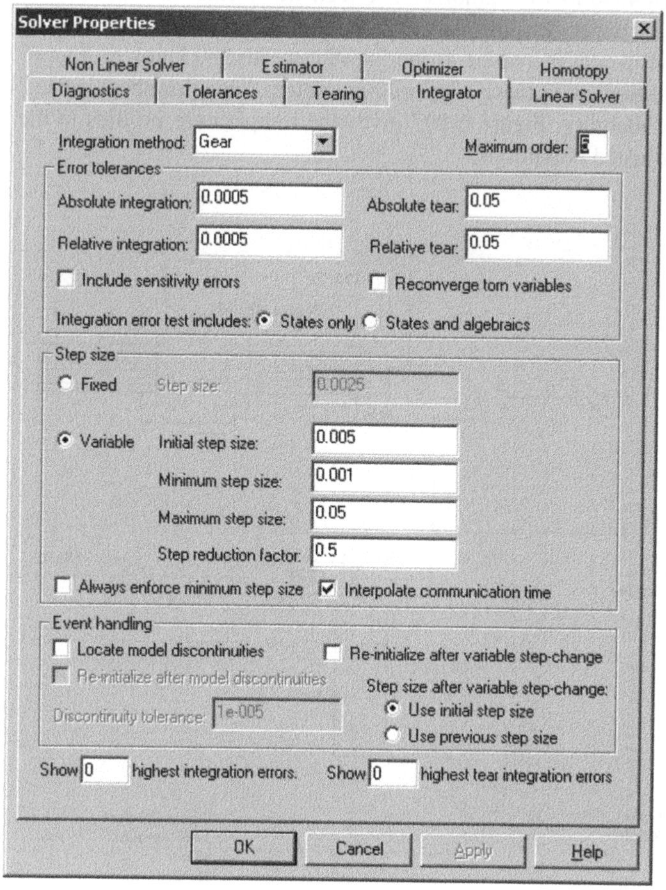

Figure 6.103 Selecting gear integration method.

vent flow occur, as shown in Figure 6.104. The explanation for this extreme difference is unclear. There are differences between the models used in Aspen Plus for plug flow reactors (a rigorous distributed system model) and the models used in Aspen Dynamics (a lumped model). A 50-lump model is used, so we would not expect such a large difference. Figure 6.105 shows the steady-state Aspen Dynamics temperature profiles in the two reactors. These should be compared with those given in Figure 6.97 from the steady-state Aspen Plus simulation.

The temperature profiles in the first reactor are significantly different. The Aspen Plus temperature profile rises very rapidly at the front of the reactor to a temperature of ~315°C. The Aspen Dynamics temperature profile rises gradually to a peak temperature of only 278°C. So the reaction rates are much smaller, and less methanol is produced.

The temperature profiles in the second reactor are also somewhat different. The Aspen Plus temperature profile drops to about 335°C at the exit. The Aspen Dynamics temperature profile drops all the way to 210°C with the same value for the coolant inlet temperature (150°C). This indicates much less reaction is occurring.

Adjustments were made in several parameters in order to increase the fresh feed. The exit temperature from the first reactor was reduced from 277 to 245°C. The exit temperature from the heater "HX3" was reduced from 150 to 110°C. These two changes increased the fresh feed from 4990 to 7415 kmol/h. The feed was further increased to 9288 kmol/h by increasing the power to the recycle compressor "K3" from 1884 to 4000 kW. The recycle flowrate increased from 34,880 to 46,380 kmol/h. Figure 6.106 gives flowsheet conditions with these adjusted parameters. This flowsheet is used for exploring dynamic controllability. Figure 6.107 gives the temperature profiles in the two reactors under these adjusted conditions.

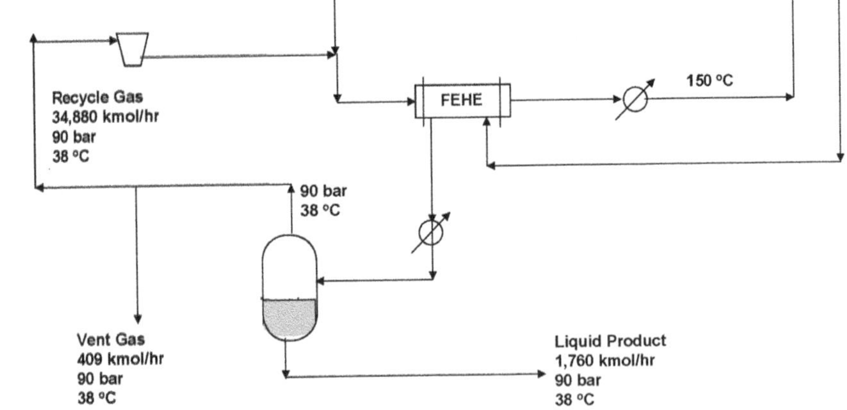

Figure 6.104 Steady state from Aspen Dynamics.

Reactor 1:

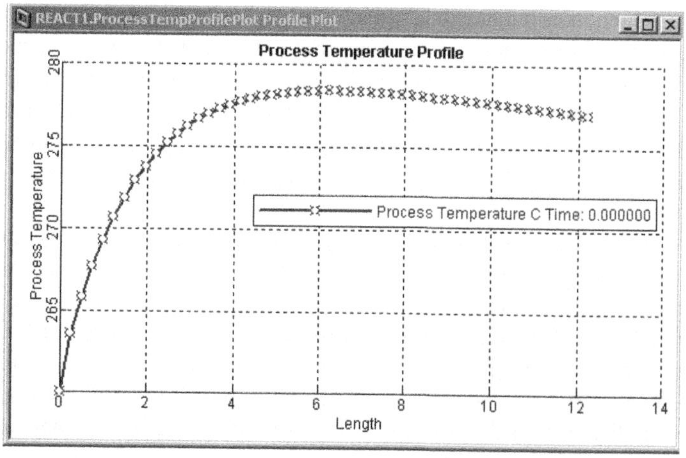

Reactor 2:

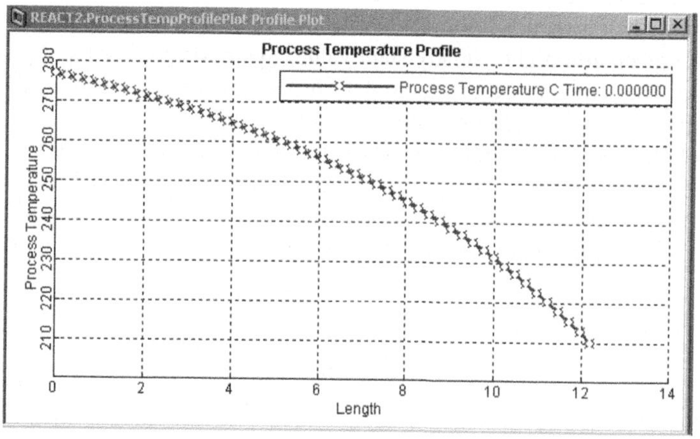

Reactor 2:

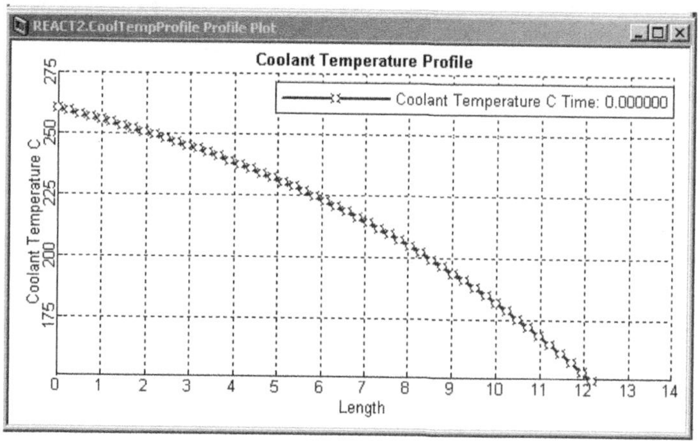

Figure 6.105 Reactor temperature profiles in steady state from Aspen Dynamics with 277°C exit temperature on first reactor.

Dynamic Control A deadtime of 1 min is inserted in the first reactor temperature controller (TCR1), which manipulates the coolant temperature. The ranges of the temperature transmitter and the coolant temperature are both 200–300°C. A deadtime of 3 min is inserted in the composition controller (CC), which manipulates the vent flowrate. The composition transmitter range is 0–30 mol% methane. Relay–feedback tests and Tyreus–Luyben settings give controller tuning constants of $K_C = 0.50$ and $\tau_I = 6.6$ min for TCR1 and $K_C = 2.5$ and $\tau_I = 51$ min for CC. Figure 6.108 shows the controller faceplates at steady-state conditions.

The process is subjected to a number of disturbances, and the control structure handles all of them quite effectively. Dynamic responses to changes in the setpoint of the temperature controller in the first reactor are shown in Figure 6.109. At 0.1 h, the setpoint is increased from 245 to 255°C. At 3 h, it is decreased to 235°C. Decreasing the temperature in the first reactor results in an increase in throughput. The synthesis gas feedrate, the product rate, and the vent rate all increase. The opposite occurs when the temperature is increased. This indicates that the reaction is equilibrium-limited, not kinetically limited. Decreasing temperature increases the equilibrium constant of exothermic reactions.

Figure 6.110 gives responses to changes in the setpoint of the methane composition controller. At 0.1 h, it is reduced from 24 to 20 mol% methane. The vent flowrate increases sharply and ends up at a higher steady-state value. Synthesis gas feed and product flows both increase. The temperature of the coolant in the first reactor decreases because the increase in the reaction rate requires more heat removal.

Figure 6.111 shows what happens when the power to the recycle compressor is changed. At 0.1 h, power is increased from 4000 to 6000 kW, and at 3 h it is decreased to 3000 kW. The recycle flowrate changes directly with compressor power, which results in corresponding changes in throughput.

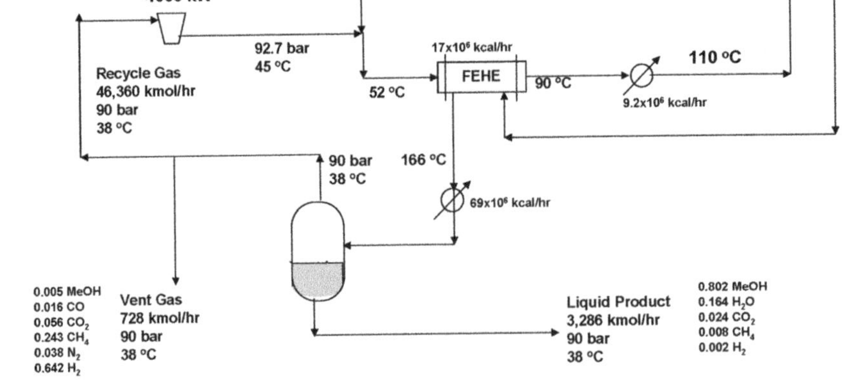

Figure 6.106 Adjusted steady state.

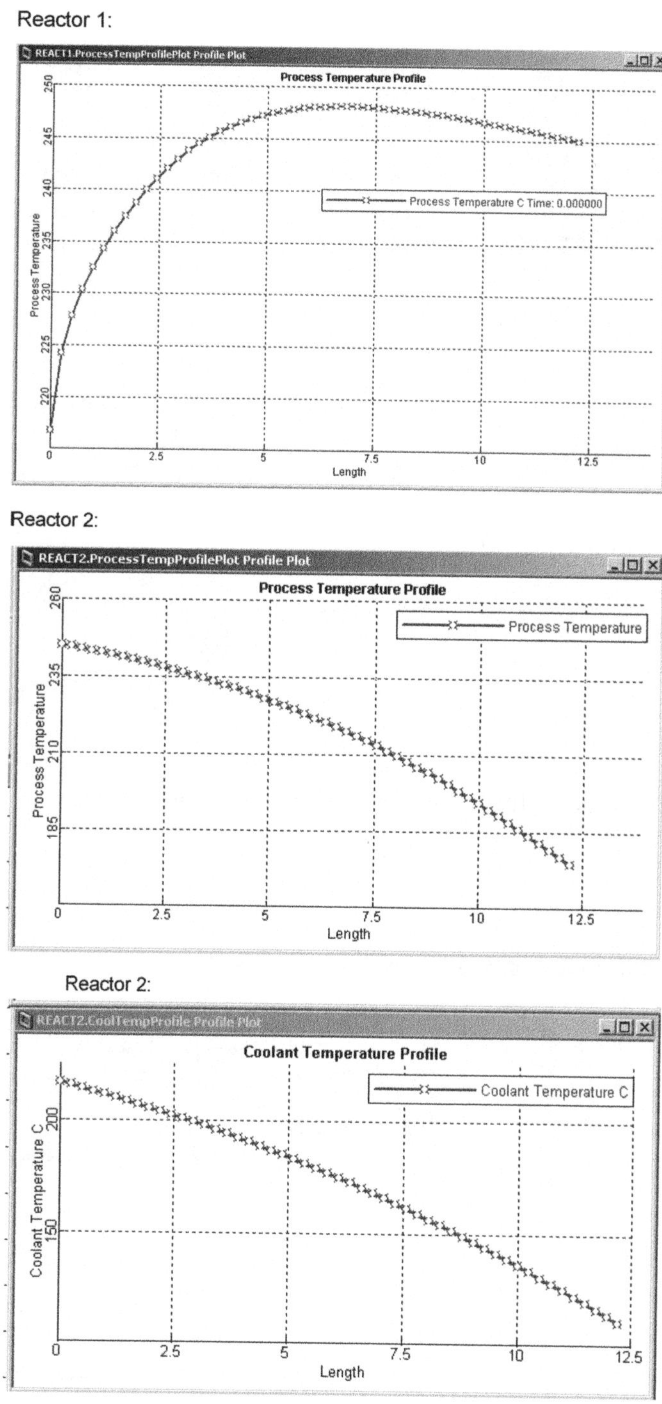

Figure 6.107 Reactor temperature profiles in adjusted steady state.

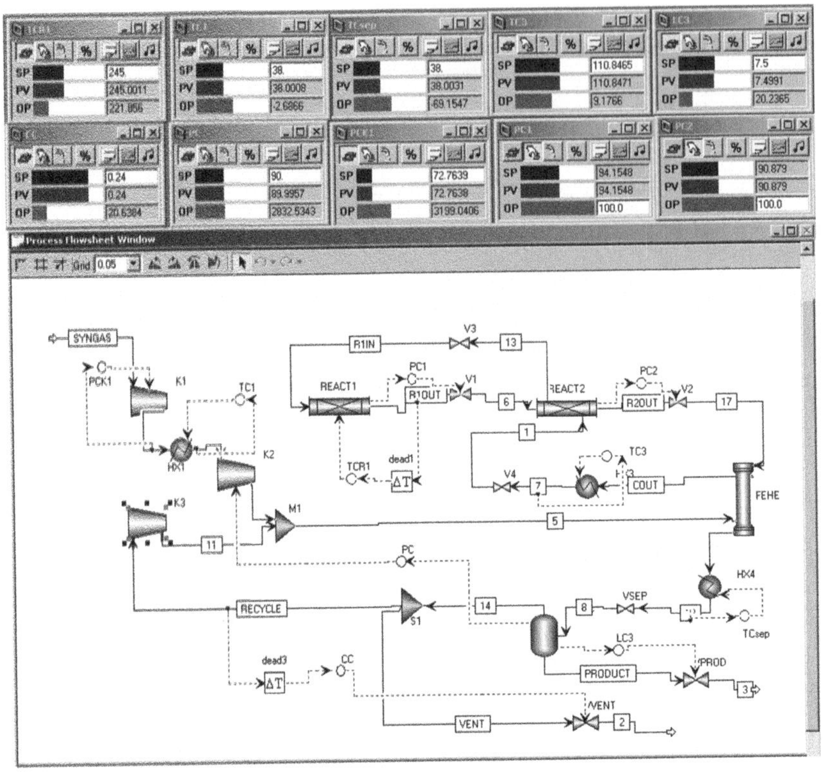

Figure 6.108 Controller faceplates and control structure.

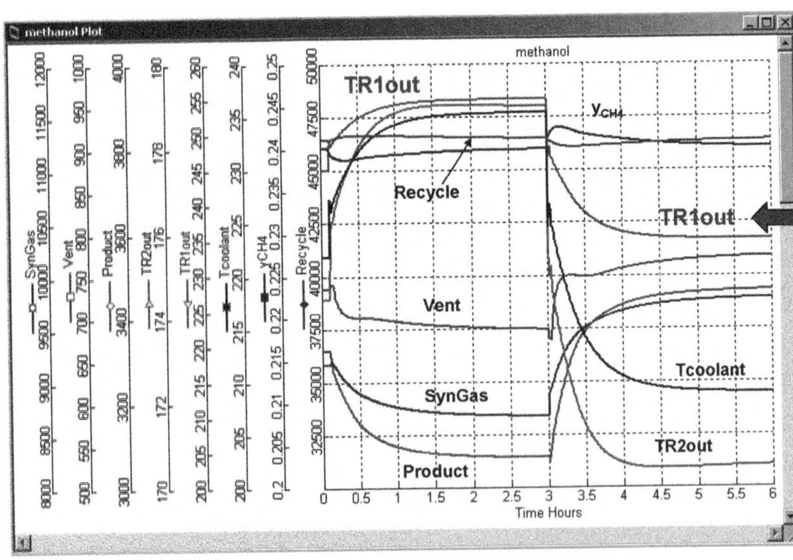

Figure 6.109 Changes in first reactor exit temperature $T_{R1,\text{out}}$.

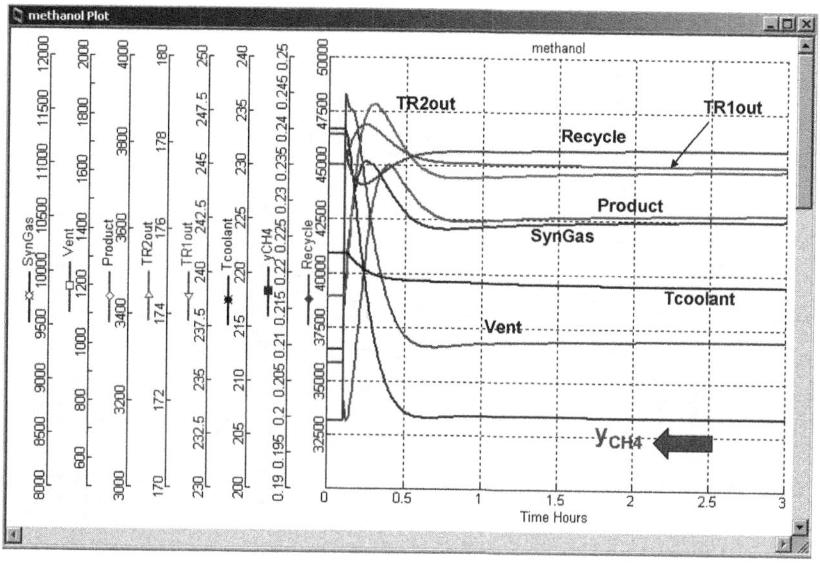

Figure 6.110 Changes in methane composition controller setpoint.

Figure 6.112 gives responses to changes in the temperature leaving the heat exchanger HX3. It is changed from 110 to 130°C at 0.1 h. Then it drops to 90°C at 3 h. The temperature of the cold stream leaving the second reactor (TR2out) changes directly with the temperature of the inlet hot stream to this reactor–heat exchanger. The higher the temperature, the lower the throughput, which again indicates equilibrium-limited reactions.

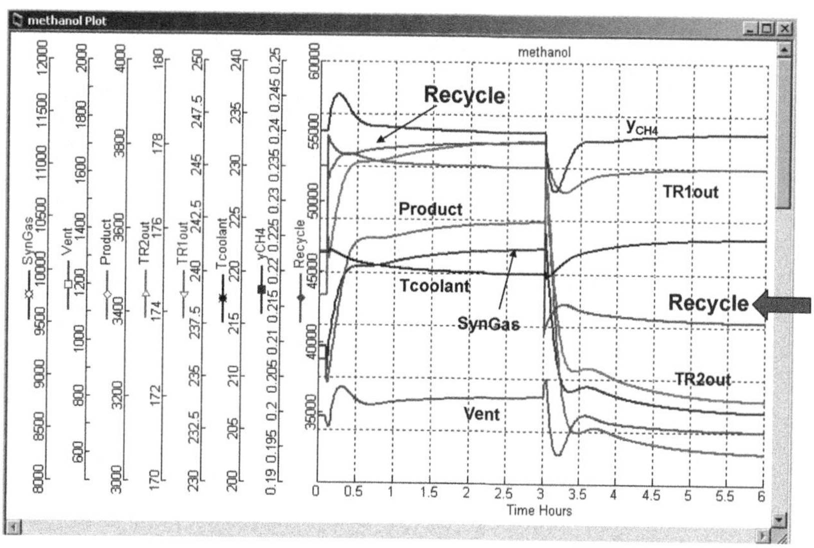

Figure 6.111 Changes in recycle compressor power.

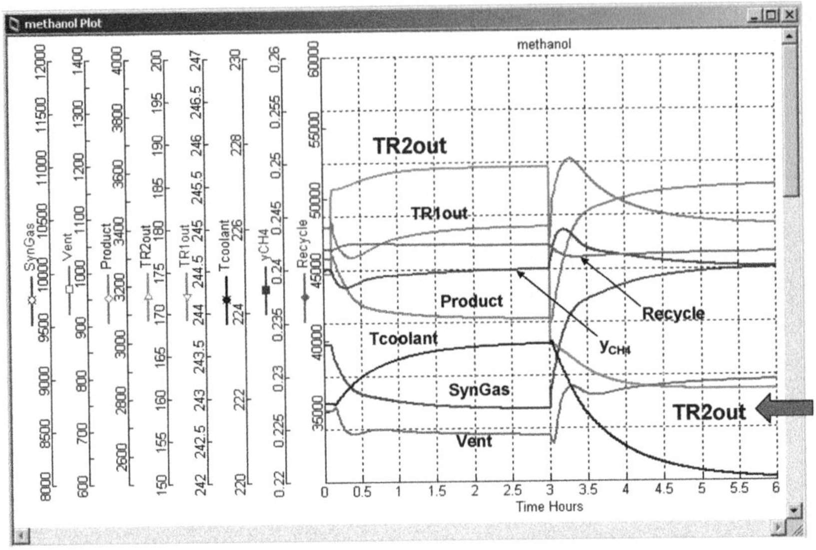

Figure 6.112 Changes in HX3 exit temperature.

6.12 CONCLUSION

The dynamics and control of a number of tubular reactor systems have been studied in this chapter. Both adiabatic and cooled tubular reactors have been explored in both isolation and a plantwide environment. Ideal systems have been studied using Matlab programs. Real chemical systems have been studied using Aspen Dynamics.

A complex methanol production process has been simulated and successfully controlled.

CHAPTER 7

HEAT EXCHANGER/REACTOR SYSTEMS

In Chapters 5 and 6, high-temperature exothermic tubular reactor systems were considered. All of these systems used feed-effluent heat exchangers (FEHE) to preheat the feed to the desired reactor inlet temperature by recovering heat from the hot reactor exit stream. Some of the systems also used a trim furnace to add additional heat if needed.

In this chapter we consider the FEHE–furnace system in more detail in order to illustrate the inherent dynamic problems with using reactor effluent for preheating the feed, despite its steady-state economic advantages.

7.1 INTRODUCTION

Because of the need to provide a desired inlet temperature in plug flow reactor (PFR) systems, the cold feedstream often needs to be heated. Also the hot effluent stream from the reactor usually needs to be cooled before sending it to the separation section. A heat exchanger network is typically used to preheat the cold feed with the hot reactor effluent.

The startup of these systems requires that some type of heater or furnace be provided to initially achieve the temperature required for the reaction to "light off." Once the reactor exit temperature becomes sufficiently higher than the reactor inlet temperature, the required preheating can usually be achieved in the FEHE, provided its area is large enough and the temperature rise through the reactor is large enough. Under these conditions, no heat input is required in the furnace. However, since this startup furnace is available, it can potentially be used as an additional manipulated variable. A comparison of these alternative process and control structures is the subject of this chapter.

Figure 7.1 shows a typical chemical process in which a feed-effluent heat exchanger is coupled with an adiabatic exothermic reactor. The heat of reaction produces a reactor

Chemical Reactor Design and Control. By William L. Luyben

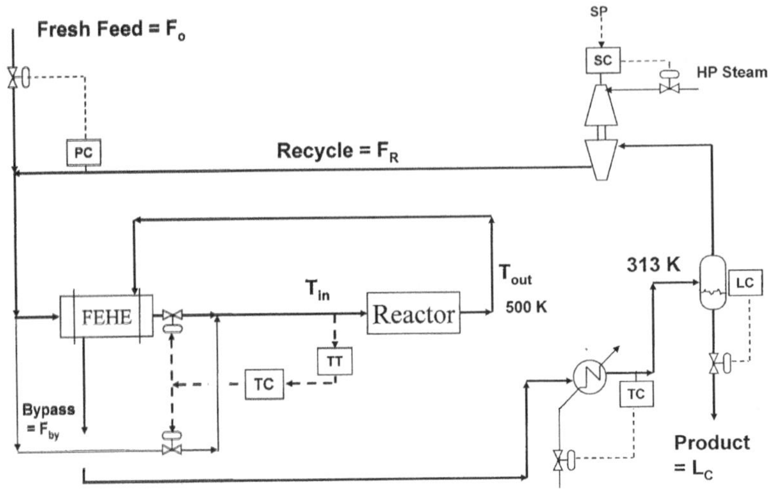

Figure 7.1 Inlet temperature control; no furnace.

effluent temperature T_{out} that is higher than the temperature of the feedstream to the reactor T_{in}. Therefore heat can be recovered from the hot stream leaving the reactor. The control objective is to maintain the reactor inlet temperature by manipulating the bypass flow F_{by} of cold material around the heat exchanger. Split-range valves are used to manipulate the bypass flowrate and the flowrate through the FEHE. A simple separation section is assumed that consists of a separator drum. All of the reactants go into the gas phase and are recycled. All of product C is removed in the liquid stream.

Figure 7.2 shows the design alternative in which a furnace is installed between the FEHE and the reactor. Furnace inlet temperature is controlled by manipulating the

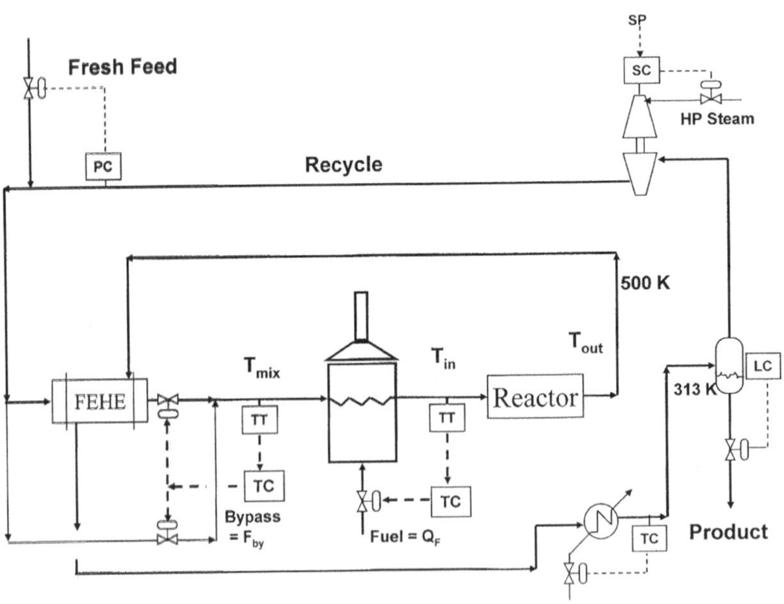

Figure 7.2 Inlet temperature control; with furnace.

flowrate of the stream F_{by} that bypasses around the heat exchanger and is blended with the gas leaving the FEHE at temperature T_{Cout}, giving a temperature of T_{mix}. Reactor inlet temperature T_{in} is controlled by manipulating heat input to the furnace Q_F.

7.2 STEADY-STATE DESIGN

The optimum steady-state design of the process involves an economic tradeoff between (1) the capital costs of the reactor vessel, catalyst, heat exchangers, furnace, and compressor and (2) the operating costs of compressor work and furnace energy. The exit temperature of the reactor is fixed at its maximum value $T_{out} = 500$ K as limited by safety, catalyst degradation, undesirable side reaction, and other factors. The highest reactor exit temperature is used at the design stage because it minimizes reactor size. At the operating stage, using the maximum reactor exit temperature maximizes capacity or minimizes recycle.

This process has two design optimization variables. The temperature rise through the reactor ΔT_R and the ratio of the furnace energy to the total energy transferred into the feedstream Q_F/Q_{total} are selected as the two variables to be adjusted to minimize the total annual cost of the process. The production rate of C is set at 0.12 kmol/s, which sets the fresh feed flowrate at 0.24 kmol/s of an equimolal mixture of reactant components A and B.

To design the heat exchanger, the heuristic is used that the minimum approach temperature differential $\Delta T_H = T_{out} - T_{Cout}$ is 25 K, which is reasonable for the temperature level in this process. This pinch temperature differential occurs at the "hot end" of the heat exchanger. An overall heat transfer coefficient $U = 0.142$ kJ s^{-1} m^{-2} K^{-1} is used in this gas–gas system.

Fixing ΔT_R sets the reactor inlet temperature T_{in} and the recycle flowrate F_R because of the energy balance around the reactor with the fixed conversion. The larger is ΔT_R, the lower is T_{in} and the smaller the recycle flowrate required to give the 500 K exit temperature. Thus increasing ΔT_R, reduces compression costs. However, the lower T_{in} results in a larger reactor.

Fixing the Q_F/Q_{total} ratio sets the furnace firing, the heat exchanger bypass flowrate, the size of the heat exchanger, and the size of the furnace. Increasing the ratio increases the capital cost of the furnace, decreases the capital cost of the heat exchanger, increases the operating cost of furnace fuel, and increases the bypass flowrate.

The design procedure is the following:

1. Specify the desired production rate and the maximum reactor exit temperature.
2. Pick a value of temperature rise through the reactor ΔT_R (to be optimized).
3. Calculate the inlet reactor temperature and the recycle flowrate from a reactor energy balance.
4. Calculate the inlet molar flowrates to the reactor.
5. Integrate down the length of the reactor until the temperature is 500 K, which also corresponds to producing the desired amount of product (0.12 kmol/s). This gives reactor size and the amount of catalyst W_{cat} required.
6. Pick a value of the Q_F/Q_{total} ratio (to be optimized).
7. Calculate the temperatures in and out of the furnace and heat exchanger, and calculate the heat transfer area.

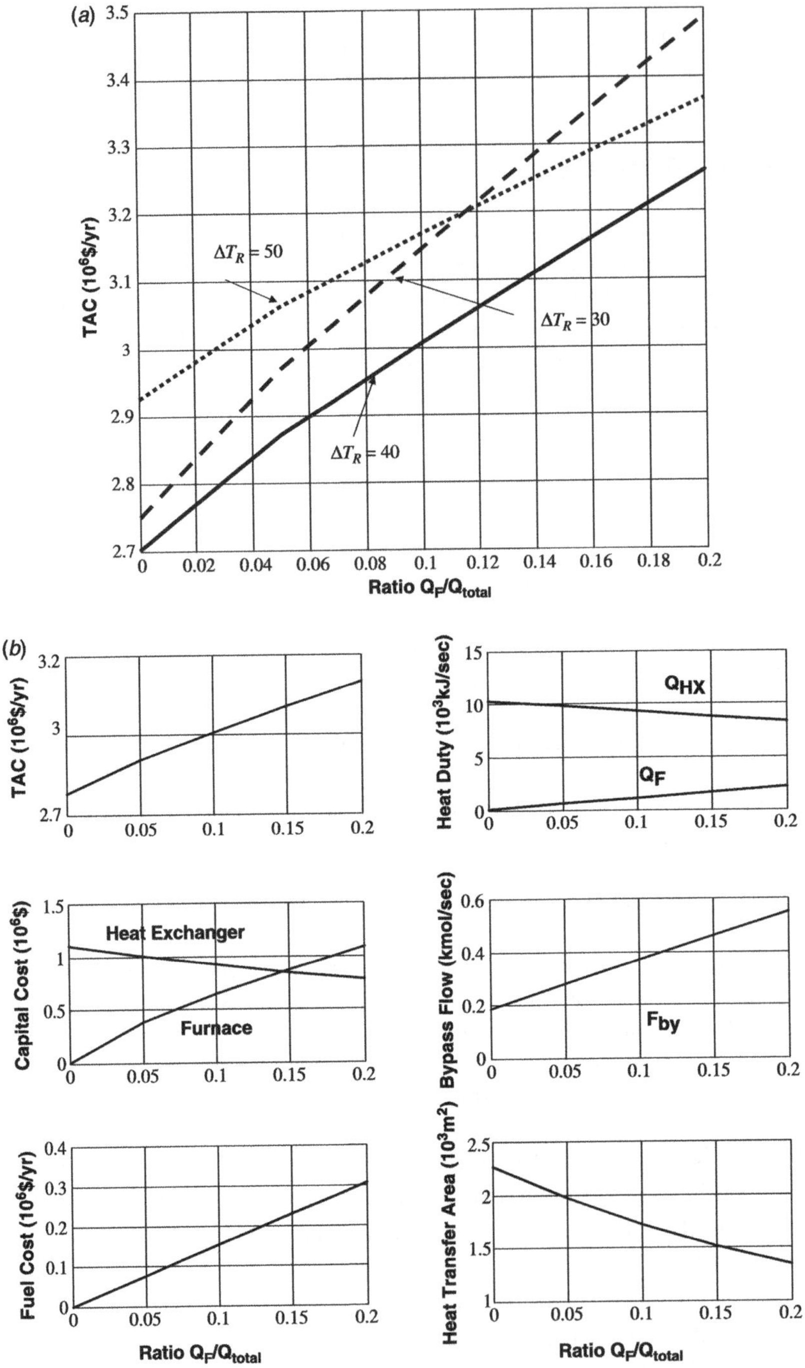

Figure 7.3 (*a*) Total annual cost; (*b*) effect of Q_F/Q_{total} for $\Delta T_R = 40$ K Case.

8. Evaluate the total annual cost TAC of the process using the economic values given in Chapter 5.
9. Vary the Q_F/Q_{total} ratio from 0 to 0.2 and repeat steps 7 and 8.
10. Vary ΔT_R until the minimum TAC is obtained.

Figure 7.3 shows results of this optimization procedure. Figure 7.3*a* shows that the minimum total annual cost occurs with $\Delta T_R = 40$ K and with no energy provided by the furnace. Figure 7.3*b* shows how the individual values vary with the Q_F/Q_{total} ratio for the $\Delta T_R = 40$ K case. Furnace capital cost, energy costs and bypass flowrate increase with larger ratios. Heat exchanger capital cost decreases. The steady-state economics favor the flowsheet in which a furnace *is not* used. As we will demonstrate in the following sections, consideration of dynamic controllability favors the flowsheet in which a furnace *is* used. Thus this process provides another important example of the ever-present interaction and conflict between steady-state economics and dynamic controllability.

7.3 LINEAR ANALYSIS

In a later section we show results of rigorous dynamic simulations of this process. However, it may be useful at this point to show the predictions of a linear model of this type of FEHE–reactor system.

Block diagrams of the linear openloop process are shown in Figure 7.4. The two alternative flowsheets are labeled "FS1," in which no furnace is used, and "FS2," in which a furnace is used. The reactor transfer function is $G_{R(s)}$, representing the adiabatic tubular reactor. The reactor by itself is openloop-stable. In Figure 7.4*a* a simple first-order lag is used for purposes of illustration:

$$G_{R(s)} = \frac{T_{\text{out}(s)}}{T_{\text{in}(s)}} = \frac{K_R}{\tau_R s + 1} \tag{7.1}$$

In this initial analysis, the dynamics of the heat exchanger are assumed to be fast compared to those of the reactor, so simple gain transfer functions are used that relate the blended temperature T_{in} in FS1 or T_{mix} in FS2 to the two inputs to the heat exchanger: bypass flowrate F_{by} and reactor exit temperature T_{out}:

$$T_{\text{in}(s)} = K_1 T_{\text{out}(s)} + K_2 F_{\text{by}(s)} \tag{7.2}$$

Later in the rigorous simulations, the dynamics of the heat exchanger are included because of the significant thermal capacitance of the metal tubes in this very large unit.

7.3.1 Flowsheet FS1 without Furnace

In this flowsheet there is only one controller, whose transfer function we call $G_{C1(s)}$:

$$G_{C1(s)} = \frac{F_{\text{by}(s)}}{E_{1(s)}} = \frac{F_{\text{by}(s)}}{T^{\text{set}}_{\text{in}(s)} - T_{\text{in}}} \tag{7.3}$$

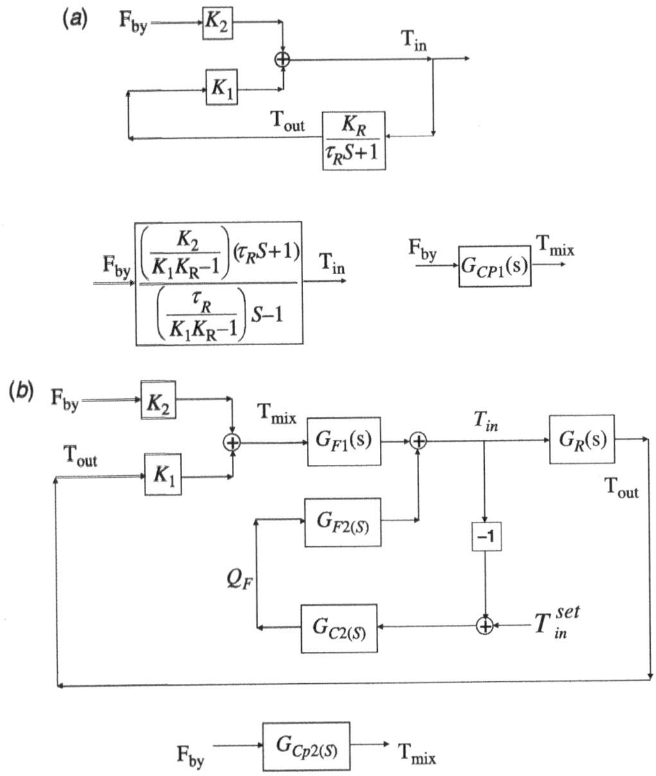

Figure 7.4 Block diagrams (*a*) without furnace (FSI); (*b*) with furnace (FS2).

Both P and PI controllers are considered below. This controller looks at the *coupled* openloop process transfer function $G_{\mathrm{CP1}(s)}$ shown in Figure 7.4*a*:

$$G_{\mathrm{CP1}(s)} = \frac{T_{\mathrm{in}(s)}}{F_{\mathrm{by}(s)}} = \left[\frac{\left(\dfrac{K_2}{K_R K_1 - 1}\right)(\tau_R s + 1)}{\left(\dfrac{\tau_R}{K_R K_1 - 1}\right) s - 1} \right] \tag{7.4}$$

The basic dynamic problem is clearly revealed by looking at the pole of this openloop transfer function. The pole is positive, meaning an openloop unstable process, if the product of the gains $K_R K_1$ is greater than one. The heat exchanger gain K_1 depends on the heat transfer area and the approach temperature differential on the hot end of the process (the temperature difference between the entering hot stream and the exiting cold stream), but it cannot be greater than unity. The reactor gain K_R has typical values ranging from 2 to 6, depending on the per-pass conversion, the concentration of reactants in the reactor inlet stream, and the heat of reaction. Therefore these systems are often openloop-unstable. Dynamic controllability degrades as reactor gain increases, as we demonstrate below.

In tuning the controller we assume that three small lags τ_m exist in the loop, so the controller sees the total openloop transfer function:

$$\left[\frac{\left(\dfrac{K_2}{K_R K_1 - 1}\right)(\tau_R s + 1)}{\left(\dfrac{\tau_R}{K_R K_1 - 1}\right)s - 1}\right]\left[\frac{1}{(\tau_m s + 1)^3}\right] \tag{7.5}$$

7.3.2 Flowsheet FS2 with Furnace

In this flowsheet the presence of the furnace provides an additional control degree of freedom, and there are now two controllers. The first $G_{C1(s)}$ controls T_{mix} by manipulating bypass flow F_{by}. The second $G_{C2(s)}$ controls T_{in} by manipulating furnace firing Q_F. The second controller sees the furnace transfer function $G_{F2(s)}$, which we assume to be a furnace first-order lag and three small lags (see Fig. 7.4*b*):

$$G_{F2(s)} = \frac{T_{\text{in}(s)}}{Q_{F(s)}} = \frac{K_{F2}}{(\tau_F s + 1)(\tau_m s + 1)^3} \tag{7.6}$$

The first controller $G_{C1(s)}$ sees a coupled "openloop" transfer function in which the second controller is nested, because the controller $G_{C2(s)}$ is on automatic. The $G_{C1(s)}$ controller is tuned by finding the ultimate gain and period and using the Tyreus–Luyben settings. The settings of this PI controller are $K_{C2} = 2.13$ and $\tau_{I2} = 1.94$.

7.3.3 Nyquist Plots

To compare the two flowsheets and the two control structures, a numerical case is used. Parameter values are $\tau_R = K_1 = K_2 = 1$ and $\tau_m = 0.1$. Several values of reactor gain K_R are explored. Figure 7.5*a* gives Nyquist plots of the openloop coupled process for the process without a furnace (FS1). The closedloop system is conditionally stable since one counterclockwise encirclement of the $(-1,0)$ point is required (because of the one openloop pole in the RHP). Therefore if the controller gain is above a maximum gain $K_{\max}$ or below a minimum gain $K_{\min}$, the closedloop system is unstable. The controller gain must be such that the $(-1,0)$ point lies between the $\omega = 0$ point and the ω_u point (the crossover or ultimate frequency) on the negative real axis. Figure 7.5*a* shows that the distance between these two points shrinks as reactor gain increases. For values of $K_R = 2, 3$, there is a stable region. However, for $K_R = 5$ it is impossible to obtain a counterclockwise encirclement, so a P controller cannot stabilize the FS1 system.

Figure 7.5*b* gives Nyquist plots for the process with the furnace (FS2). First, note that the system is now openloop-*stable* for values of reactor gain $K_R = 2, 3$, which was not the case for the FS1 flowsheet. Second, observe that even for reactor gains up to about $K_R = 8$ it is possible to use a P controller to stabilize the system. Remember that we are talking about the $G_{C1(s)}$ controller, with the $G_{C2(s)}$ controller on automatic ($K_{C2} = 2.13$ and $\tau_{I2} = 1.94$).

Figure 7.6 demonstrates the improvement obtained when PI controllers are used instead of P controllers for $G_{C1(s)}$. The controller tuning constants used are given in Table 7.1 for the two flowsheets and for two values of K_R.

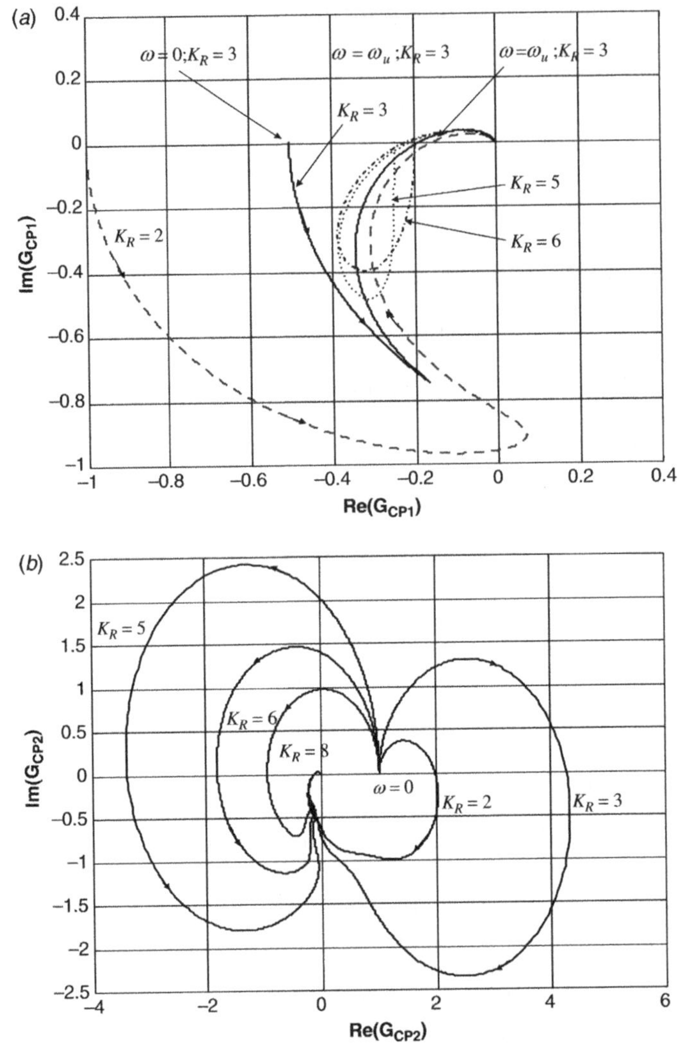

Figure 7.5 Nyquist plots (*a*) without furnace (FS1); (*b*) with furnace (FS2).

This improvement is true for both flowsheets, as a comparison of Figures 7.5 and 7.6 reveals. For example, in the process without the furnace, a P controller could not stabilize the system for $K_R = 5$. However, the PI controller can stabilize the system.

Figure 7.7 gives time-domain results for the linear model that verify the frequency-domain predictions. Figure 7.7*a* shows that control performance degrades as K_R increases without a furnace. The responses are for a unit step change in setpoint. Results from two tuning methods are shown: the Tyreus–Luyben settings and the optimum settings found from a previous study.[1]

[1]W. L. Luyben, Effect of kinetic, design and operating parameters on reactor gain, *Ind. Eng. Chem. Research* **39**, 2384 (2000).

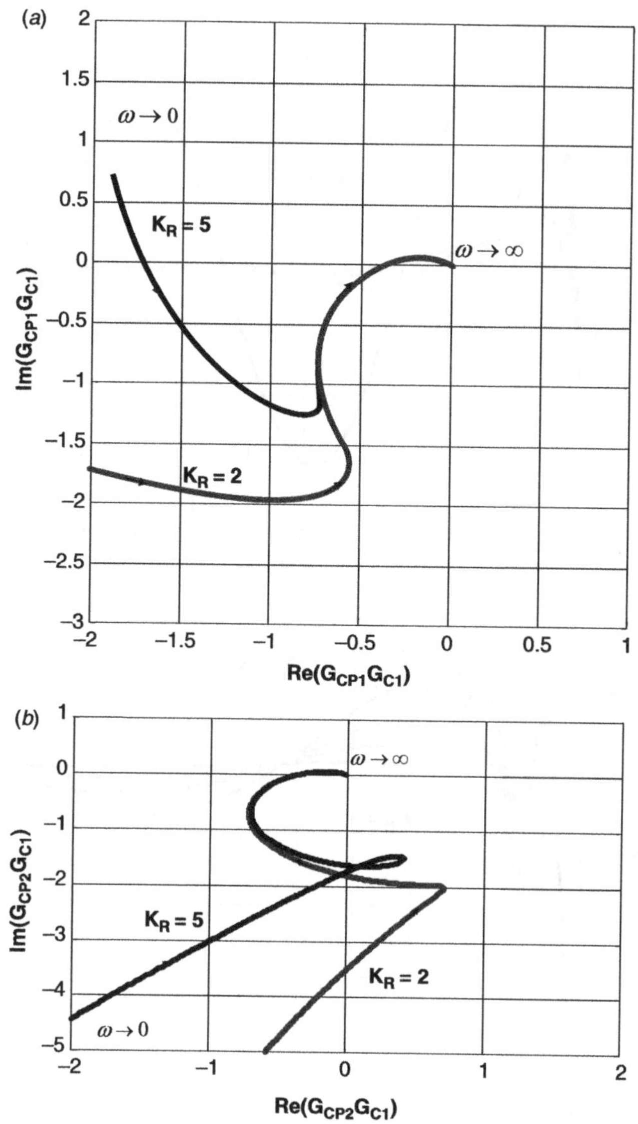

Figure 7.6 Nyquist plots (*a*) without and (*b*) with furnace (PI control).

TABLE 7.1 Controller Tuning Constants

Flowsheet	K_R	K_{C1}	τ_{I1}
FS1	2	2.05	0.876
	5	1.50	1.04
FS2	2	2.55	0.795
	5	2.56	0.801

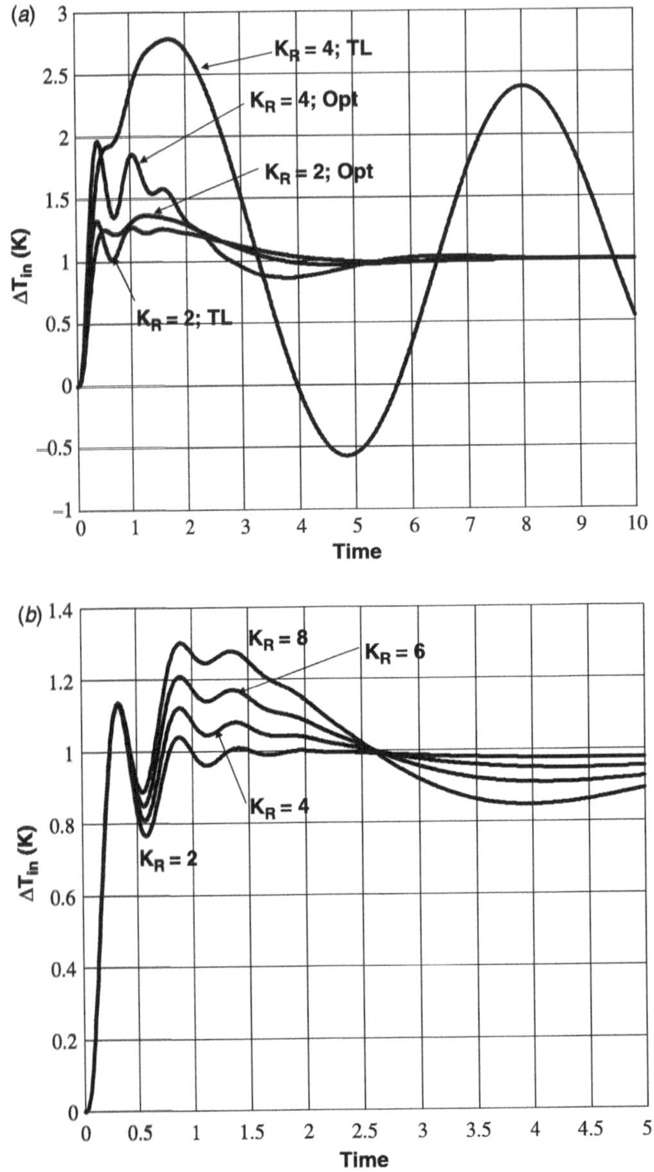

Figure 7.7 Responses (*a*) without and (*b*) with furnace.

Figure 7.7*b* shows that much larger reactor gains can be handled by the process with the furnace. The Tyreus–Luyben settings are used ($K_{C1} = 2.5$ and $\tau_{I1} = 0.80$), which were almost constant for all values of K_R.

Thus the linear analysis predicts that the dynamic controllability of the FS2 process is much better than that of the FS1 process. Even though steady-state economics do not favor using a furnace, the actual profitability of the process may be higher if a furnace is used. This is particularly true for systems that exhibit large reactor gains.

7.4 NONLINEAR SIMULATION

To verify the results of the linear analysis, we develop a nonlinear dynamic model of the process and study two different flowsheets. Figure 7.8*a* shows the FS1 flowsheet without a furnace. The FEHE area is 2261 m^2, and the FEHE bypass flowrate is 0.184 kmol/s. Figure 7.8*b* shows the FS2 flowsheet with a furnace. The FEHE area is smaller (1712 m^2), and the bypass flowrate is larger (0.365 kmol/s).

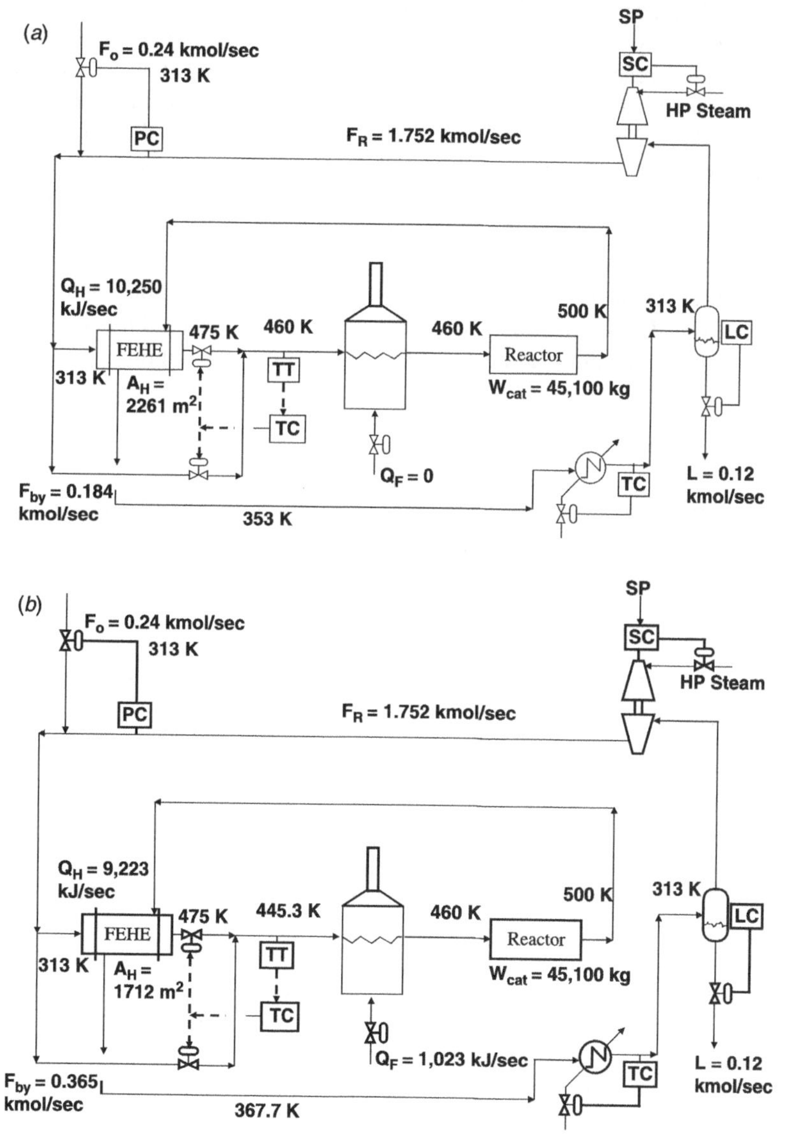

Figure 7.8 (*a*) Base case FS1 (no furnace); (*b*) Furnace supplying 10% of heat load (FS2).

7.4.1 Dynamic Model

The reaction considered is the gas-phase, irreversible, exothermic reaction $A + B \rightarrow C$ occurring in a packed tubular reactor. The reactor and the heat exchanger are both distributed systems, which are rigorously modeled by partial differential equations. Lumped-model approximations are used in this study, which capture the important dynamics with a minimum of programming complexity. There are no sharp temperature or composition gradients in the reactor because of the low per-pass conversion and high recycle flowrate.

Reactor and Gas Loop Model The number of lumps used is determined by comparing the rigorous steady-state results with models using different numbers of lumps. The rigorous steady-state reactor exit temperature is 500 K with a 460 K inlet reactor temperature. When a 10-lump model is used with the same amount of catalyst, the steady-state exit temperature is 505 K. This occurs because of the "numerical diffusion" or effective backmixing that is inherent with a lumped model. A 50-lump reactor model is used in all the simulations. It gives a steady-state reactor exit temperature of 501 K. The differences between the dynamic responses of the 10-lump and 50-lump reactors are shown in the next section.

The thermal capacitance of the gas in the reactor is assumed negligible compared to that of the solid catalyst. Therefore a single dynamic energy balance is used for each lump, and the gas temperature is assumed to be equal to the catalyst temperature

$$c_{p,\text{cat}} W_{\text{cat},n} \frac{dT_n}{dt} = F_{n-1} c_{p,n-1} T_{n-1} - F_n c_{p,n} T_n - \lambda \Re_{Cn} \tag{7.7}$$

where $W_{\text{cat},n} = W_{cat}/NR$ and NR is the number of reactor lumps. The heat capacity of the catalyst is 0.837 kJ $\text{kg}^{-1}\ \text{K}^{-1}$. The reaction rate $\Re_{Cn}$ is given by

$$\Re_{Cn} = W_{\text{cat},n} k_{n(T_n)} y_{An} y_{Bn} P^2 \tag{7.8}$$

The dynamic component balances on the gas phase in each lump assume a constant average molar density of $\rho_{\text{av}} = 1.124\ \text{kmol/m}^3$ (at 535 K and 50 bar).

Component balances for nth lump:

$$V_{\text{gas}} \rho_{\text{av}} \frac{dy_{jn}}{dt} = F_{n-1} y_{j,n-1} - F_n y_{j,n} + \upsilon_j \Re_{Cn} \tag{7.9}$$

where the stoichiometric coefficients are $\upsilon_A = \upsilon_B = -1$ and $\upsilon_C = 1$. Note that the reaction is nonequimolar, so the total molar flowrate decreases down the length of the reactor.

The total pressure in the gas loop is calculated from an approximate dynamic total mass balance. Gas enters in the fresh feed, and gas is converted into liquid component C in the condenser following the FEHE. An average density is used, assuming an average molecular weight of 17.5 kg/kmol and an average temperature of 500 K:

Total mass balance on gas loop:

$$\frac{17.5(V_{\text{FEHE}} + V_{\text{reactor}} + V_{\text{drum}})}{(0.08314)(500)} \frac{dP}{dt} = F_0(17.5) - 35\, F_{NR} y_{NR,\text{C}} \tag{7.10}$$

TABLE 7.2 Steady States from Rigorous and Lumped Dynamic Models

	FS1 Design	FS1 Dynamic	FS2 Design	FS2 Dynamic
F_0 (kmol/s)	0.24	0.248	0.24	0.248
F_{by} (kmol/s)	0.184	0.232	0.365	0.391
Q_F (kW)	0	0	1023	978
T_{mix} (K)	460	460	445.3	446
T_{in} (K)	460	460	460	460
T_{out} (K)	500	501.3	500	501.3
Area (m^2)	2261	2261	1712	1712

Total molar balance on drum:

$$\frac{dM_D}{dt} = 35F_{NR}y_{NR,C} - L_C \tag{7.11}$$

FEHE Model There are no phase changes or small pinch temperatures in the heat exchanger. Therefore a 10-lump model is used for this unit. The steady-state exit temperatures predicted by the lumped model are close to those calculated by the rigorous countercurrent model used in the steady-state design. Table 7.2 compares the steady-state conditions for the process obtained from the dynamic model with those obtained in the steady-state design.

The heat exchanger is quite large because of the low heat transfer coefficient found in these gas-phase systems. Therefore the mass of metal in the tubes is quite significant in terms of thermal capacitance. Table 7.3 gives design details of the heat exchanger. The tube diameter is 0.0254 m, length is 5 m, wall thickness is 0.000524 m, and heat capacity is 0.05 kJ kg^{-1} K^{-1}.

The FEHE is modeled by assuming the cold gas has a temperature T_{Cn} in the nth lump, into which heat is transferred at a rate Q_{Cn} from the hotter tube metal at temperature T_{Mn}:

$$Q_{Cn} = U_C A_n (T_{Mn} - T_{Cn}) \tag{7.12}$$

The hot gas flows countercurrently and has a temperature T_{Hn} in the nth lump. Heat is transferred from the hot gas into the tube metal at a rate Q_{Hn}:

$$Q_{Hn} = U_H A_n (T_{Hn} - T_{Mn}) \tag{7.13}$$

The heat transfer area in each lump is the total area divided by the number of lumps. The overall heat transfer coefficient used in the steady-state design calculations was

TABLE 7.3 FEHE Parameters (FS2 Case)

Heat transfer area (m^2)	1712
Number of tubes	4291
Volume inside tubes (m^3)	10.9
Volume of shell side (m^3)	10.9
Weight of tubes (kg)	44,500

0.142 kJ s^{-1} m^{-2} K^{-1}. The two heat transfer coefficients U_C and U_H are set 3.5 times the overall coefficient so that the exit stream temperatures of the dynamic model are approximately the same as those of the steady-state model.

Each lump of the cold and hot sides of the exchanger is described by an energy balance and three composition balances since both temperatures and compositions can change with time and axial position. Note that the cold side is always a 50/50 mixture of A and B, but the hot side can contain varying amounts of component C as conditions in the reactor change dynamically with time

$$\rho_{\text{av}} c_{pH} V_{Hn} \frac{dT_{Hn}}{dt} = F_H c_{pH}(T_{H,n+1} - T_{H,n}) - Q_{Hn} \tag{7.14}$$

$$\rho_{\text{av}} c_{pC} V_{Cn} \frac{dT_{Cn}}{dt} = F_C c_{pC}(T_{C,n+1} - T_{C,n}) - Q_{Cn} \tag{7.15}$$

$$\rho_{\text{av}} V_{Hn} \frac{dy_{H,j,n}}{dt} = F_{H,n+1} y_{H,j,n+1} - F_{Hn} y_{H,j,n} \tag{7.16}$$

where $j =$ A,B,C and $y_{H,j,n} =$ mole fraction of component j in the hot-side lump n. The stage parameter n is numbered starting from the cold inlet. An average molar gas density ρ_{av} is used in the component balances. There is flow in and out of each lump. Pressure drops are neglected. The hot- and cold-side volumes are assumed to be equal.

Each lump of the metal is described by an energy balance on the tube metal

$$c_{pM} W_n \frac{dT_{Mn}}{dt} = Q_{Hn} - Q_{Cn} \tag{7.17}$$

where the weight of tube metal in each lump is the total tube weight divided by the number of lumps.

The furnace is modeled with a simple energy balance, with the molar holdup M_F adjusted to give a 2 min time constant:

$$c_{pC} M_F \frac{dT_{\text{in}}}{dt} = Q_F + F_C c_{pC}(T_{\text{mix}} - T_{\text{in}}) \tag{7.18}$$

The flowrate into the furnace is the blended stream F_C with a temperature T_{mix}.

7.4.2 Controller Structure

The control structure has the following loops:

1. Gas loop pressure P is controlled by manipulating fresh feed flowrate F_0.
2. Liquid level in the separator is controlled by manipulating liquid flowrate.
3. Recycle gas flowrate F_R is held constant by fixing compressor speed.
4. The temperature T_{mix} of the blended stream of gas from the heat exchanger and gas bypassing the heat exchanger is controlled by manipulating bypass flowrate F_{by}.
5. In the FS2 flowsheet with the furnace, the reactor inlet temperature T_{in} is controlled by manipulating furnace heat input Q_F.

All controllers are PI except for the drum level controller, which is P only. Two 0.5-min lags are assumed in the pressure loop. Three 0.1-min lags are assumed in the temperature loops. Relay-feedback tests are conducted to get the ultimate gain and period, and the Tyreus–Luyben settings are used.

7.4.3 Results

Figure 7.9 gives the openloop responses of the heat exchanger and the reactor to step changes in inlet temperature. These results are for each of the units operating in total isolation from the rest of the plant. In Figure 7.9*a* the temperature of the hot stream entering the heat exchanger is increased 20 K at time $=0$. The temperature of the cold exit stream responds with a time constant of about 3 min. The temperature of the hot exit stream responds with a time constant of about 10 min. The temperature dynamics are due to the tube metal thermal capacitance since the holdup of gas is small. Note that the heat exchanger gain for this numerical case is $\Delta T_{C,\text{out}}/\Delta T_{H,\text{in}} = \frac{18}{20} = 0.9$.

Figure 7.9*b* gives the responses of the reactor exit temperature using the 10-lump and 50-lump models. Step changes in the reactor inlet temperature of several magnitudes are made at time zero. The 10-lump model shows more sensitivity than does the 50-lump model. Note that the reactor gain ($\Delta T_{\text{out}}/\Delta T_{\text{in}}$) increases as the magnitude of the input step size increases. This is an important effect, which we discuss in a later section when a "hotter" reaction system is considered. There is a very slight amount of "wrong way" or inverse response observed in this numerical example. This type of response can occur for certain parameter values and complicates the control problem.

During the initial simulation studies, considerable difficulty was experienced in getting the entire process simulation to stabilize. Starting with the FS1 flowsheet without the furnace, convergence of the model to a steady state from some approximate initial conditions predicted by the steady-state design was attempted. However, unless the guesses for the initial conditions were very close to the correct ones, the system would quench. Reactor inlet temperature decreased, which decreased reactor exit temperature. Despite the bypass valve closing, trying to return the inlet temperature to the setpoint, the drop in reactor exit temperature decreased the reactor inlet temperature even more. So the system shut down; pressure built up and fresh feed was cut off.

However, switching to the FS2 flowsheet with the presence of the furnace preventing the quench, the simulation was rather easily converged to a steady state. These startup experiences indicate that the FS2 flowsheet is much more robust than the FS1 flowsheet.

FS2 with Furnace Figure 7.10 illustrates the effective dynamic performance for the FS2 flowsheet. Figure 7.10*a* shows that changes in the recycle flowrate F_R of $\pm 20\%$ are easily handled. Note the interesting phenomenon that an *increase* in recycle flowrate *decreases* production rate (fresh feed flowrate F_0 decreases). This is somewhat counterintuitive until one remembers that what we are doing is holding reactor inlet temperature constant. This means that an increase in flowrate results in a decrease in the reactor exit temperature because the reactor is operating adiabatically and the increase in flow provides more thermal sink. This lowers the temperature rise in the reactor. See the lower left graph in Figure 7.10*a*. The lower temperatures in the reactor yield a smaller reaction rate.

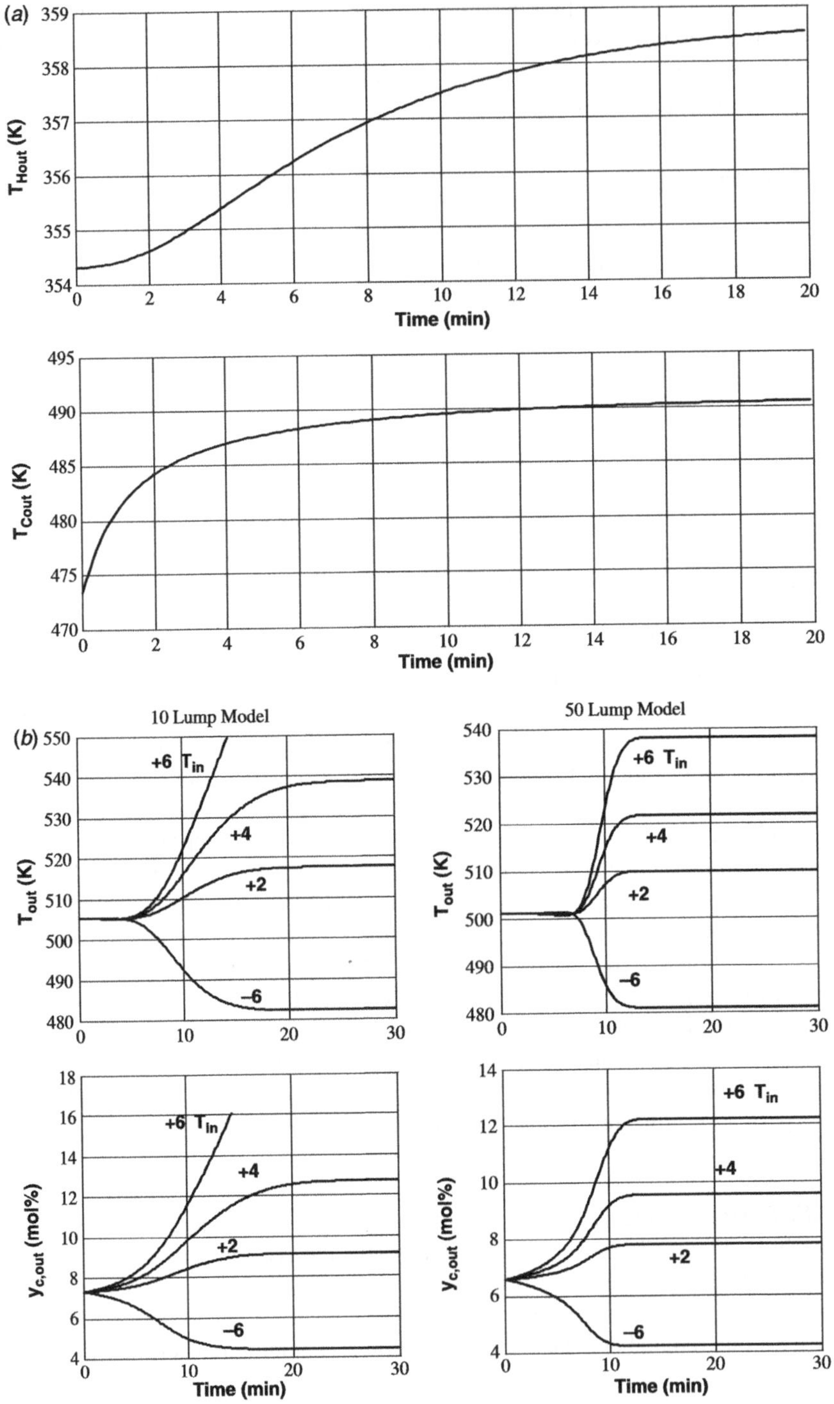

Figure 7.9 Openloop responses of (*a*) FEHE; (*b*) reactor.

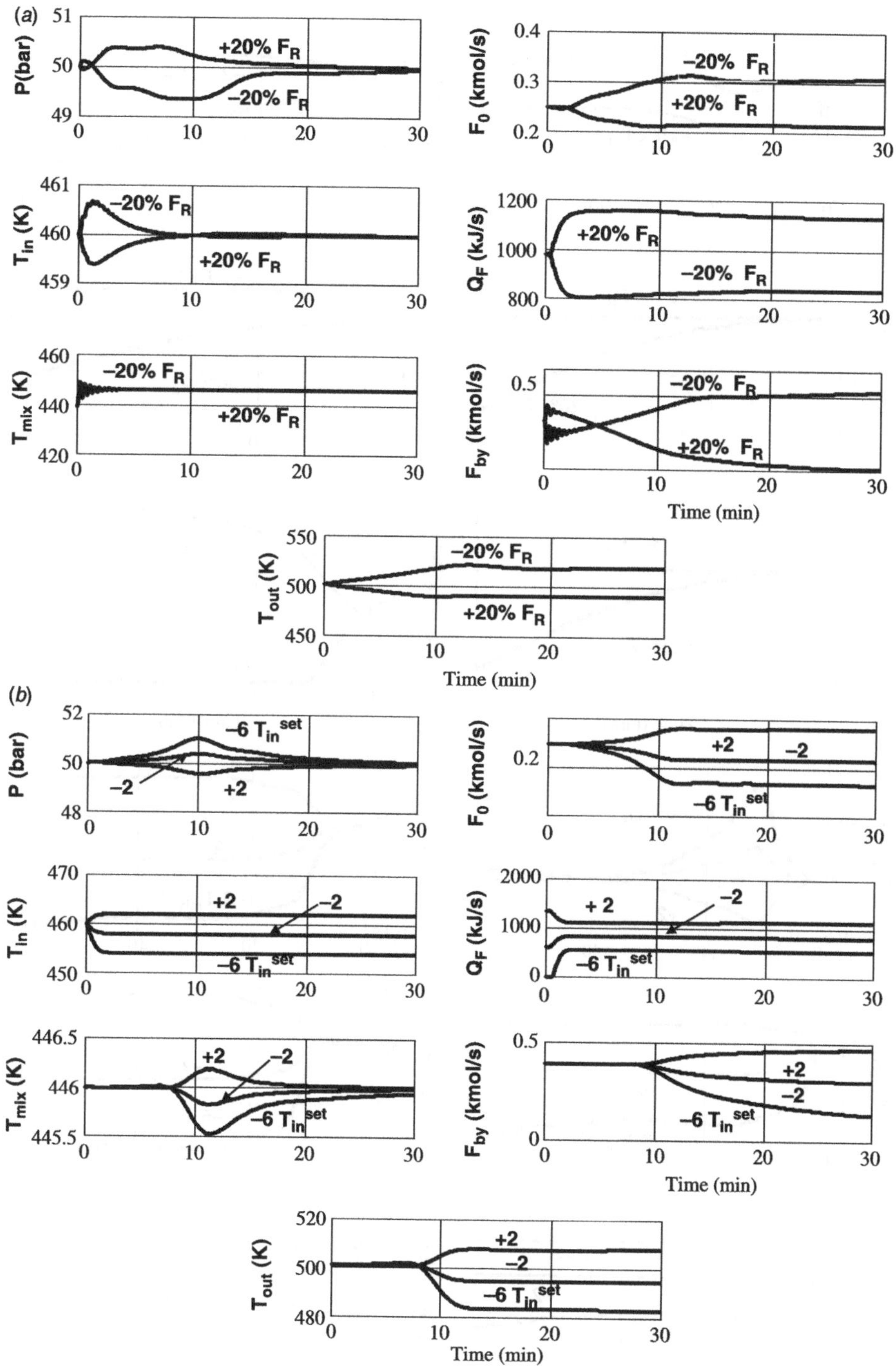

Figure 7.10 FS2 response with changes in (*a*) recycle flow and (*b*) T_{in}^{set}.

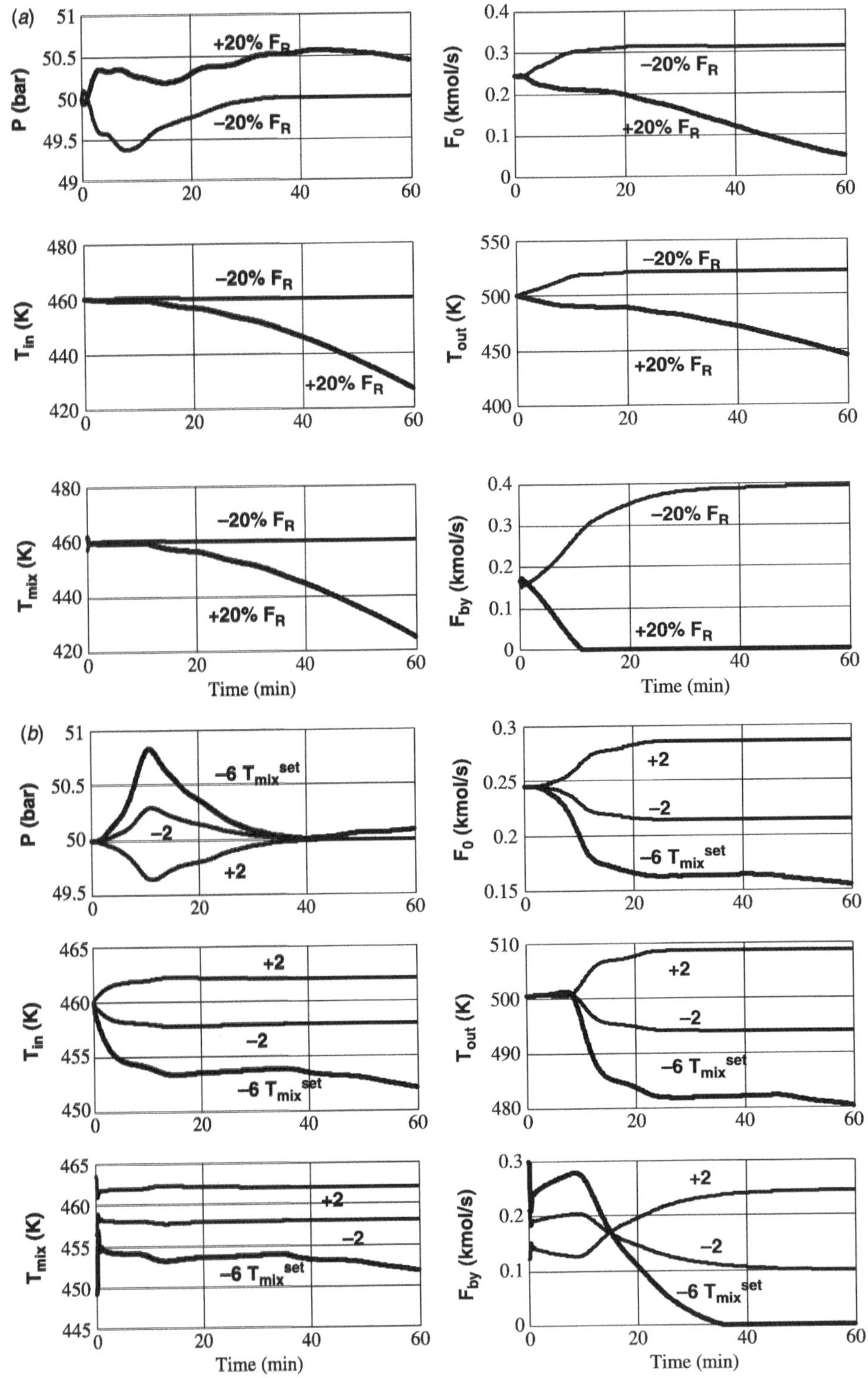

Figure 7.11 FS1 response with changes in (*a*) recycle flow and (*b*) T_{mix}^{set}.

We are considering an irreversible reaction in this book. For reversible reactions, Luyben[2] demonstrated the opposite effect of recycle flowrate on production rate, specifically, increasing recycle flowrate *increases* production rate. With reversible reactions, the impact of temperature is less important than the impact of concentrations because of the reaction equilibrium constraints.

Figure 7.10*b* demonstrates that changes in the setpoint of the reactor inlet temperature controller are handled smoothly. When reactor inlet temperature is increased, for a constant recycle flowrate, there are increases in production rate, bypass flowrate, reactor exit temperature, and furnace heat input.

FS1 without Furnace It is more difficult to operate the system without the furnace. Quenches occur because of bypass valve saturation. Figure 7.11*a* illustrates such a case. A 20% decrease in recycle flowrate is handled because the bypass valve can open more because of the higher production rate and higher reactor exit temperature. However, a 20% *increase* in recycle flowrate causes a quench. Reactor exit temperature drops, decreasing reactor inlet temperature. The bypass valve closes, but is unable to maintain the reactor inlet temperature. Reaction rate decreases, pressure increases, and fresh feed is gradually shut off. It takes about one hour for the shut down to occur. Figure 7.11*b* shows that small changes in temperature controller setpoint can be handled, but a large decrease causes the bypass valve to saturate shut, eventually leading to a quench.

Thus the absence of the furnace to provide heat input when the bypass valve is completely shut produces a process that is not as robust and not as easy to operate. Another problem with just using heat exchanger bypassing is the impact of changes in heat transfer coefficients over time due to fouling. The stability of the process will degrade as the FEHE fouls.

One might consider using a flowsheet like FS2, but setting the setpoints of the two temperature controllers at the same value (or using one temperature controller with a "split-range valve" setup). Or a low-temperature override controller could be used to prevent quenching by using furnace heat input when the bypass valve approaches the fully closed position. In all these schemes, bypass control would be used during normal operation. This would achieve the desired steady-state economic objective of using no energy. Such schemes may be possible if the heating is done using high-pressure steam or a high-temperature medium (such as Dowtherm). But if the heat is being provided by a furnace, which is the most common case because of the high temperature levels typical of many industrial reactors, these control structures are not viable. A furnace cannot be conveniently operated with no firing rate. So some minimum base load ($\sim$10% of capacity) is required.

7.5 HOT-REACTION CASE

The linear analysis in Section 7.3 predicted that the FS2 flowsheet should be able to handle systems with reactor gains up to $\sim$6. To test this prediction, kinetic parameters are modified to give reactor gains of this magnitude in a new design. Both the activation energy and the heat of reaction are increased by a factor of 1.9. The specific reaction rate at

[2]W. L. Luyben, Design and control of gas-phase reactor/recycle processes with reversible exothermic reactions, *Ind. Eng. Chem. Research* **39**, 1529 (2000).

TABLE 7.4 Optimum Design Parameters for Hot-Reaction Case

		Hot Reaction	
	Base Case Design Steady State	Design Steady State	Dynamic Steady State
E (kJ/kmol)	69,710	132,450	132,450
λ (kJ/kmol)	−23,237	−44,150	−44,150
W_{cat} (kg)	45,100	62,980	62,980
Q_F (kJ/s)	1023	2770	2667
Q_H (kJ/s)	8200	24,950	24,950
F_R (kmol/s)	1.75	4.81	4.91
F_{by} (kmol/s)	0.365	0.640	0.749
T_{in} (K)	460	470	470
T_{out} (K)	500	500	501.8
T_{mix} (K)	445.3	454.3	454.3
Area (m^2)	1712	5121	5121
TAC (10^6 \$ year^{-1})	3.01	5.53	—

500 K is kept constant by adjusting the preexponential factor as the activation energy is changed.

The optimum steady-state economic design was determined with these new kinetic parameters, and the parameters are given in Table 7.4. The FS2 flowsheet is used with a ratio $Q_F/Q_{total} = 0.1$. The impact of the kinetic parameters on the optimum design is striking. The hotter reaction requires a much larger recycle flowrate and a higher reactor inlet temperature for the same reactor exit temperature $T_{out} = 500$ K. These lead

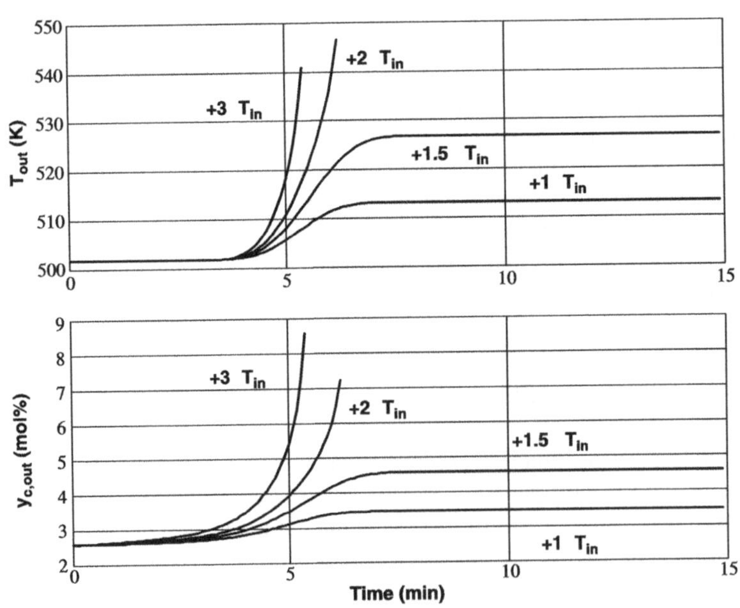

Figure 7.12 Openloop response of hot reactor.

to a larger reactor, heat exchanger, compressor, and furnace. The process is more expensive (TAC increases from \$3.01 $\times$ 10^6 to \$5.53 $\times$ 10^6 per year).

The dynamic simulations of the reactor by itself are shown in Figure 7.12. A very small increase in reactor inlet temperature produces a reactor runaway because of the high activation energy and high heat of reaction. Compare these results with those of the base case kinetic parameters shown in Figure 7.9*b*.

Figures 7.13 and 7.14 give results using the FS2 flowsheet with the furnace for this hot-reaction case. Figure 7.13 shows that a 10% decrease in recycle flowrate can be handled, but a 20% decrease produces a reactor runaway. This occurs despite the fact that the reactor inlet temperature increases only slightly (~0.5 K) during the transient. Figure 7.14 gives results for changes in the setpoint of the reactor inlet temperature controller. Rather surprisingly, inlet temperature can be increased by 2 K without a runaway. This is unexpected since the isolated reactor (Fig. 7.12) showed a runaway with a +2 K change in T_{in}. The difference may be due to the effect of pressure. In the isolated reactor simulation, pressure is held constant at 50 bar. In the simulation of the whole process, pressure drops as reactor temperature increases due to the increased consumption of reactants. Since the reaction rate depends on the square of the total pressure (P^2), the decrease in pressure lowers the reaction rates. However, a 3 K increase cannot be handled.

The new kinetic parameters drastically increase the sensitivity of the reactor to inlet temperature. The sensitivity to inlet temperature occurs because of the high activation energy and heat of reaction and because of the high reactant concentrations (low per-pass conversion). Remember that the feed to the reactor is a 50/50 molar mixture of pure reactants. There are large amounts of reactants available to fuel the reaction runaway.

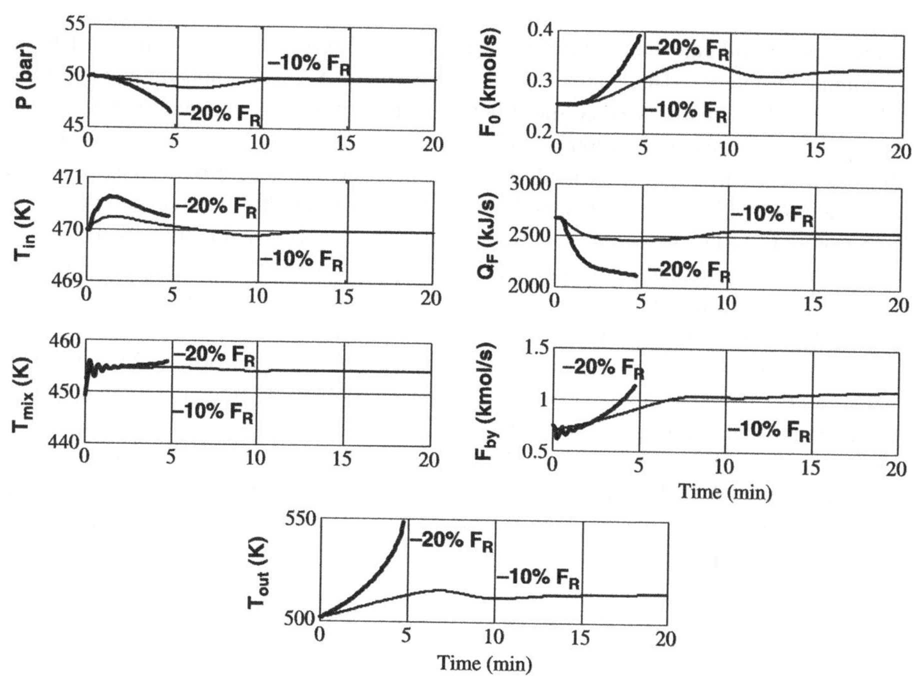

Figure 7.13 FS2 response with hot reaction; change in recycle flow.

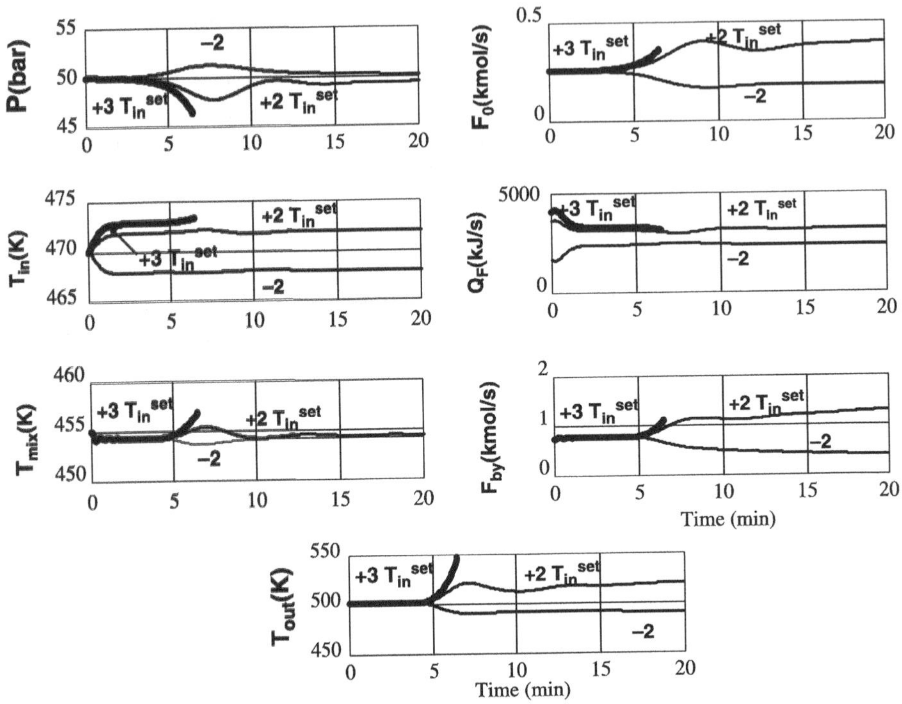

Figure 7.14 FS2 response with hot reaction; change in T_{in}^{set}.

These results indicate that a process change would probably be required to handle the dynamic problems. There are several alternatives. A cooled nonadiabatic reactor should reduce the sensitivity since more heat will be removed as temperatures increase. Probably a more practical solution would be to design for a lower concentration of one of the reactants. This mode of operation would prevent reaction runaways because the reaction rate would drop off quickly as the concentration of the limiting reactant declined. The economic penalties would include requiring a larger reactor and more recycle than in the equimolar pure reactant feed mode of operation. Alternatively, the concentrations of both reactants could be reduced by recycling an inert substance (probably product C). This would also increase reactor size and recycle flowrate.

All of these considerations again demonstrate the strong effect of process design on process control. In almost all processes there are important tradeoffs and conflicts that must be considered to produce an easy-to-operate, stable process with little product quality variability. Such a process will be more profitable in the operational stage than a process that may look better from just a steady-state standpoint. Techniques are available for quantitatively evaluating this tradeoff between steady-state economics and dynamic controllability.[3]

[3]See T. R. Elliott, M. L. Luyben, and W. L. Luyben, Application of the capacity-based economic approach to an industrial-scale process, *Ind. Eng Chem. Research* **36**, 1727 (1997).

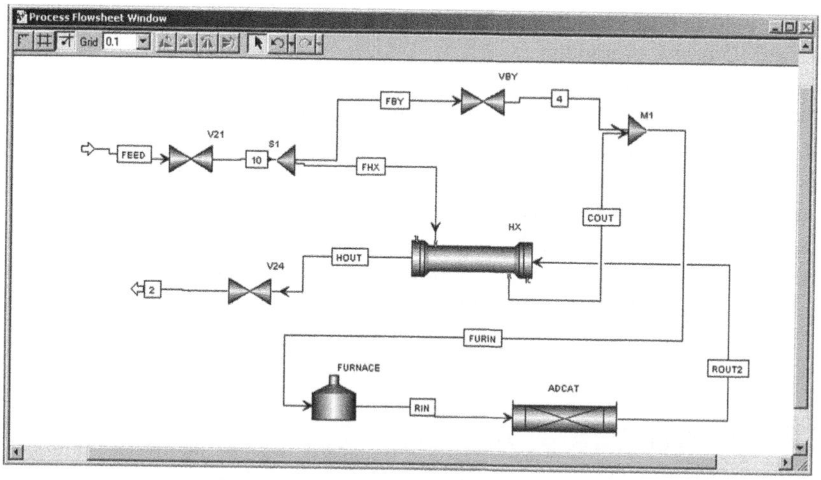

Figure 7.15 Aspen Plus flowsheet.

7.6 ASPEN SIMULATION

The simulation of a FEHE–adiabatic tubular reactor system in Aspen Plus and Aspen Dynamics involves some interesting programming issues and produces some valuable insights about the dynamics of these types of systems. For a numerical example, the chlorination of propylene considered in Chapters 5 and 6 is used. Figure 7.15 gives the steady-state Aspen Plus flowsheet. The cold feedstream is split into the bypass stream, going through valve "VBY," and the stream that is heated in the heat exchanger "HX." The heated stream leaving heat exchanger "COUT" and the bypass stream "FBY" are mixed together and sent to a furnace and then to an adiabatic tubular reactor. The reactor effluent goes to the hot side of the FEHE to preheat the feed.

The adiabatic reactor is 20 m in length and 1 m in diameter. It is packed with 15,708 kg of catalyst with a heat capacity of 500 J kg^{-1} K^{-1}. The process gas and the solid catalyst are assumed to be at the same temperature. The reactor inlet temperature is specified to be

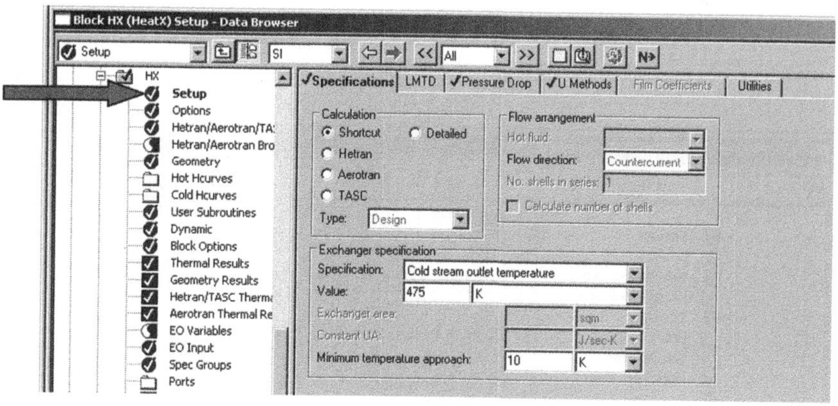

Figure 7.16 Heat exchanger specifications.

400 K. The feed is 0.025 kmol/s at 310 K and 8 atm. The feed composition is 5 mol% chlorine and 95 mol% propylene.

The heat exchanger has countercurrent flow of the cold process gas with the hot gas from the reactor exit. A 25 K difference in temperature is assumed between the reactor effluent and the cold stream leaving the FEHE at the hot end. For example, with a 5 mol% chlorine feed composition, the reactor exit temperature T_{out} is 500 K with a 400 K reactor inlet temperature T_{in}. The heat exchanger is sized so that the temperature of the exiting cold stream is 475 K. Figure 7.16 shows these specifications. Figure 7.17 shows that the pressure drops through both the cold and hot sides of the heat exchanger are 1 atm (putting a "−1" in the *Outlet pressure* box indicates a pressure drop of

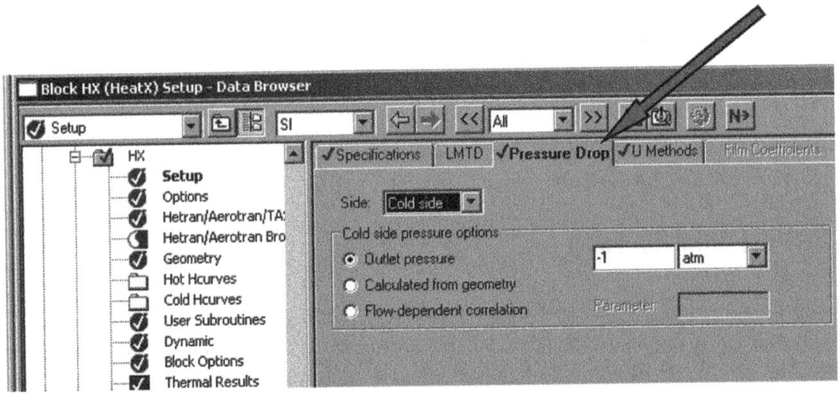

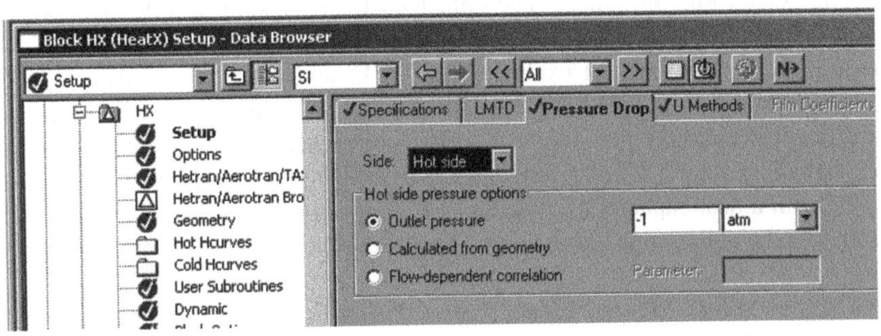

Figure 7.17 Heat exchanger pressure drops.

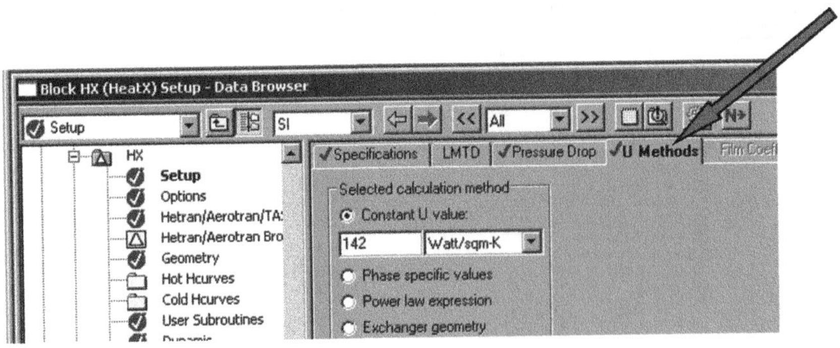

Figure 7.18 Heat exchanger *U*.

1 atm). A constant value of the overall heat transfer coefficient $U = 142\ \mathrm{W\ K^{-1}\ m^{-2}}$ is set by opening the *U Methods* page tab shown in Figure 7.18.

The dynamics of heat exchangers are modeled only very approximately in Aspen. We have to specify the inlet and outlet volumes on the tube and shell sides and the weight of metal in the tubes. These can be calculated from the heat transfer area. Using an overall heat transfer coefficient of $142\ \mathrm{J\ s^{-1}\ K^{-1}\ m^{-2}}$ in this gas–gas heat exchanger gives a heat transfer area of $21.2\ \mathrm{m^2}$ and a heat transfer rate of 174 kW. These results can be seen after the program runs by clicking the *Thermal Results* item under the heat exchanger block in the *Data Browser*.

If we assume 0.025-m-diameter tubes, 10 m in length and with a wall thickness of 0.002 m, the total tube volume will be $0.12\ \mathrm{m^3}$ and weight is 320 kg. The shell volume is assumed to be equal to the tube volume. The top window in Figure 7.19 shows that *Dynamic* is selected as the *Model type* (note that *Instantaneous* is the default), and these volumes are split between the inlet and outlet. The lower window in Figure 7.19 shows the *Equipment Heat Transfer* page tab on which the *Model equipment heat capacity* option is selected, and the weight of the tubes is distributed among the inlets and outlets of the hot and cold sides. The heat capacity of the tubes is $500\ \mathrm{J\ kg^{-1}\ K^{-1}}$. Note that the metal of the shell is not included in the metal mass because the change in its temperature would be much slower than that of the tube metal.

The amount of bypassing F_{by} is determined by blending the 475 K stream coming from the FEHE with the 310 K bypass stream to achieve the 400 K reactor inlet temperature. The bypass flowrate is 0.0119 kmol/s, and the flow through the heat exchanger is

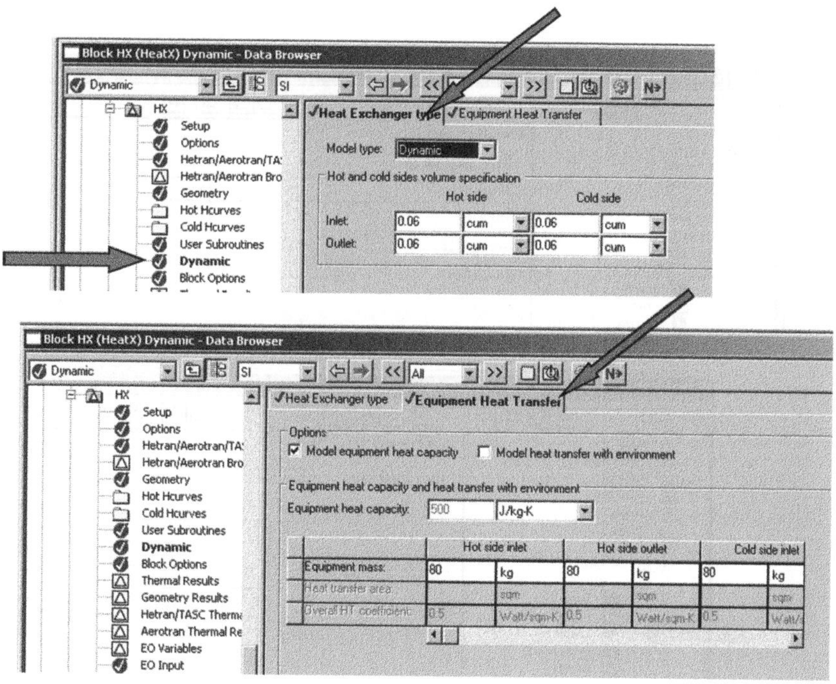

Figure 7.19 Heat exchanger dynamics.

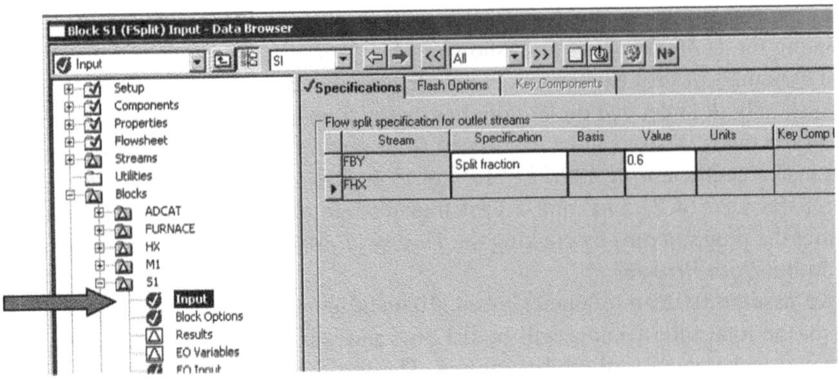

Figure 7.20 Splitter setup.

0.0131 kmol/s. This split is determined by the specification made for the splitter "S1" shown in the flowsheet given in Figure 7.15. Either the flowrate of one of the streams or the split fraction can be specified. Figure 7.20 shows that the fraction bypassed is initially set at 0.6. This fraction is adjusted, as discussed in the next section, by a design spec to achieve the desired 400 K temperature of the mixed stream.

A furnace is located after the FEHE. Its heat input is normally zero in the steady-state design because the reactor effluent is hot enough to provide all the heat required for preheating. Figure 7.21 shows its setup with no heat added under steady-state conditions.

The furnace will be used in the control scheme to provide additional heat if required as conditions changed. The dynamic results given in the next section show that using

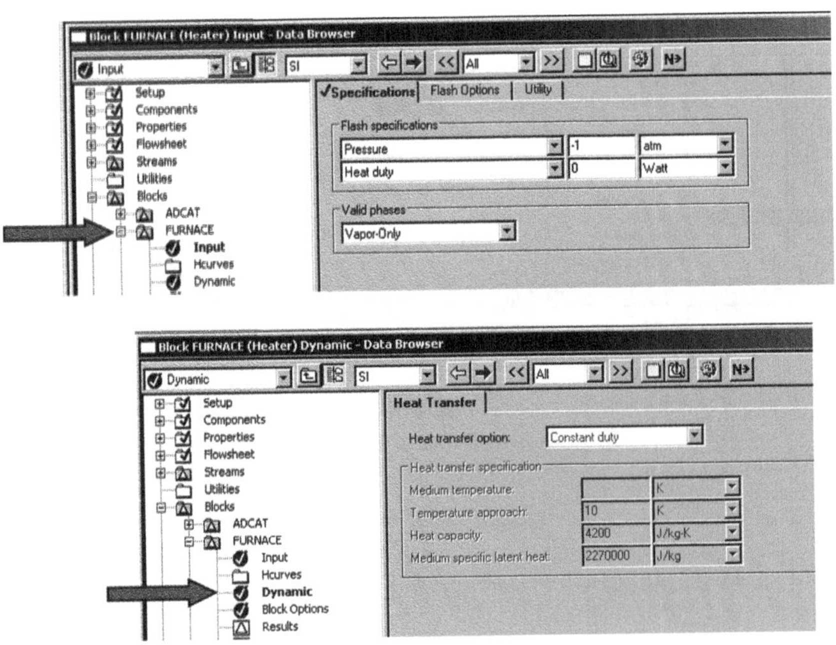

Figure 7.21 Furnace setup.

a furnace in addition to using bypassing improves the dynamic control of the process by permitting the system to handle larger disturbances. Bypassing prevents temperatures from becoming too hot and producing runaways. Furnace heating prevents temperatures from becoming too cold and producing reactor quenches. Reaction rates get so small that conversion drops to the point that the reactor exit temperature can no longer preheat the feed. Thus using both bypassing and furnace heating provide more control muscle.

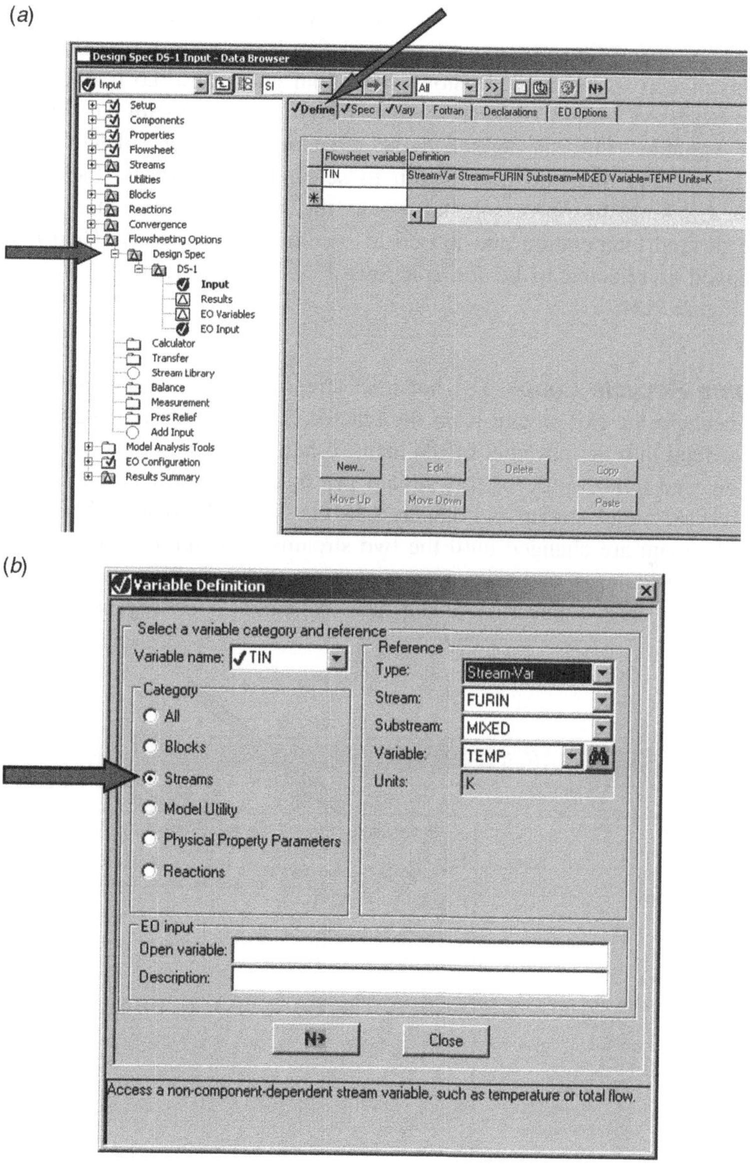

Figure 7.22 (*a*) Defining flowsheet variable in design spec; (*b*) editing variable.

7.6.1 Aspen Plus Steady-State Design

There are two sources of complexity in this flowsheet. The first is the split of the feed between the bypass stream and the stream going through the heat exchanger. The second is the recycle stream from the reactor back to the heat exchanger. The final flowsheet is shown in Figure 7.15, but to get to this converged flowsheet, these issues must be handled.

Design Spec to Set Flow Split The cold stream leaving the heat exchanger "COUT" and the bypass stream "FBY" are mixed together and become stream "FURIN." The feed split should be adjusted to produce a 400 K temperature of this stream. A flowsheet *Design Spec*, which is under *Flowsheeting Options* in the *Data Browser* window, is used to achieve this objective. Figure 7.20 shows the initial guess of a 0.6 fraction of the total feed being bypassed. Figure 7.22*a* shows the *Design Spec* item with the *Define* page tab open. The variable "TIN" is defined to be the temperature of "FURIN" by clicking the *New* button and then the *Edit* button to select the stream and the property (see Fig. 7.22*b*. Clicking the *Spec* page tab opens the window shown in Figure 7.23, on which the desired temperature and the error tolerance are specified. Then the variable to be changed is selected to be the fraction of "FBY" in the splitter block "S1" as shown in Figure 7.24.

Converging Recycle Loop The hot inlet stream to the heat exchanger is initially unknown because the exit stream from the reactor is not known. A temporary stream is specified to feed into the hot inlet of the heat exchanger. Initial guesses of the flowrate, composition, and temperature of this stream are made. The simulation is run, and the resulting reactor outlet stream is compared to the guessed stream. The parameters of the guessed stream are changed until the two streams have similar properties. Then the guessed stream is deleted, and the reactor outlet stream "ROUT2" (see Fig. 7.15) is connected to the hot inlet connection of the heat exchanger. Convergence is aided if this

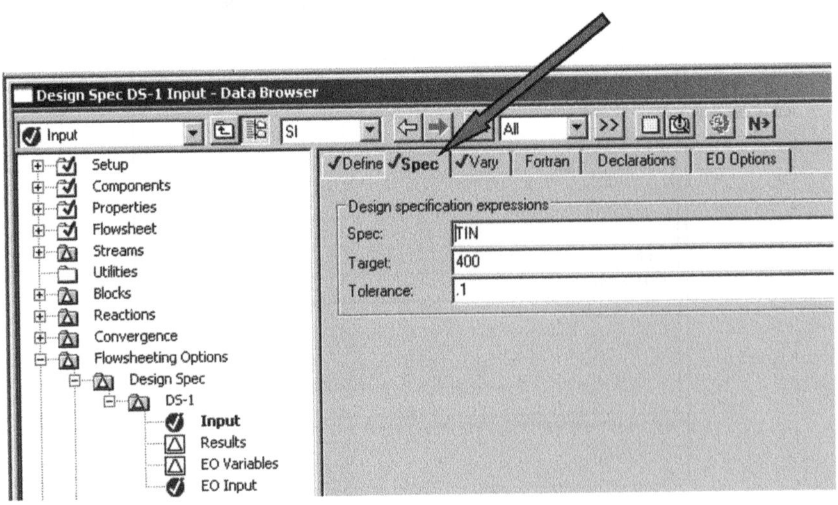

Figure 7.23 Specifying desired value.

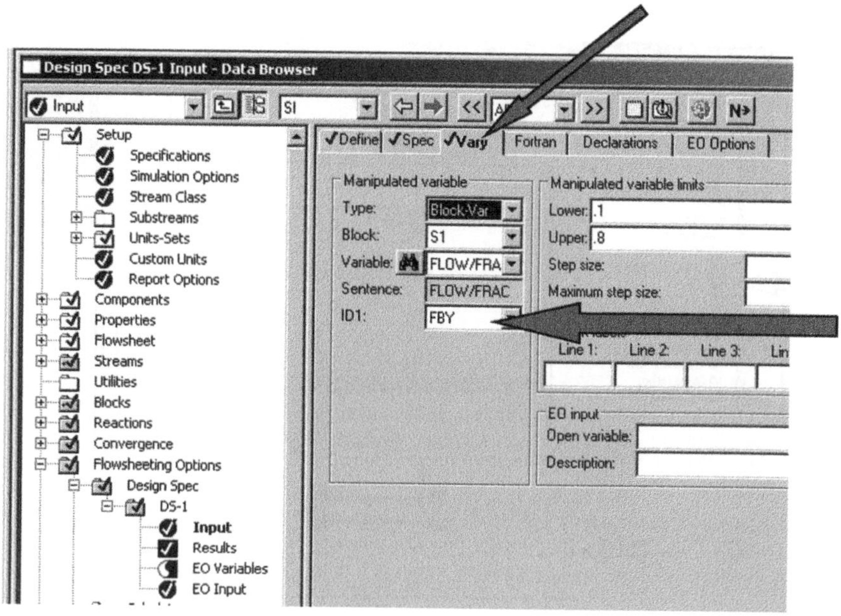

Figure 7.24 Selecting the fraction in splitter as variable.

stream is specified as a *Tear* stream under the *Convergence* item in the *Data Browser* window as shown in Figure 7.25.

Then the program is run and the resulting reactor exit temperature is found (with the inlet temperature set at 400 K). Then the temperature of the cold stream leaving the heat exchanger "COUT" is specified to be 25 K less than the reactor exit temperature, and the program is run again. Figure 7.26 gives results from the two iterative loops. The Wegstein method is used in "$OLVER12" to converge the recycle flow "ROUT2." The secant method is used in "$OLVER05" to converge the flow split. Steady-state temperature and composition profiles in the reactor are shown in Figure 7.27. Results for the heat exchanger are shown in Figure 7.28. The calculated area is 21.2 m^2.

The results presented above are for a feed composition of 5 mol% chlorine. If the feed composition is changed to 10 mol% chlorine, the reactor temperature and composition profiles change, as does the heat exchanger design. With a 400 K reactor inlet temperature, the reactor outlet temperature increases to 587 K because of the greater heat generated. The temperature of stream "COUT" is set at 562 K ($\Delta T = 25$ K at the hot end of the heat exchanger). The required heat transfer area in the heat exchanger changes from 21.1 to 13.8 m^2 because of the larger temperature driving force at the cold end. The heat transfer rate is slightly lower (169 vs. 174 kW). The bypass flow is larger (0.0169 vs. 0.0119 kmol/s).

The shapes of the temperature profiles in the two cases are quite different. Figure 7.27 shows that when the feed contains 5 mol% chlorine, it takes almost the entire length of the reactor to achieve a high conversion of chlorine. When the feed contains 10 mol% chlorine, the temperatures increase more quickly and climb to higher values as the gas flows down through the reactor because of the higher concentration of the limiting reactant. Figure 7.29 shows that the chlorine is essentially all consumed by the time the

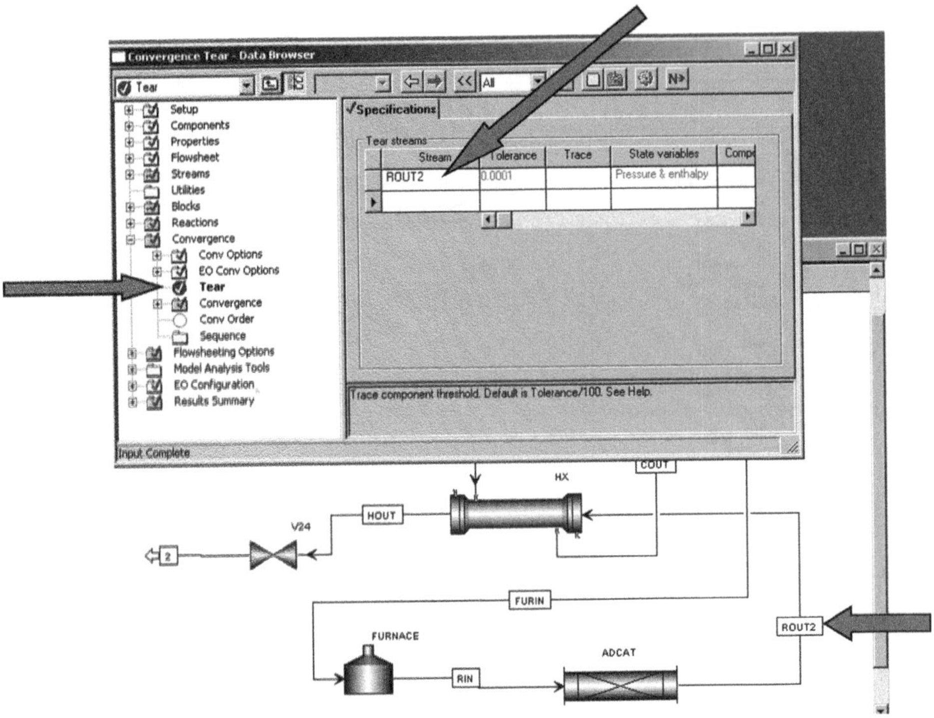

Figure 7.25 Tear stream.

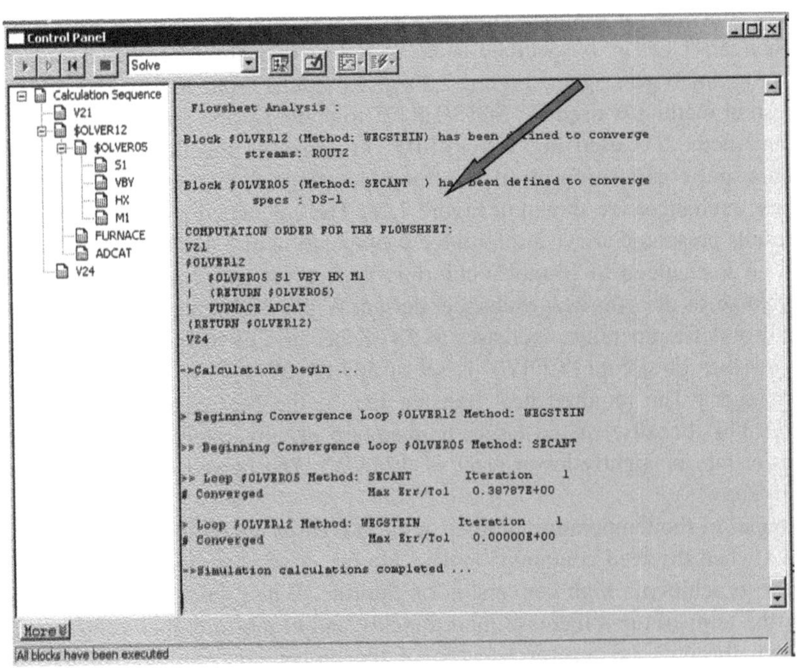

Figure 7.26 Convergence.

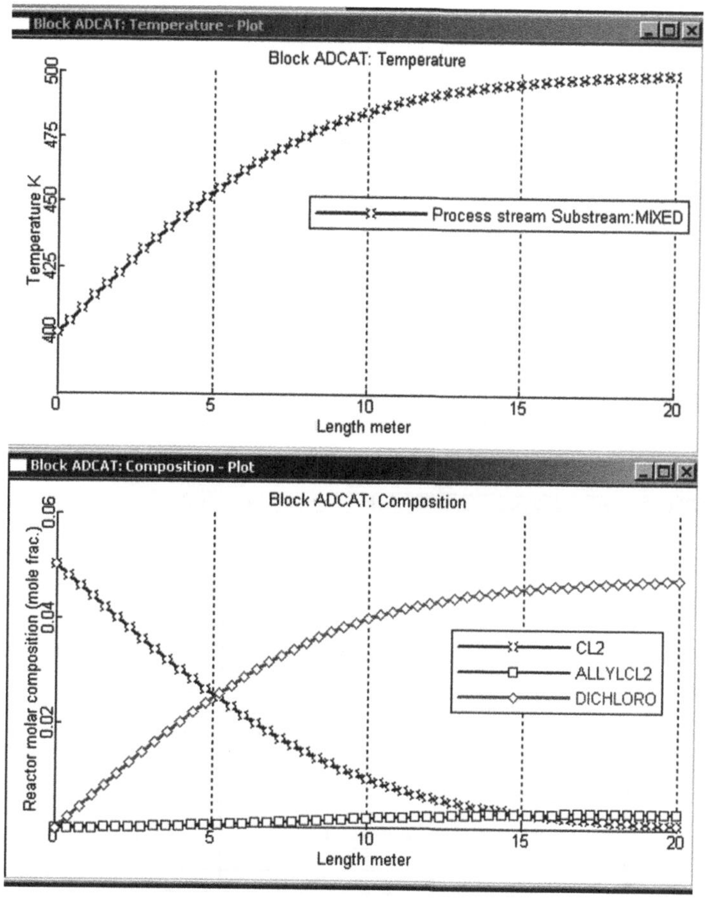

Figure 7.27 Reactor temperature and composition profiles (5 mol% Cl_2 feed).

gas gets about halfway down the reactor. We will compare the dynamics of these two cases in the next section.

7.6.2 Aspen Dynamics Control

The file is pressure checked and exported into Aspen Dynamics. A flow controller is installed on the feed. A pressure controller is installed that holds reactor pressure by adjusting valve "V24" in the process discharge line. Figure 7.30 shows the flowsheet with these controllers installed.

Control with Only Bypass The important control loop in this process is the temperature controller that manipulates the bypass flow to control the temperature of the mixed hot and cold streams. The controller is direct acting (an increase in temperature opens the bypass valve). A 1-min deadtime is inserted in the loop, and a relay–feedback test is run that gives Tyreus–Luyben settings $K_C = 0.48$ and $\tau_I = 4.0$ min. The temperature transmitter span is 350–450 K.

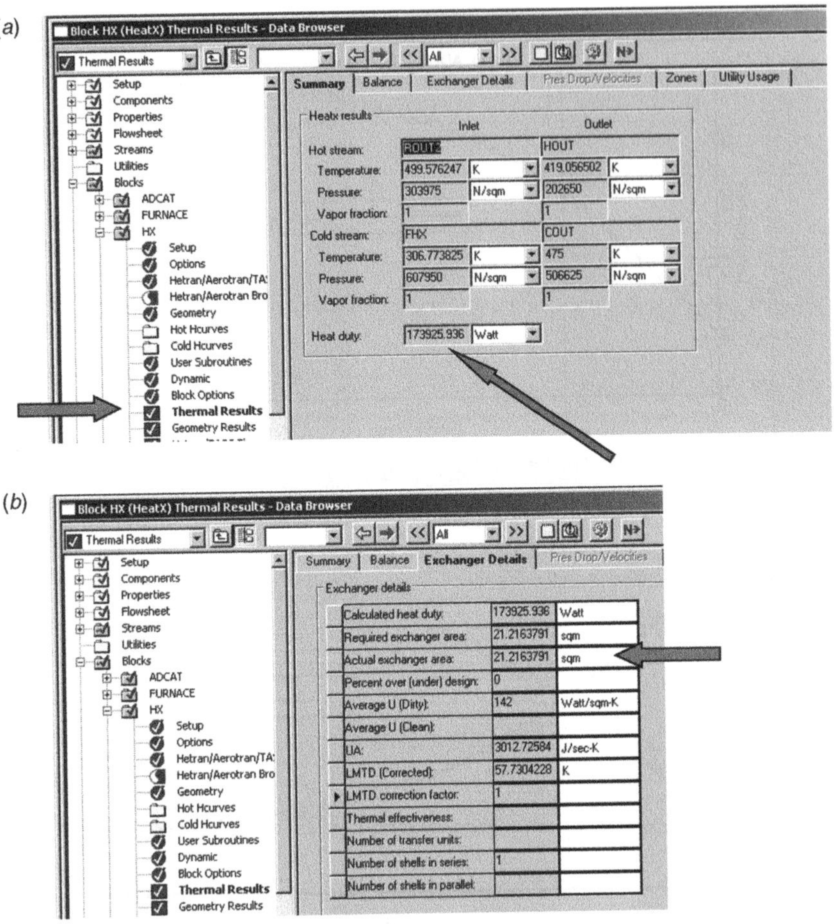

Figure 7.28 Heat exchanger results.

The performance of this control structure, which does not use the furnace, is shown in Figure 7.31. At 0.1 hours, the feed composition is changed from 5 to 7.5 mol% chlorine. The reactor outlet temperature climbs because of the increase in reaction heat generation. The hotter gas entering the FEHE raises the temperature of exit stream, which raises the temperature of the mixture. The temperature controller increases the bypass flow to hold the reactor inlet temperature at 400 K.

At 4 h, the feed composition is dropped to 2.5 mol% chlorine. Temperatures decrease sharply. The valve in the bypass line is driven completely shut at about 5.7 h, but the reactor inlet temperature cannot be maintained at the desired 400 K and drops to 393 K. The reactor outlet temperature drops from 500 to 438 K because of the reduction in reactant in the feed. Figure 7.32 shows the temperature profile at this new steady state.

If the reactions had higher activation energies and were more sensitive to temperature, the reaction rates could become so small with the low reactor temperatures that the process could "quench" (with nothing reacting and all temperatures dropping to the feed temperature).

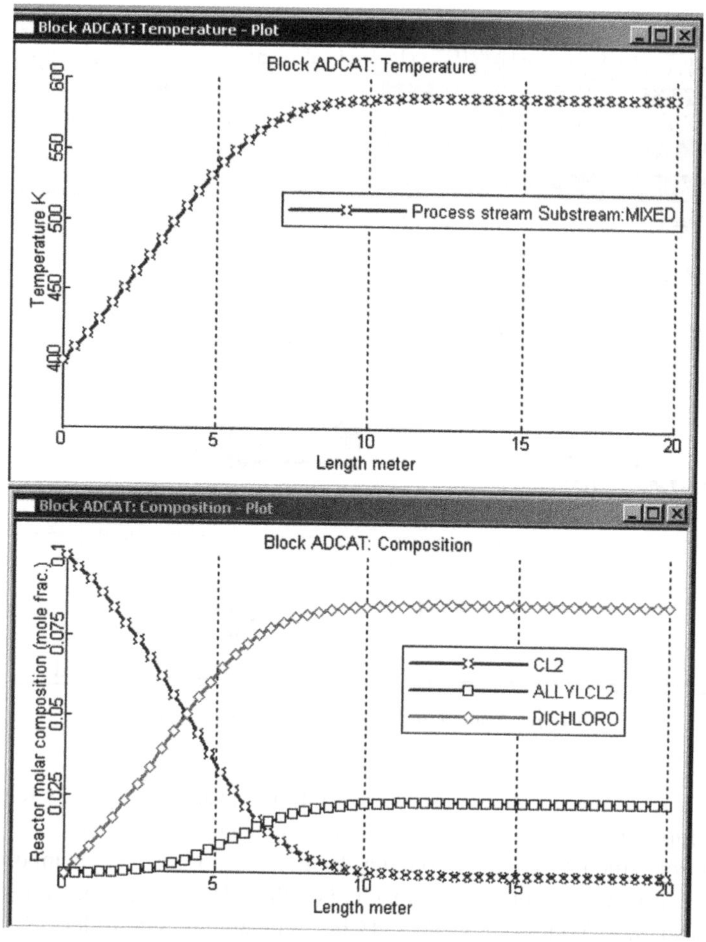

Figure 7.29 Reactor profiles for 10% feed composition.

Figure 7.33 gives results for changes in feed flowrate from 0.025 to 0.03 kmol/s as 0.1 h and from 0.03 to 0.02 kmol/s as 2 h. These disturbances are handled without saturating the bypass valve.

Control with Both Bypass and Furnace A second temperature controller is installed to control the temperature of the stream leaving the furnace by manipulating the heat input in the furnace (see Fig. 7.34). The setpoint of this temperature controller is set at 401 K. The steady-state heat input is 3.65 kW, which is only 2% of the heat transferred in the FEHE (174 kW). With a 1-min deadtime and a maximum heat input of 20 kW, a relay–feedback test gives Tyreus–Luyben settings $K_C = 4.0$ and $\tau_I = 14.5$ min.

Figure 7.35 gives results for the same scenario of feed composition disturbance used previously. When feed chlorine composition is increased, both the furnace heat input and the bypass flow respond. The furnace heat input Q_F drops to zero for about 0.7 h. When the composition of chlorine in the feed is decreased, the bypass flow goes to zero, but the furnace heat input increases and holds the reactor inlet temperature T_{in} at

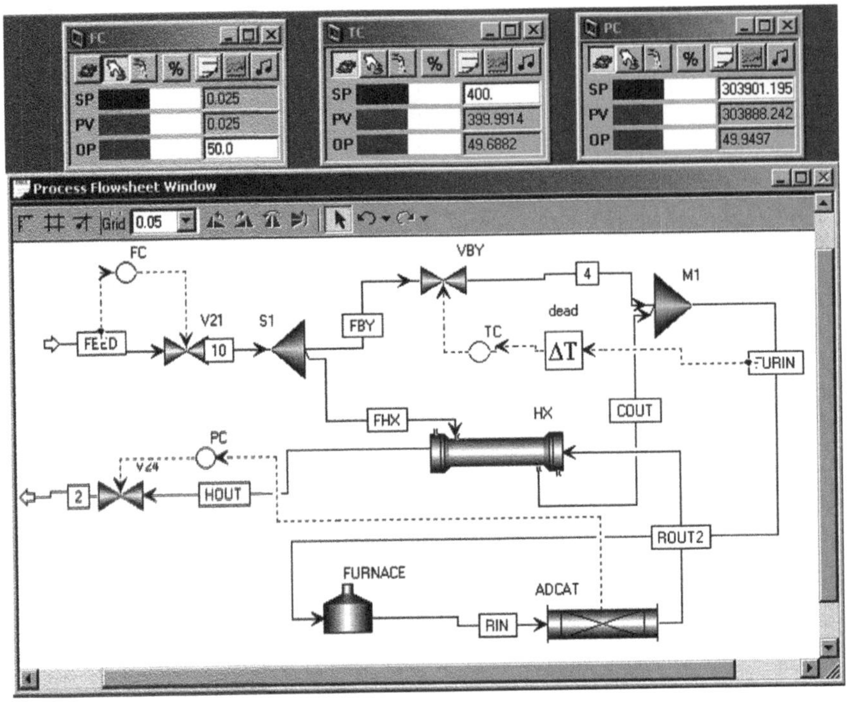

Figure 7.30 Control structure and faceplates.

400 K. The reactor outlet temperature drops to 450 K instead of the 438 K that occurs without the furnace. Thus the chances of a reactor quench are reduced.

This control structure is tested on the 10 mol% chlorine feed composition case. The tuning constants do not chance much. For the "TCmix" temperature controller, $K_C = 0.32$ and $\tau_I = 5.3$ min. For the "TCin" temperature controller, $K_C = 3.9$ and

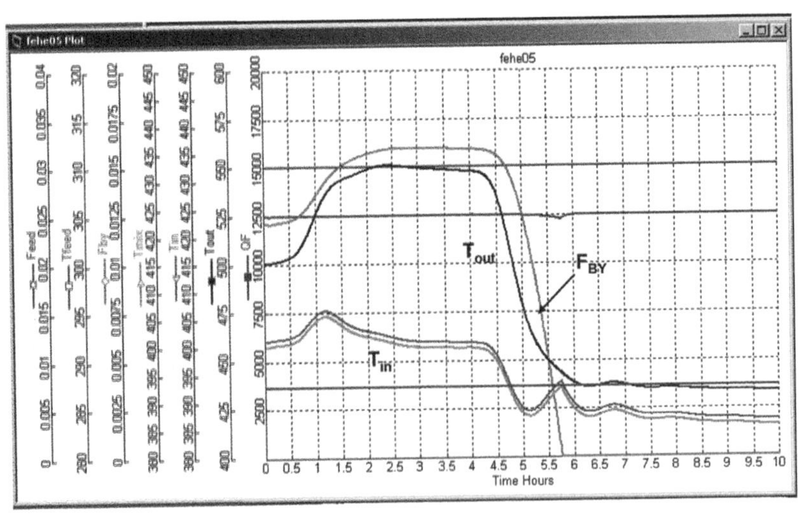

Figure 7.31 Feed composition disturbances.

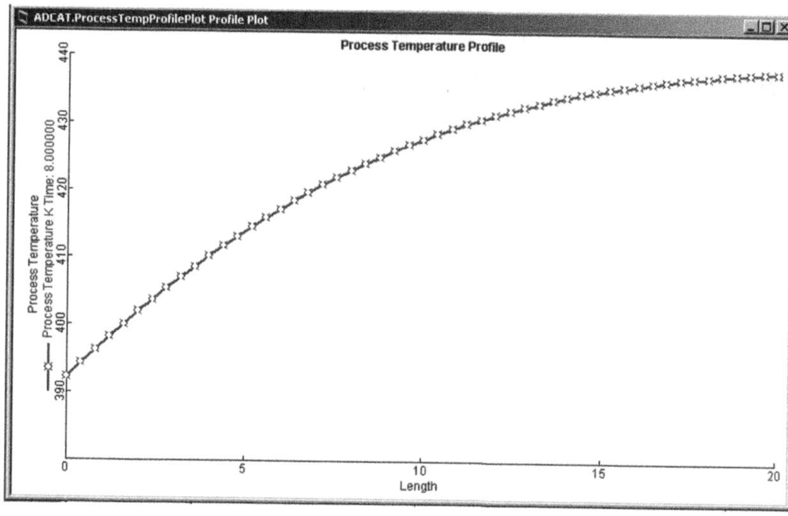

Figure 7.32 Temperature profile at the new steady state.

$\tau_I = 13$ min. Results for feed composition changes from 10 to 15 mol% chlorine at 0.1 h and from 15 to 5 mol% at 5 h are given in Figure 7.36. Effective dynamic control is achieved in the face of these large disturbances.

Figure 7.36 shows an interesting plumbing problem. Note that the feed flowrate cannot be maintained when the bypass flowrate is reduced at about 6.5 h. The OP signal from the feedflow controller goes to 100%, but the flow is not maintained at 0.025 kmol/s. This occurs because the flowrate through the heat exchanger increases, which causes the pressure drop through the heat exchanger to increase. There is less pressure drop available over the feedflow control valve.

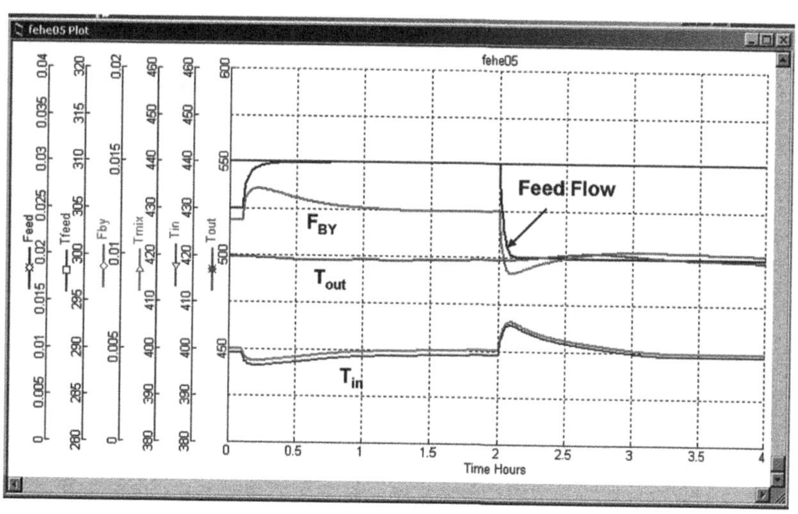

Figure 7.33 Feed flowrate disturbances.

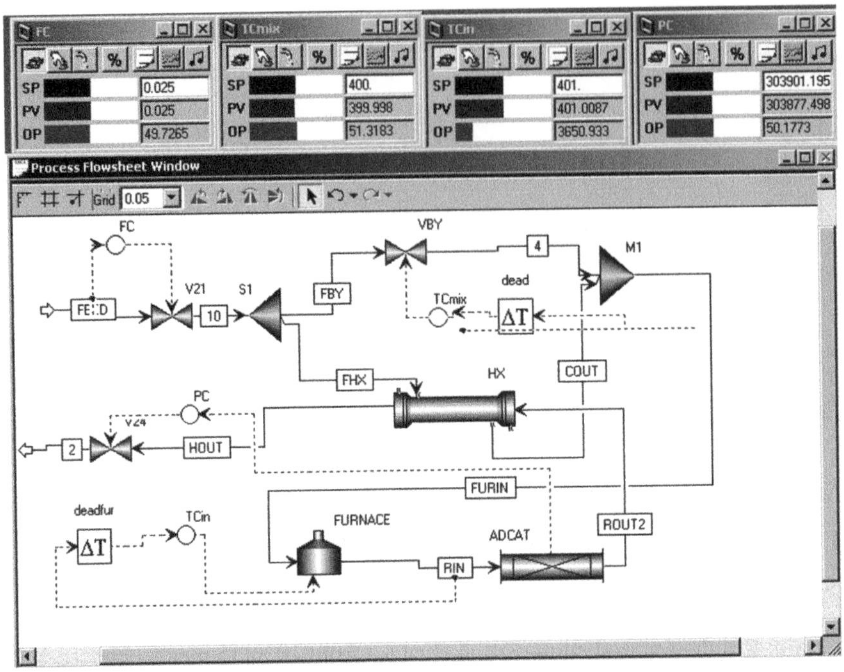

Figure 7.34 Control structure with furnace used.

The process considered in this numerical example has a high conversion of the limiting reactant and a significant temperature rise in the reactor. The result is a design with a significant fraction of the feed bypassed around the FEHE. With other reactions and other designs that result in a small fraction of the feed being bypassed, the use of a furnace to prevent reactor quenching becomes much more important.

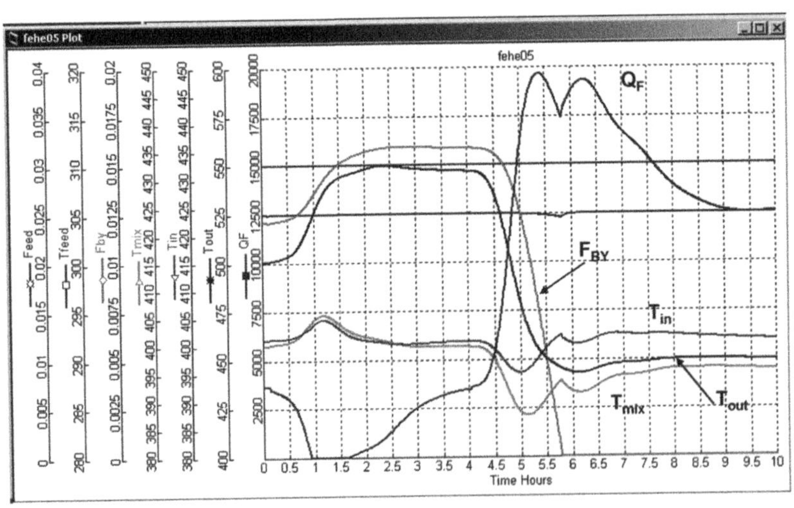

Figure 7.35 Feed composition disturbances with furnace.

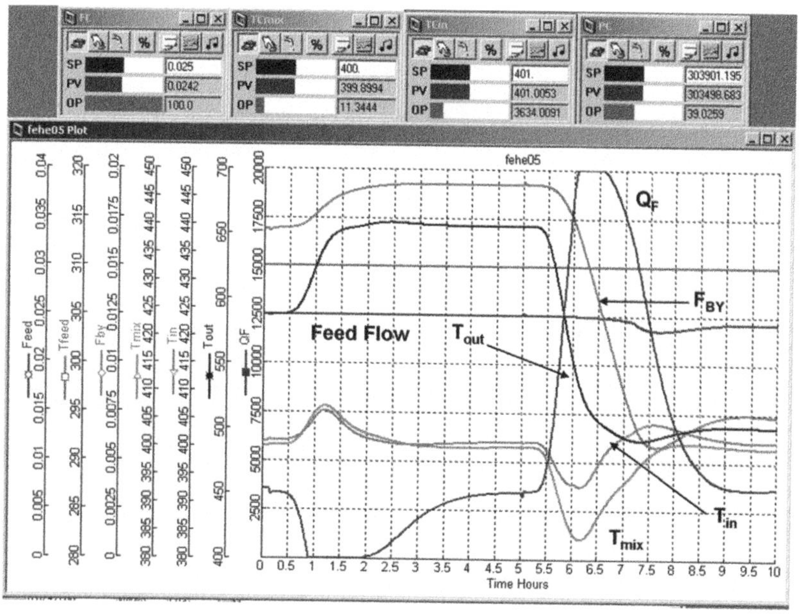

Figure 7.36 Feed composition disturbances with furnace for 10% case.

7.7 CONCLUSION

This chapter has two alternative structures for feed preheating. Both use a feed effluent heat exchanger, but one also uses a furnace. Steady-state economics favor use of only a heat exchanger. Dynamic controllability favors the use of both a heat exchanger and a furnace.

Completely independent preheating and cooling system on the reactor feed and effluent streams would provide very effective dynamic control but would be very uneconomical because of high energy consumption. The use of some coupled heating and cooling represents a compromise between the two extremes of a completely independent heating/cooling system and complete dependence on reaction heat for feed preheating. This study has provided some indication of the degree of interdependence that can be tolerated before control problems occur. The impact of kinetic and design parameters have also been illustrated.

CHAPTER 8

CONTROL OF SPECIAL TYPES OF INDUSTRIAL REACTORS

In this final chapter we take a brief look a several industrially important reactors that are not the ideal CSTR, plug flow tubular, or batch. These reactors have some unique features in terms of both steady-state design and control. A brief discussion of the process and the control problems is presented for each of these reactors, and some useful references are provided for the reader desiring more in-depth understanding.

8.1 FLUIDIZED CATALYTIC CRACKERS

Fluidized-bed reactors are used in many industries when heterogeneous reactions take place involving solids and gases. One of the earliest and most important applications is fluid catalytic cracking, which is extensively used in petroleum refineries around the world. Another application is coal gasification, which is becoming increasingly important as industry switches from imported petroleum and natural gas to coal as a primary source of chemical raw material feedstocks and energy.

Almost half of the world's gasoline is produced in fluid catalytic cracking units. In addition, much of the petrochemical industry is based on the ethylene, propylene, and other unsaturated components produced in catalytic crackers. The high-molecular-weight, high-boiling, saturated gasoil separated in the crude-oil fractionators is fed to catalytic cracking units in which the heavy material is cracked at high temperature on a solid silica–alumina catalyst to produce a wide range of lower-molecular-weight, low-boiling products. The gasoline cuts provide high-octane material for blending into the gasoline pool. The light hydrocarbons contain a variety of unsaturated olefins and aromatics that are widely used in the production of a host of chemical products.

Chemical Reactor Design and Control. By William L. Luyben

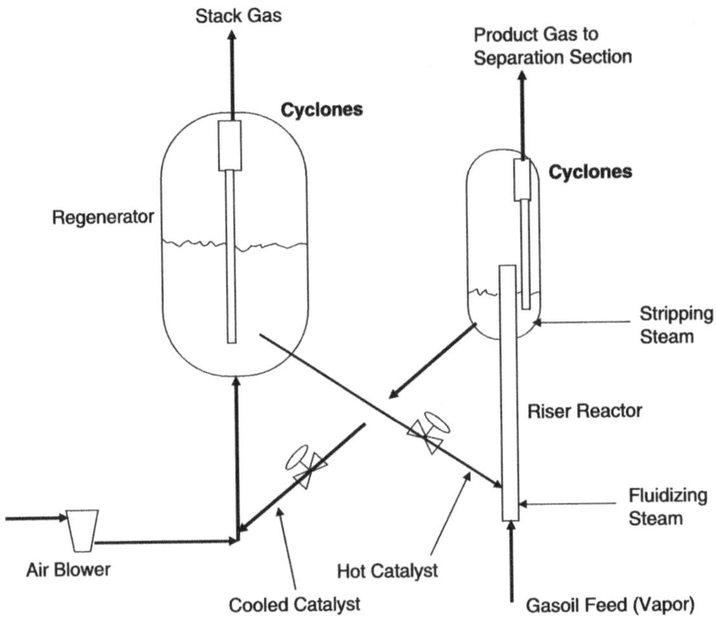

Figure 8.1 FCCU process.

Chemical reactions occur in two fluidized-bed reactors, as sketched in Figure 8.1. The hydrocarbon feed is contacted with catalyst in the "reactor" in which cracking occurs and coke is formed on the catalyst. The catalyst is then fed to a second fluidized-bed reactor, the "regenerator," along with air, which burns the coke off the catalyst. There is a very large circulation of catalyst between the two vessels (catalyst-to-oil weight ratio of 6.4). The cracking reactions are highly endothermic, and the required heat is provided by the sensible heat of the hot catalyst coming from the regenerator. The combustion reactions occurring in the regenerator are highly exothermic. A critical design–control issue is maintaining the heat balance of the unit such that the two heat requirements are closely matched.

8.1.1 Reactor

The gasoil feed is preheated and vaporized in an extensive heat exchanger network and a fired furnace. It is then combined with a large stream of hot solid catalyst and some steam at the bottom of a riser. The hydrocarbon vapor and steam provide the motive force to convey the solid and gas phases upward together at high velocity. This type of fluidized bed is called a "transport" reactor (co-current flow of solid and gas phases). The gasoil cracks to form lighter hydrocarbons and coke (mostly carbon), which is deposited on the solid catalyst particles. The very large endothermic heat of reaction of the cracking reactions is provided by the sensible heat of the hot catalyst. The reactor operates at about 1000°F, and the hot catalyst enters from the regenerator at about 1250°F. The hot gases leave the top of the reactor through cyclones to reject any catalyst particles and go to the separation section of the plant (fractionator, compressor, absorber, and light-end distillation columns). The solid catalyst, which now contains a significant amount of coke, is steam-stripped to remove most of the residual hydrocarbons.

8.1.2 Regenerator

The second reaction vessel in a catalytic cracker is called the "regenerator." The solid catalyst from the reactor is combined with a compressed air stream from an air blower, and the solid and gas phases flow upward into a bed of fluidized solid catalyst. The early designs used a "bubbling bed" reactor in which the velocity in the bed is slightly above the minimum fluidization velocity. More recent designs use a transport fluidized-bed reactor. A typical air-to-oil weight ratio is 0.54.

The oxygen in the air burns the carbon and any residual entrained hydrocarbons off the catalyst, producing carbon monoxide, carbon dioxide, and water. Catalytic coolers are usually installed in the regenerator vessel to provide heat removal capacity when needed for balancing the energy requirements in the unit. These coolers generate steam for use elsewhere in the refinery. The combustion gases flow through cyclones and leave as stack gas.

There is incomplete burning of the CO to CO_2 in the dense phase of the catalyst bed in the regenerator. However, in the dilute phase above the bed, this reaction proceeds further. Since the $CO + \frac{1}{2}O_2 \rightarrow CO_2$ reaction is very exothermic, there is an increase in temperature between the catalyst bed and the stack gas. This is called "afterburning." If the stack gas temperature gets too high, there may be thermal damage to the cyclones.

An extensive description of the process is given in by Upson et al.[1]

8.1.3 Control Issues

The control of the coupled reactor–regenerator is challenging because of the interaction between the two vessels and the "neat" operation in terms of energy. The temperatures in both vessels must be controlled at levels that are just below the metallurgical limits of the equipment materials.

A variety of conventional and advanced control structures have been used or proposed for catalytic cracking units. The most authoritative work is presented by Shinnar and coworkers.[2] They discuss both modeling and control structure issues in detail and show that the 2×2 control structure provides effective control: $T_{reactor}$ controlled by F_{cat} and T_{regen} controlled by F_{air}.

Other advanced control structures have been studied, including nonlinear control, fuzzy logic control and MPC. A recent review of control studies is given by Uygun and coworkers.[3]

In current industrial practice, reactor (or riser) temperature is usually controlled by the flowrate of hot catalyst fed to the reactor from the regenerator. A slide valve in the

[1]L. L. Upson, F. S. Rosser, C. L. Helmer, P. Palmas, L. E. Bell, W. J. Reagan, and B. W. Hedrick, Fluid catalytic cracking (FCC) units, Regeneration, *Kirk-Othmer Encyclopedia of Chemical Technology*, 5th Ed., Vol. 11, John Wiley & Sons, 2001.

[2]R. Shinnar, A. Arbel, and I. H. Rinard, Dynamics and control of fluidized catalytic crackers. 3. Designing the control system: Choice of manipulated and measured variables for partial control, *Ind. Eng. Chem. Res.*, **35**(7), 2215–2233 (1996).

[3]Ö. Uygun, H. Taşkin, C. Kubat, and S. Arslankaya, FUZZYFCC: Fuzzy logic control of a fluid catalytic cracking unit (FCCU) to improve dynamic performance, *Computers & Chemical Engineering*, **30**(5), 850–863 (2006).

catalyst transfer line adjusts this flowrate. Since high catalyst-to-oil ratios favor high conversion, the feed preheating temperature can be reduced to increase catalyst flow. A minimum feed temperature is set by requiring that the material be completely vaporized. If the slide valve on the transfer line becomes wide open, reactor temperature can be controlled with feed preheat.

The temperature in the regenerator is conventionally controlled by the flowrate of air from the blower. However, many catalytic cracking units run at maximum airflow as limited by the air blower. In this situation catalyst cooling rates in the regenerator can be used to hold regenerator temperature. However, if air is at a maximum constraint, the amount of coke that can be burned is limited. Therefore, feed flowrate may have to be reduced or feed composition altered (feed more light gasoil and less heavy material).

The catalyst bed level in the reactor is controlled by adjusting a slide valve in the catalyst return line from the reactor to the regenerator. Since the total amount of catalyst in both vessels is essentially constant, the level in the regenerator is not controlled. The thermal capacitance in the system is mostly in the solid catalyst. The catalyst holdup in the regenerator is much larger than in the reactor, so the dynamic response of the reactor is faster than that of the regenerator.

Disturbances include changes in the composition of the feed and changes in throughput. Since crude oil is a naturally occurring raw material, its composition changes from source to source.

8.2 GASIFIERS

With the rapid increase in the price of imported petroleum and the political instability in many regions of crude-oil production, the incentive to use domestically available coal supplies has increased in recent years. Gasifiers are used in several types of coal-consuming processes to produce gas that can be used for fuel in a combustion turbine. If the gasifier is "*air* blown" (where air, coal, and water are the main feeds), the gas contains nitrogen, so a low calorific value fuel gas is produced. If the gasifier is *oxygen*-blown, the product is synthesis gas that contains much less nitrogen (there is some nitrogen in the coal) and therefore can be use as source of fuel or chemical feedstocks.

The Department of Energy has a very active program to develop a coal-based, low-emissions "FutureGen" process for the production of electric power, hydrogen, and other chemicals. A vital part of this process is a fluidized-bed gasifier, in which the coal is partially oxidized using oxygen to produce synthesis gas. The heat required for the endothermic reforming reactions is provided by the partial oxidation of some of the coal. The synthesis gas produced is a mixture of hydrogen, carbon monoxide, and carbon dioxide, which can be used as feed to a combustion turbine for generating electric power, or it can by used as a chemical feedstock for the production of a wide variety of chemical products.

To produce more hydrogen than is in the original synthesis gas, a water–gas shift reactor is used downstream of the gasifier in which the reaction $CO + H_2O \rightarrow CO_2 + H_2$ takes place. Then the H_2 is separated from the CO_2 using membranes or adsorption methods (PSA). To cut greenhouse gas emissions, the hope is that the CO_2 can be sequestered in underground or undersea reservoirs. The hydrogen can to used directly as fuel in a combustion turbine or fuel cell to generate power with nonpolluting water as the product, or it can used for as a chemical feedstock (e.g., hydrotreating of

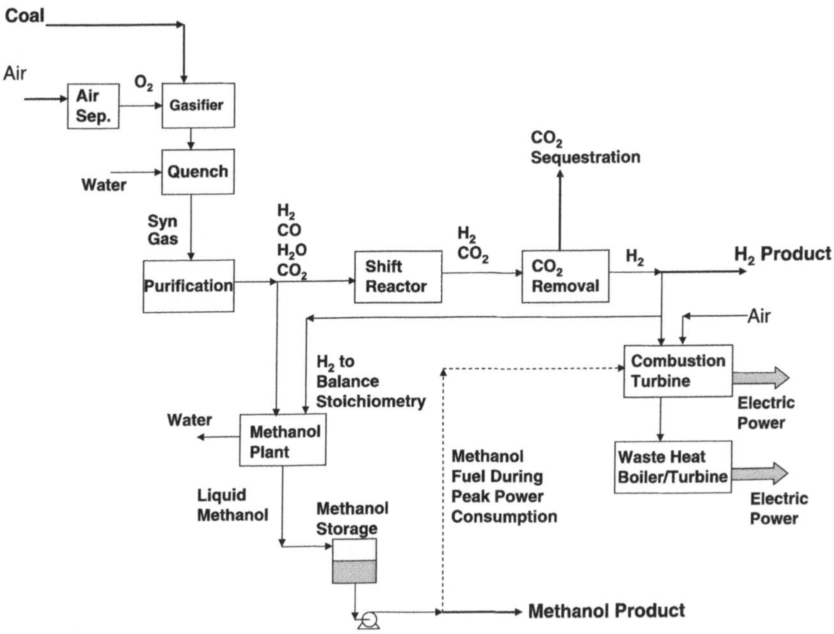

Figure 8.2 Conceptual flowsheet of hybrid FutureGen/Methanol process.

petroleum products to reduce sulfur content). The hydrogen could also be used in transportation vehicles.

Figure 8.2 gives a schematic of one possible configuration of this type of future power/chemical process. Air is fed to an air separation unit (cryogenic or membrane) to produce an oxygen-rich steam that is fed with the coal to the gasifier. The high-temperature gas leaving the gasifier is quenched and fed to a purification unit for the removal of sulfur and nitrogen compounds. The synthesis gas can be used for the production of chemicals, such as in methanol production. Likewise, the hydrogen can be used for the production of chemicals (e.g., ammonia for fertilizer), or it can provide pollution-free fuel for a combustion turbine or fuel cell.

In addition to coal, gasifiers can be used on a variety of feedstocks. Almost anything that will burn can be gasified. There are many types of gasifiers. Some are fluidized beds, and others are moving beds. The coal fed to the gasifier can be dry or a coal/water slurry.

A comprehensive and readable discussion of many types of gasifiers is given in a text by Higman and van der Burgt.[4] Shadle et al.[5] provide a detailed description of many aspects of the coal gasification processes: historical, technological, and economic. A paper by Xu et al.[6] presents an interesting discussion of selecting combinations of bubbling and transport fluidized beds.

[4]C. Higman and M. van der Burgt, *Gasification*, Elsevier, 2003.
[5]L. J. Shadle, D. A. Berry, and M. Syamlal, Coal conversion processes, gasification, *Kirk-Othmer Encyclopedia of Chemical Technology*, 5th Ed., John Wiley & Sons, 2001.
[6]G. Xu, T. Murakami, T. Suda, Y. Matsuzawa, and H. Tani, The superior technical choice for dual fludized bed gasification, *Ind. Eng. Chem. Research*, **45**, 2281–2286 (2003).

The dominant control issue in coal gasification is maintaining gasifier temperature in the face of varying coal composition (carbon and water content) and frequent changes in the demand for electric power. The very high temperature and the severe conditions in the vessel make temperature measurement difficult and unreliable. As a result, inferential methods for estimating temperature are often used. The methane composition of the gas gives a fairly estimate of the temperature; low temperature gives low conversion and high methane compositions.

Some of the practical issues of measuring temperatures and gas compositions in gasifiers are discussed by Higman and van der Burgt (see Footnote 4). Gasifier control is discussed by Dixon,[7] who also presented the Alstom benchmark challenge problem that provides a nonlinear model of an air-blown coal gasifier in Matlab for other researches to test new control methods. The use of nonlinear model predictive control on this process is presented by Al Seyab and Cao.[8]

8.3 FIRED FURNACES, KILNS, AND DRIERS

Fired furnaces, kilns, and driers are special types of chemical reactor in which a combustible fuel (gas, oil, coal, wood, etc.) is burned in the presence of air to provide heat at a high temperature level. Generating high-pressure steam in power plants, providing heat in process units (distillation columns, reactors with endothermic reactions, etc.), producing lime, and smelting ore are common examples.

The control issues with the fuel combustion units involve supplying the required energy in a safe and efficient manner. This energy demand usually changes dynamically as dictated by the needs of the process.

For a typical and important example, let us consider the steam-generating boiler in an electrical power plant. The consumption of steam by the turbine driving the generator

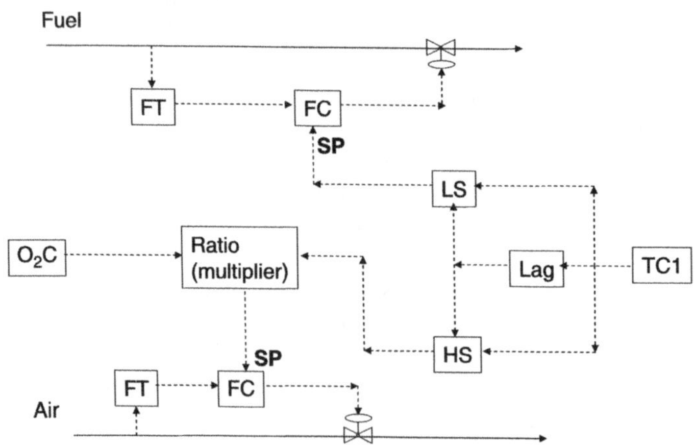

Figure 8.3 Firing controls.

[7]R. Dixon, Advanced gasifier control, *Comput. Control. Eng. J. IEE*, **10**(3), 93–96 (1999).

[8]R. K. Al Seyab and Y. Cao, Nonlinear model predictive control of the ALSTOM gasifier, *J. Process Control*, **16**, 795–808 (2006).

varies with the demand for electric power. The pressure of the high-pressure steam supplied to the turbine is controlled by changing the flowrate of the fuel. Of course, the air flowrate also needs to be changed as the fuel flowrate changes. If too much air is added to the furnace, the percent excess oxygen in the stack gas will be high, and fuel will be wasted. If too little air is added, the furnace will smoke and violate air pollution regulations. In extreme cases, the fuel may build up in the furnace and present a potential risk of explosion.

In a typical furnace control system the fuel flow is manipulated to hold temperature or pressure in the process. The air is ratioed to the fuel. This ratio is adjusted to maintain a reasonable oxygen composition in the stack gas.

In addition, there is a need to handle the dynamic changes in load in a safe manner. If there is an increase in demand, the airflow should be increased *before* fuel flow. If there is a decrease in demand, the airflow should be decreased *after* the fuel flow. This is achieved by the use of some simple dynamic lags and selectors, as illustrated in Figure 8.3.

8.4 PULP DIGESTERS

One of the critical units in the production of paper is a reactor called a "digester." In the kraft process this reactor is a two-phase tubular reactor in which the lignin that binds the wood chips together is broken down through a combination of chemical and thermal effects. The "white liquor" (aqueous solution of sodium hydroxide and hydrosulfide) and solid wood chips flow countercurrently in some zones and co-currently in others. The residence time of the pulp is about 10 h.

There are many control challenges in this process. These include strong nonlinearity, distributed system, long deadtimes, and a feedstock that varies significantly because of its biological source. The key variable is kappa number (degree of delignification), which cannot be measured online, so it must be estimated from secondary measurements.

The papers by Doyle and coworkers[9–10] present a thorough treatment of the control of this type of reactor.

The Weyerhaeuser digester problem, which is used in these studies, is presented in another paper.[11]

8.5 POLYMERIZATION REACTORS

The polymer industry experienced very rapid growth over the last five decades. A vast variety of products are produced for a myriad of applications. The heart of a polymer process is the reactor. Because there are many types of polymerization reactions, there a wide variety of types of polymer reactors, both batch and continuous.

An in-depth discussion of polymerization reactors is far beyond the scope of this book. Several volumes would be required to thoroughly cover the subjects of both

[9] P. A. Wisnewski and F. J. Doyle, Control structure selection and model predictive control of the Weyerhaeuser digester problem, *J. Process Control*, **8**, 487–495 (1998).

[10] F. J. Doyle and F. Kayihan, Reaction profile control of the continuous pulp digester, *Chem. Eng. Sci.*, **54**, 2679–2688 (1999).

[11] F. Kayihan, M. S. Gelormina, E. M. Hanczyc, F. J. Doyle, and Y. A. Arkun, A Kamyr continuous digester model for identification and controller design, *Proc. IFAC World Congress*, San Francisco, CA 1996, pp. 37–42.

design and control of the many, many types of polymerization reactors with their varied chemistry and chemical compounds. Therefore we will limit ourselves to a brief presentation of some of the unique problems and provide some references where more details can be found.

Some polymerization reactions are highly exothermic, so the problems of temperature control, which are the major emphasis of this book, are important in these systems. However, beyond the issue of temperature control, polymer reactors must produce a product with the desired properties. The final polymer product properties, such as viscosity, molecular weight distribution, particle size, and composition, are important for consistent performance of the polymer. These properties depend on more than just temperature and few can be measured online.[12]

According to Ray,[13] "One of the greatest difficulties in achieving quality control of the polymer product is that the actual customer specifications may be in terms of nonmolecular parameters such as tensile strength, crack resistance, temperature stability, color, clarity, adsorption capacity for plasticizer, etc. The quantitative relationship between these product-quality parameters and reactor operating conditions may be the least understood area of polymerization reaction engineering."

Many polymerization reactions occur in batch reactors, and the product properties result from an integral average of reaction conditions. Even if the polymerization reactor is continuous, grade transitions are frequently required, which cause significant dynamic problems.[14]

Suffice it to say that the control of polymerization reactors is a most challenging and complex problem.

8.6 BIOCHEMICAL REACTORS

Bioreactors need to provide an environment in which the microorganisms grow, multiply, and produce the desired product. Their control is challenging because of significant process variability, the need for a sterile environment, the complexity of biological systems, and relatively few online measurements. The control system must provide the right concentration of nutrients (carbon, nitrogen, oxygen, phosphorous, etc.), remove any toxic metabolic products (e.g., CO_2), and maintain internal cellular parameters (e.g., temperature and pH). Alford discusses the control issues of biochemical reactors.[15]

One of the most important biochemical reactors is the fermentor. We provided a simple example of a batch fermentor in Chapter 4. But there are many other types including continuous, batch, and fed-batch. There are some other useful references in the literature.[16–17]

[12] J. R. Richards and J. P. Congalidis, Measurement and control of polymerization reactors, *Comput. Chem. Eng.*, **30**, 1447–1463 (2006).

[13] W. H. Ray, Polymerization reactor control, *IEEE Control Sys. Mag.*, 3–8 (Aug. 1986).

[14] M. Asteausuain, A. Bandoni, C. Sarmoria, and A. Brandolin, Simultaneous process and control system design for grade transition in styrene polymerization, *Chem. Eng. Sci.*, **61**, 3362–3378 (2006).

[15] J. S. Alford, Biprocess control: Advances and challenges, *Comput. Chem. Eng.*, **30**, 1464–1475 (2006).

[16] W. Bequette, *Process Control: Modeling, Design, and Simulation*, Prentice-Hall, 2003.

[17] A. Cinar, S. J. Parulekar, C. Undey, and G. Birol, *Batch Fermentation: Modeling, Monitoring, and Control*, Marcel Dekker, 2003.

8.7 SLURRY REACTORS

Multiple phases exist in slurry reactors. The solid phase is typically a catalyst. The liquid phase is a reactant or product. There can also be a gas phase, usually one of the reactants. The design issues involve separating the phases and recycling catalyst and reactant back to the reactor.

The polymerization of propylene is a typical example. A paper by De Wolf et al.[18] describes the process and the control issues. Liquid propylene and gaseous hydrogen are fed to a slurry reactor. The standard control scheme controls the level in the reactor by the addition of propylene, controls the temperature in the reactor by the heat removal system, and controls the hydrogen composition in the vapor space above the reactor. The important product quality variable is the melt index, which is a measure of the average chain length. The hydrogen composition in the slurry phase, where it acts as a chain transfer agent, affects the melt index. However, it is only practical to measure the composition of the vapor phase, and this is affected by other process variables. The authors study the use of a "soft sensor" (a linear Kalman filter) to estimate the slurry phase hydrogen composition and a model predictive controller.

8.8 MICROSCALE REACTORS

The design and control of very small chemical reactors (reactors on a chip) have received considerable attention in recent years, particularly in academic research. The paper by Kothare[19] provides a broad review of work in this area.

Quoting Kothare, "Microchemical systems are a new generation of miniature chemical systems that carry out chemical reactions and separations in precisely fabricated three dimensional microreactor configurations in the size range of a few microns to a few hundred microns. Typical microchemical systems combine fluid handling and reaction capabilities with electronic sensing and actuation . . ." and ". . . are fabricated using integrated circuit manufacturing techniques . . . ." A variety of fluid-handling devices have been developed, for example, micropumps and microvalves, which can be used in these chip devices.

Many potential applications are under study. Miniature chemical reactors could be used for portable applications in which they provide advantages of rapid startup and shutdown and of increased safety (intensification by requiring only small quantities of hazardous materials). The development of chip-scale chemical and biological analysis systems has the potential to reduce the time and cost associated with conventional laboratory methods. These devices could be used as portable analysis systems for detection of hazardous chemicals in air and water. There is considerable interest in using a microreactor to provide in situ production of hydrogen for small-scale fuel-cell power applications by conducting a reformation reaction from some liquid hydrocarbon raw material (e.g., methanol).

[18]S. De Wolf, R. L. E. Cuypers, L. C. Zullo, B. J. Vos, and B. J. Bax, Model predictive control of a slurry polymerization reactor, *Comput. Chem. Eng.*, **20**(Suppl.), S955–S961 (1996).

[19]M. V. Kothare, Dynamics and control of integrated microchemical systems with application to micro-scale fuel processing, *Comput. Chem. Eng.*, **30**, 1725–1734 (2006).

INDEX

Printed and bound by CPI Group (UK) Ltd, Croydon, CR0 4YY

Printed and bound by CPI Group (UK) Ltd, Croydon, CR0 4YY
18/09/2026
14965012-0001